FACILITIES PLANNING

THIRD EDITION

JAMES A. TOMPKINS
Tompkins Associates, Inc.

JOHN A. WHITE
University of Arkansas

YAVUZ A. BOZER
University of Michigan

J. M. A. TANCHOCO
Purdue University

JOHN WILEY & SONS, INC.

ACQUISITIONS EDITOR   Wayne Anderson
ASSOCIATE EDITOR   Jennifer Welter
MARKETING MANAGER   Katherine Hepburn
SENIOR PRODUCTION EDITOR   Valerie A. Vargas
SENIOR DESIGNER   Kevin Murphy
COVER DESIGNER   Pamela King, Tompkins Associates
PRODUCTION MANAGEMENT SERVICES   Suzanne Ingrao/Ingrao Associates

This book was set in 10/12 Garamond Light by TechBooks and printed and bound by Hamilton Printing Company. The cover was printed by Phoenix Color Corp.

This book is printed on acid-free paper. ∞

To order books or for customer service please, call 1(800)-CALL-WILEY (225-5945).

ISBN 0-471-41389-5

ISBN 0-471-38937-4 (WIE)

Printed in the United States of America

10 9 8 7 6 5 4 3 2

PREFACE

The third edition of *Facilities Planning* is a comprehensive guide assembling new and updated information on the operations around which we are planning facilities today. The purpose of this book is to continue the creativity, rigor, and design aspects of facilities planning while rejecting the cookbook and checklist approaches to this discipline.

Because it impacts so many activities—handling and maintenance costs, employee morale, operating costs, capital investment, facility management, adapting to change, and satisfying future requirements—the importance of facilities planning will continue to grow and challenge the engineering professional. Rapid changes in production techniques and equipment will continue as well. With so many challenges, it is more important than ever for new professionals to have the right tools when entering the marketplace.

The size of a typical investment in new facilities makes this field particularly important. We have reexamined the role of facilities planning within the supply chain. It is now imperative that the facilities planner assist his or her company in progressing through the six levels of supply chain excellence: *Business as Usual, Link Excellence, Visibility, Collaboration, Synthesis,* and *Velocity.* Successful enterprises know that there is no more "business as usual." Facilities planning is no longer just a science, but a strategy for navigating a competitive global economy. Every entity must insist on the highest return on investment—not only to prosper, but also to survive.

In this edition, we endeavor to continue the engineering design process and application of quantitative tools as a foundation for doing appropriate and successful facilities planning. We have added more coverage of cost justification, safety, and cellular manufacturing. To support these topics and others, we have given more detailed real-world examples and problems concerning current facilities planning practices and deleted nonrelevant and dated material. We have added new photography of the industry's latest material-handling equipment and updated the drawings and diagrams to better reflect the practices taking place today. The instructor will find improved problem sets, including more quantitative problems, and a greater variety of helpful questions. We have mentioned valuable software applications (like VisFactory) and included that information in the Instructor Solutions Manual and on the Web site. The chapter titles, order, and content have been changed to provide clarity for the instructor and the student, and the references cited were revised to offer opportunity for additional study.

This text is organized into the following instruction methods:

- Defining Requirements
  - Strategic Facilities Planning
  - Product, Process, and Schedule Design
  - Flow, Space, and Activity Relationships
  - Personnel Requirements

- Developing Alternatives: Concepts and Techniques
  - Material Handling
  - Layout Planning Models and Design Algorithms

- Developing Alternatives: Functions
  - — Warehouse Operations
  - — Manufacturing Operations
  - — Facilities Systems
- Developing Alternatives: Quantitative Approaches
  - — Quantitative Facilities Planning Models
  - — Evaluating, Selecting Preparing, Presenting, Implementing, and Maintaining
- Evaluating and Selecting the Facilities Plan
  - — Preparing, Presenting, Implementing, and Maintaining Facilities Plans

The instructor can easily rearrange the material as best suited for his or her method of instruction.

Many people have influenced the development of this text. Deserving special mention are: Edward H. Frazelle and Jaime Trevino who contributed their expertise as authors and editors of the previous editions of *Facilities* Planning, and Marvin H. Agee, James M. Apple, Robert P. Davis, Marc Goetschalckx, Hugh D. Kinney, Leon F. McGinnis, Benoit Montreuil, Colin L. Moodie, James M. Moore, Richard Muther, Ruddell Reed, Jr., Jerry D. Smith, John D. Spain, Richard A. Wysk, and Kazuho Yoshimoto, whose professionalism and expertise have influenced our thinking on the subject. Also, we thank Pamela King for the cover design. Finally, the contributions of numerous graduate students, administrative assistants, and other professionals within our organizations are significant and appreciated.

Finally, we thank our families for their patience, support, and encouragement while we worked to create a teaching tool that reflects our collective vision of facilities planning excellence.

James A. Tompkins
John A. White
Yavuz A. Bozer
J. M. A. Tanchoco

CONTENTS

Part One

DEFINING
REQUIREMENTS

1

INTRODUCTION

Facilities planning has taken on a whole new meaning in the past 10 years. In the past, facilities planning was primarily considered to be a science. In today's competitive global marketplace, facilities planning is a strategy. Governments, educational institutions, and businesses no longer compete against one another individually. These entities now align themselves into cooperatives, organizations, associations, and, ultimately, synthesized supply chains to remain competitive by bringing the customer into the process.

The subject of facilities planning has been a popular topic for many years. In spite of its long heritage, it is one of the most popular subjects of current publications, conferences, and research. The treatment of facilities planning as a subject has ranged from checklist, cookbook-type approaches to highly sophisticated mathematical modeling. In this text, we intend to employ a practical approach to facilities planning, taking advantage of empirical and analytical approaches using both traditional and contemporary concepts. It should be noted that facilities planning, as addressed in this text, has broad applications. For example, the contents of this book can be applied equally to the planning of a new hospital, an assembly department, an existing warehouse, or the baggage department of an airport. Whether the activities in question occur in the context of a hospital, production plant, warehouse, airport, retail store, school, bank, office, or any portion of these facilities, the material presented in this text should be useful in planning. It is important to recognize that contemporary facilities planning considers the facility as a dynamic entity and that a key requirement for a successful facilities plan is its adaptability and its ability to become suitable for new use.

1.1 **FACILITIES PLANNING DEFINED**

The facilities we plan today must help an organization achieve Supply Chain Excellence. Supply Chain Excellence is a process with six steps, or levels. These steps are Business as Usual, Link Excellence, Visibility, Collaboration, Synthesis, and Velocity.

Business as Usual is when a company works hard to maximize its individual functions. The goal of individual departments, such as finance, marketing, sales, purchasing, information technology, research and development, manufacturing, distribution, and human resources, is to be the best department in the company. Organizational effectiveness is not the emphasis. Each organizational element attempts to function well within its individual silo.

Only after one's link achieves performance excellence can they begin to pursue Supply Chain Excellence. To achieve *Link Excellence,* companies must tear down the internal boundaries until the entire organization functions as one. Companies usually have numerous departments and facilities, including plants, warehouses, and distribution centers (DCs). If an organization hopes to pursue Supply Chain Excellence, it must look within itself, eliminate and blur any boundaries between departments and facilities, and begin a never-ending journey of continuous improvement. It must have strategic and tactical initiatives at the department, plant, and link levels for design and systems.

Supply Chain Excellence requires everyone along the supply chain to work together. Everyone in the supply chain cannot work together, however, if they cannot see one another. *Visibility,* the third level of Supply Chain Excellence, brings to light all links in the supply chain. It minimizes supply chain surprises because it provides the information links need to understand the ongoing order status. It could be considered the first real step toward Supply Chain Excellence.

Through visibility, organizations come to understand their roles in a supply chain and are aware of the other links. An example is an electronics company with a Web site that allows its customers to view circuit boards and then funnels information about those customers to suppliers. Visibility thus requires sharing information so that the links understand the ongoing order status and thus minimize supply chain surprises.

Once a supply chain achieves visibility, it can move to *Collaboration,* the fourth level of Supply Chain Excellence. Through collaboration, the supply chain can determine how best to meet the demands of the marketplace. The supply chain works as a whole to maximize customer satisfaction while minimizing inventories. Collaboration is achieved through the proper application of technology and true partnerships. Various collaboration technologies exist, and, as with visibility software, the supply chain must choose the right technology or combination of technologies if it hopes to collaborate properly. True partnerships require total commitment from all the links in the supply chain and are based on trust and a mutual desire to work as one for the benefit of the supply chain.

After collaboration is in place, the supply chain then must pursue the continuous improvement process of *Synthesis.* Synthesis is the unification of all supply chain links to form a whole. It creates a complete pipeline from a customer perspective. The results of synthesis are:

- *Increased ROA.* This is achieved by maximizing inventory turns, minimizing obsolete inventory, maximizing employee participation, and maximizing continuous improvement.

- *Improved customer satisfaction.* This is achieved because synthesis creates companies that are responsive to the customer's needs through customization. They understand value-added activity. They also understand the issue of flexibility and how to meet ever-changing customer requirements. They completely comprehend the meaning of high quality and strive to provide high value.

- *Reduced costs.* This is achieved by scrutinizing transportation costs, acquisition costs, distribution costs, inventory carrying costs, reverse logistics costs, packaging costs, and so on and continually searching for ways to drive costs down.
- *An integrated supply chain.* This is achieved by using partnerships and communication to integrate the supply chain and focus on the ultimate customer.

Synthesis is not achieved overnight. It takes time to take the links of a supply chain and remove the boundaries between them. However, if all links are visible and all collaborate, then synthesis is within reach.

Velocity is synthesis at the speed of light. It is the embodiment of the statement, "Ride fast or you will fall." Today's business environment demands speed. The Internet has created immediate orders, and customers expect their products to arrive almost as quickly. Synthesis with speed creates multilevel networks that meet these demands—these are complex entities that can meet the demands of today's economy through a combination of partnerships, flexibility, and robust design methods.

Facilities are critical components of the multilevel networks necessary for Supply Chain Excellence. Each organization in the supply chain should therefore plan facilities with their supply chain partners in mind. Proper facilities planning along the supply chain ensures that the product will be manufactured and shipped to the satisfaction of the ultimate customer. Therefore, all facilities in the supply chain have the following characteristics:

- *Flexibility.* Flexible facilities are able to handle a variety of requirements without being altered.
- *Modularity.* Modular facilities are those with systems that cooperate efficiently over a wide range of operating rates.
- *Upgradability.* Upgraded facilities gracefully incorporate advances in equipment systems and technology.
- *Adaptability.* This means taking into consideration the implications of calendars, cycles, and peaks in facilities use.
- *Selective operability.* This means understanding how each facility segment operates and allows contingency plans to be put in place.

Creating these facilities requires a holistic approach. The elements of this approach are:

- *Total integration*—the integration of material and information flow in a true, top-down progression that begins with the customer.
- *Blurred boundaries*—the elimination of the traditional customer/supplier and manufacturing/warehousing relationships, as well as those among order entry, service, manufacturing, and distribution.
- *Consolidation*—the merging of similar and disparate business entities that results in fewer and stronger competitors, customers, and suppliers. Consolidation also includes the physical merging of sites, companies, and functions.
- *Reliability*—the implementation of robust systems, redundant systems, and fault-tolerant systems to create very high levels of uptime.
- *Maintenance*—a combination of preventive maintenance and predictive maintenance. Preventive maintenance is a continuous process that minimizes future maintenance problems. Predictive maintenance anticipates potential problems by sensing the operations of a machine or system.

- *Economic progressiveness*—the adoption of innovative fiscal practices that integrate scattered information into a whole that may be used for decision making.

In this regard, for a facilities planner, the notion of continuous improvement for Supply Chain Excellence must be an integral element of the facilities planning cycle. The continuous improvement facilities planning cycle shown in Figure 1.1 details this concept. Whether you are involved in planning a new facility or planning to update an existing facility, the subject matter should be of considerable interest and benefit.

Facilities planning determines how an activity's tangible fixed assets best support achieving the activity's objective. For a manufacturing firm, facilities planning involves the determination of how the manufacturing facility best supports production. In the case of an airport, facilities planning involves determining how the airport facility is to support the passenger–airplane interface. Similarly, facilities planning for a hospital determines how the hospital facility supports providing medical care to patients.

It is important to recognize that we do not use the term *facilities planning* as a synonym for such related terms as *facilities location, facilities design, facilities layout,* or *plant layout*. As depicted in Figure 1.2, it is convenient to divide a facility into its location and its design components.

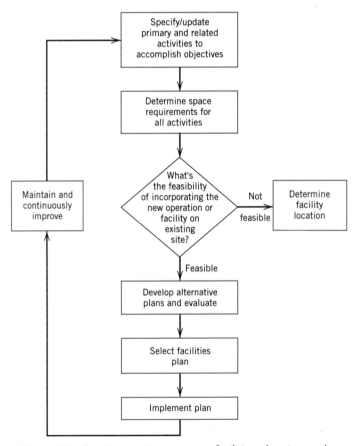

Figure 1.1 Continuous improvement facilities planning cycle.

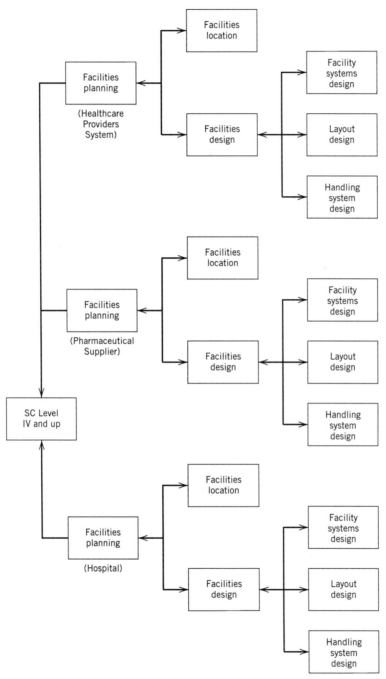

Figure 1.2 Facilities planning as part of Supply Chain Excellence. Continuous improvement of each operation within each supply chain link takes an organization through the first three levels of Supply Chain Excellence. To move to levels 4, 5, and 6, the links must collaborate, as illustrated above, to synthesize their operations and continue to improve the chain.

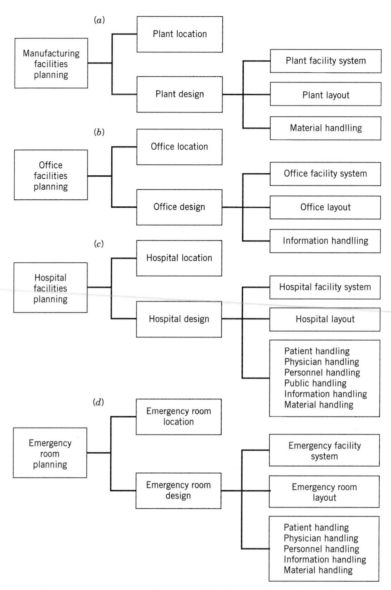

Figure 1.3 Facilities planning for specific types of facilities. (*a*) Manufacturing plant. (*b*) Office. (*c*) Hospital. (*d*) Emergency room.

The *location* of the facility refers to its placement with respect to customer, suppliers, and other facilities with which it interfaces. Also, the location includes its placement and orientation on a specific plot of land.

The design *components* of a facility consist of the facility systems, the layout, and the handling system. The facility systems consist of the structural systems, the atmospheric systems, the enclosure systems, the lighting/electrical/communication systems, the life safety systems, and the sanitation systems. The layout consists of all equipment, machinery, and furnishings within the building envelope; and the handling system consists of the mechanisms needed to satisfy the required facility interactions. The facility systems for a manufacturing facility may include the envelope (structure and enclosure elements), power, light, gas, heat, ventilation, air conditioning, water, and sewage needs. The layout consists of the production areas, production-related or support areas, and personnel areas within the building. The handling system consists of the materials, personnel, information, and equipment-handling systems required to support production.

Determining how the *location* of a facility supports meeting the facility's objective is referred to as *facilities location*. The determination of how the design components of a facility support achieving the facility's objectives is referred to as *facilities design*. Therefore, facilities planning may be subdivided into the subjects of facilities location and facilities design. Facilities location addresses the macro issues, whereas facilities design looks at the microelements.

The general terms facilities planning, facilities location, facilities design, facility systems design, layout design, and handling system design are utilized to indicate the breadth of the applicability of this text. In Figure 1.3, the facilities planning hierarchy is applied to a number of different types of facilities. It is because of its breadth of application that we employ a unified approach to facilities planning.

1.2 SIGNIFICANCE OF FACILITIES PLANNING

According to the U.S. Census, U.S. businesses invested $1.038 trillion in capital goods in 1999. Of that money, $320.8 billion was spent on structures, with $297.4 billion (92.7%) being spent on new structures in 1999.

Since 1955, approximately 8% of the gross national product (GNP) has been spent annually on new facilities in the United States. Table 1.1 indicates the typical expenditures, in percentage of GNP, for major industry groupings. The size of the investment in new facilities each year makes the field of facilities planning important. As stated previously, contemporary facilities planning must include the notion of continuous improvement in the design approach. The importance of adaptability, as a key design criterion, is evidenced by the ever-increasing performance of previously

Table 1.1 *Percentage of the Gross National Product (GNP) by Industry Grouping Typically Expended on New Facilities Between 1955 and Today*

Industry	GNP Percentage
Manufacturing	3.2
Mining	0.2
Railroad	0.2
Air and other transportation	0.3
Public utilities	1.6
Communication	1.0
Commercial and other	1.5
All industries	8.0

Source: U.S. Bureau of Census.

purchased facilities, which are modified each year and require replanning. For these reasons, it seems reasonable to suggest that over $250 billion will be spent annually in the United States alone on facilities that will require planning or replanning.

Although the annual dollar volume of the facilities planned or replanned indicates the scope of facilities planning, it does not appear that adequate planning is being performed. Based on our collective experience, it appears that there exists a significant opportunity to improve the facilities planning process as practiced in industry today.

To stimulate your thoughts on the breadth of the facilities planning opportunities, consider the following questions:

1. What impact does facilities planning have on handling and maintenance costs?
2. What impact does facilities planning have on employee morale, and how does employee morale impact operating costs?
3. In what do organizations invest the majority of their capital, and how liquid is their capital once invested?
4. What impact does facilities planning have on the management of a facility?
5. What impact does facilities planning have on a facility's capability to adapt to change and satisfy future requirements?

Although these questions are not easily answered, they tend to highlight the importance of effective facilities planning. As an example, consider the first question. Between 20 and 50% of the total operating expenses within manufacturing is attributed to material handling. Furthermore, it is generally agreed that effective facilities planning can reduce these costs by at least 10 to 30%. Hence, if effective facilities planning were applied, the annual manufacturing productivity in the United States would increase approximately three times more than it has in any year in the past 15 years.

It is difficult to make similar projections for the other sectors of our economy. However, there is reason to believe that facilities planning will continue to be one of the most significant fields of the future. It represents one of the most promising areas for increasing our rate of productivity improvement. The "reindustrialization of the United States" and the relative performances of Asian, European, and other nations' industries in competition with the United States are strongly related to a need for improved facilities planning.

Economic considerations force a constant reevaluation and recognition of existing systems, personnel, and equipment. New machines and processes render older models and methods obsolete. Facilities planning must be a continuing activity in any organization that plans to keep abreast of developments in its field.

With the rapid changes in production techniques and equipment that have taken place in the recent past and those that are expected in the future, very few companies will be able to retain their old facilities or layouts without severely damaging their competitive position in the marketplace. Productivity improvements must be realized as quickly as they become available for implementation.

One of the most effective methods for increasing plant productivity and reducing costs is to reduce or eliminate all activities that are unnecessary or wasteful. A facilities design should accomplish this goal in terms of material handling, personnel and equipment utilization, reduced inventories, and increased quality.

If an organization continually updates its production operations to be as efficient and effective as possible, then there must be continuous relayout and rearrangement.

Only in very rare situations can a new process or piece of equipment be introduced into a system without disrupting ongoing activities. A single change may have a significant impact on integrated technological, management, and personnel systems, resulting in suboptimization problems that can be avoided or resolved only through the redesign of the facility.

Employee health and safety is an area that has become a major source of motivation behind many facilities planning studies. In 1970, the Occupational Safety and Health Act became law and brought with it a far-reaching mandate: "to assure so far as possible every working man and woman in the nation safe and healthful working conditions and to preserve our human resources."

Because the act covers nearly every employer in a business affecting commerce that has 10 or more employees, it has had and will continue to have a significant impact on the structure, layout, and material handling systems of any facility within its scope. Under the law, an employer is required to provide a place of employment free from recognized hazards and to comply with occupational safety and health standards set forth in the act.

Because of these stringent requirements and attendant penalties, it is imperative during the initial design phase of a new facility or the redesign and revamping of an existing facility to give adequate consideration to health and safety norms and to eliminate or minimize possible hazardous conditions within the work environment.

Equipment and/or processes that may create hazards to workers' health and safety must be in areas where the potential for employee contact is minimal. By incorporating vital health and safety measures into the initial design phase, the employer may avoid fines for unsafe conditions and losses in money and manpower resulting from industrial accidents.

Energy conservation is another major motivation for the redesign of a facility. Energy has become an important and expensive raw material. Equipment, procedures, and materials for conserving energy are introduced to the industrial marketplace as fast as they can be developed. As these energy-conserving measures are introduced, companies should incorporate them into their facilities and manufacturing process.

These changes often necessitate changes in other aspects of the facility design. For example, in some of the energy-intensive industries, companies have found it economically feasible to modify their facilities to use the energy discharged from the manufacturing processes to heat water and office areas. In some cases, the addition of ducting and service lines has forced changes in material flows and the relocation of in-process inventories. In one large office complex, the basic facility design was altered so that many rooms could be heated by the excess energy from the computing equipment.

If a company is going to retain a competitive edge today, it must reduce its consumption of energy. One method of doing this is to modify and relayout facilities or redesign material handling systems and manufacturing processes to accommodate new energy-saving measures.

Other factors that motivate investment in new facilities or the alteration of existing facilities are community considerations, fire protection, security, and the Americans With Disabilities Act (ADA) of 1990. Community rules and regulations regarding noise, air pollution, and liquid and solid waste disposal are frequently cited as reasons for the installation of new equipment that requires modification of facilities and systems operating policies.

One of the most significant challenges to facilities planners today is how to make the facility "barrier free" in compliance with the ADA. The enactment of this legislation has resulted in a significant increase in the alteration of existing facilities and has radically shaped the way facilities planners approach planning and design. The act impacts all elements of the facility, from parking space allocation and space design, ingress and egress ramp requirements, and restroom layout to drinking-fountain rim heights. Companies are aggressively spending billions of dollars to comply with the law, and those involved with facilities planning must be the leaders in pursuing the required changes.

On nearly a daily basis in any large city newspaper, reports of fires that destroy a whole facility can be found. In many instances, these fires can be attributed to poor housekeeping or poor facilities design. Companies are now carefully seeking modifications to existing material handling systems, storage systems, and manufacturing processes to lower the risk of fire.

Pilferage is yet another major and growing problem in many industries today. Several billion dollars' worth of merchandise is stolen annually from manufacturing companies in the United States. The amount of control designed into material handling, flow of materials, and design of the physical facility can help reduce losses to a firm.

1.3 OBJECTIVES OF FACILITIES PLANNING

As previously mentioned, facilities planning must be done within the context of the supply chain to maintain a strategic competitive advantage. Just as SCS is driven by customer satisfaction, so too should customer satisfaction be the primary objective of facilities planning. This will ensure that the other objectives are in alignment with what drives the enterprise, namely revenues and profits from customers. Many entities lose sight of the importance their customers have to their existence. Looking at customers as an internal element of the supply chain allows the focus to sustain itself indefinitely. Too many companies, governmental agencies, educational institutions, and services become so focused on the other internal elements and issues that the primary end-customer focus is lost. Many cannot properly define who their primary end customers are and fail as a result. The term business-to-business (B2B) should be viewed as B2B2B2B2B2C with the "C" representing the customer. By incorporating the primary end customer into the supply chain and building the communication links and other infrastructure, the primary end customer is now a part of the entire supply chain, as it should be. As a result, the facilities planning process will take place with this primary end customer as the focus.

The facilities planning objectives are to:

- Improve customer satisfaction by being easy to do business with, conforming to customer promises, and responding to customer needs.
- Increase return on assets (ROA) by maximizing inventory turns, minimizing obsolete inventory, maximizing employee participation, and maximizing continuous improvement.
- Maximize speed for quick customer response.
- Reduce costs and grow the supply chain profitability.

- Integrate the supply chain through partnerships and communication.
- Support the organization's vision through improved material handling, material control, and good housekeeping.
- Effectively utilize people, equipment, space, and energy.
- Maximize return on investment (ROI) on all capital expenditures.
- Be adaptable and promote ease of maintenance.
- Provide for employee safety and job satisfaction.

It is not reasonable to expect that one facility design will be superior to all others for every objective listed. Some of the objectives conflict. Hence, it is important to evaluate carefully the performance of each alternative, using each of the appropriate criteria.

1.4 FACILITIES PLANNING PROCESS

The facilities planning process is best understood by placing it in the context of a facility life cycle. Although a facility is planned only once, it is frequently replanned to synchronize the facility and its constantly changing objectives. The facilities planning and replanning processes are linked by the continuous improvement facilities planning cycle shown in Figure 1.1. This process continues until a facility is torn down. The facility is continuously improved to satisfy its constantly changing objectives.

Even though facilities planning is not an exact science, it can be approached in an organized, systematic way. The traditional engineering design process can be applied to facilities planning as follows:

1. Define the problem.
   - *Define (or redefine) the objective of the facility.* Whether planning a new facility or the improvement of an existing facility, it is essential that the product(s) to be produced and/or service(s) to be provided be specified quantitatively. Volumes or levels of activity are to be identified whenever possible. The role of the facility within the supply chain must also be defined.
   - *Specify the primary and support activities to be performed in accomplishing the objective.* The primary and support activities to be performed and requirements to be met should be specified in terms of the operations, equipment, personnel, and material flows involved. Support activities allow primary activities to function with minimal interruption and delay. As an example, maintenance is a support activity for manufacturing.
2. Analyze the problem.
   - *Determine the interrelationships among all activities.* Establish whether and how activities interact with or support one another within the boundaries of the facility and how this is to be undertaken. Both quantitative and qualitative relationships should be defined.
3. *Determine the space requirements for all activities.* All equipment, material, and personnel requirements must be considered when calculating space requirements for each activity. Generate alternative designs.

- *Generate alternative facilities plans.* The alternative facilities plans will include both alternative facilities locations and alternative designs for the facility. The facilities design alternatives will include alternative layout designs, structural designs, and material handling system designs. Depending on the particular situation, the facility location decision and the facility design decision can be decoupled.

4. Evaluate the alternatives.

- *Evaluate alternative facilities plans.* On the basis of accepted criteria, rank the plans specified. For each, determine the subjective factors involved and evaluate whether and how these factors will affect the facility or its operation.

5. Select the preferred design.

- *Select a facilities plan.* The problem is to determine which plan, if any, will be the most acceptable in satisfying the goals and objectives of the organization. Most often, cost is not the only major consideration when evaluating a facilities plan. The information generated in the previous step should be utilized to arrive at the final selection of a plan.

6. Implement the design.

- *Implement the facilities plan.* Once the plan has been selected, a considerable amount of planning must precede the actual construction of a facility or the layout of an area. Supervising installation of a layout, getting ready to start up, actually starting up, running, and debugging are all part of the implementation phase of facilities planning.

- *Maintain and adapt the facilities plan.* As new requirements are placed on the facility, the overall facilities plan must be modified accordingly. It should reflect any energy-saving measures or improved material handling equipment that becomes available. Changes in product design or mix may require changes in handling equipment or flow patterns that, in turn, require an updated facilities plan.

- *Redefine the objective of the facility.* As indicated the first step, it is necessary to identify the products to be produced or services to be provided in specific, quantifiable terms. In the case of potential modifications, expansions, and so on for existing facilities, all recognized changes must be considered and integrated into the layout plan.

A novel approach to contemporary facilities planning is the Winning Facilities Planning process, as shown in Figure 1.4. A more detailed explanation of the Winning Facilities Planning process is shown in Table 1.2.

The Model of Success referred to in Figure 1.4 presents a clear direction for where a business is headed. Experience has shown that in order for the facilities plan to be successful, not only a clear understanding of the vision is needed but also the mission, the requirements of success, the guiding principles, and the evidence of success. It is the total of these five elements (Vision, Mission, Requirements of Success, Guiding Principles, and Evidence of Success) that forms an organization's Model of Success.

The definitions of these five elements are:

1. *Vision:* A description of where you are headed

2. *Mission:* How to accomplish the vision

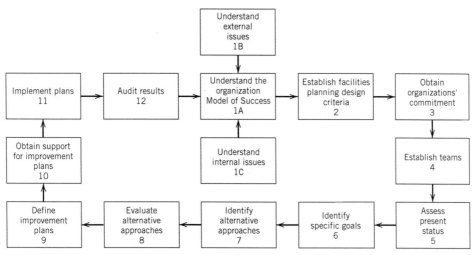

Figure 1.4 Winning Facilities Planning process. *Source:* Tompkins [3].

3. *Requirement of Success:* The science of your business
4. *Guiding Principles:* The values to be used while pursuing the vision
5. *Evidence of Success:* Measurable results that will demonstrate when an organization is moving toward its vision

Table 1.2 *Explanation of Winning Facilities Planning Process*

Step	Function	Comment
1A	Understand the organization model of success.	This requires an education program for all levels of an organization. Understanding an organization's Model of Success is a prerequisite for successful facilities planning.
1B	Understand external issues.	This requires external outreach via professional society involvement, participation in trade shows, seminars, and conferences, and reading magazines and books. A coordinated effort is required if external issues are to be well understood.
1C	Understand internal issues.	A winning organization must understand not only the Model of Success, but also the organization's business plan, resources, and constraints, and the objectives of the overall supply chain. A prerequisite to winning is understanding a company's future.
2	Establish facilities planning design criteria.	To implement improvements, an organization must have focus. This step requires that management determine the facilities planning design criteria.
3	Obtain organizational commitment.	Management must make a clear commitment to implement the justified improvements consistent with the facilities planning design criteria. This commitment must be uncompromised.
4	Establish teams.	Teams having a broad-based representation and the ability to make decisions should be established for each design requirement. These teams must be uncompromised.

Table 1.2 *continued*

Step	Function	Comment
5	Assess present status.	This assessment will result in the baseline against which improvements will be measured. Both quantitative and qualitative factors should be assessed.
6	Identify specific goals.	Identify clear, measurable, time-related goals for each design criterion, for example, "Reduce raw material inventory to $300,000 by June 1."
7	Identify alternative approaches.	The creative process of identifying alternative systems, procedures, equipment, or methods to achieve the specified goals. The investigation of all feasible alternatives.
8	Evaluate alternative approaches.	The economic and qualitative evaluation of the identified alternatives. The economic evaluation should adhere to corporate guidelines while estimating the full economic benefit of pursuing each alternative.
9	Define improvement plans.	Based upon the evaluation done in step 8, select the best approach. Define a detailed implementation and cash flow schedule.
10	Obtain support for improvement plans.	Sell the improvement plans to management. Document the alternatives, the evaluation, and the justification. Help management visualize the improved operation.
11	Implement plans.	Oversee development, installation, soft load, startup, and debug. Train operators and assure proper systems utilization. Stay with effort until results are achieved.
12	Audit results.	Document actual systems operation. Compare results with the specified goal and anticipated performance. Identify and document discrepancies. Provide appropriate feedback.

Source: Tompkins [3].

To help people understand where their organization is headed, it is often useful to illustrate the first four elements of the Model of Success in graphical form as shown in Figure 1.5. This graphical representation is often called the winning circle and is viewed as the organization's bull's eye.

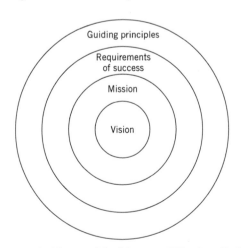

Figure 1.5 The Model of Success "Winning Circle."

Table 1.3 *Comparison of the Engineering Design Process, Facilities Planning Process, and Winning Facilities Planning Process*

Phase	The Engineering Design Process	The Facilities Planning Process	The Wining Facilities Planning Process
Phase I	Define problem.	1. Define or redefine objective of the facility. 2. Specify primary and support activities.	1A Understand the organization Model of Success. 1B Understand external issues. 1C Understand internal issues. 2. Establish facilities planning design criteria. 3. Obtain organizational commitment.
Phase II	Analyze the problem. Generate alternatives. Evaluate the alternatives. Select the preferred design.	3. Determine the inter-relationships. 4. Determine space requirements. 5. Generate alternative facilities plan. 6. Evaluate alternative facilities plan. 7. Select a facilities plan.	4. Establish teams. 5. Assess present status. 6. Identify specific goals. 7. Identify alternative approaches. 8. Evaluate alternative approach. 9. Define improvement plans. 10. Obtain support for improvement plans.
Phase III	Implement the design.	8. Implement the plan. 9. Maintain and adopt the facilities plan. 10. Redefine the objective of the facility	11. Implement plans. 12. Audit results

In Table 1.3, the traditional engineering design process and the Winning Facilities Planning process are compared. The first phases of the facilities planning process involve either the initial definition of the objectives of a new facility or the updating of an existing facility. These first phases are undertaken by the people charged with overall responsibility for facilities planning and management of the facility.

The second phase of the facilities planning process is assessing the present status, identifying specific goals, identifying alternative approaches, evaluating alternative approaches, defining improvement plans, and obtaining support for improvement. The final phase consists of implementing the plans and auditing the results. In applying the facilities planning concepts, an iterative process is often required to develop satisfactory facilities plans. The iterative process might involve considerable overlap, backtracking, and cycling through the analysis, generation, evaluation, and selection steps of the engineering design process.

At this point, a word of caution seems in order. You should not infer from our emphasis on a unified approach to facilities planning that the process of replanning a pantry in a cafeteria is identical to planning a new manufacturing facility. The scope of a project does affect the intensity, magnitude, and thoroughness of the study. However, the facility planning process described above and depicted in Figure 1.6 should be followed.

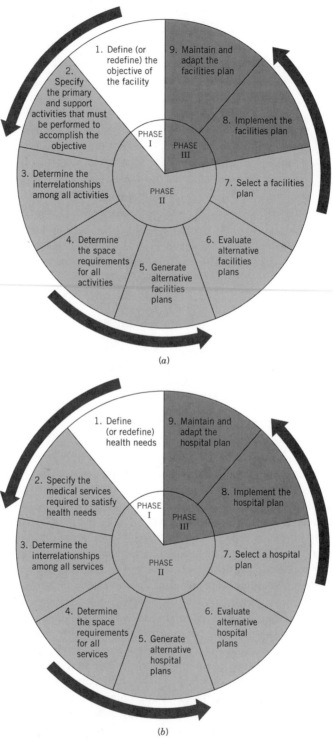

Figure 1.6 The facilities planning process. (*a*) General and manufacturing facilities. (*b*) Hospital facilities.

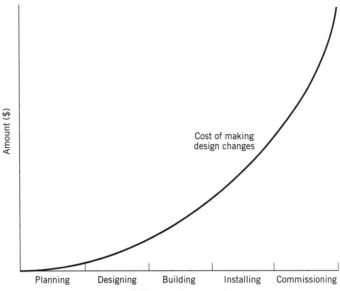

Figure 1.7 Cost of design changes during a project.

1.5 STRATEGIC FACILITIES PLANNING[1]

While it is true that the concerns of facilities planning are the location and the design of the facility, there exists another primary responsibility—*planning!* The importance of planning in facilities planning cannot be overemphasized, for it is this emphasis that distinguishes the activities of the facilities planner from the facilities designer and the facilities "locator."

Dwight D. Eisenhower said, "The plan is nothing, but planning is everything." As an indication of its importance in facilities planning, consider the process of planning and designing a manufacturing facility, building it, and installing and using the equipment. As shown in Figure 1.7, the costs of design changes increase exponentially as a project moves beyond the planning and designing phases.

The term **strategic planning** appears to have originated in the military. Webster defines strategy as "the science and art of employing the armed strength of a belligerent to secure the objects of a war." Today, the term is frequently used in politics, sports, investments, and business. Our concern is with the latter usage.

Based on Webster's definition of strategy, *business* strategies can be defined as *the art and science of employing the resources of a firm to achieve its business objectives.* Among the resources available are marketing resources, manufacturing resources, and distribution resources. Hence, marketing strategies, manufacturing strategies, and distribution strategies can be developed to support the achievement of the business objectives.

---

[1]A condensed version of the material presented here was published in *Industrial Engineering*. See reference 15 at the end of this chapter.

Recall that facilities planning was defined as determining how a firm's resources (tangible fixed assets) best support achieving the business objectives. In a real sense, facilities planning is itself a strategic process and must be an integral part of overall corporate strategy.

Historically, the development of corporate strategies has been restricted to the presidential level in many companies. Furthermore, business strategies tended to be limited to a consideration of such issues as acquisition, finance, and marketing. Consequently, decisions were often made without a clear understanding of the impact on manufacturing and distribution or on such support functions as facilities, material handling, information systems, and purchasing.

As an illustration, suppose an aggressive market plan is approved without the realization that manufacturing capacity is inadequate to meet the plan. Furthermore, suppose the lead-times required to achieve the required capacity are excessive. As a result, the market plan will fail because the impact of the plan on people, equipment, and space was not adequately comprehended. A Winning Facilities Plan must consider integrating all elements that will impact the plan. An example of the accumulating benefits that can result from integrating operations is shown in Figure 1.8.

Business Week, Industry Week, Time, Fortune, and other business publications have focused on the competitiveness of America. This attention reflects the growing

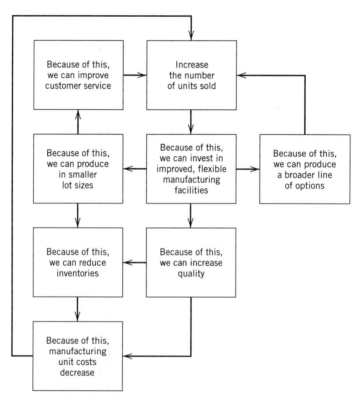

Figure 1.8 Synergistic benefit of Winning Manufacturing on an integrated manufacturing-marketing team. *Source:* Tompkins [1].

awareness in the business community of the importance of improved manufacturing productivity and technology. General Electric, Tektronix, Texas Instruments, and Westinghouse, among others, have expanded the strategic planning process to include the development of manufacturing strategies. Top management is recognizing the need to become involved in manufacturing decision making.

As noted by Skinner [14, p. 25], "When companies fail to recognize the relationship between manufacturing decisions and corporate strategy, they may become saddled with seriously noncompetitive production systems that are expensive and time-consuming to change." He goes on to state, "Manufacturing affects corporate strategy, and corporate strategy affects manufacturing. Even in an apparently routine operating area such as a production scheduling system, strategic considerations should outweigh technical and conventional industrial engineering factors invoked in the name of 'productivity'" [14, p. 27].

Manufacturing strategies must be integrated with the other elements of overall business strategy. Each element of the overall organization should have a strategic plan that supports the objectives of the firm. Within the manufacturing and distribution strategies, the storage, movement, protection, and control of material must be addressed.

1.6 DEVELOPING FACILITIES PLANNING STRATEGIES

The process of effectively translating objectives into actions can take place only if the power of the individuals inside an organization is unleashed. Team-based implementation of company objectives will ensure that all members of the organization are involved in its achievement.

As noted in the previous section, strategies are needed for such functions as marketing, manufacturing, distribution, purchasing, facilities, material handling, and data processing/information systems, among others. It is important to recognize that each functional strategy is multidimensional. Namely, each must support or contribute to the strategic plan for the entire organization. Furthermore, each must have its own set of objectives, strategies, and tactics.

As previously stated, one method used to ensure that the objectives are effectively translated into action is the Model of Success. The Model of Success is effective because it is lateral rather than hierarchical in its approach. With the traditional top-down approach, only a handful of people are actively involved in ensuring that the objectives are met by driving these goals and plans into action. The lateral structure of the Model of Success communicates to everyone in an organization where the organization is headed.

The facilities planning process can be improved in a number of ways. Three potential dimensions for improvement are illustrated in Figure 1.9. Suppose the objective is to increase the size of the box shown. One approach is to make it taller by focusing on the physical aspects of facilities planning, for example, buildings, equipment, and people. Another approach is to make the box wider by focusing on control aspects of facilities planning, for example, space standards, materials

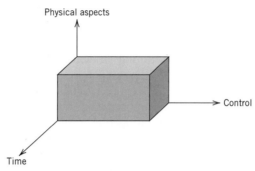

Figure 1.9   Three dimensions for improvement.

control, stock locator systems, and productivity measures. While it is possible to make the box taller and wider, we must not overlook the benefits provided by the third dimension: time. To make the box deeper requires time for planning. The old adage, "There's never time to do it right, but there's always time to do it over" has been repeatedly demonstrated with respect to facilities planning. Sufficient lead-time is needed to do it right!

Another way to improve this process is to do it in the context of Supply Chain Synthesis, a process that is well defined, integrated, and based on continuous improvement for maximized supply chain performance. It also harnesses the energy of change and has no information delays.

The facilities planning process should also be well defined as to how each function fits, interacts, and integrates. Otherwise, critical information will be lost or an important link will be missing and all will be lost. An excellent visualization of this is to consider your supply chain as if it were Y2K. No matter how much money was spent on upgrading computer systems so that they were Y2K compliant, the failure of one little chip somewhere could have brought down power grids, telecommunications systems, and banking systems. Those who wrote the files to ready systems for Y2K had to keep this fact in mind. Fortunately, they did and little or no effects of the changeover were felt.

The facilities planning process should be integrated and not allow selfishness. This includes eliminating silos and focusing all functions on customer satisfaction. To eliminate silos, we synthesize the whole from its origination point to the ultimate customer. The result is a focus on continuous improvement.

In the facilities planning process, everyone involved should understand the energy of change and have a desire to harness this energy for the competitive advantage of the total pipeline. This involves courage and innovation. By harnessing change we can turn it into an asset. Instead of thinking, "I want to improve my function," you may have to think, "Tradeoffs might be what are needed to improve the facilities planning process and create the ideal facility." The facilities planning process should not accept information delays. It requires true partnerships and an integration of information. To meet today's demands for speed, everyone involved in the process must do the right thing and let everyone else know what they are doing quickly. Communication is critical, robust, and simultaneous.

Facilities planning should be a continuous improvement process focused on achieving total performance excellence with the objectives presented earlier. Because all parties involved in the plan focus on these objectives, facilities planning excellence will be achieved.

A number of internal functional areas tend to have a significant impact on facilities planning, including marketing, product development, manufacturing, production and inventory control, human resources, and finance. Marketing decisions affect the location of facilities and the handling system design. For example, material handling will be affected by decisions related to unit volume, product mix, packaging, service levels for spares, and delivery times.

Product development and design decisions affect processing and materials requirements, which in turn affect layout and material handling. Changes in materials, component shapes, product complexity, number of new part numbers and package sizes introduced (due to a lack of standardization in design), stability of product design, and the number of products introduced will affect the handling, storage, and control of materials. Manufacturing decisions will have an impact on both facilities location and facilities design. Decisions concerning the degree of vertical integration, types and levels of automation, types and levels of control over tooling and work-in-process, plant sizes, and general-purpose versus special-purpose equipment can affect the location and design of manufacturing and support facilities.

Production planning and inventory control decisions affect the layout and handling system. Lot size decisions, production scheduling, in-process inventory requirements, inventory turnover goals, and approaches used to deal with seasonal demand affect the facilities plan.

Human resources and finance decisions related to capital availability, labor skills and stability, staffing levels, inventory investment levels, organizational design, and employee services and benefits will impact the size and design of facilities, as well as their number and location. Space and flow requirements will be affected by financial and human resources decisions. In turn, they have an impact on the storage, movement, protection, and control of material.

For the facilities plan to support the overall strategic plan, it is necessary for facilities planners to participate in the development of the plan. Typically, facilities planners tend to *react* to the needs defined by others, rather than participate in the decision making that creates the needs. A proactive rather than a reactive role for facilities planning is recommended. The Model of Success approach will ensure that facilities planners are on board, focusing on the overall direction of the company.

Close coordination is required in developing facilities plans to support manufacturing and distribution. Manufacturing/facilities planning and distribution/facilities planning interfaces are especially important. As the manufacturing plan addresses automatic load/unload of machines, robotics, group technology, transfer lines, flexible manufacturing systems, numerically controlled machines, just-in-time and computer-integrated manufacturing, alternative storage systems for tooling and work-in-process, real-time inventory control, shop floor control, and waste handling/removal systems, the facilities plan must support changes in manufacturing technology. Likewise, the facilities plan must support a distribution plan that addresses automatic palletizers, shrinkwrap/stretchwrap, automatic identification, automatic loading and unloading of vehicles and trailers, automated storage and retrieval of unit loads and small parts, and automated guided-vehicle systems.

It is also important for the level of manufacturing/distribution technology in use to be assessed objectively and compared with the state-of-the-art. Five- and 10-year technology targets should be identified and an implementation plan developed to facilitate the required evolution.

Developing contingency plans by asking numerous "what-if" questions is another important element of the process. By asking such questions, an uncertainty envelope can be developed for facility requirements. Also, in translating market projections to requirements for facilities, it is important to consider learning-curve effects, productivity improvements, technological forecasts, and site-capacity limits.

The following 10 issues may have a long-range impact on the strategic facilities plan:

1. Number, location, and sizes of warehouses and/or distribution centers
2. Centralized versus decentralized storage of supplies, raw materials, work-in-process, and finished goods for single- and multibuilding sites, as well as single- and multisite companies
3. Acquisition of existing facilities versus design of modern factories and distribution centers of the future
4. Flexibility required because of market and technological uncertainties
5. Interface between storage and manufacturing
6. Level of vertical integration, including "subcontract versus manufacture" decisions
7. Control systems, including material control and equipment control, as well as level of distributed processing
8. Movement of material between buildings and between sites, both inbound and outbound
9. Changes in customers' and suppliers' technology as well as a firm's own manufacturing technology and material movement, protection, storage, and control technology
10. Design-to-cost goals for facilities

1.7 EXAMPLES OF INADEQUATE PLANNING

Numerous examples exist of situations where inadequate planning was being performed. The following actual situations are presented to illustrate the need for improved planning.

- A large consumer products company decided to allow each of its acquisitions to remain independent, thus requiring the management of many duplicate logistics organizations. The organizations consisted of duplicate planning functions, execution systems, and facility locations. After poor performance, the management team soon began to question the rationale of the separate organizations.

- A major manufacturer in the Midwest made a significant investment in storage equipment for a parts distribution center. The selection decision was based on the need for a "quick fix" to a pressing requirement for increased space utilization. The company soon learned that the "solution" would not provide the required throughput and was not compatible with long-term needs.

- An electronics manufacturer was faced with rapid growth. Management received proposals that required approximately equivalent funding for large warehouses at two sites having essentially the same storage and throughput requirements. Management questioned the rationale for one "solution" being a high-rise Automated Storage/Retrieval System and the other being a low-rise warehouse with computer-controlled industrial trucks.

- Another firm installed miniload systems at two sites. One system was designed for random storage, the other for dedicated storage. The storage and throughput requirements were approximately the same for the two systems; however, different suppliers had provided the equipment and software. Management raised the questions: Why are they different? And which is best?

- A textile firm installed a large high-rise AS/RS for one of its divisions. The amount and size of the product to be stored subsequently changed. Other changes in technology were projected. The system became obsolete before it was operational.

- An engine manufacturer was planning to develop a new site. Decisions had not been made concerning which products would be off-loaded to the new site, nor what effect the off-load would have on requirements for moving, protecting, storing, and controlling material.

- An electronics manufacturer was planning to develop a new site. The facilities planners and architects were designing the first building for the site. No projections of space and throughput had been developed since decisions had not been made concerning the occupant of the building.

- The mission for a major military supply center was changed in order that additional bases could be serviced. The throughput, storage, and control requirements for the new customers were significantly different from those for which the system was originally designed. However, no modifications to the system were funded.

- A manufacturer of automotive equipment acquired the land for a new manufacturing plant. The manufacturing team designed the layout, and the architect began designing the facility before the movement, protection, storage, and control system was designed.

- An aerospace-related manufacturer implemented group technology in its process planning and converted to manufacturing cells in a machining department. No analyses had been performed to determine queue or flow requirements. Subsequent analyses showed the manufacturing cells were substantially less efficient as a result of their impact on movement, protection, storage, and control of work-in-process [14].

- A large consumer products company decided to allow each of its acquisitions to remain independent, thus requiring the management of many duplicate logistics organizations. The organizations consisted of duplicate planning functions, execution systems and facility locations. After poor performance, the management team soon began to question the rationale of the separate organizations.

- An established brick-and-mortar retailer began accepting orders through its Web site. The volume of orders received during the holiday season peak could not be processed by its distribution center. Gift certificates had to be mailed to all of the customers whose orders weren't delivered by Christmas. A study conducted after the new year showed that poor configuration of storage racks, ineffective replenishment processes, lack of proper product slotting, and material handling equipment that could not efficiently process the variety of the products' attributes created a situation that forced the entire fulfillment operation to grind to a halt.

In practically every case the projects were interrupted and significant delays were incurred because proper facilities planning had not been performed. These examples emphasize once more the importance of providing adequate leadtimes for planning.

The previous list of examples of inadequate facilities planning could possibly create a false impression that no one is doing an adequate planning job. Such is not the case; several firms have recognized the need for strategic facilities planning and are doing it.

A major U.S. airline developed 10-year and 20-year facilities plans to facilitate decision making regarding fleet size and mix. Maintenance and support facilities requirements were analyzed for wide-body and mid-sized aircraft. The impact of route planning, mergers and acquisitions, and changes in market regions to include international flights were considered in developing the plan.

The airline industry operates in a dynamic environment. Governmental regulations and attitudes toward business are changeable, energy costs and inflationary effects are significant, and long leadtimes are required for aircraft procurement. For new-generation aircraft, an airline company might negotiate procurement conditions, including options, eight years before taking delivery of the airplane.

1.8 SUMMARY

Facilities planning is a process that is dynamic over time. The methodology continues to change as technology evolves and new approaches are developed. The focus at the current time is on the customer and the view that all components of a supply chain must band together to plan the facilities that will successfully support all of the activities of the supply chain.

Facilities planning:

- Determines how an activity's tangible fixed assets should contribute to meeting the activity's objectives
- Consists of facilities location and facilities design
- Is part art and part science
- Can be approached using the engineering design process
- Is a continuous process and should be viewed from a life-cycle perspective
- Represents one of the most significant opportunities for cost reduction and productivity improvement

Strategic facilities planning is needed to support competition from a supply chain versus supply chain point of view. No longer is the focus of strategic facilities planning only internal. The focus now is on how our facilities planning process supports the entire supply chain from basic raw materials to the final customer. If the facilities planning process does not support the entire supply chain, it is at a disadvantage. Other supply chains may be able to leverage themselves into an advantage by focusing on the customer and on the big picture, rather than simply one location or one company. Moving forward, this focus on the entire supply chain will grow even stronger, and those companies and those supply chains that do not realize this fact will no longer exist.

BIBLIOGRAPHY

1. Cullinane, T. P., and Tompkins, J. A., "Facility Layout in the 80's: The Changing Conditions," *Industrial Engineering,* vol. 12, no. 9, pp. 34–42, September 1980.
2. Holmes, W. G., *Plant Location,* McGraw-Hill, New York, 1930.
3. Tompkins, J. A., *Winning Manufacturing: The How To Book of Successful Manufacturing,* IIE, Norcross, GA, 1989.
4. Material Handling Industry of America, *ProMat 93 Proceedings,* Charlotte, NC, vol. 6, pp. 75–83.
5. Apple, J. M., Jr., "Strategic Planning for Material Handling Improvements," presentation to the AIIE/MHI Seminar, San Frncisco, March 1980.
6. Collins, J.D., "Material Handling System Concept," presentation to the 26th Annual Material Handling Management Course, American Institute of Industrial Engineers, Mt. Pocono, PAP, June 1979.
7. Collins, J.D., "Strategic Planning for Material Handling and Storage," presentation to the Computer Integrated Manufacturing: Productivity for the 1980s, A Joint Industry–DoD Manufacturing Technology Workshop, Society of Manufacturing Engineers, Detroit, MIP, September 1979.
8. Geoffrion, A. M., Graves, G. W., and Lees, S. J., "Strategic Distribution System Planning: A Status Report," Chapter 7 in A. Hax (ed.), *Studies in Operations Management,* North-Holland, Amsterdam, New York, 1978.
9. Kinney, H. D., "A Totally Intergrated Material Management System," presentation to the 1980 Annual Conference of the Internatinal Material Management Society, Cincinnti, OH, June 1980.
10. McCord, A. R., "TI's Preparations for the Challenges of the Eighties," a presentation to members of the Institute de L'Enterprise given at the University of Texas at Austin, March 1979.
11. Phipps, C. H., "Strategic Development in a Technology Intensive Company: Texas Instruments OST System," presented to U.S. Army Corporate planners, May 1979.
12. Radford, K. J., *Strategic Planning: An Analytical Approach,* Reston Publishing, Reston, VAP, 1980.
13. Rothschild, W. E., "How to Ensure the Countinued Growth of Strategic Planning," *The Journal of Business Strategy,* vol. 1, no. 1, pp. 11–18, Summer 1980.
14. Skinner, W., *Manufacturing in the Corporate Strategy,* Wiley, New York, 1978.
15. Tomkins, J. A., *Winning Manufacturing: The How To Book of Successful Manufacturing,* IIE, Norcross, GA, 1989.
16. White, J. A., "The Planning Process Need Not Be Painful," presentation to the National Material Handling Show and Seminar, The Material Handling Institute, Chicago, ILP, June 1978.
17. White, J. A., "SEP—Strategic Facilities Planning," *Modern Materials Handling,* pp. 23–25, November 1978.

18.   White, J. A., *Yale Management Guide to Productivity,* Eaton Corp., Philadelphia, PAP, September 1979.

19.   White, J. A., "Strategic Facility Planning," *Industrial Engineering,* vol 2, no. 9, pp. 94–100, September 1980.

20.   *Measuring Productivity in Physical Distribution: A 40 Billion Dollar Goldmine,* National Council of Physical Distribution Management, Chicago, IL, 1978.

PROBLEMS

1.1   Describe both the planning and operating activities required to conduct a professional football game from the point of view of the
   a. visiting team's football coach.
   b. home team quarterback.
   c. manager of refreshment vending.
   d. ground crew manager.
   e. stadium maintenance manager.

1.2   List 10 components of a football stadium facility.

1.3   Describe the activities that would be involved in the (a) facilities location, (b) facilities design, and (c) facilities planning of an athletic stadium. Consider baseball, football, soccer, and track and field.

1.4   Assume you are on a job interview and you have listed on your resume a career interest in facilities planning. The firm where you are interviewing is a consulting firm that specializes in problem solving for transportation, communication, and the service industries. React to the following statement directed to you by the firm's personnel director: "Facilities planning is possibly of interest to a firm involved in manufacturing, but it is not clear that our customers have needs in this area sufficient for you to pursue your field of interest."

1.5   What criterion should be utilized to determine the optimal facilities plan?

1.6   Evaluate the facilities plan for your campus and list potential changes you would consider if you were asked to replan the campus. Why would you consider these?

1.7   Chart the facilities planning process for:
   a. a bank.
   b. a university campus.
   c. a warehouse.
   d. a consulting and engineering office.

1.8   Describe the procedure you would follow to determine the facilities plan for a new library on your campus.

1.9   Is facilities planning ever completed for an enterprise? Why or why not?

1.10   With the aid of at least three references, write a paper on the industrial engineer and architect's roles in the planning of a facility.

1.11   Consider the definition of industrial engineering approved by the Institute of Industrial Engineers. Discuss the extent to which the definition applies to facilities planning.

1.12   Read three papers on strategic planning published in the *Harvard Business Review,* summarize the material, and relate it to strategic facilities planning.

1.13   Develop a list of strategic issues that must be addressed in performing facilities planning for:
   a. an airport.
   b. a community college.

c. a bank.

d. a grocery store chain.

e. a soft drink bottler and distributor.

f. a library.

g. an automobile dealership.

h. a shopping center developer.

i. a public warehousing firm.

j. a professional sports franchise.

1.14 Develop a set of responses to the following "reasons" for not doing strategic facilities planning:

a. There are more critical short-term problems to be solved.

b. The right people internally are too busy to be involved in the project.

c. The future is too hard to predict, and it will probably change anyway.

d. Nobody really knows what alternatives are available and which ones might apply.

e. Technology is developing very rapidly; any decisions we make will be obsolete before they can be implemented.

f. The return on investment in strategic planning is hard to measure.

1.15 What is the impact of facilities planning on the competitiveness of manufacturing facilities?

1.16 What are the implications of strategic planning on your personal career planning?

1.17 What are the impacts of automation on facilities planning?

1.18 What are the issues to be addressed in strategic planning for warehousing/distribution? What are the cost and customer services implications?

1.19 Explain the impacts of SCS on facilities planning.

1.20 What are the differences between strategic planning and contingency planning?

1.21 How does the issue of time impact the process of facilities planning?

2

PRODUCT, PROCESS, AND SCHEDULE DESIGN

2.1 INTRODUCTION

Based on the presentation given in Chapter 1, the facilities planning process for manufacturing and assembly facilities can be listed as follows:

1. Define the products to be manufactured and/or assembled.
2. Specify the required manufacturing and/or assembly processes and related activities.
3. Determine the interrelationships among all activities.
4. Determine the space requirements for all activities.
5. Generate alternative facilities plans.
6. Evaluate the alternative facilities plans.
7. Select the preferred facilities plan.
8. Implement the facilities plan.
9. Maintain and adapt the facilities plan.
10. Update the products to be manufactured and/or assembled and redefine the objective of the facility.

The facilities planning process will be greatly impacted by the business strategic plan and the concepts, techniques, and technologies to be considered in the manufacturing and assembly strategy.

Before the 1980s, most American companies were not considering small lot purchasing and production, multiple receiving docks, deliveries to the points of use, decentralized storage, areas, cellular manufacturing, flat organizational structures, "pull" approaches with kanbans, team decision making, electronic data interchange, process quality, self-rework, multifunctional "associates," and many other concepts that are drastically changing the facilities planning process and the final facilities plan.

Among the questions to be answered before alternative facility plans can be generated are the following:

1. What is to be produced?
2. How are the products to be produced?
3. When are the products to be produced?
4. How much of each product will be produced?
5. For how long will the products be produced?
6. Where are the products to be produced?

The answers to the first five questions are obtained from product design, process design, and schedule design, respectively. The sixth question might be answered by facilities location determination or it might be answered by schedule design when production is to be allocated among several existing facilities.

Answering the sixth question has become much more complicated in recent years. Currently, many firms have global production strategies and utilize combinations of contract manufacturing and contract assembly. The textile industry, for example, has undergone tremendous change, with global sourcing occurring for yarn and textile production as well as for garment assembly. Few domestic sewing operations continue to exist in the United States.

The automobile is another example of global sourcing, resulting in the final product being called a *world car*; engines, powertrains, bodies, electronic assemblies, seating, and tires may well be manufactured in different countries. Similar conditions exist for the production of home appliances, computers, and televisions, with subassemblies and components being produced around the world.

Product designers specify what the end product is to be in terms of dimensions, material composition, and perhaps, packaging. The process planner determines how the product will be produced. The production planner specifies the production quantities and schedules the production equipment. The facilities planner is dependent on timely and accurate input from product, process, and schedule designers to carry out his task effectively.

In this chapter we focus on the product, process, and schedule (PP&S) design functions as they relate to facilities planning. The success of a firm is dependent on having an efficient production system. Hence, it is essential that product designs, process selections, production schedules, and facilities plans be mutually supportive. Figure 2.1 illustrates the need for close coordination among the four functions.

Frequently, organizations create teams with product, process, scheduling, and facilities design planners and with personnel from marketing, purchasing, and accounting to address the design process in an integrated, simultaneous, or concurrent way. Customer and supplier representatives are often involved in this process. These teams are referred to as *concurrent* or *simultaneous engineering* teams and reduce the design cycle time, improve the design process, and eliminate engineering changes. Companies implementing this integrated approach have reported impressive results in cost, quality, productivity, sales, customer satisfaction, delivery time, inventories, space and handling requirements, and building size, among others [14].

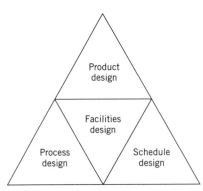

Figure 2.1   Relationship between product, process, and schedule (PP&S) design and facilities planning.

Product, process, schedule, and facilities design decisions are not made independently and sequentially. A clear vision is needed of what to do and how to do it (including concepts, techniques, and technologies to consider). For example, management commitment to the use of multiple receiving docks, smaller lot sizes, decentralized storage areas, open offices, decentralized cafeterias, self-managing teams, and focused factories will guide the design team in the generation of the best alternatives to satisfy business objectives and goals and make the organization more competitive.

In the case of an existing facility with ongoing production operations, a change in the design of a product, the introduction of a new product, changes in the processing of products, and modifications to the production schedule can occur without influencing the location or design of the facility.

In many cases, however, changes in product, process, and schedule design of current manufacturing facilities will require layout, handling, and/or storage modifications (especially if the design change requires more or less inventories, space, people, offices, and machines).

In this chapter, we are concerned with the interactions between facilities planning and PP&S design. Hence, we assume either a new facility or a major expansion/modification to an existing facility is being planned.

The *seven management and planning tools* methodology presented in Section 2.5 illustrates the integration of the facilities design process for a given scenario.

2.2   PRODUCT DESIGN

Product design involves both the determination of which products are to be produced and the detailed design of individual products. Decisions regarding the products to be produced are generally made by top management based on input from marketing, manufacturing, and finance concerning projected economic performance. In other instances, as illustrated in Chapter 1, the lead times to plan and build facilities, in the face of a dynamic product environment, might create a situation in which it is not possible to accurately specify the products to be produced in a given facility.

The facilities planner must be aware of the degree of uncertainty that exists concerning the mission of the facility being planned, the specific activities to be performed and the direction of those activities. As an illustration, a major electronics firm initially designed a facility for semiconductor manufacture. Before the facility was occupied, changes occurred in space requirements and another division of the company was assigned to the facility; the new occupant of the site used the space for manufacturing and assembling consumer electronic products. As the division grew in size, many of the manufacturing and assembly operations were off-loaded to newly developed sites and the original facility was converted to predominantly an administrative and engineering site.

Depending on the type of products being produced, the business philosophy concerning facilities, and such external factors as the economy, labor availability and attitudes, and competition, the occupants of a facility might change frequently or never change at all. Decisions must be made very early in the facilities planning process regarding the assumptions concerning the objectives of the facility.

If it is decided that the facility is to be designed to accommodate changes in occupants and mission, then a highly flexible design is required and very general space will be planned. On the other hand, if it is determined that the products to be produced can be stated with a high degree of confidence, then the facility can be designed to optimize the production of those particular products.[1]

The design of a product is influenced by aesthetics, function, materials, and manufacturing considerations. Marketing, purchasing, industrial engineering, manufacturing engineering, product engineering, and quality control, among others, will influence the design of the product. In the final analysis, the product must meet the needs of the customer.

This challenge can be accomplished through the use of Quality Function Deployment (QFD) [1]. QFD is an organized planning approach to identify customer needs and to translate the needs to product characteristics, process design, and tolerances required.

Benchmarking can also be used to identify what the competition is doing to satisfy the needs of your customer or to exceed the expectations of your customer [8]. It can also be used to identify best practices and the best of the best organizations.

Through QFD and benchmarking, product designers can focus their work on customer needs being met marginally or not at all (compared to the competition and to the best organizations).

Detailed operational specifications, pictorial representations, and prototypes of the product are important inputs for the facilities planner. Exploded assembly drawings, such as that given in Figure 2.2, are quite useful in designing the layout and handling system. These drawings generally omit specifications and dimensions, although they are drawn to scale.

As an alternative to the exploded assembly drawing, a photograph can be used to show the parts properly oriented. Such a photograph is given in Figure 2.3. Photographs and drawings allow the planner to visualize how the product is assembled, provide a reference for part numbers, and promote clearer communications during oral presentations.

---

[1]Minor changes in product design and the addition of similar products to the product family would be included in this scenario.

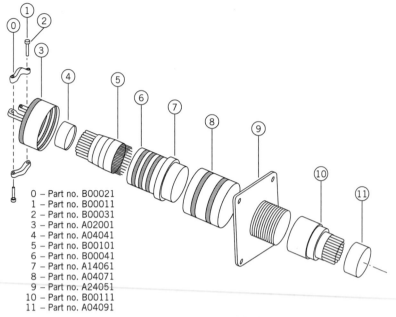

0 – Part no. B00021
1 – Part no. B00011
2 – Part no. B00031
3 – Part no. A02001
4 – Part no. A04041
5 – Part no. B00101
6 – Part no. B00041
7 – Part no. A14061
8 – Part no. A04071
9 – Part no. A24051
10 – Part no. B00111
11 – Part no. A04091

Figure 2.2   Exploded assembly drawing.

Detailed component part drawings are needed for each component part. The drawings should provide part specifications and dimensions in sufficient detail to allow part fabrication. Examples of component part drawings are given in Figures 2.4 and 2.5. The combination of exploded assembly drawings and component part drawings fully documents the design of the products.

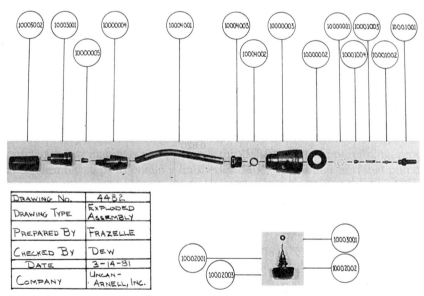

Figure 2.3   Exploded parts photograph.

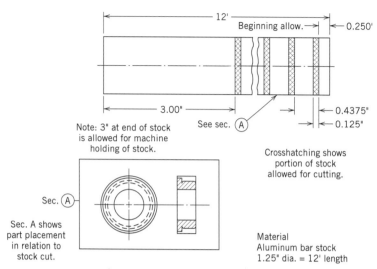

Figure 2.4 Component part drawing of a plunger.

The drawings can be prepared and analyzed with *computer aided design* (CAD) systems. CAD is the creation and manipulation of design prototypes on a computer to assist the design process of the product. A CAD system consists of a collection of many application modules under a common database and graphics editor. The blending of computers and the human ability to make decisions enable us to use CAD systems in design, analysis, and manufacturing [9].

During the facilities design process, the computer's graphics capability and computing power allow the planner to visualize and test ideas in a flexible manner. The CAD system also can be used for area measurement, building and interior design, layout of furniture and equipment, relationship diagramming, generation of block and detailed layouts, and interference checks for process oriented plants [2].

In addition to CAD, *concurrent engineering* (CE) can be used to improve the relationship between the function of a component or product and its cost. Concurrent engineering provides a simultaneous consideration in the design phase of life cycle factors such as product, function, design, materials, manufacturing processes,

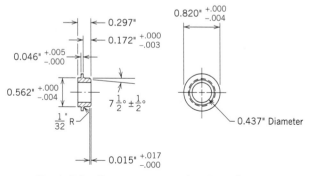

Figure 2.5 Component part drawing of a seat.

testability, serviceability, quality, and reliability. As a result of this analysis, a less expensive but functionally equivalent product design might be identified. Concurrent engineering is important because it is at the design stage that many of the costs of a product are specified. It has been estimated that more than 70% of a product's manufacturing cost is dictated by design decisions [14].

In addition to the design tools we have described, many other tools can be used to help the product-process designer in the evaluation and selection of the best product and process design combination [10, 21, 23].

2.3   PROCESS DESIGN

The process designer or process planner is responsible for determining *how* the product is to be produced. As a part of that determination, the process planner addresses *who* should do the processing; namely, should a particular product, sub-assembly, or part be produced in-house or subcontracted to an outside supplier or contractor? The "make-or-buy" decision is part of the process planning function.

In addition to determining whether a part will be purchased or produced, the process designer must determine how the part will be produced, which equipment will be used, and how long it will take to perform the operation. The final process design is quite dependent on input from both the product and schedule designs.

Identifying Required Processes

Determining the scope of a facility is a basic decision and must be made early in the facilities planning process. For a hospital whose objective is to serve the health needs of a community, it may be necessary to limit the scope of the facility by not including in the facility a burn-care clinic, specific types of diagnostic equipment, and/or a psychiatric ward. The excluded services, although needed by the community, may not be feasible for a particular hospital. Patients requiring care provided elsewhere would be referred to other hospitals. Similarly, the scope of a manufacturing facility must be established by determining the processes that are to be included within the facility. The extremes for a manufacturing facility may range from a vertically integrated firm that purchases raw materials and proceeds through a multitude of refining, processing, and assembly steps to obtain a finished product, to another firm that purchases components and assembles finished products. Therefore, it is obvious that the scope and magnitude of activities within a manufacturing facility are dependent on the decisions concerning the level of vertical integration. Such decisions are often referred to as "make-or-buy" decisions or "sourcing" decisions.

Large corporations have downsized large facilities and broken them into small business units that keep only processes that are economically feasible. Small business units operate with low overhead, low management levels, and frequently with self-managing operator teams. Buildings for this type of organization are smaller and management functions and offices are usually decentralized.

Make-or-buy decisions are typically managerial decisions requiring input from finance, industrial engineering, marketing, process engineering, purchasing, and perhaps human resources, among others. A brief overview of the succession of questions leading to make-or-buy decisions is given in Figure 2.6. The input to the facilities planner is a listing of the items to be made and the items to be purchased. The listing often takes the form of a parts list or a bill of materials.

The parts list provides a listing of the component parts of a product. In addition to make-or-buy decisions, a parts list includes at least the following:

1. Part numbers
2. Part name

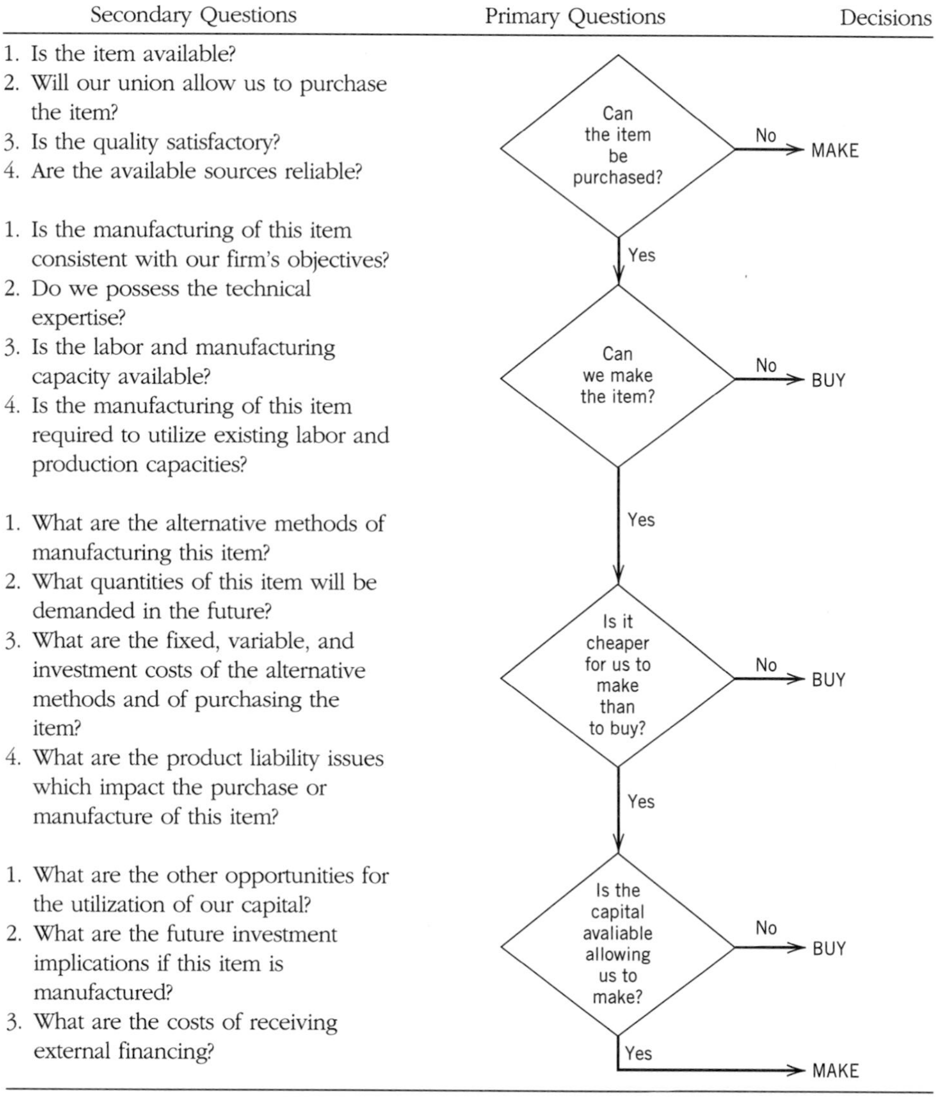

Secondary Questions	Primary Questions	Decisions
1. Is the item available? 2. Will our union allow us to purchase the item? 3. Is the quality satisfactory? 4. Are the available sources reliable?	Can the item be purchased? — No	MAKE
	Yes	
1. Is the manufacturing of this item consistent with our firm's objectives? 2. Do we possess the technical expertise? 3. Is the labor and manufacturing capacity available? 4. Is the manufacturing of this item required to utilize existing labor and production capacities?	Can we make the item? — No	BUY
	Yes	
1. What are the alternative methods of manufacturing this item? 2. What quantities of this item will be demanded in the future? 3. What are the fixed, variable, and investment costs of the alternative methods and of purchasing the item? 4. What are the product liability issues which impact the purchase or manufacture of this item?	Is it cheaper for us to make than to buy? — No	BUY
	Yes	
1. What are the other opportunities for the utilization of our capital? 2. What are the future investment implications if this item is manufactured? 3. What are the costs of receiving external financing?	Is the capital avaliable allowing us to make? — No	BUY
	Yes	MAKE

Figure 2.6 The make-or-buy decision process.

PARTS LIST

Company T. W., Inc.   Prepared by J. A.
Product Air Flow Regulator   Date

Part No.	Part Name	Drwg. No.	Quant./ Unit	Material	Size	Make or Buy
1050	Pipe plug	4006	1	Steel	.50″ × 1.00″	Buy
2200	Body	1003	1	Aluminum	2.75″ × 2.50″ × 1.50″	Make
3250	Seat ring	1005	1	Stainless steel	2.97″ × .87″	Make
3251	O-ring	—	1	Rubber	.75″ dia.	Buy
3252	Plunger	1007	1	Brass	.812″ × .715″	Make
3253	Spring	—	1	Steel	1.40″ × .225″	Buy
3254	Plunger housing	1009	1	Aluminum	1.60″ × .225″	Make
3255	O-ring	—	1	Rubber	.925″ dia.	Buy
4150	Plunger retainer	1011	1	Aluminum	.42″ × 1.20″	Make
4250	Lock nut	4007	1	Aluminum	.21″ × 1.00″	Buy

Figure 2.7   Parts list for an air flow regulator.

3.   Number of parts per product

4.   Drawing references

A typical parts list is given in Figure 2.7.

A bill of materials is often referred to as a structured parts list, as it contains the same information as a parts list plus information on the structure of the product. Typically, the product structure is a hierarchy referring to the level of product assembly. Level 0 usually indicates the final product; level 1 applies to subassemblies and components that feed directly into the final product; level 2 refers to the subassemblies and components that feed directly into the first level, and so on. A bill of materials is given in Figure 2.8 and is illustrated in Figure 2.9 for the product described in the parts list in Figure 2.7.

Selecting the Required Processes

Once a determination has been made concerning the products to be made "inhouse," decisions are needed as to how the products will be made. Such decisions are based on previous experiences, related requirements, available equipment, production rates, and future expectations. Therefore, it is not uncommon for different processes to be selected in different facilities to perform identical operations. However, the selection procedure used should be the same. In Figure 2.10, the procedure for process selection is given.

BILL OF MATERIALS

Company T. W., Inc.   Prepared by J. A.
Product Air Flow Regulator   Date

Level	Part No.	Part Name	Drwg. No.	Quant./ Unit	Make or Buy	Comments
0	0021	Air flow regulator	0999	1	Make	
1	1050	Pipe plug	4006	1	Buy	
1	6023	Main assembly	—	1	Make	
2	4250	Lock nut	4007	1	Buy	
2	6022	Body assembly	—	1	Make	
3	2200	Body	1003	1	Make	
3	6021	Plunger assembly	—	1	Make	
4	3250	Seat ring	1005	1	Make	
4	3251	O-ring	—	1	Buy	
4	3252	Plunger	1007	1	Make	
4	3253	Spring	—	1	Buy	
4	3254	Plunger housing	1009	1	Make	
4	3255	O-ring	—	1	Buy	
4	4150	Plunger retainer	1011	1	Make	

Figure 2.8 Bill of materials for an air flow regulator.

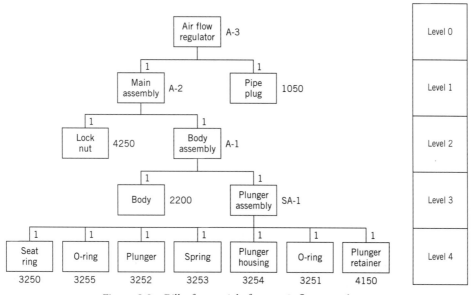

Figure 2.9 Bill of materials for an air flow regulator.

PROCESS IDENTIFICATION

Define elemental operations	Step 1
Identify alternative process for each operation	Step 2
Analyze alternative processes	Step 3
Standardize processes	Step 4
Evaluate alternative processes	Step 5
Select processes	Step 6

Figure 2.10   Process selection procedure.

Input into the process selection procedure is called **process identification.** Process identification consists of a description of what is to be accomplished. For a manufactured product, process identification consists of a parts list indicating what is to be manufactured, component part drawings describing each component, and the quantities to be produced.

Computer aided process planning (CAPP) can be used to automate the manual planning process. There are two types of CAPP Systems: *variant and generative.* In a variant CAPP, standard process plans for each part family are stored within the computer and called up whenever required. In generative process planning, process plans are generated automatically for new components without requiring the existing plans. Selection of these systems basically depends on product structure and cost considerations. Typically, variant process planning is less expensive and easier to implement.

Since process planning is a critical bridge between design and manufacturing, CAPP systems can be used to test the different alternative routes and interact with the facility design process. The input of a CAPP system is commonly a three dimensional model from a CAD database including information related to tolerances and special features. Based on these inputs, manufacturing leadtime and resource requirements can also be determined [2].

The facilities planner will not typically perform process selection. However, an understanding of the overall procedure provides the foundation for the facilities plan. Step 1 of the procedure involves the determination of the operations required to produce each component. In order to make this determination, alternative forms of raw materials and types of elemental operations must be considered. Step 2 involves the identification of various equipment types capable of performing elemental operations. Manual, mechanized, and automated alternatives should be considered. Step 3 includes the determination of unit production times and equipment utilizations for various elemental operations and alternative equipment types. The utilizations are inputs into step 4 of the procedure. Step 5 involves an economic evaluation of alternative equipment types. The results of the economic evaluation along with intangible factors such as flexibility, versatility, reliability, maintainability, and safety serve as the basis for step 6.

Computer programs that perform time and cost evaluations of different process combinations have also been developed to aid process planners [10, 21].

The outputs from the process selection procedure are the processes, equipment, and raw materials required for the in-house production of products. Output is generally given in the form of a **route sheet.** A route sheet should contain at least the data given in Table 2.1. Figure 2.11 is a route sheet for the production given in part in Table 2.1.

Table 2.1   *Route Sheet Data Requirements*

Data	Production Example
Component name and number	Plunger housing—3254
Operation description and number	Shape, drill, and cut off—0104
Equipment requirements	Automatic screw machine and appropriate tooling
Unit times	Setup time—5 hr operation time—.0057 hr per component
Raw material requirements	1 in. dia. × 12 ft. aluminum bar per 80 components

Sequencing the Required Processes

The only process selection information not yet documented is the method of assembling the product. An **assembly chart,** given in Figure 2.12, provides such documentation. The easiest method of constructing an assembly chart is to begin with the completed product and to trace the product disassembly back to its basic components. For example, the assembly chart given in Figure 2.12 would be constructed by beginning in the lower right-hand corner of the chart with a finished air flow regulator. The first disassembly operation would be to unpackage the air flow regulator (operation A-4). The operation that precedes packaging is the inspection of the air flow regulator. Circles denote assembly operation; inspections are indicated on assembly charts as squares. Therefore, in Figure 2.12, a square labeled I-1 immediately precedes operation A-4. The first component to be disassembled from the air flow regulator is part number 1050, the pipe plug, indicated by operation A-3. The lock nut is then disassembled, followed by the disassembly of the body assembly (the subassembly made during subassembly operation SA-1) and the body. The only remaining steps required to complete the assembly chart are the labeling of the circles and lines of the seven components flowing into SA-1.

Although route sheets provide information on production methods and assembly charts indicate how components are combined, neither provides an overall understanding of the flow within the facility. However, by superimposing the route sheets on the assembly chart, a chart results that does give an overview of the flow within the facility. This chart is an **operation process chart.** An example of an operation process chart is given in Figure 2.13.

To construct an operation process chart, begin at the upper right side of the chart with the components included in the first assembly operation. If the components are purchased, they should be shown as feeding horizontally into the appropriate assembly operation. If the components are manufactured, the production methods should be extracted from the route sheets and shown as feeding vertically into the appropriate assembly operation. The operation process chart may be completed by continuing in this manner through all required steps until the product is ready for release to the warehouse.

The operation process chart can also include materials needed for the fabricated components. This information can be placed below the name of the component. Additionally, operation times can be included in this chart and placed to the left of

ROUTE SHEET

Company Produce	A.R.C., Inc. Air Flow Regulator		Part Name	Plunger Housing	Part No. 3254	Prepared by	J. A. Date

Oper. No.	Operation Description	Machine Type	Tooling	Dept.	Set-up Time (hr.)	Operation Time (hr.)	Materials or Parts Description
0104	Shape, drill, cut off	Automatic screw machine	.50 in. dia. coller, feed fingers, cir. form tool, .45 in. dia. center drill, .129 in. twist drill, finish spiral drill, cut off blade		5	.0057	Aluminum 1.0 in. dia. × 12 ft.
0204	Machine slot and thread	Chucker	.045 in. slot saw, turret slot attach. 3/8–32 thread chaser		2.25	.0067	
0304	Drill 8 holes	Auto. dr. unit (chucker)	.078 in. dia. twist drill		1.25	.0038	
0404	Deburr and blow out	Drill press	Deburring tool with pilot		.5	.0031	
SA1	Enclose subassembly	Dennison hyd. press	None		.25	.0100	

Figure 2.11 Route sheet for one component of the air flow regulator.

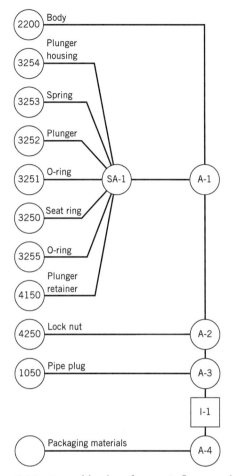

Figure 2.12 Assembly chart for an air flow regulator.

operations and inspections. A summary of the number of operations and inspections and operation times can be provided below the chart.

The operation process chart can be complemented with transportations, storages, and delays (including distances and times) when the information is available. Such a chart is referred to as a **flow process chart** by some and a **process chart** by others [3].

Assembly charts and operation process charts may be viewed as analog models of the assembly process and the overall production process, respectively. As noted previously, circles and squares represent time and horizontal connections represent sequential steps in the assembly of the product.

Notice in the assembly chart that components have been identified with a four-digit code starting with 1, 2, 3, and 4. Furthermore, subassemblies (SA) and assemblies (A) have been identified with letters and numbers. The same identification approach has been used in the operation process chart. Additionally, fabrication operations have been represented with a four-digit code starting with 0.

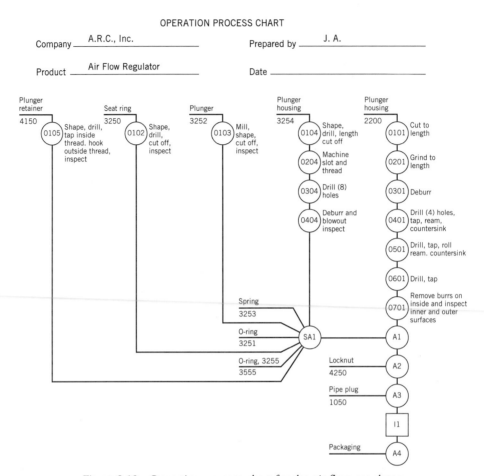

OPERATION PROCESS CHART

Company ___A.R.C., Inc.___   Prepared by ___J. A.___

Product ___Air Flow Regulator___   Date _____

Figure 2.13 Operation process chart for the air flow regulator.

A second viewpoint (from graph and network theory) is to interpret the charts as network representations, or more accurately, tree representations of a production process. A variation of the network viewpoint is to treat the assembly chart and the operation process chart as special cases of a more general graphical model, the **precedence diagram.** The precedence diagram is a directed network and is used in project planning; critical path diagrams and PERT charts are examples of precedence diagrams.

A precedence diagram for the air flow regulator is given in Figure 2.14. The precedence diagram shows part numbers on the arcs and denotes operations and inspections by circles and squares, respectively. A procurement operation, 0100, is used in Figure 2.14 to initiate the process.

Notice in the precedence diagram that purchased parts and materials that do not require modifications are placed on the top and bottom part of the diagram so that they can be inserted in the center part of the diagram when needed (packaging materials, pipe plug, lock nut, spring, and O-rings). The center part of the diagram is then used to include purchased materials and/or components that require some work before being assembled (body, plunger housing, plunger, seat ring, and

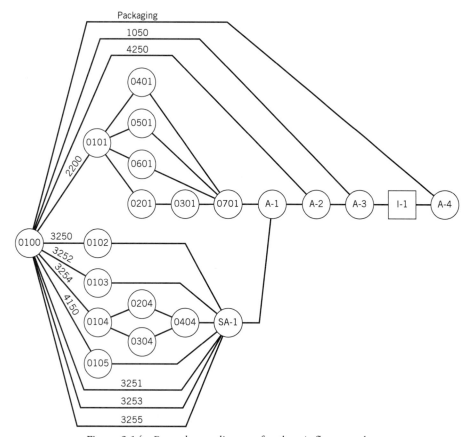

Figure 2.14 Precedence diagram for the air flow regulator.

plunger retainer). Fabrication and assembly operations are placed in the center part of the diagram.

The precedence diagram representation of the operations and inspections involved in a process can be of significant benefit to the facilities planner. It establishes the precedence relationships that must be maintained in manufacturing and assembling a product. No additional constraints are implicitly imposed; no assumptions are made concerning which parts move to which parts; no material handling or layout decisions are implicit in the way the precedence diagram is constructed. Unfortunately, the same claims cannot be made for the assembly chart and operation process chart.

Recall the instructions we gave for constructing the assembly chart: "The easiest method of constructing an assembly chart is to begin with the completed product and to trace the product disassembly back to its basic components." Implicit in such an instruction is that *the sequence used to disassemble the product should be reversed to obtain the sequence to be used to assemble the product.* Just as there are alternative disassembly sequences that can be used, there are also alternative assembly sequences. *The assembly chart and the operation process chart depict a single sequence.* The particular sequence used can have a major impact on space and handling system requirements.

Notice operations 0101, 0201, 0301, 0401, 0501, 0601, and 0701 are not shown in series in Figure 2.14, even though they were so represented on the operation process chart. Hence, there exists some latitude in how the product is assembled.

Unfortunately, there doesn't exist a mechanism to show the possibility of alternative processing sequences on the operation process chart.

In order to motivate further our concerns regarding the misuse of the operation process chart in layout planning, consider the processes involved in manufacturing an axle for an over-the-road tractor. Using the advice typically provided in texts that describe the construction of operation process charts, the axle itself should be shown at the extreme right side of the chart. Subassemblies, components, and purchased parts would be shown sequentially feeding into the axle until a finished assembly was produced.

By observing the operation process chart one might be tempted to develop an assembly line for the axle (assuming sufficient quantities are to be produced). The axle would be moved along the assembly line and subassemblies, components, and purchased parts would be attached to it. Using such an approach, *space and handling equipment requirements for the line would be based on the largest component part in the assembly.*

As an alternative, after performing the last processing operation on the axle it could remain stationary and all subassemblies could flow to it. No further movement of the axle would have to occur.

Because of the limitations of the assembly chart and the operation process chart, we recommend a precedence diagram be constructed first. Based on the precedence diagram, alternative assembly charts and operation process charts should then be constructed.

Other techniques to generate and evaluate assembly sequences have been explored. These techniques consider the assembly according to the relationship among parts instead of the order in which parts will be assembled. With these approaches, liaison sequences are generated instead of assembly sequences.

The original *Liaison Sequence Analysis* (LSA) method for assembly sequencing was proposed by Bourjault [4]. Bourjault's method is an algorithmic approach that uses the relationships between parts to generate all possible assembly sequences.

The second LSA method is a modification of Bourjault's technique developed by Whitney and DeFazio [22]. It reduces the number of questions while keeping the algorithmic nature of the original approach.

Improvements to LSA are currently being addressed. For example, Silva [20] developed the assembly sequences generator system (ASGS) to produce all the physically possible assembly sequences for a given functional product design. ASGS also considers the factors affecting the decision-making process to select the best assembly sequence.

Group technology is having an impact on product and process design. Group technology refers to the grouping of parts into families and then making design decisions based on family characteristics. Groupings are typically based on part shapes, part sizes, material types, and processing requirements. In those cases where there are thousands of individual parts, the number of families might be less than 100. Group technology is an aggregation process that has been found to be quite useful in achieving standardized part numbers, standard specifications of purchased parts, for example, fasteners and standardized process selection.

The importance of the process design or process plan in developing the facilities plan cannot be overemphasized. Furthermore, it is necessary that the process planner understands the impact of process design decisions on the facilities plan. Our experience indicates that process planning decisions are frequently made without such an understanding. As an example, it is often the case that alternatives exist in both the selection of the processes to be used and their sequence of usage. The final choice should be based on interaction between schedule design and facilities planning. Unfortunately, the choice is often based on "We've always done it this way" or "The computer indicated this was the best way."

In many companies the process selection procedure has become sufficiently routine that CAPP systems have been developed. The resulting standardization of process selection has yielded considerable labor savings and reductions in production lead times. At the same time, standardization in process selection might create disadvantages for schedule and facility design. If such a situation occurs, a mechanism should exist to allow exceptions.

The facilities planner must take the initiative and participate in the process selection decision process to ensure that "route decisions" are not unduly constraining the facilities plan. Many of the degrees of freedom available to the facilities planner can be affected by process selection decisions.

2.4 SCHEDULE DESIGN

Schedule design decisions provide answers to questions involving how much to produce and when to produce. Production quantity decisions are referred to as lot size decisions; determining when to produce is referred to as production scheduling. In addition to how much and when, it is important to know how long production will continue; such a determination is obtained from market forecasts.

Schedule design decisions impact machine selection, number of machines, number of shifts, number of employees, space requirements, storage equipment, material handling equipment, personnel requirements, storage policies, unit load design, building size, and so on. Consequently, schedule planners need to interface continuously with marketing and sales personnel and with the largest customers to provide the best information possible to facilities design planners.

To plan a facility, information is needed concerning production volumes, trends, and the predictability of future demands for the products to be produced. The less specificity provided regarding product, process, and schedule designs, the more general purpose will be the facility plan. The more specific the inputs from product, process, and schedule designs, the greater the likelihood of optimizing the facility and meeting the needs of manufacturing.

Marketing Information

A facility that produces 10,000 television sets per month should differ from a facility that produces 1000 television sets per month. Likewise, a facility that produces

Table 2.2 *Minimum Market Information Required for Facilities Planning*

Product or Service	First Year Volume	Second Year Volume	Fifth Year Volume	Tenth Year Volume
A	5000	5000	8000	10,000
B	8000	7500	3000	0
C	3500	3500	3500	4,000
D	0	2000	3000	8,000

10,000 television sets for the first month and increases production 10% per month thereafter should not be considered the same as a facility that produces 10,000 television sets per month for the foreseeable future. Lastly, consider a facility that produces 10,000 television sets per month for the next 10 years versus one that produces 10,000 television sets per month for 3 months and is unable to predict what product or volume will be produced thereafter; they too should differ.

As a minimum, the market information given in Table 2.2 is needed. Preferably, information regarding the dynamic value of demands to be placed on the facility is desired. Ideally, information of the type shown in Table 2.3 would be provided. If such information is available, a facilities plan can be developed for each demand state and a facility designed with sufficient flexibility to meet the yearly fluctuations in product mix. By developing facilities plans annually and noting the alterations to the plan, a facilities master plan may be established [16]. Unfortunately, information of the type given in Table 2.3 is generally unavailable. Therefore, facilities typically are planned using deterministic data. The assumptions of deterministic data and known demands must be dealt with when evaluating alternative facilities plans.

In addition to the volume, trend, and predictability of future demands for various products, the qualitative information listed in Table 2.4 should be obtained. Additionally, the facilities planner should solicit input from marketing as to why market trends are occuring. Such information may provide valuable insight to the facilities planner.

An Italian economist, Pareto, observed that 85% of the wealth of the world is held by 15% of the people. Surprisingly, his observations apply to several aspects of facilities planning. For example, *Pareto's law* often applies to the product mix of a facility. That is, 85% of the production volume is attributed to 15% of the product line. Such a situation is depicted by the **volume-variety chart** or **Pareto chart** given in Figure 2.15. This chart suggests that the facilities plan should consist of a mass production area for the 15% of high-volume items and a job shop arrangement for the remaining 85% of the product mix. By knowing this at the outset, development of the facilities plan may be significantly simplified.

This volume-variety information is very important in determining the layout type to use. As described in Chapter 3, identifying the layout type will impact among others the type of material handling alternatives, storage policies, unit loads, building configuration, and the location of receiving/shipping docks.

Pareto's law may not always describe the product mix of a facility. Figure 2.16 represents the product mix for a facility where 15% of the product line represents approximately 25% of the production volume; Pateto's law clearly does not apply. However, knowing that Pareto's law does not apply is valuable information; because if no product dominates the production flow, a general job shop facility is suggested.

Table 2.3 *Market Analysis Indicating the Stochastic Nature of Future Requirements for Facilities Planning*

Product or Service	Demand State	First Year Probability	First Year Volume	Second Year Probability	Second Year Volume	Fifth Year Probability	Fifth Year Volume	Tenth Year Probability	Tenth Year Volume
A	Pessimistic	.3	3000	.2	3500	.1	5500	.1	7,000
	Most likely	.5	5000	.6	5500	.8	8000	.9	10,000
	Optimistic	.2	6000	.2	6500	.1	9500		
B	Pessimistic	.1	7000						
	Most likely	.6	8000	.7	7000	.9	3000	1.0	0
	Optimistic	.3	8500	.3	8000	.1	3500		
C	Pessimistic	.2	2000	.2	2000	.2	2000	.2	2,000
	Most likely	.7	4000	.7	4000	.7	4000	.6	4,000
	Optimistic	.1	4500	.1	4500	.1	4500	.2	5,000
D	Pessimistic			.1	1500	.1	2500	.2	7,000
	Most likely	1.0	0	.9	2000	.8	3000	.6	8,000
	Optimistic					.1	3500	.2	9,000
Confidence level or degree of certainty		90%		85%		70%		59%	

49

Table 2.4   *Valuable Information that Should Be Obtained from Marketing and Used by a Facilities Planner*

Information to Be Obtained from Marketing	Facilities Planning Issues Impacted by the Information
Who are the consumers of the product?	1. Packaging 2. Susceptibility to product changes 3. Susceptibility to changes in marketing strategies
Where are the consumers located?	1. Facilities location 2. Method of shipping 3. Warehousing systems design
Why will the consumer purchase the product?	1. Seasonability 2. Variability in sales 3. Packaging
Where will the consumer purchase the product?	1. Unit load sizes 2. Order processing 3. Packaging
What percentage of the market does the product attract and who is the competition?	1. Future trends 2. Growth potential 3. Need for flexibility
What is the trend in product changes?	1. Space allocations 2. Materials handling methods 3. Need for flexibility

Process Requirements

Process design determines the specific equipment types required to produce the product. Schedule design determines the number of each equipment type required to meet the production schedule.

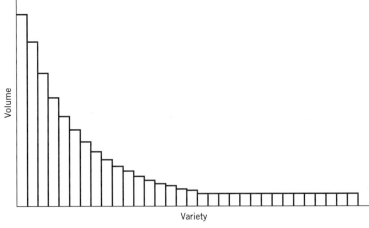

Figure 2.15   Volume-variety chart for a facility where Pareto's law is applicable.

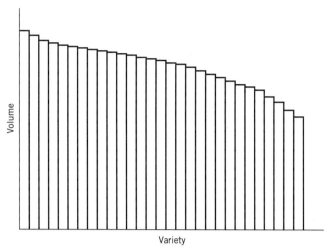

Figure 2.16   Volume-variety chart for a facility where Pareto's law is not applicable.

Specification of process requirements typically occurs in three phases. The first phase determines the quantity of components that must be produced, including scrap allowance, in order to meet the market estimate. The second phase determines the equipment requirements for each operation, and the third phase combines the operation requirements to obtain overall equipment requirements.

Scrap Estimates

The market estimate specifies the annual volume to be produced for each product. To produce the required amount of product, the number of units scheduled through production must equal the market estimate plus a scrap estimate. Hence, production capacity must be planned for the production of scrap. Otherwise, when scrap is produced the market estimate will not be met. Scrap is the material waste generated in the manufacturing process due to geometric or quality considerations. For example, scrap due to geometry is generated when a rectangular steel plate is used to create circular components or when rolls of fabric are used to make shirts. Scrap due to bad quality might be caused by mistakes in machining or assembly operations. Ideally, a company should strive for continuous improvement to achieve zero scrap. An estimate of the percentage of scrap to be incurred from each operation must be made. This may be based on historical data or estimated from similar operations. In general, the more automated a process, the less scrap is produced; the looser the part tolerance, the less scrap is produced; the bigger the number of certified suppliers, the less scrap is generated; the more quality at the source and prevention techniques are applied, the less scrap is generated; and the higher the grade of material, the less scrap is produced.

Let P_k represent the percentage of scrap produced on the kth operation, O_k the desired output of nondefective product from operation k, and I_k the production input to operation k. It follows that, on the average

$$O_k = I_k - P_k I_k \tag{2.1}$$

or

$$O_k = I_k(1 - P_k)$$

Hence,

$$I_k = \frac{O_k}{1 - P_k} \tag{2.2}$$

Thus, the expected number of units to start into production for a part having n operations is

$$I_1 = \frac{O_n}{(1 - P_1)(1 - P_2) \cdots (1 - P_n)} \tag{2.3}$$

where in this case O_n is the market estimate.

Example 2.1

A product has a market estimate of 97,000 components and requires three processing steps (turning, milling, and drilling) having scrap estimates of $P_1 = 0.04$, $P_2 = 0.01$, and $P_3 = 0.03$. The market estimate is the output required from step 3. Therefore,

$$I_3 = \frac{97,000}{1 - 0.03} = 100,000$$

Assuming no damage between operations 2 and 3 and an inspection operation to remove all rejects, the output of good components from operation 2 (O_2) may be equated to the input to operation 3 (I_3). Therefore, the number of components to start into operation 2 (I_2) is

$$I_2 = \frac{100,000}{1 - 0.01} = 101,000$$

Likewise, for operation 1:

$$I_1 = \frac{101,010}{1 - 0.04} = 105,219$$

The calculations are identical to:

$$I_1 = \frac{97,000}{(1 - 0.03)(1 - 0.01)(1 - 0.04)} = 105,219$$

The amout of raw material and processing on operation 1 is not to be based on the market estimate of 97,000 components, but on 105,219 components, as summarized in Table 2.5.

Table 2.5 *Summary of Production Requirements for Example 2.1*

Operation	Production Quantity Scheduled (units)	Expected Number of Good Units Produced
Turning	105,219	101,010
Milling	101,010	100,000
Drilling	100,000	97,000

Reject Allowance Problem

As suggested by the previous analysis, random events influence the number of acceptable units produced. The calculations performed using Equations 2.1 through 2.3 are reasonable when high-volume production is occurring. However, when producting small batches the use of average values is less appropriate. As an example, consider a foundry that produces small numbers of custom-designed castings. If conditions are such that the foundry has only one chance to produce the number of castings required, then the probability of a casting being good should be considered when determining the batch size to be produced.

In determining how many castings to produce, the following questions come to mind:

1. How much does it cost to produce a good casting? How much for a bad casting?

2. How much revenue is generated from a good casting? How much from a bad casting?

3. What is the probability distribution for the number of good castings resulting from a production lot?

If answers are available for these questions, then a determination can be made regarding the number of castings to schedule in order to, say, maximize the expected profit or achieve a desired confidence level of not producing fewer good castings than are needed. Determining the number of additional units to allow when scheduling low-volume production where rejects randomly occur is called the *reject allowance problem* [22].

To facilitate a formulation of the reject allowance problem, let

X = random variable representing the number of good units produced
$p(x)$ = probability of producing exactly x good units
Q = quantity of units to produce
$C(Q, x)$ = cost of producing Q units, of which exactly x are good units
$R(Q, x)$ = revenue from producing Q units, of which exactly x are good units
$P(Q, x)$ = profit from producing Q units, of which exactly x are good units
= $R(Q, x) - C(Q, x)$
$E[P(Q)]$ = expected profit from producing Q units

$$= \sum_{x=0}^{Q} P(Q, x)\, p(x)$$

The expected profit from producing Q units can be determined as follows:

$$E[P(Q)] = \sum_{x=0}^{Q} \{R(Q, x) - C(Q, x)\}\, p(x)$$

If it is desired to maximize expected profit, the value of Q that maximizes Equation 2.4 can be determined by enumerating over various values of Q. For most cost and revenue formulations, Equation 2.4 is a concave function; therefore, unimodal search routines can be used. See [22] for necessary and sufficient conditions for Q to be the optimum production quantity when X is binomially distributed.

Table 2.6 *Probability Distributions for the Number of Good Castings (x) Out of Q*

# Good Castings (x)	Number of Castings Produced (Q)										
	20	21	22	23	24	25	26	27	28	29	30
12	0.05	0.00	0.00	0.00	0.00	0.00	0.00	0.00	0.00	0.00	0.00
13	0.05	0.05	0.00	0.00	0.00	0.00	0.00	0.00	0.00	0.00	0.00
14	0.05	0.05	0.05	0.00	0.00	0.00	0.00	0.00	0.00	0.00	0.00
15	0.05	0.05	0.05	0.05	0.00	0.00	0.00	0.00	0.00	0.00	0.00
16	0.10	0.05	0.05	0.05	0.05	0.00	0.00	0.00	0.00	0.00	0.00
17	0.10	0.10	0.05	0.05	0.05	0.05	0.00	0.00	0.00	0.00	0.00
18	0.15	0.10	0.10	0.05	0.05	0.05	0.05	0.00	0.00	0.00	0.00
19	0.20	0.15	0.10	0.10	0.05	0.05	0.05	0.05	0.00	0.00	0.00
20	0.25	0.20	0.15	0.10	0.10	0.05	0.05	0.05	0.05	0.00	0.00
21	0.00	0.25	0.20	0.15	0.10	0.10	0.05	0.05	0.05	0.05	0.00
22	0.00	0.00	0.25	0.20	0.15	0.10	0.10	0.05	0.05	0.05	0.05
23	0.00	0.00	0.00	0.25	0.20	0.15	0.10	0.10	0.05	0.05	0.05
24	0.00	0.00	0.00	0.00	0.25	0.20	0.15	0.10	0.10	0.05	0.05
25	0.00	0.00	0.00	0.00	0.00	0.25	0.20	0.15	0.10	0.10	0.05
26	0.00	0.00	0.00	0.00	0.00	0.00	0.25	0.20	0.15	0.10	0.10
27	0.00	0.00	0.00	0.00	0.00	0.00	0.00	0.25	0.20	0.15	0.10
28	0.00	0.00	0.00	0.00	0.00	0.00	0.00	0.00	0.25	0.20	0.15
29	0.00	0.00	0.00	0.00	0.00	0.00	0.00	0.00	0.00	0.25	0.20
30	0.00	0.00	0.00	0.00	0.00	0.00	0.00	0.00	0.00	0.00	0.25

Example 2.2

A foundry produces castings to order. An order for 20 custom-designed castings has been received. The casting process costs $1,100 per unit scheduled. If a casting is not sold, it has a recycle value of $200. The customer has indicated a willingness to pay $2,500 per casting for 20 acceptable castings—no more, no less! Based on historical records, the probability distributions given in Table 2.6 have been estimated. How many castings should be scheduled for production to maximize expected profit? What is the probability of losing money at this production level?

The revenue and cost functions can be given as follows:

$$R(Q, x) = \begin{cases} \$200Q & x < 20 \\ \$2,500(20) + \$200(Q - 20) & 20 \leq x \leq Q \end{cases}$$

$$C(Q, x) = \$1,100Q \qquad 0 \leq x \leq Q$$

$$P(Q, x) = \begin{cases} -\$900Q & x < 20 \\ \$46,000 - \$700Q & 20 \leq x \leq Q \end{cases}$$

Therefore, the expected profit can be given as follows:

$$E[p(Q)] = -\sum_{x=0}^{19} 900Q\, p(x) + \sum_{x=20}^{Q} (46,000 - 700Q)\, p(x)$$

The expected profit expression can be shown to reduce to

$$E[p(Q)] = \sum_{x=20}^{Q} (46,000 + 200Q)\, p(x) - 900Q$$

Table 2.7 *Profit from Producing Q Castings, with Exactly x Being Good*

# Good Castings	Number of Castings Scheduled										
	20	21	22	23	24	25	26	27	28	29	30
12	−$18,000	−$18,900	−$19,800	−$20,700	−$21,600	−$22,500	−$23,400	−$24,300	−$25,200	−$26,100	−$27,000
13	−$18,000	−$18,900	−$19,800	−$20,700	−$21,600	−$22,500	−$23,400	−$24,300	−$25,200	−$26,100	−$27,000
14	−$18,000	−$18,900	−$19,800	−$20,700	−$21,600	−$22,500	−$23,400	−$24,300	−$25,200	−$26,100	−$27,000
15	−$18,000	−$18,900	−$19,800	−$20,700	−$21,600	−$22,500	−$23,400	−$24,300	−$25,200	−$26,100	−$27,000
16	−$18,000	−$18,900	−$19,800	−$20,700	−$21,600	−$22,500	−$23,400	−$24,300	−$25,200	−$26,100	−$27,000
17	−$18,000	−$18,900	−$19,800	−$20,700	−$21,600	−$22,500	−$23,400	−$24,300	−$25,200	−$26,100	−$27,000
18	−$18,000	−$18,900	−$19,800	−$20,700	−$21,600	−$22,500	−$23,400	−$24,300	−$25,200	−$26,100	−$27,000
19	−$18,000	−$18,900	−$19,800	−$20,700	−$21,600	−$22,500	−$23,400	−$24,300	−$25,200	−$26,100	−$27,000
20	$32,000	$31,300	$30,600	$29,900	$29,200	$28,500	$27,800	$27,100	$26,400	$25,700	$25,000
21	$0	$31,300	$30,600	$29,900	$29,200	$28,500	$27,800	$27,100	$26,400	$25,700	$25,000
22	$0	$0	$30,600	$29,900	$29,200	$28,500	$27,800	$27,100	$26,400	$25,700	$25,000
23	$0	$0	$0	$29,900	$29,200	$28,500	$27,800	$27,100	$26,400	$25,700	$25,000
24	$0	$0	$0	$0	$29,200	$28,500	$27,800	$27,100	$26,400	$25,700	$25,000
25	$0	$0	$0	$0	$0	$28,500	$27,800	$27,100	$26,400	$25,700	$25,000
26	$0	$0	$0	$0	$0	$0	$27,800	$27,100	$26,400	$25,700	$25,000
27	$0	$0	$0	$0	$0	$0	$0	$27,100	$26,400	$25,700	$25,000
28	$0	$0	$0	$0	$0	$0	$0	$0	$26,400	$25,700	$25,000
29	$0	$0	$0	$0	$0	$0	$0	$0	$0	$25,700	$25,000
30	$0	$0	$0	$0	$0	$0	$0	$0	$0	$0	$25,000

The profits resulting from various combinations of Q and x are shown in Table 2.7. (Zero profit values are shown for infeasible combinations of Q and x to simplify the calculation of expected profit using Excel's SUMPRODUCT function.) The vector products of columns from Tables 2.6 and 2.7 are obtained using the SUMPRODUCT function and yield the expected profit values in Table 2.8. The maximum expected profit ($26,400) results from scheduling 28 castings to be produced. From Table 2.6, when 28 castings are produced there is a zero probability of losing money; the penalty for not producing at least 20 good units is so severe that the minimum number of units having a zero probability of producing less than 20 good units is optimum.

Equipment Fractions

The quantity of equipment required for an operation is referred to as the equipment fraction. The equipment fraction may be determined for an operation by dividing the total time required to perform the operation by the time available to complete the operation. The total time required to perform an operation is the product of the standard time for the operation and the number of times the operation is to be performed. For example, if it takes ½ hr to duplicate a paper and if six papers are to be duplicated in 2 hr, it clearly follows that 1.5 duplicating machines are needed to complete the operation. Whether or not 1.5 duplicating machines are actually adequate to duplicate the required papers depends on the following questions.

1. Are the papers actually being duplicated according to the ½ hr per paper standard?
2. Is the duplicating machine available when needed during the 2-hr period?
3. Are the standard time, the number of papers, and the time the equipment is actually available known with certainty and fixed over time?

The first question may be handled by dividing the standard time by the historical efficiency of performing the operation. The second question may be handled by multiplying the time the equipment is available to complete an operation by the historical reliability factor for the equipment. The reliability factor is the percentage of time the equipment is actually producing and is not down because of malfunctioning or planned maintenance.

The third question dealing with the uncertainty and time-varying nature of machine fraction variables can be an important factor in determining process requirements. If considerable uncertainty and variation over time exists, it may be useful to consider using probability distributions, instead of point estimates for the parameters, and utilize a probabilistic machine fraction model. Typically, such models are not utilized and the approach taken is to use a deterministic model and plan the facility to provide sufficient flexibility to handle changes in machine fraction variables.

Table 2.8   *Expected Profit from Producing Q Castings*

Number of Castings Scheduled (Q)										
20	21	22	23	24	25	26	27	28	29	30
−$5,500	$3,690	$10,440	$14,720	$19,040	$20,850	$22,680	$24,530	$26,400	$25,700	$25,000

The following deterministic model can be used to estimate the equipment fraction required:

$$F = \frac{SQ}{EHR} \qquad (2.5)$$

where

F = number of machines required per shift
S = standard time (minutes) per unit produced
Q = number of units to be produced per shift
E = actual performance, expressed as a percentage of standard time
H = amount of time (minutes) available per machine
R = reliability of machine, expressed as percent "uptime"

Additionally, equipment requirements are a function of the following factors:

- Number of shifts (the same machine can work in more than one shift).
- Setup times (if machines are not dedicated, the longer the setup, the more machines needed).
- Degree of flexibility (customers may require small lot sizes of different products delivered frequently—extra machine capacity will be required to handle these requests).
- Layout type (dedicating manufacturing cells or focused factories to the production of product families may require more machines).
- Total productive maintenance (will increase machine up time and improve quality, thus fewer machines will be needed).

Example 2.3

A machined part has a standard machinery time of 2.8 min per part on a milling machine. During an 8-hr shift 200 units are to be produced. Of the 480 min available for production, the milling machine will be operational 80% of the time. During the time the machine is operational, parts are produced at a rate equal to 95% of the standard rate. How many milling machines are required?

For the example, $S = 2.8$ min per part, $Q = 200$ units per shift, $H = 480$ min per shift, $E = 0.95$, and $R = 0.80$. Thus,

$$F = \frac{2.8(200)}{0.95(480)(0.80)} = 1.535 \text{ machines per shift}$$

Specifying Total Equipment Requirements

The next step in determining process requirements is to combine the equipment fractions for identical equipment types. Such a determination is not necessarily straightforward. Even if only one operation is to be performed on a particular equipment type, overtime and subcontracting must be considered. If more than one operation is to be run on a particular equipment type, several alternatives must be considered.

Table 2.9   *Total Equipment Requirement Specification Example*

Operation Number	Equipment Fraction	Next Highest Whole Number
109	1.1	2
206	2.3	3
274	0.6	1

Example 2.4

The machine fractions for an ABC drill press are given in Table 2.9. No drill press operator, overtime, or subcontracting is available for any operation on the ABC drill press. It may be seen that a minimum o four and a maximum of six machines are required. How many should be purchased? The answer is either four, five, or six. With no further information, a specific recommendation cannot be made. Information on the cost of the equipment, the length of machine setups, the cost of in-process inventories, the cost and feasibility of over-time, production, and/or setups, the expected future growth of demand, and several other qualitative factors must be analyzed to reach a decision.

Machine Assignment Problem

Chapter 4 addresses facility requirements for personnel, including employee parking, locker rooms, restrooms, food services, and so forth. It is assumed that a determination of the number of people to be employed in the facility already has been made. Typically, such decisions are not a part of the facilities planning process. However, the combination of product, process, and schedule design decisions significantly influences the number of employees involved in producing the product. In this section, we consider how decisons regarding the assignment of machines to operators can affect the number of employees. Specifically, we consider a situation involving the assignment of operators to semiautomatic production equipment. For purposes of this discussion, it is assumed the machines are identical. In contrast to the reject allowance problem, it is assumed that the times required to load and unload each machine are constant, the automatic machining time is constant, and the "independent" time required for the operator to travel between machines, prepare parts for maching, and inspect and pack parts are constant.

To illustrate the situation under consideration, see Figure 2.17, which depicts the activities of an operator and three machines. The chart is called a **human-machine chart** or a **multiple activity chart,** since it shows the activities of one or more poeple and one or more machines. Such charts can be used to analyze multiple activity relationships when nonidentical machines are being tended by one or more operatiors. The multiple activity chart can prove useful in analyzing the activities or each operator and each machine during "transient" and "steady-state" conditions.

Example 2.5

For the situation illustrated in Figure 2.17, it takes 0.5 minute to travel between machines, 1.0 minute to load a machine, 1.0 minute to unload a machine, 6 minutes of automatic machine time, and 0.5 minute to inspect and pack a finished part. As shown, the analysis

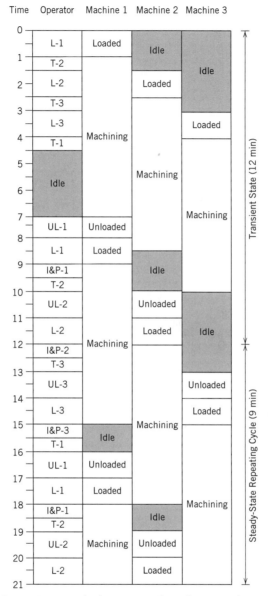

Figure 2.17 Multiple Activity Chart for Example 2.5.

begins with each machine empty and the operator standing in front of Machine 1 (M-1). The operator loads M-1, walks to M-2, loads M-2, walks to M-3, loads M-3, walks to M-1, unloads M-1, loads M-1, inspects and packs the part removed from M-1, travels to M-2, and so forth. As shown in Figure 2.18, it takes 12 minutes for the operator and 3 machines to achieve a steady-state condition; thereafter, a repeating cycle of 9 minutes in duration occurs. (In other words, if nothing interrupts the activities of the operator and the three machines, the 9-minute cycle will repeat indefinitely.)

Under conditions similar to those depicted by Figure 2.17, a deterministic model can be developed to determine the optimum number of machines to assign to an individual

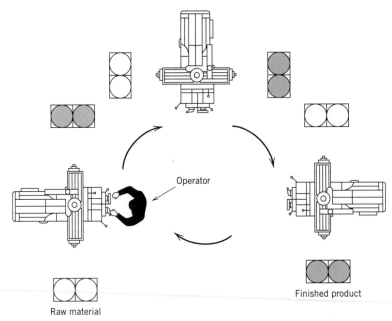

Figure 2.18 Assignment of three machines to one operator. (Reprinted with permission from [23].)

operator. To facilitate the development of the model, let

a = concurrent activity time (e.g., loading and unloading a machine)
b = independent operator activity time (e.g., walking, inspecting, packing)
t = independent machine activity time (e.g., automatic machining time)
n' = ideal number of identical machines to assign an operator
m = number of identical machines assigned an operator
T_c = repeating cycle time
I_o = idle operator time during a repeating cycle
I_m = idle time for each machine during a repeating cycle

Excluding idle time, each machine cycle requires $a + t$ minutes to complete a cycle. Likewise, the operator devotes $a + b$ minutes to each machine during a cycle. Hence, an ideal assignment is

$$n' = (a + t)/(a + b) \qquad (2.6)$$

For the example, a equals 2 minutes, b equals 1 minute, and t equals 6 minutes. Therefore, n' equals 2.67 machines. Since a fractional number of machines cannot be assigned to an operator, consider what will happen if some integer number of machines, m, is assigned. The work content for the operator will total $m(a + b)$, while a machine cycle will be $(a + t)$ in duration. The repeating cycle will be the larger of the two and the difference in the two will be idle time. If $m > n'$, then the repeating cycle will be $m(a + b)$ in duration; if $m < n'$, then the repeating cycle will be $(a + t)$ in duration. Therefore,

$$T_c = \begin{cases} (a + t) & m \le n' \\ m(a + b) & m > n' \end{cases} \qquad (2.7)$$

$$I_m = \begin{cases} 0 & m \le n' \\ T_c - (a + t) & m > n' \end{cases} \tag{2.8}$$

$$I_o = \begin{cases} T_c - m(a + b) & m \le n' \\ 0 & m > n' \end{cases} \tag{2.9}$$

For the example, with $m = 3$ and $n' = 2.67$, the repeating cycle is $3(2 + 1)$, or 9 minutes as observed in Figure 2.17. The idle time for each machine during a repeating cycle equals $9 - (2 + 6)$, or 1 minute, again as observed in Figure 2.17. As shown in Figure 2.18, the work cell must be designed carefully to provide adequate access for incoming and outgoing material, machine maintenance, and operator movement.

If we wish to determine the cost per unit produced by an m machine assignment, the following notation will be helpful

C_o = cost per operator-hour
C_m = cost per machine-hour
ϵ = C_o / C_m
$TC(m)$ = cost per unit produced based on an assignment of m machines per operator

The cost per hour of a combination of m machines and an operator totals $C_o + mC_m$. Assuming each machine produces one unit during a repeating cycle, the cost per unit produced during a repeating cycle can be determined as follows:

$$TC(m) = \begin{cases} (C_o + mC_m)(a + t)/m & m \le n' \\ (C_o + mC_m)(a + b) & m > n' \end{cases} \tag{2.10}$$

To minimize $TC(m)$ when $m \le n'$, m should be made a large as possible; to minimize $TC(m)$ when $m > n'$, m should be made as small as possible. If n' is integer valued, n' will minimize $TC(m)$. If n' is not integer valued then either n or $n + 1$ will minimize $TC(m)$ where n is defined to be the integer portion of n'.

To facilitate the determination, let Φ represent the ratio of $TC(n)$ to $TC(n + 1)$. Hence, from Equation 2.10,

$$\Phi = TC(n)/TC(n + 1)$$
$$= \frac{(C_o + nC_m)(a + t)}{n[C_o + (n + 1)C_m](a + b)} \tag{2.11}$$

which reduces to

$$\Phi = \frac{\epsilon + n}{\epsilon + n + 1} \frac{n'}{n} \tag{2.12}$$

If $\Phi < 1$, then $TC(n) < TC(n + 1)$ and n machines should be assigned; if $\Phi > 1$, then $TC(n + 1) < TC(n)$ and $n + 1$ machines should be assigned; if $\Phi = 1$, then either n or $n + 1$ machines should be assigned.

For the example, suppose C_o equals $15 per hour and C_m equals $50 per hour. Therefore, ϵ equals 0.30 and Φ equals 0.929. Since $\Phi < 1$, then 2 machines should be assigned to an operator. The problems at the end of the chapter explore various aspects of the machine assignment problem, such as assigning machines to operators if, say, a total of 11 machines are required to meet the daily production schedule or there is uncertainty regarding the value of C_m.

2.5   FACILITIES DESIGN

Once the product, process, and schedule design decisions have been made, the facilities planner needs to organize the information and generate and evaluate layout, handling, storage, and unit load design alternatives. As explained in Chapter 1 and Section 2.1, the facilities planner must be aware of top management's objectives and goals to maximize the impact of the facilities design effort on such objectives and goals. Some typical business objectives include breakthroughs in production cost, on-time delivery, quality, and lead time.

Some tools frequently used by quality practitioners (e.g., Pareto chart) can be very useful in facilities planning efforts. More recently, the *seven management and planning tools* have gained acceptance as a methodology for improving planning and implementation efforts in general [5]. The tools have their roots in post-WWII operations research work and the total quality control (TQC) movement in Japan.

In the mid 1970s a committee of engineers and scientists in Japan refined and tested the tools as an aid for process improvement, as proposed by the Deming cycle. In 1950 Dr. W. E. Deming proposed a model for continuous process improvement that involves four steps: planning and goal setting, doing or execution, checking or analysis, and specifying corrective actions (Plan–Do–Check–Act).

The seven management and planning tools are the **affinity diagram,** the **interrelationship digraph,** the **tree diagram,** the **matrix diagram,** the **contingency diagram,** the **activity network diagram,** and the **prioritization matrix.** Each is described below and illustrated with examples related to facilities design.

Affinity Diagram

The affinity diagram is used to gather verbal data, such as ideas and issues, and organize it into groupings. Suppose we are interested in generating ideas for reducing manufacturing leadtime. In a brainstorming session, the issues are written down on "post-it" notes and grouped on a board or wall. Each group then receives a heading. An affinity diagram for reducing manufacturing leadtime is presented in Figure 2.19. The headings selected were: facilities design, equipment issues, quality, set-up time, and scheduling.

Interrelationship Diagraph

The interrelationship diagram is used to map the logical links among related items, trying to identify which items impact others the most. The term *digraph* is employed because the graph uses directed arcs. Suppose we want to study the relationship between the items in Figure 2.19 under facilities design. The interrelationships are presented in Figure 2.20. Note that this graph helps us understand the logical sequence of steps for the facilities design. The efforts must be initiated with the formation of product families.

	Issues in reducing manufacturing leadtime			
Facilities design	Equipment issues	Quality	Set-up time	Scheduling
1. form product families	1. operator certification program	1. provide training on how to use process documentation	1. provide documentation on set-up procedures	1. provide visibility to daily product sequence
2. assign families to cells	2. sit technicians closer to production	2. implement successive inspection with feedback	2. locate fixtures and tooling close to machines	2. do not authorize products for which the needed parts are not available
3. assign raw mtl's to their point of use	3. monitor breakdowns to predict future occurrences	3. develop mistake-proof devices	3. provide training so operators can participate	3. negotiate frequent and smaller lots to customers
4. keep receiving and shipping close to production	4. recruit enough technicians per shift	4. develop capabilities for monitoring key machine parameters	4. provide information on daily sequence	

Figure 2.19 Affinity diagram example for reducing manufacturing leadtime.

Tree Diagram

The tree diagram is used to map in increasing detail the actions that need to be accomplished in order to achieve a general objective. Assuming that we want to construct a tree diagram for the formation of product families, the tree is presented in Figure 2.21. Note that the same exercise can be performed for each item in the interrelationship digraph.

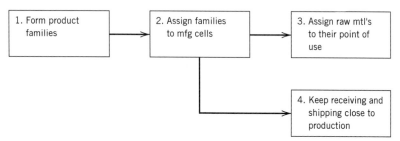

Figure 2.20 Interrelationship digraph for facilities design.

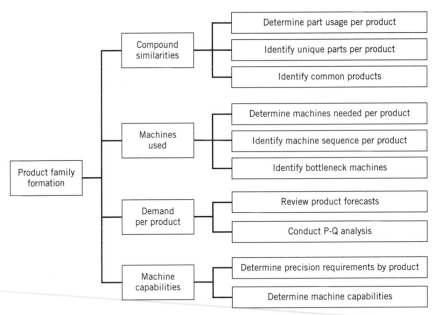

Figure 2.21   Tree diagram for the formation of product families.

Matrix Diagram

The matrix diagram organizes information such as characteristics, functions, and tasks into sets of items to be compared. A simple application of this tool is the design of a table in which the participants and their role within the small teams are defined. This tool provides visibility to key contacts on specific issues and helps identify individuals who are assigned to too many teams. Table 2.10 assumes that three teams are formed in response to the actions listed in the tree diagram (Figure 2.21). The teams focus on (1) part usage, (2) machine usage and capability, and (3) demand forecasts. Note that additional teams will be formed for the other activities identified in the interrelationship digraph (Figure 2.20). In the table, team leaders and coordinators have been identified since they might carry a heavier workload and it is desirable for them to have less involvement in other efforts.

Table 2.10   *Matrix Diagram for Team Participation*

Team\Participants	Joe	Mary	Jerry	Lou	Linda	Daisy	Jack
Part usage team	P	C	P	L			P
Machine use & cap team	L		C				P
Demand forecast team				P	C	L	

Note. L: Team Leader
     C: Team Coordinator
     P: Team Participant

Contingency Diagram

The contingency diagram, formally known as process decision program chart, maps conceivable events and contingencies that might occur during implementation. It is particularly useful when the project being planned consists of unfamiliar tasks. The benefit of preventing or responding effectively to contingencies makes it worthwhile to look at these possibilities during the planning phase.

Activity Network Diagram

The activity network diagram is used to develop a work schedule for the facilities design effort. This diagram is synonymous to the critical path method (CPM) graph. It can also be replaced by a Gantt chart [6] and if a range is defined for the duration of each activity, the Program Evaluation and Review Technique (PERT) chart can also be used. The important message is that a well thought out time table is needed to understand the length of the facilities design project. This timetable can be developed after the actions on the tree diagram (Figure 2.21) have been evaluated with the prioritization matrix. An example of an activity network diagram for a production line expansion is illustrated in Figure 2.22.

After understanding the magnitude of the facilities design effort, it might be necessary to form several small teams to work concurrently on the various tasks. In a participative environment, production, material handling, and representatives from support functions are included in the facilities design project. They have the best understanding of how the plant processes are being executed. With their participation, more ideas are generated, there is significantly less resistance to change in the organization, and the end result is improved. If a team has varying levels of understanding about facilities design, the team interaction should be started with education and training.

Team activities can also be planned, and a weekly schedule can be prepared, assigning one or two work sessions per team. Scheduling the work sessions helps team members in allocating time for the effort. A typical weekly work schedule for the teams defined above is presented in Table 2.11.

Prioritization Matrix

In developing facilities design alternatives it is important to consider:

(a) Layout characteristics
  - total distance travelled
  - manufacturing floor visibility
  - overall aesthetics of the layout
  - ease of adding future business
(b) Material handling requirements
  - Use of current material handling equipment
  - investment requirements on new equipment
  - space and people requirements

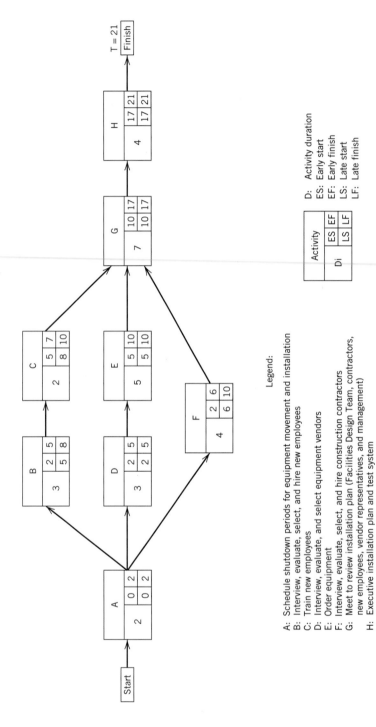

Legend:

A: Schedule shutdown periods for equipment movement and installation
B: Interview, evaluate, select, and hire new employees
C: Train new employees
D: Interview, evaluate, and select equipment vendors
E: Order equipment
F: Interview, evaluate, select, and hire construction contractors
G: Meet to review installation plan (Facilities Design Team, contractors, new employees, vendor representatives, and management)
H: Executive installation plan and test system

D: Activity duration
ES: Early start
EF: Early finish
LS: Late start
LF: Late finish

Figure 2.22   Activity network diagram example for a production line expansion facilities design project.

Table 2.11   *Weekly Timetable for Team Work Sessions*

Time\Day	Mon	Tue	Wed	Thu	Fri
8–10 A.M.					
10–12 P.M.	Parts		Parts		
1–3 P.M.		Mach's		Demand	Mach's
3–5 P.M.				Demand	

(c)   Unit load implied
- impact on WIP levels
- space requirements
- impact on material handling equipment

(d)   Storage strategies
- space and people requirements
- impact on material handling equipment
- human factor risks

(e)   Overall building impact
- estimated cost of the alternative
- opportunities for new business.

The prioritization matrix can be used to judge the relative importance of each criterion as compared to each other. Table 2.12 presents the prioritization of the criteria for the facilities design example. The criteria are labeled to help in building a table with weights.

A. Total distance travelled   G. Space requirements
B. Manufacturing floor visibility   H. People requirements
C. Overall aesthetics of the layout   I. Impact on WIP levels

Table 2.12   *Prioritization Matrix for the Evaluation of Facilities Design Alternatives*

	A	B	C	D	E	F	G	H	I	J	K	Row totals (%)
A	1	5	10	5	1	1	1	1	1	5	1	32. (9.9)
B	1/5	1	5	1/5	1/5	1/10	1/5	1/5	1/10	1/5	1/5	7.6 (2.4)
C	1/10	1/5	1	1/10	1/10	1/10	1/5	1/5	1/10	1/10	1/10	2.3 (0.7)
D	1/5	5	10	1	1/5	1/5	1/5	1/5	1/10	1/5	1/10	17.4 (5.4)
E	1	5	10	5	1	1	5	5	1/5	1	1/5	34.4 (10.7)
F	1	10	10	5	1	1	5	5	1	1	1	41. (12.7)
G	1	5	5	5	1/5	1/5	1	5	1/5	1/5	1/5	23. (7.1)
H	1	5	5	5	1/5	1/5	5	1	1/10	1/5	1/5	22.9 (7.1)
I	1	10	10	10	5	1	5	10	1	1	5	59. (18.3)
J	1/5	5	10	5	1	1	5	5	1	1	5	39.2 (12.2)
K	1	5	10	10	5	1	5	5	1/5	1/5	1	43.4 (13.5)
Column total	7.7	56.2	86.	51.3	14.9	6.8	32.6	37.6	5.	10.1	14.	322.2 Grand total

Table 2.13 *Prioritization of Layout Alternatives Based on WIP Levels*

WIP Levels	Layout					Row totals (%)
	P	Q	R	S	T	
P	1	5	1	1/10	1/5	7.3 (9.9)
Q	1/5	1	1/5	1/10	1/10	1.6 (2.2)
R	1	5	1	10	5	22. (30.0)
S	10	10	1/10	1	1/5	21.3 (29.0)
T	5	10	1/5	5	1	21.2 (28.9)
Column Total	17.2	31	2.5	16.2	6.5	73.4 Grand total

D. Ease of adding future business
E. Use of current MH equipment
F. Investment in new MH equipment

J. Human factor risks
K. Estimated cost of alternative

The weights typically used to compare the importance of each pair of criteria are

1 = equally important
5 = significantly more important
10 = extremely more important

1/5 = significantly less impotant
1/10 = extremely less important

Note that the values in cells (i, j) and (j, i) are reciprocals. The resulting relative importance is presented in the last column in parenthesis. For this application, the most important criterion for facilities design selection is the impact on WIP levels (weight = 18.3), followed by the estimated cost of the solution (weight = 13.5).

This same methodology can be employed to compare all facilities design alternatives in each weighted criterion. For example, suppose five layout alternatives are generated, namely, P, Q, R, S, and T. Table 2.13 presents the ranking of the layout alternatives based on the impact of WIP levels criterion.

If we construct a similar table for the remaining ten criteria, we will be able to evaluate each layout alternative using the eleven criteria to identify the best layout. The format of this final table is presented in Table 2.14. The last column is computed as in Tables 2.12 and 2.13. The row totals (represented by Σ) are added to obtain the grand total, after which the percentages (%P,..., %T) are determined.

Table 2.14 *Ranking of Layouts by All Criteria*

	Criteria											Row totals (%)
	A	B	C	D	E	F	G	H	I	J	K	
P									$.099 \times .183 = .018$			Σ (%P)
Q									$.022 \times .183 = .004$			Σ (%Q)
R									$.300 \times .183 = .055$			Σ (%R)
S									$.290 \times .183 = .053$			Σ (%S)
T									$.289 \times .183 = .053$			Σ (%T)
Column									.183			Grand Total

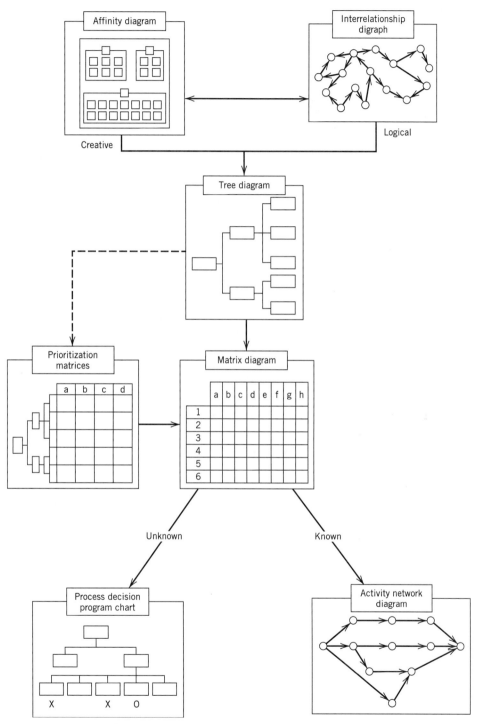

Figure 2.23 Logical application sequence of the seven management and planning tools.

These percentages tell us the relative goodness of each layout alternative. These results should be presented to plant management to facilitate final decisions regarding the layout.

The examples presented in this section are a subset of the possible uses of the seven management and planning tools. Interestingly, all the tools can be used to support an integrated facilities design process or they can also be used independently to support facilities design teams in the planning or resolution of specific issues. Figure 2.23 presents a graphical representation on the logical sequence when the tools are used in an integrated fashion. Figure 2.24 provides a flowchart of the application of the tools to a facilities design planning project. To summarize, Figure 2.25 presents a possible scenario for a facilities design team using the tools.

2.6 SUMMARY

Product, process, and schedule design decisions can have a significant impact on both the investment cost for a facility and the cost-effective performance of the activities assigned to the facility. The decisions made concerning product design, process planning, production schedules, and facilities planning must be jointly determined in order to obtain an integrated production system that achieves the firm's business objectives.

Even though facilities planners are not typically involved in product, process, and schedule design, it is important for the facilities planner to be familiar with these activities. Rather than simply reacting to requirements established by the product, process, and schedule design process, the facilities planner should be proactive and influence requirements definitions. Rather than being a passive observer, the facilities planner should be an active participant in decision making who influences the degrees of freedom for planning the facility.

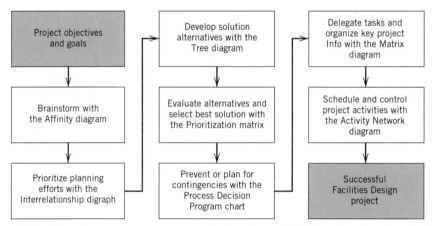

Figure 2.24 How the seven management and planning tools facilitate the planning of a facilities design project.

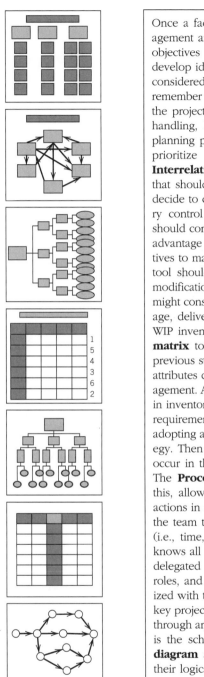

Once a facilities design team, trained to use the seven management and planning tools, has properly defined the project objectives and goals, it should use the **Affinity diagram** to develop ideas and document important issues that need to be considered for the project planning effort. It is important to remember that these tools are used at the planning phase of the project. At this step issues and ideas concerning material handling, layout, inventory control policies, and production planning policies may arise. The next step for the team is to prioritize its planning efforts through the use of the **Interrelationship digraph.** This tool will identify the issues that should be considered first. For instance, the team might decide to concentrate its efforts on modifying current inventory control policies. Once the team has identified where it should concentrate its efforts and in what order, it should take advantage of the **Tree diagram** to develop solution alternatives to materialize an idea or resolve an important issue. This tool should allow the team to map all possibilities. For the modification of current inventory control policies the team might consider strategies like decentralization of inventory storage, delivery of parts to their points of use and reduction of WIP inventory. The team should then use the **Prioritization matrix** to evaluate the different strategies developed in the previous step and select the best solution based on evaluation attributes defined in coordination with the organization's management. Attributes for the example could be: estimated impact in inventory reduction, estimated impact in reduction of space requirements, and cost of solution. The team might find that adopting a decentralization of inventory policy is the best strategy. Then the team should consider contingencies that could occur in the process of implementing the identified solution. The **Process decision program chart** should be used for this, allowing the team to include preventive and corrective actions in its implementation plan. Being well prepared allows the team to make the most effective use of limited resources (i.e., time, people, money, and equipment). Once the team knows all it needs to do and what it will require, tasks can be delegated and information about schedules, team member roles, and other important project information may be organized with the use of the **Matrix diagram.** This is important as key project information needs to be shared and communicated through an effective means. The last step of the planning phase is the scheduling phase and for this the **Activity Network diagram** should be used. All tasks are carefully mapped in their logical sequence and dates are identified for key project milestones. This is useful for scheduling and controlling the execution of the facilities design project.

Figure 2.25   A facilities design team using the seven management and planning tools.

BIBLIOGRAPHY

1. Akao, Y. (ed.), *QFD: Integrating Customer Requirements into Product Design,* Productivity Press, Cambridge, MA, 1990.
2. Amirouche, F. M. L., *Computer-Aided Design and Manufacturing,* Prentice Hall, NJ, 1993.
3. Barnes, R. M., *Motion and Time Study Design and Measurement of Work,* John Wiley, New York, 1980.
4. Bourjault, A., *Contribution a une Approache Methodologique de l'Assemblage Automatise: Elaboration Automatique des Sequences Operatories,* Thesis for Grade de Docteur en Sciences Physiques at l'Universite de Franche Comte, November 12, 1984.
5. Brassard, M., *The Memory Jogger Plus,* GOAL/QPC, Methuen, MA, 1989.
6. Buffa, E. S., *Modern Production Management,* John Wiley, New York, 1977.
7. Burbidge, J. L., *The Introduction of Group Technology,* John Wiley, New York, 1975.
8. Camp, R. C., *Benchmarking: The Search for Industry Best Practices that Lead to Superior Performance,* Quality Press, White Plains, NY, 1989.
9. Chang, T., Wysk, R. A., Wang, H., *Computer Aided Manufacturing,* Prentice Hall, Englewood Cliffs, NJ, 1991.
10. El-Hadidi, T., *Simultaneous Evaluation Tool for Printed Circuit Boards,* Master Thesis in Industrial Engineering, Department of Industrial Engineering, North Carolina State University, Raleigh, NC, August 1991.
11. Francis, R. L., McGinnis, L. F., Jr., and White, J. A., *Facility Layout and Location: An Analytical Approach,* 2ed, Prentice Hall, Englewood Cliffs, NJ, 1992.
12. Hales, H. L., *Computer-Aided Facilities Planning,* Marcel Dekker, New York, 1984.
13. Ham, I., *Introduction to Group Technology,* Pennsylvania State University, University Park, PA, 1975.
14. Hartley, J. R., *Concurrent Engineering,* Productivity Press, Cambridge, 1992.
15. Klein, C., *Generation and Evaluation of Assembly Sequence Alternatives,* Master of Science Thesis, MIT Mech. Eng. Dept., MIT, Boston, 1986.
16. Morris, W. T., *Engineering Economic Analysis,* Reston Publishing, Reston, VA., 1976.
17. Ranson, G. M., *Group Technology,* Pergamon, New York, 1970.
18. Shore, R. H., and Tompkins, J. A., *Designing Flexible Facilities,* North Carolina State Univ., Raleigh, NC, 1977.
19. Shore, R. H., and Tompkins, J. A., "Flexible Facilities Design," *AIIE Transactions,* vol. 12, no. 2. pp. 200–205, June 1980.
20. Silva, D., *Concurrent Product and Process Design for Assembly,* Master Project in Integrated Manufacturing Systems Engineering, Integrated Manufacturing Systems Engineering Institute, North Carolina State University, Raleigh, NC, August 1990.
21. Subramaniam, S., *Design Evaluation and Cost Estimation Expert System,* Master Thesis in Industrial Engineering, Department of Industrial Engineering, North Carolina State Univ., Raleigh, NC, 1991.
22. White, J. A., "On Absurbing Markov Chains and Optimim Batch Production Quantities," *AIIE Transactions,* vol. 2, no. 1, March 1970, pp. 82–88.
23. White, J. A., Case, K. E., Pratt, D. B., and Agee, M. H., *Principles of Engineering Economic Analysis,* 4th ed., John Willy New York, 1998.
24. Whitney, D. E., DeFazio, T. L., Gustavson, R. E., Graves, S. C., Abell, T., Cooprider, K., and Pappu, S., *Tools for Strategic Product Design, First Report, CSDL-R-2115,* MIT and The Charles Stark Draper Laboratory, Cambridge, MA, November 1988.
25. Zorowski, C. F., *PDM—A Product Assemblability Merit Analysis Tool,* 1986 National Design Engineering Technical Conference, ASME, Columbus, OH, 1986.

PROBLEMS

2.1 You have been asked to plan a facility for a hospital. Explain in detail your first step.

2.2 Why is it important to integrate product, process, quality, scheduling, and facilities design decisions? Who should be involved in this integration? Which techniques are available to support this integrated approach?

2.3 Identify from papers and books at least five companies that have designed or redesigned their facilities considering new manufacturing approaches (i.e., multiple receiving docks, decentralized storage areas, cellular manufacturing, kanbans). For each company, describe in detail the methodology employed, approaches used, and results obtained.

2.4 Identify from papers and books at least five companies that have used benchmarking and/or quality function deployment to identify competition and customer needs. For each company, describe in detail the methodology employed and results obtained.

2.5 Identify from papers and books at least two companies that are implementing concurrent engineering techniques for integrated product and process design using the team approach. For each company, explain the team composition, techniques used, methodology employed, and results obtained. Did they consider facilities design in the product and process integration?

2.6 Identify at least five CAD systems available to support the facilities design process. Prepare a comparison of these systems and recommend the best one.

2.7 Develop a bill of materials, an assembly chart, and an operation process chart for a cheese hamburger and a taco with everything on it. Identify the components that are purchased and the ones that are prepared internally.

2.8 State three differences between the assembly chart and the operation process chart.

2.9 Part X requires machining on a milling machine (operations A and B are required). Find the number of machines required to produce 3000 parts per week. Assume the company will be operating 5 days per week, 18 hours per day. The following information is known:

Operation	Standard Time	Efficiency	Reliability	Scrap
A	3 min	95%	95%	2%
B	5 min	95%	90%	5%

Note. The milling machine requires tool changes and preventive maintenance after every lot of 500 parts. These changes require 30 min.

2.10 Form a team and apply the seven management and planning tools for planning the redesign of any of the following:
a. a bank.
b. a kindergarden.
c. a hospital.
d. a cafeteria.
e. a laundry facility.
f. a passenger area from a local airport, or
g. a bookstore.

2.11 Given the following, what are the machine fractions for machines A, B, and C to produce parts X and Y?

	Machine A	Machine B	Machine C
Part X standard time	0.15 hr	0.25 hr	0.1 hr
Part Y standard time	0.10 hr	0.10 hr	0.15 hr
Part X scrap estimate	5%	4%	3%

Part Y scrap estimate	5%	4%	3%
Historical efficiency	90%	90%	95%
Reliability factor	90%	90%	95%
Equipment availability	1600 hr/yr	1600 hr/yr	1600 hr/yr

Part X routing is machine A, then B, and then C; 110,000 parts are to be produced per year. Part Y routing is machine B, then A, and then C; 250,000 parts are to be produced per year. Setup times for parts X and Y are 20 min and 40 min, respectively.

2.12 During one 8-hr shift, 750 nondefective parts are desired from a fabrication operation. The standard time for the operation is 15 min. Because the machine operators are unskilled, the actual time it takes to perform the operation is 20 min and, on the average, one-fifth of the parts that begin fabrication are scrapped. Assuming that each of the machines used for this operation will not be available for 1 hr each shift, determine the number of machines required.

2.13 Part A is produced in machines 1 and 2. Part B is produced in machines 3 and 4. Part A and B are assembled in workstation 5 to create C. Assembly C is painted in process 6. Given the scrap percentages P1, ..., P6 for each operation and the desired output 06 for C, determine an equation that can be used to calculate the required input 11 to machine 1.

2.14 A foundry has received an order for 20 custom-designed castings. The casting process costs $700 per unit scheduled. If a casting is good, then it is machined to specifications at an added cost of $500 per unit. If a casting is not sold, it has a recycle value of $300. The customer has indicated a willingness to pay $2,000 per casting for 20 acceptable castings; the customer has also agreed to pay $1,500 each for 1 or 2 additional castings. However, the customer is unwilling to purchase fewer than 20 or more than 22 castings. Based on historical records, the following probability distributions have been estimated. How many castings should be scheduled for production to maximize expected profit? What is the probability of losing money?

# Good Castings	Number of Castings Scheduled										
	20	21	22	23	24	25	26	27	28	29	30
12	0.10	0.05	0.00	0.00	0.00	0.00	0.00	0.00	0.00	0.00	0.00
13	0.10	0.10	0.05	0.05	0.05	0.05	0.00	0.00	0.00	0.00	0.00
14	0.10	0.10	0.10	0.05	0.05	0.05	0.05	0.00	0.00	0.00	0.00
15	0.10	0.10	0.10	0.10	0.05	0.05	0.05	0.00	0.00	0.00	0.00
16	0.10	0.10	0.10	0.10	0.05	0.05	0.05	0.05	0.00	0.00	0.00
17	0.10	0.10	0.10	0.10	0.10	0.05	0.05	0.05	0.05	0.00	0.00
18	0.20	0.10	0.10	0.10	0.10	0.05	0.05	0.10	0.05	0.05	0.05
19	0.10	0.15	0.10	0.10	0.10	0.10	0.10	0.10	0.10	0.05	0.05
20	0.10	0.10	0.15	0.10	0.10	0.10	0.10	0.10	0.10	0.10	0.05
21	0.00	0.10	0.10	0.10	0.10	0.10	0.10	0.10	0.10	0.10	0.10
22	0.00	0.00	0.10	0.10	0.10	0.10	0.10	0.10	0.10	0.10	0.10
23	0.00	0.00	0.00	0.10	0.10	0.10	0.10	0.10	0.10	0.10	0.10
24	0.00	0.00	0.00	0.00	0.10	0.10	0.10	0.10	0.10	0.10	0.10
25	0.00	0.00	0.00	0.00	0.00	0.10	0.10	0.10	0.10	0.10	0.10
26	0.00	0.00	0.00	0.00	0.00	0.00	0.05	0.05	0.10	0.10	0.10
27	0.00	0.00	0.00	0.00	0.00	0.00	0.00	0.05	0.05	0.10	0.10
28	0.00	0.00	0.00	0.00	0.00	0.00	0.00	0.00	0.05	0.05	0.05
29	0.00	0.00	0.00	0.00	0.00	0.00	0.00	0.00	0.00	0.05	0.05
30	0.00	0.00	0.00	0.00	0.00	0.00	0.00	0.00	0.00	0.00	0.05

2.15 A foundry produces castings to order. An order for 20 special castings has been received. Since the casting process is highly variable, not all castings produced are good. The cost of producing each casting is $550; the additional cost of finishing a good casting is $125. If a casting is not good, it is recycled at a value of $75; excess good castings are not finished, but are recycled at a value of $75. The customer has agreed to accept 15, 16, 17, 18, 19, or 20 castings at a price of $1,250, each. If fewer than 15 good castings are produced, none will be purchased by the customer. Probability distributions for the number of good castings produced in a batch of varying sizes are given below. How many castings should be scheduled in order to maximize expected profit?

# Good Castings	Number of Castings Scheduled															
	15	16	17	18	19	20	21	22	23	24	25	26	27	28	29	30
5	0.05	0.00	0.00	0.00	0.00	0.00	0.00	0.00	0.00	0.00	0.00	0.00	0.00	0.00	0.00	0.00
6	0.05	0.05	0.00	0.00	0.00	0.00	0.00	0.00	0.00	0.00	0.00	0.00	0.00	0.00	0.00	0.00
7	0.05	0.05	0.05	0.00	0.00	0.00	0.00	0.00	0.00	0.00	0.00	0.00	0.00	0.00	0.00	0.00
8	0.05	0.05	0.05	0.05	0.00	0.00	0.00	0.00	0.00	0.00	0.00	0.00	0.00	0.00	0.00	0.00
9	0.05	0.05	0.05	0.05	0.05	0.00	0.00	0.00	0.00	0.00	0.00	0.00	0.00	0.00	0.00	0.00
10	0.05	0.05	0.05	0.05	0.05	0.05	0.00	0.00	0.00	0.00	0.00	0.00	0.00	0.00	0.00	0.00
11	0.10	0.05	0.05	0.05	0.05	0.05	0.05	0.00	0.00	0.00	0.00	0.00	0.00	0.00	0.00	0.00
12	0.10	0.10	0.05	0.05	0.05	0.05	0.05	0.05	0.00	0.00	0.00	0.00	0.00	0.00	0.00	0.00
13	0.15	0.10	0.10	0.05	0.05	0.05	0.05	0.05	0.05	0.00	0.00	0.00	0.00	0.00	0.00	0.00
14	0.15	0.15	0.10	0.10	0.05	0.05	0.05	0.05	0.05	0.05	0.00	0.00	0.00	0.00	0.00	0.00
15	0.20	0.15	0.15	0.10	0.10	0.05	0.05	0.05	0.05	0.05	0.05	0.00	0.00	0.00	0.00	0.00
16	0.00	0.20	0.15	0.15	0.10	0.10	0.05	0.05	0.05	0.05	0.05	0.05	0.00	0.00	0.00	0.00
17	0.00	0.00	0.20	0.15	0.15	0.10	0.10	0.05	0.05	0.05	0.05	0.05	0.05	0.00	0.00	0.00
18	0.00	0.00	0.00	0.20	0.15	0.15	0.10	0.10	0.05	0.05	0.05	0.05	0.05	0.05	0.00	0.00
19	0.00	0.00	0.00	0.00	0.20	0.15	0.15	0.10	0.10	0.05	0.05	0.05	0.05	0.05	0.05	0.00
20	0.00	0.00	0.00	0.00	0.00	0.20	0.15	0.15	0.10	0.10	0.05	0.05	0.05	0.05	0.05	0.05
21	0.00	0.00	0.00	0.00	0.00	0.00	0.20	0.15	0.15	0.10	0.10	0.05	0.05	0.05	0.05	0.05
22	0.00	0.00	0.00	0.00	0.00	0.00	0.00	0.20	0.15	0.15	0.10	0.10	0.05	0.05	0.05	0.05
23	0.00	0.00	0.00	0.00	0.00	0.00	0.00	0.00	0.20	0.15	0.15	0.10	0.10	0.05	0.05	0.05
24	0.00	0.00	0.00	0.00	0.00	0.00	0.00	0.00	0.00	0.20	0.15	0.15	0.10	0.10	0.05	0.05
25	0.00	0.00	0.00	0.00	0.00	0.00	0.00	0.00	0.00	0.00	0.20	0.15	0.15	0.10	0.10	0.05
26	0.00	0.00	0.00	0.00	0.00	0.00	0.00	0.00	0.00	0.00	0.00	0.20	0.15	0.15	0.10	0.10
27	0.00	0.00	0.00	0.00	0.00	0.00	0.00	0.00	0.00	0.00	0.00	0.00	0.20	0.15	0.15	0.10
28	0.00	0.00	0.00	0.00	0.00	0.00	0.00	0.00	0.00	0.00	0.00	0.00	0.00	0.20	0.15	0.15
29	0.00	0.00	0.00	0.00	0.00	0.00	0.00	0.00	0.00	0.00	0.00	0.00	0.00	0.00	0.20	0.15
30	0.00	0.00	0.00	0.00	0.00	0.00	0.00	0.00	0.00	0.00	0.00	0.00	0.00	0.00	0.00	0.20

2.16 In Problem 2.15, suppose the castings are produced independently with the probability of an individual casting being good being equal to 0.88. Using the binomial distribution to generate the probabilities, determine the optimum number of castings to produce. What is the probability of losing money on the transaction? [Use Excel's binomial distribution function to determine the probability of x good castings out of n produced, with the probability of a casting being good equal to p: =BINOMDIST (x, n, p, FALSE).

2.17 A foundry receives an order for four custom-designed castings. The customer will pay $30,000 for each of four good castings. The customer will accept neither fewer nor more than four castings. It will cost $15,000 to produce each casting. Each casting is produced independently; the probability of a casting being good is estimated to be 0.90.

   a. How many castings should be produced?
   b. What is the probability of losing money on the transaction?
   c. It has been argued that the estimate of cost to produce each casting is not very accurate. Analyze the sensitivity of the optimum batch production quantity to errors in estimating the values of the cost parameters; determine how much the cost of

producing each casting has to decrease in order to increase by one the number of castings to produce.

d. It has been argued that the estimate of the probability of a casting being good is inaccurate. Analyze the sensitivity of the optimum batch production quantity to changes in the probability.

2.18 A job shop has received an order for high-precision formed parts. The cost of producing each part is estimated to be $65,000. The customer requires that either 8, 9, or 10 parts be supplied. Each good part sold will produce revenue of $100,000. However, if fewer than 8 good parts are produced, none will be purchased; if more than 10 good parts are produced, the excess will not be purchased. The probability of an individual part being acceptable equals 0.85. Determine the expected profits for batch production quantities of 10, 11, and 12. For each, determine the probability of losing money on the transaction. Of the three choices, which is least preferred? Why?

2.19 A semiconductor wafer fabrication facility received an order for specially designed prototype semiconductor wafers. The cost of producing each wafer is estimated to be $20,000. The customer agrees to pay $150,000 for 3 good wafers, $200,000 for 4 good wafers, and $250,000 for 5 good wafers. Other than the 3, 4, or 5 good wafers, all other wafers, good or bad, must be destroyed. To obtain the contract, the wafer fab offers to pay the customer $100,000 if at least 3 good wafers are not produced. Each wafer is produced independently; the probability of a wafer being acceptable is estimated to be 0.75. Determine the number of wafers to produce, as well as the probability of losing money on the transaction.

2.20 A firm has received an order for 25 die cast parts made from precious metals. The parts will sell for $5000 each; it will cost $2500 to produce an individual part. The probability of a part meeting final inspection equals 0.98. Parts not sold, good and bad, can be recycled at a value of $1000. A penalty clause in the contract results in the firm having to pay the customer $500 per unit short. Determine the number of parts to be scheduled for production to maximize expected profit.

2.21 An order for 45 castings has been received. Each casting is produced independently, with a 0.85 probability of an individual casting being good. The customer will accept as few as 40 and as many as 50 castings at a unit sales price of $2000; each casting costs $800 to produce. (*Hint:* The optimum lot size is at least 55 castings.)

a. Determine the lot size that maximizes expected profit.

b. Determine the economic lot size assuming a 0.98 probability of a good unit.

2.22 In Example 2.5, suppose 11 machines are to be assigned. Which of the following assignments will minimize the total cost over a common time period: {2, 2, 2, 2, 3} or {2, 2, 2, 2, 2, 1}? Is there another combination that will be less expensive?

2.23 In Example 2.5, suppose the cost per machine hour is unknown. For what range of values of C_m will the optimum assignment remain the same? Justify your answer.

2.24 In Example 2.5, suppose an operator is assigned 3 machines, the work shift begins at 8:00 A.M., a 15-minute break occurs at 9:45 A.M., a 30-minute lunch break occurs at 11:45 A.M., a 15-minute break occurs at 2:00 P.M., and the shift ends at 4:00 P.M. When breaks occur, all machines are stopped. Parts cannot be partially machined when machines are stopped. Hence, transient conditions exist when machines are started and when machines are stopped. How many parts are produced per shift? Under steady-state conditions, how many parts would be produced during an 8-hour period?

2.25 Suppose 5 identical machines are to be used to produce two different products. The operating parameters for the two products are as follows: $a_1 = 2$ min; $a_2 = 2.5$ min; $b_1 = 1$ min; $b_2 = 1.5$ min; $t_1 = 6$ min; and $t_2 = 8$ min. The cost parameters are the same for each operator-machine combination: C_o = \$15/hr and C_m = \$50/hr. Determine the method of assigning operators to machines that minimizes the cost per unit produced.

2.26 Semiautomatic mixers are used in a paint plant. It takes 6 minutes for an operator to load the appropriate pigments and paint base into a mixer. Mixers run automatically and then automatically dispense paint into 50-gallon drums. Mixing and unloading take 30 minutes to complete. Mixers are cleaned automatically between batches; it takes 4 minutes to clean each mixer. Between batches, an operator places empty drums in the magazine to position them for filling; it requires 6 minutes to load the drums into the magazine. Filled drums are transported automatically by conveyor to a test area before being stored. Mixers are located close enough for travel between mixers to be negligible.

    a. What is the maximum number of mixers that can be assigned to an operator without creating idle time for the mixers?

    b. If C_o = \$12/hr and C_m = \$25/hr, what assignment of mixers to operators will minimize the cost per batch produced?

2.27 Use a multiple activity chart to illustrate how one operator can tend Machine A, Machine B, and Machine C during a repeating cycle, based on the following data: a_A = 2 min; a_B = 2.5 min; a_C = 3 min; b_A = 1 min; b_B = 1 min; b_C = 1.5 min; t_A = 7 min; t_B = 8 min; and t_C = 9 min. What is the length of the repeating cycle?

2.28 It takes 3 minutes to load and 2 minutes to unload a machine. Inspection, packing, and travel between machines totals 1 minute. Machines run automatically for 20 minutes. Operators cost \$12 per hour; machines cost \$30 per hour.

    a. What is the maximum number of machines that can be assigned an operator without creating machine idle time during a repeating cycle?

    b. What assignment minimizes the cost per unit produced?

    c. If 4 machines are assigned an operator, what will be the cost per unit produced?

    d. For what range of values for concurrent activity is the optimum assignment equal to 4?

    e. If 15 machines are needed to meet the production requirement, how should they be assigned to operators in order to minimize cost per unit produced?

2.29 Cases of product are conveyed to automatic palletizers for palletizing. Depending on the particular product and carton dimensions, the palletizer has to be reprogrammed by an operator. Typically, 25 pallet loads are completed before reprogramming is needed. The reprogramming requires 4 minutes. The palletizer operates automatically for 40 minutes. The palletizer operator also must restock the machine with empty pallets; this can be done at any time during the last 10 minutes of the palletizer's automatic run time; restocking pallets requires 5 minutes. Travel time between palletizers, plus data entry in the computer system by the operator, requires 3 minutes. What is the maximum number of palletizers one operator can tend without creating idle time for the palletizer?

2.30 A 5-aisle, automated storage and retrieval system (AS/RS) with a storage/retrieval (S/R) machine in each aisle is installed in a distribution center. Full pallet loads are stored and retrieved by the S/R machines. The input/output (I/O) point is at the end of the aisle and consists of a simple pickup and deposit (P/D) station. Pallet loads to be stored are retrieved from an inbound conveyor, delivered to the P/D station, and placed on the P/D station by lift truck; likewise, pallet loads brought to the P/D station by the S/R machines are removed by the lift truck and transported on an outbound conveyor.

    It takes 0.5 minute to travel between the retrieval point of the inbound conveyor and the P/D station, 0.5 minute to travel between the P/D station and the deposit point of the outbound conveyor, and 0.5 minute to travel between the deposit point of the outbound conveyor and the retrieval point of the inbound conveyor; travel times include picking up and placing the pallet load. The time required for the S/R machine to pick up a load at the P/D station, travel into the aisle, store the load, and return to the P/D station equals 4 minutes. How many lift truck operators are required to service the 5-aisle AS/RS and not create idle time on the part of the S/R machines?

2.31 Carousel conveyors are used for storage and order picking for small parts. The conveyors rotate clockwise or counterclockwise, as necessary, to position storage bins at the storage and retrieval point. The conveyors are closely spaced, such that the operator's travel time between conveyors is negligible. The conveyor rotation time for each item equals 1 minute; the time required for the operator to retrieve an item after the conveyor stops rotating equals 0.25 minute. How many carousel conveyors can one operator tend without creating idle time on the part of the conveyors?

2.32 Choose a simple recipe consisting of no more than ten ingredients. Examine carefully how the recipe is made. Develop a parts list, a bill of materials, a route sheet, an assembly chart, an operations process chart, and a precedence diagram for this recipe so that someone could follow the recipe without additional instructions.

2.33 Take the parts list, bill of materials, route sheet, assembly chart, operations process chart, and precedence diagram from Problem 2.32 and give it to another individual unfamiliar with the recipe. Have this individual decompose these charts into a written form of the recipe. Examine how close this derived recipe is to the original.

2.34 In Figure 2.19, the second item under scheduling is "do not authorize products for which the needed parts are not available." Construct an affinity diagram of actions to provide visibility to this issue.

2.35 Construct an interrelationship digraph for the results of Problem 2.34.

2.36 Develop a tree diagram to identify the specifics on developing an operator certification program (first item under equipment issues in Figure 2.19).

2.37 In a facilities design lab assignment, your instructor might be expecting specific deliverables (select one of your lab assignments). For such deliverables, identify the activities that must be performed. Construct a matrix diagram with deliverables (in rows) and activities (in columns). Using some symbols (i.e., star = extremely important, circle = important, triangle = less important, empty = no impact), specify the impact of each activity on the deliverables. An activity might impact more than one deliverable.

2.38 Develop a contingency diagram for the field trips to nearby industries that the local IIE student chapter might be organizing for the year.

2.39 Develop an implementation plan, using the activity network diagram, for your facilities design final project.

2.40 Prepare a summary of the computer-based tools that can be employed in the design of products and processes.

3

FLOW, SPACE, AND ACTIVITY RELATIONSHIPS

3.1 INTRODUCTION

In determining the requirements of a facility, three important considerations are flow, space, and activity relationships. **Flow** depends on lot sizes, unit load sizes, material handling equipment and strategies, layout arrangement, and building configuration. **Space** is a function of lot sizes, storage system, production equipment type and size, layout arrangement, building configuration, housekeeping and organization policies, material handling equipment, and office, cafeteria, and restroom design. **Activity relationships** are defined by material or personnel flow, environmental considerations, organizational structure, continuous improvement methodology (teamwork activities), control issues, and process requirements.

As indicated in previous chapters, facilities planning is an iterative process. The facilities planning team or the facilities planner needs to interact not only with product, process, and schedule designers but also with top management to identify alternative issues and strategies to consider in the analysis (i.e., lot sizes, storage-handling strategies, office design, organizational structure, environmental policy).

The facilities planner also needs to continually investigate the impact of modern manufacturing approaches on flow, space, and activity relationships. For example, concepts like decentralized storage, multiple receiving docks, deliveries to points of use, decentralized management and support functions, quality at the source, cellular manufacturing, lean organizational structures, and small lot purchasing and production could challenge traditional activity relationships and reduce flow and space requirements.

Some of the traditional activity relationships to be challenged are centralized offices, a centralized storage area, a single receiving area, and a centralized rework area. Flow requirements are reduced with external and internal deliveries to points of use, storage of inventories in decentralized storage areas close to points of use, movement of material controlled by pull strategies with kanbans, and manufacturing cells. Less space is required for inventories; production, storage, and handling equipment; offices; parking lots; and cafeterias.

Different procedures, forms, and tables to determine flow, space, and activity relationships are covered in this chapter. Additionally, layout types are explained in Section 3.2 (including a consideration of cellular manufacturing), the logistics system is described in Section 3.3, and visual management is covered in Section 3.7. Activity relationships are considered in Section 3.3. Sections 3.4 through 3.6 address flow relationships. Space relationships are the subject of Section 3.7.

3.2   DEPARTMENTAL PLANNING

To facilitate the consideration of flow, space, and activity relationships, it is helpful to introduce the subject of departmental planning. At this point in the facilities planning process, we are not so concerned with organizational entities. Rather, we are interested in forming **planning departments.** Planning departments can involve production, support, administrative, and service areas (called production, support, administrative, and service planning departments).

Production planning departments are collections of workstations to be grouped together during the facilities layout process. The formulation of organizational units should parallel the formation of planning departments. If for some reason the placement of workstations violates certain organizational objectives, then modifications should be made to the layout.

As a general rule, planning departments may be determined by combining workstations that perform "like" functions. The difficulty with this general rule is the definition of the term "like." "Like" could refer to workstations performing operations on similar products or components or to workstations performing similar processes.

Depending on product volume-variety, production planning departments can also be classified as product, fixed materials location, product family (or group technology), and process planning departments (see Figure 3.1). As examples of production planning departments that consist of a combination of workstations performing operations on similar products or components, consider engine block production line departments, aircraft fuselage assembly departments, and uniform flat sheet metal departments. As described in Chapter 6, these are referred to as product planning departments because they are formed by combining workstations that produce similar products or components. **Product planning departments** may be further subdivided by the characteristics of the products being produced.

Suppose a large, stable demand for a standardized product, like an engine block, is to be met by production. In such a situation, the workstations should be combined into a planning department so that all workstations required to produce the product are combined. The resulting product planning department may be referred to as **production line department.**

Next, suppose a low, sporadic demand exists for a product that is very large and awkward to move, for example, an aircraft fuselage. The workstations should

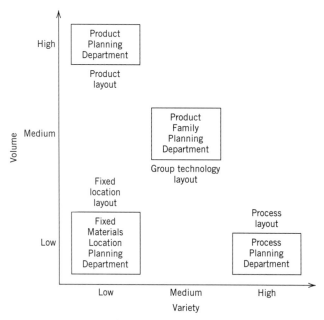

Figure 3.1 Volume-variety layout classification.

be combined into a planning department that includes all workstations required to produce the product and the staging area. This type of product planning department may be referred to as a **fixed materials location department.**

A third type of product planning department may be identified when there exists a medium demand for a medium number of similar components. Similar components form a family of components that, in group technology terminology, may be produced via a "group" of workstations. The combination of the group of workstations results in a product planning department that may be referred to as a **product family department.**

Examples of planning departments based on the combination of workstations containing "similar" processes are metal cutting departments, gear cutting departments, and hobbing departments. Such planning departments are referred to as **process departments** because they are formed by combining workstations that perform "similar" processes.

The difficulty in defining process departments is in the interpretation of the word "similar." For example, in a facility specializing in the production of gears, gear hobbing, gear shaping, and shaft turning might not be considered similar and each might be grouped into their own planning departments. However, in a facility producing mechanical switching mechanisms, these same processes might be grouped into two, not three, planning departments: a gear cutting department, containing similar gear hobbing and shaping processes, and a turning department. Even more extreme, in a furniture facility, all metalworking might take place in a metalworking planning department. Therefore, the same three processes might be seen to be similar and grouped into a single process planning department. The determination of which workstations are to be considered similar depends on not only the workstations but also the relationships among workstations and between workstations and the overall facility.

Most facilities consist of a mixture of product and process planning departments. For example, in a facility consisting of mainly process planning departments producing a large variety of rather unrelated products, the detailed placement of individual workstations within a process department might be based on a product planning department philosophy. (As an illustration, all painting activities might be grouped together in a painting process department. However, the layout of the painting department can consist of a painting line designed on the basis of a product planning department philosophy). Conversely, in a facility consisting mainly of product planning departments producing a few, high-volume, standard products, it would not be surprising to find several "specialized" components produced in process planning departments.

A systematic approach should be used in combining workstations into departments. Each product and component should be evaluated and the best approach determined for combining workstations into planning departments. Table 3.1 summarizes the bases for combining workstations into planning departments.

Support, administrative, and service planning departments include offices and areas for storage, quality control, maintenance, administrative processes, cafeterias, restrooms, lockers, and so on. Traditionally, support, administrative, and service planning departments have been treated as "process" departments (similar activities performed within a certain area).

Organizations using modern manufacturing approaches are combining production, support, administrative, and service planning departments to create integrated production-support-administrative-service planning departments. For example, a manufacturing cell dedicated to the production of a family of parts and with dedicated support and administrative personnel and services (e.g., maintenance, quality, materi-

Table 3.1 *Procedural Guide for Combining Workstations in Planning Departments*

If the Product Is	The Type of Planning Department Should Be	And the Method of Combining Workstations into Planning Departments Should Be
Standardized and has a large stable demand.	Production line, product department.	Combine all workstations required to produce the product.
Physically large, awkward to move, and has a low sporadic demand	Fixed materials location, product department	Combine all workstations required to produce the product with the area required for staging the product
Capable of being grouped into families of similar parts that may be produced by a group of workstations	Product family, product department	Combine all workstations required to produce the family of products
None of the above	Process department	Combine identical work stations into initial planning departments and attempt to combine similar initial planning departments without obscuring important interrelationships within departments

als, engineering, tooling, purchasing, management, vending machines, restrooms, lockers) could be an integrated planning department.

Many companies train their operators in most of the support, administrative, and management functions so that they can become autonomous. In these cases, the operators (called technicians or associates) plus a facilitator-coordinator can manage the operation of the manufacturing cell with minimum external support.

Activity relationships and flow and space requirements in a facility with self-managing teams will be totally different than in a facility with traditional production, support, and administrative planning departments (less material, personnel, tooling, and paperwork flow requirements and less space needs).

Many companies using modern manufacturing approaches are converting their facilities to combinations of product and product family (group technology) planning departments. Group technology layouts are combined with *just-in-time* (JIT) concepts in cellular manufacturing arrangements. A more detailed explanation of group technology and cellular manufacturing follows.

Manufacturing Cells

Product family or group technology departments aggregate medium volume-variety parts into families based on similar manufacturing operations or design attributes. Machines required to manufacture the part family are grouped together to form a "cell."

Cellular manufacturing [1, 2, 3, 10, 31, 46] involves the use of *manufacturing cells.* The manufacturing cells can be formed in a variety of ways, with the most popular involving grouping of machines, employees, materials, tooling, and material handling and storage equipment to produce families of parts. Cellular manufacturing became quite popular in the late 1900s and is often associated with just-in-time (JIT) [19, 39, 40, 45], total quality management (TQM) [8, 13, 21, 22, 23, 24, 32, 38, 52, 53, 54], and lean manufacturing concepts and techniques [37, 51, 55]. (For additional information on lean manufacturing, explore the Web site for the Lean Enterprise Institute, http://www.lean.org/.)

Successful implementation of manufacturing cells requires addressing selection, design, operation, and control issues. Selection refers to the identification of machine and part types for a particular cell. Cell design refers to layout and production and material handling requirements. Operation of a cell involves determining lot sizes, scheduling, number of operators, type of operators, and type of production control (push vs. pull). Finally, control of a cell refers to the methods used to measure the performance of the cell.

Several approaches have been proposed to address selection issues of manufacturing cells. The most popular approaches are classification and coding, production flow analysis, clustering techniques, heuristic procedures, and mathematical models [1, 2, 5, 7, 14, 16, 17, 25, 26, 28, 30, 31, 44, 46, 58].

Classification is the grouping of parts into classes or part families based on design attributes and coding is the representation of these attributes by assigning numbers or symbols to them.

Production flow analysis [5] is a procedure for forming part families by analyzing the operation sequences and the production routing of a part or component through the plant.

Clustering methodologies are used to group parts together so they can be processed as a family [7, 14, 16, 20, 26, 28, 31, 46, 58]. This methodology lists parts and machines in rows and columns, and interchanges them based on some criterion like similarity coefficients. For example, the *direct clustering algorithm* (DCA) [7] forms clustered groups based on sequentially moving rows and columns to the top and left.

Several heuristic procedures have been developed for the formation of cells [2, 17, 20, 25, 46]. One developed by Ballakur and Steudel [2] assigns machines to cells based on work load factors and assigns parts to cells based on the percentage of operations of a part processed within a cell.

Several other mathematical models have also been developed [1, 30, 46, 47]. One proposed by Al-Qattan [1] to form machine cells and families of parts is based on the branch and bound method.

Singh and Rajamani [46] present a wide range of algorithms for forming manufacturing cells. Among those they consider are bond energy (BEA), rank-order clustering (ROC and ROC2), modified rank-order clustering (MODROC), direct clustering (DCA), cluster identification (CIA), single linkage clustering (SLC), and linear cell clustering (LCC) algorithms. In addition, they consider the use of mathematical programming approaches, including formulating the cell formation problem as a p-median covering problem, an assignment problem, a quadratic assignment problem, and a nonlinear optimization problem. Although of less practical interest, they also show how simulated annealing, genetic algorithms, and neural networks can be used to address the problem of cell formation.

It is important to note that the formation of cells is seldom the responsibility of the facilities planner. Instead, it is typically performed by the manufacturing engineer in conjunction with the production planner. Cell formation, inventory control, demand forecasting, assembly line balancing, and a host of other subjects are of great interest and significance to facilities planning. However, they are seldom (if ever) found in the domain of the facilities planner. For that reason, at most, we mention them briefly, if at all.

Because cellular manufacturing is increasing in importance and application and because it can significantly impact the facilities layout, we have chosen to introduce the subject through the following examples. For illustrative purposes, we will limit our treatment to the use of the direct clustering algorithm developed by Chan and Milner [7]. The DCA methodology is based on a machine-part matrix in which 1 indicates that the part requires processing by the indicated machine; a blank indicates the machine is not used for the particular part. The DCA methodology consists of the following steps:

Step 1. Order the rows and columns. Sum the 1s in each column and in each row of the machine-part matrix. Order the rows (top to bottom) in descending order of the number of 1s in the rows and order the columns (left to right) in ascending order of the number of 1s in each. Where ties exist, break the ties in descending numerical sequence.

Step 2. Sort the columns. Beginning with the first row of the matrix, shift to the left of the matrix all columns having a 1 in the first row. Continue the process row-by-row until no further opportunity exists for shifting columns.

Step 3. Sort the rows. Column-by-column, beginning with the leftmost column, shift rows upward when opportunities exist to form blocks of 1s. (It

Machine #

Part #	1	2	3	4	5	# of 1s
1	1		1			2
2	1					1
3		1		1	1	3
4	1		1			2
5		1				1
6				1	1	2
# of 1s	3	2	2	2	2	

Figure 3.2 Machine-part matrix for Example 3.1.

should be noted that performing the column and row sortation is facilitated by using spreadsheets, such as Excel.)

Step 4. Form cells. Look for opportunities to form cells such that all processing for each part occurs in a single cell.

Example 3.1

Consider the machine-part matrix shown in Figure 3.2 for a situation involving 6 parts to be processed; 5 machines are required. As noted above, the entries in the matrix indicate the machine-part combination that is required; for example, part 1 requires machining by machines 1 and 3.

Applying step 1 of the direct clustering algorithm, as shown in Figure 3.3, the rows are ranked in descending order of the number of 1s and ties are broken in descending numerical sequence. The row-ordered sequence of the part numbers is {3, 6, 4, 1, 5, 2}. Likewise, the columns are arranged in ascending order of the number of 1s, with ties broken in descending numerical order; the resulting column-ordered sequence of the machine numbers is {5, 4, 3, 2, 1}. The ordered machine-part matrix is shown in Figure 3.3.

Step 2 involves sorting the columns to move toward the left all columns having a 1 in the first row, which represents part 3. Since the columns for machines 5 and 4 are already located to the left of the matrix, only the column for machine 2 can be shifted. That is the only column shift required for this example. The resulting column-sorted machine-part matrix is depicted in Figure 3.4.

Step 3 consists of sorting the rows by moving upward rows having a 1 in the first column that are not already located as far toward the top of the matrix as possible. Since

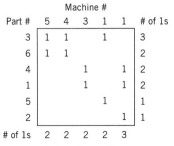

Machine #

Part #	5	4	3	1	1	# of 1s
3	1	1		1		3
6	1	1				2
4			1		1	2
1			1		1	2
5			1			1
2					1	1
# of 1s	2	2	2	2	3	

Figure 3.3 Ordered machine-part matrix.

Machine #

Part #	5	4	2	3	1	# of 1s
3	1	1	1			3
6	1	1				2
4				1	1	2
1				1	1	2
5			1			1
2					1	1
# of 1s	2	2	2	2	3	

Figure 3.4 Column-sorted machine-part matrix.

Machine #

Part #	5	4	2	3	1	# of 1s
3	1	1	1			3
6	1	1				2
5			1			1
4				1	1	2
1				1	1	2
2					1	1
# of 1s	2	2	2	2	3	

Figure 3.5 Row-sorted machine-part matrix.

Machine #

Part #	5	4	2	3	1
3	1	1	1		
6	1	1			
5			1		
4				1	1
1				1	1
2					1

Figure 3.6 Formation of two cells.

none can be shifted further for either machines 5 or 4, the first row to be moved is that for part 5, based on its processing requirement with machine 2. The resulting row-sorted machine-part matrix is shown in Figure 3.5.

In this case, as shown in Figure 3.6, the machines can be grouped into 2 cells, with parts 3, 5 and 6 being processed in a cell made up of machines 2, 4, and 5 and with parts 1, 2, and 4 being processed in a cell consisting of machines 1 and 3. Unfortunately, it is not always the case that cells can be formed without conflicts existing, as illustrated in Example 3.2.

Example 3.2

Consider the machine-part matrix shown in Figure 3.7. Applying the DCA methodology results in the ordered machine-part matrix shown in Figure 3.8. Notice that no further improvement will occur by performing step 2 or step 3. Also, notice that because machine 2 is needed for parts 3 and 5, a conflict exists; alternatively, we could say that because part 5 requires machines 2 and 3, a conflict exists. As shown in Figure 3.9a, two cells can be formed, one consisting of machines 4 and 5 and the other consisting of machines 1, 2, and 3, with the machining required for part 3 on machine 2 to be resolved. Alternatively, as shown in Figure 3.9b, machines 2, 4, and 5 can form a cell and machines 1 and 3 can form another cell; in this case, the machining of part 5 on machine 2 must be resolved. Finally, as shown in Figure 3.9c, the cellular formation as given in Figure 3.9b can be used, but with part 5 assigned to the cell consisting of machines 2, 4, and 5; as shown, the processing of part 5 on machine 3 would need to be resolved for this cellular formation.

Examining Figure 3.9a, a possible solution comes to mind. Depending on the facility, if machines 2 and 3 can be located relatively close to one another, albeit in

Machine #

Part #	1	2	3	4	5	# of 1s
1	1		1			2
2	1					1
3		1		1	1	3
4	1		1			2
5		1	1			2
6				1	1	2
# of 1s	3	2	3	2	2	

Figure 3.7 Machine-part matrix.

Machine #

Part #	5	4	2	3	1	# of 1s
3	1	1	1			3
6	1	1				2
5			1	1		1
4				1	1	2
1				1	1	2
2					1	1
# of 1s	2	2	2	2	3	

Figure 3.8 Ordered machine-part matrix.

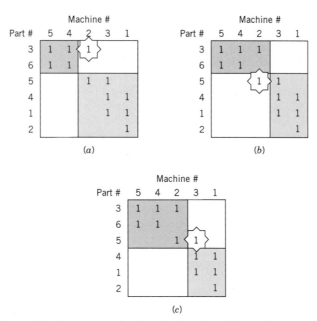

Figure 3.9 Formation of cells with "bottleneck" machine 2 or 3.

different cells, then part 5 could be processed by machines on the "boundaries" of
the cells.

Another option is to duplicate machine 2 and place it in each cell, as shown in
Figure 3.10a. Alternatively, as shown in Figure 3.10b, machine 3 could be duplicated and
placed in each cell. The tradeoff between having a part travel to both cells versus having
to duplicate a machine depends on many factors, not the least of which is the overall
utilization of the machine to be duplicated. For example, if the processing requirements for
parts 3 and 5 are such that multiple machines of type 2 are required, then the conflict
involving the formation of cells disappears or is minimized; likewise, if the volume of
processing required for part 5 fully utilizes machine 3, then providing another machine 3
to process parts 2 and 4 is a natural means of resolving the conflict.

The situation depicted in Example 3.2 points out a weakness of several of the cell
formation algorithms. Namely, the simplest ones do not take into account machine
utilization and the possibilities of multiple machines of a given type being required.

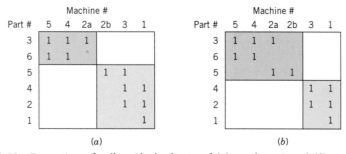

Figure 3.10 Formation of cells with duplicate of (*a*) machine 2 and (*b*) machine 3.

Example 3.3

Consider the machine-part matrix for a situation involving 13 parts and 26 machines given in Figure 3.11. Applying the DCA algorithm yields the results depicted in Figures 3.12 through 3.14. As with the previous example, step 3 was not required, since shifting rows would not improve the formation of cells. From Figure 3.14, it is evident that only 2 "pure" cells can be formed, due to machines 1 and 3. However, if multiple machines are needed, such that Figure 3.14c is feasible, then 3 "pure" cells can be formed for this example. As noted in the previous example, an alternative approach is to form the cells as shown in Figure 3.14b and locate machines 1 and 3 at the boundary between the cells A and B to minimize material handling between cells.

   Using cellular manufacturing terminology, machines 1 and 3 are called "bottle-neck" machines since they bind two cells together. When bottleneck conditions exist, as discussed previously, one can attempt to minimize the disruptive effects of having parts from other cells intrude on neighboring cells by locating bottleneck machines at the boundary between cells. Alternatively, the parts that require processing by the bottleneck machines could be reexamined to determine if alternative processing approaches can be used. Perhaps parts can be redesigned so that other machines can

Part #												Machine #															# of 1s
	1	2	3	4	5	6	7	8	9	10	11	12	13	14	15	16	17	18	19	20	21	22	23	24	25	26	
1	1		1		1			1	1		1						1	1				1					9
2	1		1		1			1	1		1						1	1				1					9
3	1		1		1			1	1		1						1	1				1					9
4	1		1		1			1	1		1						1	1				1					9
5	1		1			1		1	1		1						1	1				1					9
6	1		1			1		1	1		1						1	1				1					9
7	1		1			1		1	1		1						1	1				1					9
8		1		1			1			1		1							1	1			1				8
9		1		1			1			1		1							1	1			1				8
10	1		1										1	1	1	1					1			1	1	1	10
11	1		1										1	1	1	1					1			1	1	1	10
12	1		1										1	1	1	1					1			1	1	1	10
13	1		1										1	1	1	1					1			1	1	1	10
# of 1s	11	2	11	2	4	3	2	7	7	2	7	2	4	4	4	4	7	7	2	2	4	7	2	4	4	4	

Figure 3.11   Machine-part matrix.

Part #												Machine #															# of 1s
	23	20	19	12	10	7	4	2	6	26	25	24	21	16	15	14	13	5	22	18	8	17	11	9	3	1	
13										1	1	1	1	1	1	1	1								1	1	10
12										1	1	1	1	1	1	1	1								1	1	10
11										1	1	1	1	1	1	1	1								1	1	10
10										1	1	1	1	1	1	1	1								1	1	10
7									1										1	1	1	1	1	1	1	1	9
6									1										1	1	1	1	1	1	1	1	9
5									1										1	1	1	1	1	1	1	1	9
4																		1	1	1	1	1	1	1	1	1	9
3																		1	1	1	1	1	1	1	1	1	9
2																		1	1	1	1	1	1	1	1	1	9
1																		1	1	1	1	1	1	1	1	1	9
9	1	1	1	1	1	1	1	1																			8
8	1	1	1	1	1	1	1	1																			8
# of 1s	2	2	2	2	2	2	2	2	3	4	4	4	4	4	4	4	4	4	7	7	7	7	7	7	11	11	

Figure 3.12   Ordered machine-part matrix.

Part #	3	1	26	25	24	21	16	15	14	13	22	18	17	11	9	8	6	5	23	20	19	12	10	7	4	2	# of 1s
13	1	1	1	1	1	1	1	1	1	1																	10
12	1	1	1	1	1	1	1	1	1	1																	10
11	1	1	1	1	1	1	1	1	1	1																	10
10	1	1	1	1	1	1	1	1	1	1																	10
7	1	1									1	1	1	1	1	1	1										9
6	1	1									1	1	1	1	1	1	1										9
5	1	1									1	1	1	1	1	1	1										9
4	1	1									1	1	1	1	1	1		1									9
3	1	1									1	1	1	1	1	1		1									9
2	1	1									1	1	1	1	1	1		1									9
1	1	1									1	1	1	1	1	1		1									9
9																			1	1	1	1	1	1	1	1	8
8																			1	1	1	1	1	1	1	1	8
# of 1s	11	11	4	4	4	4	4	4	4	4	7	7	7	7	7	7	7	3	4	2	2	2	2	2	2	2	

Figure 3.13 Column-sorted machine-part matrix.

be used; in the best of all possible worlds, the part would be redesigned for processing by machines already assigned to the cell! If no better alternative is available, then the possibility of outsourcing the processing of the part should be considered.

In the example, every part processed in the first cell is also processed in the second cell. Hence, it is unlikely that the conflicts can be resolved by redesigning or outsourcing certain manufacturing steps for the parts. As noted previously, the viability of adding multiple machines of types 1 and 3 should be explored.

The cellular manufacturing system can be designed once the cells have been formed. The system can be either decoupled or integrated. Typically, a decoupled cellular manufacturing system uses a storage area to store part families after a cell has finished operating on them. Whenever another cell or department is to operate on the parts, they are retrieved from the storage area. Thus, the storage area acts as a decoupler making the cells and departments independent of each other. Unfortunately, this leads to excessive material handling and poor responsiveness.

To eliminate such inefficiencies, many companies use an integrated approach to the design and layout of cellular manufacturing systems. Here, cells and departments are linked through the use of kanbans or cards.

Shown in Figure 3.15, production cards (POK) are used to authorize production of more components or subassemblies and withdrawal cards (WLK) are used to authorize delivery of more components, subassemblies, parts, and raw materials.

To understand what kanbans are, the motivation behind their development needs to be discussed. Traditionally, when a workstation completes its set of operations, it pushes its finished parts onto the next workstation irrespective of its need for those parts. This is referred to as "push" production control. For a situation where the supplying workstation operates at a rate faster than its consuming workstation, parts will begin to build up. Eventually, the consuming workstation will be overwhelmed with work.

To prevent this from happening, it would make sense if the supplying workstation did not produce any parts until its consuming workstation requested parts. This "pull" production control is typically called *kanban*. Kanban means signal and commonly uses cards to signal the supplying workstation that its consuming workstation requests more parts.

Machine #

Part #	3	1	26	25	24	21	16	15	14	13	22	18	17	11	9	8	6	5	23	20	19	12	10	7	4	2
13	1	1	1	1	1	1	1	1	1	1																
12	1	1	1	1	1	1	1	1	1	1																
11	1	1	1	1	1	1	1	1	1	1																
10	1	1	1	1	1	1	1	1	1	1																
7	1	1									1	1	1	1	1	1	1									
6	1	1									1	1	1	1	1	1	1									
5	1	1									1	1	1	1	1	1	1									
4	1	1				Cell A					1	1	1	1	1	1		1								
3	1	1									1	1	1	1	1	1		1								
2	1	1									1	1	1	1	1	1		1								
1	1	1									1	1	1	1	1	1		1				Cell B				
9																			1	1	1	1	1	1	1	1
8																			1	1	1	1	1	1	1	1

(a)

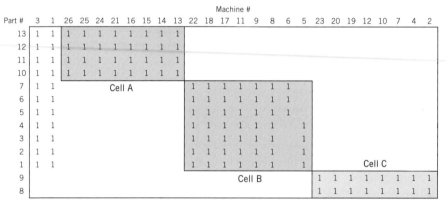

(b)

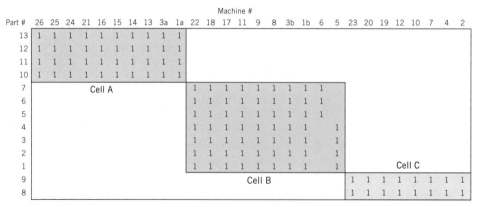

(c)

Figure 3.14   Final solution to Example 3.3.

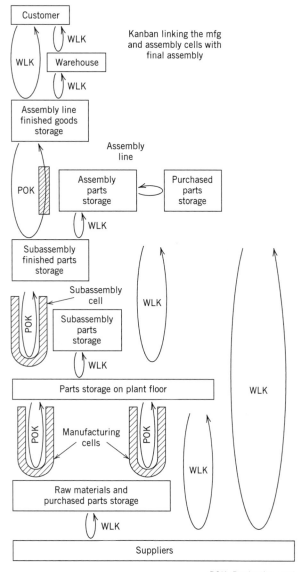

Figure 3.15   Integrated cellular manufacturing system.

The next phase in the design of a cellular manufacturing system is the layout of each cell. Figure 3.16 illustrates an assembly cell layout at the Hewlett-Packard, Greely Division [4]. The U-shaped arrangement of workstations significantly enhances visibility since the workers are aware of everything that is occurring within the cell. Notice that the agenda easel ensures that all workers know what are the daily production requirements of the cell. Materials flow from workstation to workstation via kanbans. Also, red and yellow lights or *andons* are used to stop production whenever a workstation has a problem. Problems as they occur are tabulated on the "Problem" display. This helps the workers by indicating potential problems that might arise.

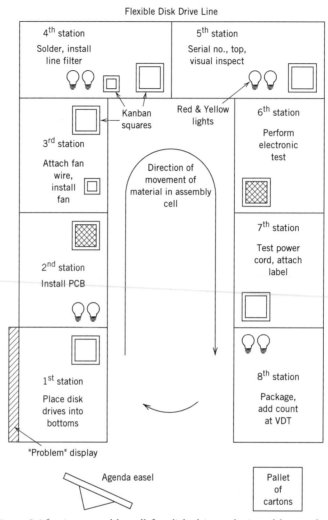

Figure 3.16 An assembly cell for disk drives, designed by workers at Hewlett-Packard, Greely Division.

3.3 ACTIVITY RELATIONSHIPS

Activity relationships provide the basis for many decisions in the facilities planning process. The primary relationships considered are

1. Organizational relationships, influenced by span of control and reporting relationships
2. Flow relationships, including the flow of materials, people, equipment, information, and money
3. Control relationships, including centralized versus decentralized materials control, real time versus batch inventory control, shop floor control, and levels of automation and integration

4.   Environmental relationships, including safety considerations and temperature, noise, fumes, humidity, and dust

5.   Process relationships other than those considered above, such as floor loadings, requirements for water treatment, chemical processing, and special services

Several relationships can be expressed quantitatively; others must be expressed qualitatively. Flow relationships, for example, are typically expressed in terms of the number of moves per hour, the quantity of goods to be moved per shift, the turnover rate for inventory, the number of documents processed per month, and the monthly expenditures for labor and materials.

Organizational relationships are usually represented formally by an organization chart. However, informal organizational relationships often exist and should be considered in determining the activity relationships for an organization. As an example, quality control may seem to have a limited organizational relationship with the receiving function in a warehouse; however, because of their requirement to interact closely, informal organizational relationships generally develop between the two functions. Organizational restructurings based on modern manufacturing approaches and motivated by increased international competition are decentralizing and relocating functions. The facilities design planner needs to be aware of such possibilities.

Flow relationships are quite important to the facilities planner, who views flow as the movement of goods, materials, energy, information, and/or people. The movement of refrigerators from the manufacturer through various levels of distribution to the ultimate customer is an important flow process. The transmission of sales orders from the sales department to the production control department is an example of an information flow process. The movement of patients, staff, and visitors through a hospital are examples of flow processes involving people.

The situations described are discrete flow processes where individual, discrete items move through the flow process. A continuous flow process differs from a discrete flow process in that movement is perpetual. Examples of continuous flow processes would include the flow of electricity, chemicals flowing through a processing facility, and oil flowing through a pipeline. Although many of the concepts described in this text are applicable to continuous flow processes, the primary emphasis is on discrete flow processes.

A flow process may be described in terms of the **subject** of flow, the **resources** that bring about flow, and the **communications** that coordinate the resources. The subject is the item to be processed. The resources that bring about flow are the processing and transporting facilities required to accomplish the required flow. The communications that coordinate the resources include the procedures that facilitate the management of the flow process. The perspective adopted for a flow process depends on the breadth of subjects, resources, and communications that exist in a particular situation.

If the flow process being considered is the **flow of materials into a manufacturing facility,** the flow process is typically referred to as a **materials management system.** The subjects of material management systems are the materials, parts, and supplies purchased by a firm and required for the production of its product. The resources of material management systems include:

1.   The production control and purchasing functions

2.   The vendors

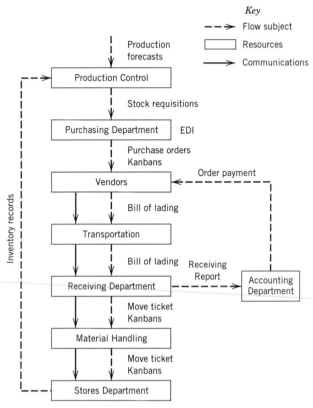

Figure 3.17 Material management system.

3.   The transportation and material handling equipment required to move the materials, parts, and supplies
4.   The receiving, storage, and accounting functions

The communications within material management systems include production forecasts, inventory records, stock requisitions, purchase orders, bills of lading, move tickets, receiving reports, kanbans, electronic data interchange (EDI), and order payment. A schematic of the material management system is given in Figure 3.17.

If the flow of materials, parts, and supplies ***within* a manufacturing facility** is to be the subject of the flow process, the process is called the **material flow system.** The type of material flow system is determined by the makeup of the activities or planning departments among which materials flow. As noted previously, there are four types of production planning departments.

1.   Production line departments
2.   Fixed materials location departments
3.   Product family departments
4.   Process departments

Typical material flow systems for each department type are shown in Figure 3.18. (Additional discussion regarding these layout alternatives is provided in Chapter 6,

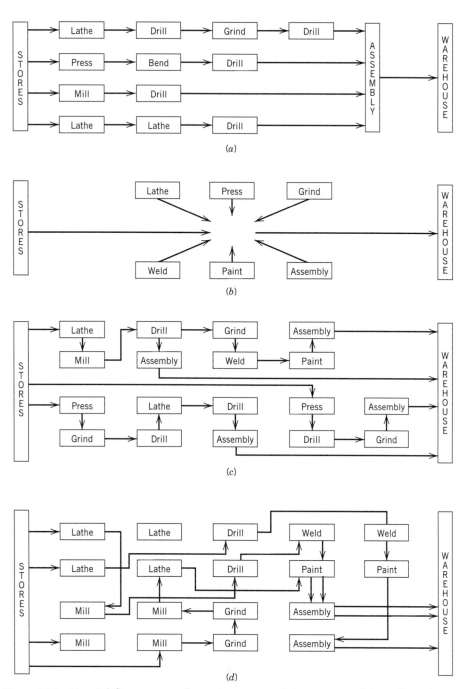

Figure 3.18   Material flow systems for various types of departments. (*a*) Product planning departments. (*b*) Fixed materials location planning departments. (*c*) Product family planning departments. (*d*) Process planning departments.

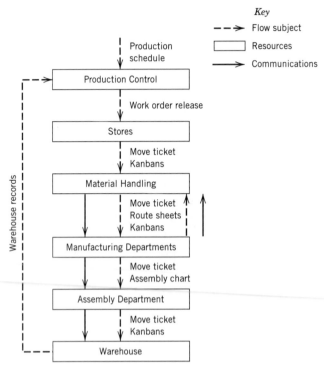

Figure 3.19 Material flow system.

Sections 6.2 and 6.3.) The subjects of material flow systems are the materials, parts, and supplies used by a firm in manufacturing its product. The resources of material flow systems include:

1. The production control and quality control departments
2. The manufacturing, assembly, and storage departments
3. The material handling equipment required to move materials, parts, and supplies
4. The warehouse

Communication within the material flow system includes: production schedules, work order releases, move tickets, kanbans, bar codes, route sheets, assembly charts, and warehouse records. A schematic of the material flow system is given in Figure 3.19.

If the **flow of products from a manufacturing facility** is to be the subject of the flow, the flow process is referred to as the **physical distribution system.** The subject of physical distribution systems are the finished goods produced by a firm. The resources of physical distribution systems include:

1. The customer
2. The sales and accounting departments and warehouses
3. The material handling and transportation equipment required to move the finished product
4. The distributors of the finished product

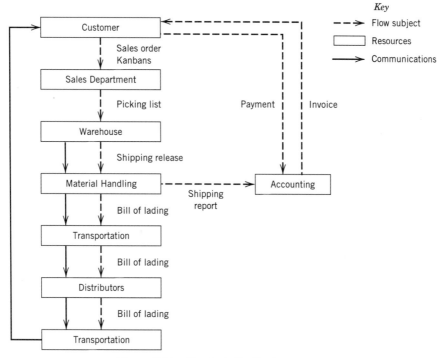

Figure 3.20 Physical distribution system.

The communications within the physical distribution system include: sales orders, packing lists, shipping reports, shipping releases, kanbans, EDI invoices, and bills of lading. A schematic of the physical distribution system is given in Figure 3.20.

The material management, material flow, and physical distribution system may be combined into one overall flow system. Such an overall flow process is referred to as the logistics system. A schematic of the **logistics system** is given in Figure 3.21.

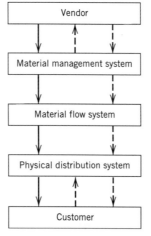

Figure 3.21 Logistics system.

Modern manufacturing approaches are impacting the logistics system in different ways. For example, some suppliers are locating facilities closer to the customer to deliver smaller lot sizes; customers are employing electronic data interchange systems and kanbans to request materials just-in-time; customers and suppliers are using continuous communication technologies with transportation system operators to prevent contingencies; products are being delivered to multiple receiving docks; products are being received in decentralized storage areas (supermarkets) at the points of use; in many cases, no receiving inspection is being performed (suppliers have been certified) and no paperwork is needed; production operators are retrieving materials from the supermarkets when needed; products are being moved short distances in manufacturing cells and/or product planning arrangements; and simpler material handling and storage equipment alternatives are being employed to receive, store, and move materials (production operators perform retrieval and handling operations from supermarkets and among processes). These changes are creating efficient logistics systems with shorter lead times, lower cost, and better quality.

_3.4_ FLOW PATTERNS

The macroflow considerations of material management, material flow, physical distribution, and logistics are of value to the facilities planner in that they define the overall flow environment within which movement takes place. Within the overall flow environment, a critical consideration is the pattern of flow. Patterns of flow may be viewed from the perspective of flow within workstations, within departments, and between departments.

Flow Within Workstations

Motion studies and ergonomics considerations are important in establishing the flow within workstations. For example, flow within a workstation should be simultaneous, symmetrical, natural, rhythmical, and habitual. Simultaneous flow implies the coordinated use of hands, arms, and feet. Hands, arms, and feet should begin and end their motions together and should not be idle at the same instant except during rest periods. Symmetrical flow results from the coordination of movements about the center of the body. The left and right hands and arms should be working in coordination. Natural flow patterns are the basis for rhythmical and habitual flow patterns. Natural movements are continuous, curved, and make use of momentum. Rhythmical and habitual flow implies a methodical, automatic sequence of activity. Rhythmical and habitual flow patterns also allow for reduced mental, eye and muscle fatigue, and strain.

Flow Within Departments

The flow pattern within departments is dependent on the type of department. In a product and/or product family department, the flow of work follows the product flow. Product flows typically follow one of the patterns shown in Figure 3.22 End-to-end, back-to-back, and odd-angle flow patterns are indicative of product departments

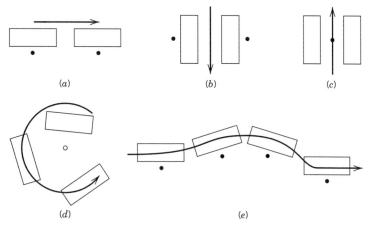

Figure 3.22 Flow within product departments. (*a*) End-to-end. (*b*) Back-to-back. (*c*) Front-to-front. (*d*) Circular. (*e*) Odd-angle.

where one operator works at each workstation. Front-to-front flow patterns are used when one operator works on two workstations and circular flow patterns are used when one operator works on more than two workstations.

In a process department, little flow should occur between workstations within departments. Flow typically occurs between workstations and aisles. Flow patterns are dictated by the orientation of the workstations to the aisles. Figure 3.23 illustrates three workstation-aisle arrangements and the resulting flow patterns. The determination of the preferred workstation-aisle arrangement pattern is dependent on the interactions among workstation areas, available space, and size of the materials to be handled.

Diagonal flow patterns are typically used in conjunction with one-way aisles. Aisles that support diagonal flow patterns often require less space than aisles with either parallel or perpendicular workstation-aisle arrangements. However, one-way aisles also result in less flexibility. Therefore, diagonal flow patterns are not utilized often.

Flow within workstations and within departments should be enriched and enlarged to allow the operators not only to use their muscles but also their minds. Multifunctional operators can work on more than one machine if needed and can get involved in support and continuous improvement functions like quality, basic maintenance, material handling, record keeping, performance measurement tracking, and teamwork. This means that flow and location of materials, tools, paperwork, and quality verification devices should be considered in an integrated way.

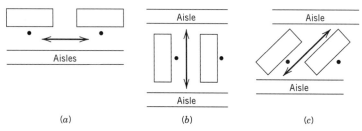

Figure 3.23 Flow within process departments. (*a*) Parallel. (*b*) Perpendicular. (*c*) Diagonal.

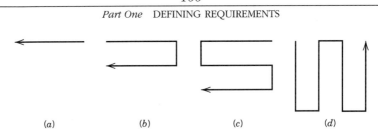

| (a) | (b) | (c) | (d) |

Figure 3.24 General flow patterns. (*a*) Straight-line. (*b*) U-shaped. (*c*) S-shaped. (*d*) W-shaped.

Flow Between Departments

Flow between departments is a criterion often used to evaluate overall flow within a facility. Flow typically consists of a combination of the four general flow patterns shown in Figure 3.24 An important consideration in combining the flow patterns shown in Figure 3.24 is the location of the entrance and exit. As a result of the plot plan or building construction, the location of the entrance (receiving department) and exit (shipping department) is often fixed at a given location and flow within the facility conforms to these restrictions. A few examples of how flow within a facility may be planned to conform to entrance and exit restrictions are given in Figure 3.25.

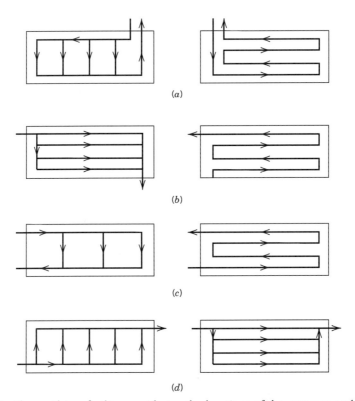

Figure 3.25 Flow within a facility considering the locations of the entrance and exit.
(*a*) At the same location. (*b*) On adjacent sides. (*c*) On the same side but at opposite ends.
(*d*) On opposite sides.

An important design issue in Just-In-Time facilities is the determination of the appropriate number of receiving/shipping docks and decentralized storage areas (supermarkets) and their location. Each combination of number and location of receiving/shipping docks and supermarkets should be analyzed in detail considering integrated layout-handling alternatives to identify flow-time-cost-quality impact.

3.5 FLOW PLANNING

Planning effective flow involves combining the flow patterns given in Section 3.4 with adequate aisles to obtain a progressive movement from origination to destination. Effective flow within a facility includes the progressive movement of materials, information, or people *between departments*. Effective flow within a department involves the progressive movement of materials, information, or people *between work-stations*. Effective flow within a workstation addresses the progressive movement of materials, information, or people *through the workstation.*

As noted, effective flow planning is a hierarchical planning process. The effective flow within a facility is contingent upon effective flow between departments. Such flow depends on effective flow within departments, which depends on effective flow within workstations. This hierarchy is shown in Figure 3.26 Planning for effective flow within the hierarchy requires the consideration of flow patterns and flow principles.

Morris [33] defines a principle as "simply a loose statement of something which has been noticed to be sometimes, but not always, true." The following principles have been observed to frequently result in effective flow: maximize directed flow paths, minimize flow, and minimize the costs of flow.

A directed flow path is an uninterrupted flow path progressing directly from origination to destination. An uninterrupted flow path is a flow path that *does not intersect with other paths.* Figure 3.27 illustrates the congestion and undesirable intersections that may occur when flow paths are interrupted. A directed flow path progressing from origination to destination is a flow path with no backtracking. As can be seen in Figure 3.28 backtracking increases the length of the flow path. The

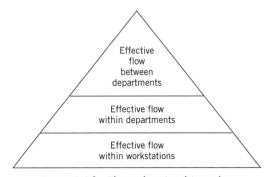

Figure 3.26 Flow planning hierarchy.

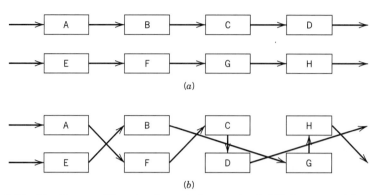

Figure 3.27   The impact of interruptions on flow paths. (*a*) Uninterrupted flow paths. (*b*) Interrupted flow paths.

principle of minimizing flow represents the work simplification approach to material flow. The work simplification approach to material flow includes:

1.   Eliminating flow by planning for the delivery of materials, information, or people directly to the point of ultimate use and eliminate intermediate steps
2.   Minimizing multiple flows by planning for the flow between two consecutive points of use to take place in as few movements as possible, preferably one
3.   Combining flows and operations wherever possible by planning for the movement of materials, information, or people to be combined with a processing step

The principle of minimizing the cost of flow may be viewed from either of the following two perspectives.

1.   Minimize manual handling by minimizing walking, manual travel distances, and motions.
2.   Eliminate manual handling by mechanizing or automating flow to allow workers to spend full time on their assigned tasks.

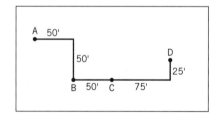

Flow Path A – B – C – D
    (50' + 50') + 50' + (75' + 25') = 250 feet

Flow Path A – B – A – C – D
    (50' + 50') + (50' + 50') + (50' + 50') + 50' + (75' + 25') = 450 feet

Backtrack Penalty

Figure 3.28   Illustration of how backtracking impacts the length of flow paths.

3.6 MEASURING FLOW

Flow among departments is one of the most important factors in the arrangement of departments within a facility. To evaluate alternative arrangements, a measure of flow must be established. Flows may be specified in a quantitative manner or a qualitative manner. Quantitative measures may include pieces per hour, moves per day, or pounds per week. Qualitative measures may range from an absolute necessity that two departments be close to each other to a preference that two departments not be close to each other. In facilities having large volumes of materials, information, and people moving between departments, a quantitative measure of flow will typically be the basis for the arrangement of departments. On the contrary, in facilities having very little actual movement of materials, information, and people flowing between departments, but having significant communication and organizational interrelations, a qualitative measure of flow will typically serve as the basis for the arrangement of departments. Most often, a facility will have a need for both quantitative and qualitative measures of flow and both measures should be used.

A chart that can be useful in flow measurement is the mileage chart shown in Figure 3.29. Notice the diagonal of the mileage chart is blank, since the question "How far is it from New York to New York?" makes little sense. Furthermore, the mileage chart is a symmetric matrix. (Distance charts do not have to be symmetric; if one-way aisles or roads are used, distance between two points will seldom be symmetric.) In Figure 3.29, it is 963 miles from Boston to Chicago and also 963 miles from Chicago to Boston. When this occurs, the format of the mileage chart is often changed to a triangular matrix as shown in Figure 3.30.

Quantitative Flow Measurement

Flows may be measured quantitatively in terms of the amount moved between departments. The chart most often used to record these flows is a from-to chart. As

From \ To	Atlanta, GA	Boston, MA	Chicago, IL	Dallas, TX	New York, NY	Pittsburgh, PA	Raleigh, NC	San Francisco, CA
Atlanta, GA		1037	674	795	841	687	372	2496
Boston, MA	1037		963	1748	206	561	685	3095
Chicago, IL	674	963		917	802	452	784	2142
Dallas, TX	795	1748	917		1552	1204	1166	1753
New York, NY	841	206	802	1552		368	489	2934
Pittsburgh, PA	687	561	452	1204	368		445	2578
Raleigh, NC	372	685	784	1166	489	445		2843
San Francisco, CA	2496	3095	2142	1753	2934	2578	2843	

Figure 3.29 Mileage chart.

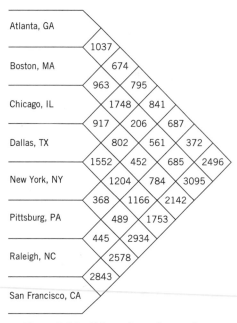

Figure 3.30 Triangular mileage chart.

can be seen in Figure 3.31, a from-to chart resembles the mileage chart given in Figure 3.29. The from-to chart is a square matrix, but is seldom symmetric. The lack of symmetry is because there is no definite reason for the flows from stores to milling to be the same as the flows from milling to stores.

A from-to chart is constructed as follows:

1. List all departments down the row and across the column following the over-all flow pattern. For example, Figure 3.32 shows various flow patterns that result in the departments being listed as in Figure 3.31.

2. Establish a measure of flow for the facility that accurately indicates equivalent flow volumes. If the items moved are equivalent with respect to ease of movement, the number of trips may be recorded in the from-to chart. If the items

From \ To	Store	Milling	Turning	Press	Plate	Assembly	Warehouse
Stores		12	6	9	1	4	
Milling					7	2	
Turning		3			4		
Press					3	1	1
Plate		3	1			4	3
Assembly	1						7
Warehouse							

Figure 3.31 From-to chart.

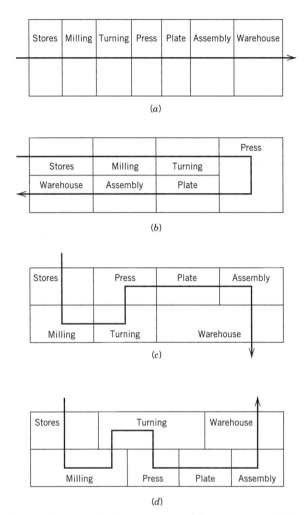

Figure 3.32 Flow patterns indicating the order of flow given in (*a*) Straight-line flow. (*b*) U-shaped flow. (*c*) S-shaped flow. (*d*) W-shaped flow.

moved vary in size, weight, value, risk of damage, shape, and so on, then some common unit of measure may be established so that the quantities recorded in the from-to chart represent the proper relationships among the volumes of movement.

3. Based on the flow paths for the items to be moved and the established measure of flow, record the flow volumes in the from-to chart.

Example 3.4

A firm produces three components. Components 1 and 2 have the same size and weight and are equivalent with respect to movement. Component 3 is almost twice as large and moving two units of either component 1 or 2 is equivalent to moving 1 unit of component 3.

From \ To	A	C	B	D	E
A		① 30 ③2(7) = 14	②12		
		44	12	0	0
C			①30	③2(7) = 14	
	0		30	14	0
B				①30 ②12	③2(7) = 14
	0	0		42	14
D			③2(7) = 14		①30 ②12
	0	0	14		42
E	0	0	0	0	

Figure 3.33 From-to chart for Example 3.4. The circled numbers represent component numbers and the number following the circled numbers indicates the volume of equivalent flows for the component.

The departments included in the facility are A, B, C, D, and E. The overall flow path is A-B-C-D-E. The quantities to be produced and the component routings are as follows:

	Production Quantities	
Component	(per day)	Routing
1	30	A-C-B-D-E
2	12	A-B-D-E
3	7	A-C-D-B-E

The first step in producing the from-to chart is to list the departments in order of the overall flow down the rows and across the columns. Then by considering each unit of component 3 moved to be equivalent to two moves of components 1 and 2, the flow volumes may be recorded as shown in Figure 3.33.

Notice that flow volumes below the diagonal represent backtracking and the closer the flow volumes are to the main diagonal, the shorter will be the move in the facility. The moves below the diagonal in the from-to chart given in Figure 3.31 are turning to milling, plate to milling, plate to turning, and assembly to stores. If these moves are traced on the flow paths shown in Figure 3.32 it may be seen that they are all counter to the overall flow pattern. The diagonal 1 units above the main diagonal in Figure 3.31 include the moves: stores to milling, press to plate, plate to assembly, and assembly to warehouse. These moves may be seen in Figure 3.32a to be between adjacent departments, or departments one department away. The diagonal 2 units above the main diagonal in Figure 3.31 include the moves: stores to turning, turning to plate, and press to assembly. These moves may be seen in Figure 3.32a to be of length two departments away along the overall flow path. In a similar manner, the fifth diagonal units above the diagonal includes the move stores to assembly, and the fifth diagonal unit below the diagonal includes the move assembly to stores; these moves are five departments apart along the overall flow path, with the move from assembly to stores being counter to the direction of flow.

Table 3.2 *Closeness Relationship Values*

Value	Closeness
A	Absolutely necessary
E	Especially important
I	Important
O	Ordinary closeness okay
U	Unimportant
X	Undesirable

Qualitative Flow Measurement

Flows may be measured qualitatively using the closeness relationships values developed by Muther [34] and given in Table 3.2. The values may be recorded in conjunction with the reasons for the closeness value using the relationship chart given in Figure 3.34.

A relationship chart may be constructed as follows:

1. List all departments on the relationship chart.

2. Conduct interviews or surveys with persons from each department listed on the relationship chart and with the management responsible for all departments.

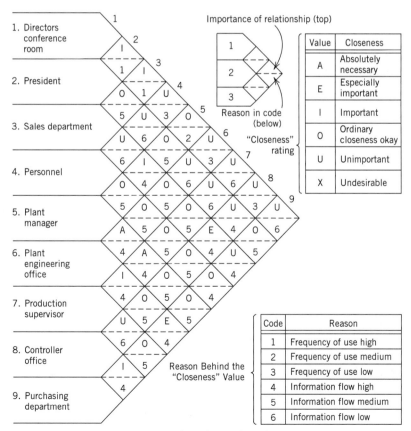

Figure 3.34 Relationship chart.

3.   Define the criteria for assigning closeness relationships and itemize and record the criteria as the reasons for relationship values on the relationship chart.

4.   Establish the relationship value and the reason for the value for all pairs of departments.

5.   Allow everyone having input to the development of the relationship chart an opportunity to evaluate and discuss changes in the chart.

It is important that this procedure be followed in developing a relationship chart. If, instead of the facilities planner synthesizing the relationship among departments as described above, the department heads are allowed to assign the closeness relationships with other departments, inconsistencies may develop. The inconsistencies follow from the form of the chart. The relationship chart by definition requires that the relationship value between departments A and B be the same as the relationship value between departments B and A. If individual relationship values were assigned by department heads and the head of department A said the relationship with B was unimportant (U), and the head of department B said the relationship with A was of ordinary importance (O), an inconsistency would exist. It is best to avoid these inconsistencies by having the facilities planner assign relationship values based on input from the important parties and then have the same parties evaluate the final result.

It is important to emphasize the difference between relationship values U and X. Two departments can be placed adjacent to each other with a relationship value U but they cannot be placed adjacent to each other with a relationship value X (environmental, safety, and facilities constraints). Departments with relationship values U do not gain (or lose) anything by being close to each other.

A cautionary word is needed relative to developing activity relationships. If there are n departments, for example, then $n(n-1)/2$ pairwise combinations must be considered. Hence, in the case of 10 departments, there are 45 pairwise combinations to consider. Likewise, for 50 activities there are 1,225 pairwise combinations to consider. If there are a large number of activities, then an iterative approach should be used. For example, activities that are most likely to be located together can be grouped into pseudodepartments. Once their relative locations are determined, then activity relationship charts can be developed for each "department."

From a practical perspective, it is expected that more than half the pairwise combinations of activities will have an activity relationship of U. If the cells having U relationships are left blank, then the resulting chart will be quite sparse. It is reasonable to expect less than 5% of the pairwise combinations to have A activity relationships, less than 5% to have X relationships, less than 12% to have either A or E relationships, less than 25% to have either A, E, or I relationships, and less than 40% to have A, E, I, or O relationships. Even with a high degree of sparseness, the number of pairwise combinations can become unmanageable. Hence, caution must be used when dealing with a large number of activities.

From a facilities planning perspective, activity relationships are often translated into **proximity requirements.** For example, if two activities have a strong, positive relationship, then they are typically located close together if not adjacent to one another. Likewise, if two activities have a strong, negative relationship they are typically separated and located far apart from each other.

It should be emphasized that activity relationships can frequently be satisfied in ways other than through physical separation. As an example, information relationships may be satisfied with communication links that include live television hookups, computer ties, pneumatic tube delivery systems, and so on. Likewise, noisy areas can be enclosed, fumes can be vented, and other environmental relationships can be dealt with using special facilities, rather than using distance separation.

Due to the multiplicity of relationships involved, it is advisable to construct separate relationship charts for each *major* relationship being measured. For example, different relationship charts might be constructed for material flow, personnel flow, information flow, organizational, control, environmental, and process relationships. A comparison of the resulting relationship charts will indicate the need for satisfying the relationships by using nondistance-related solutions.

3.7 SPACE REQUIREMENTS

Perhaps the most difficult determination in facilities planning is the amount of space required in the facility. The design year for a facility is typically 5 to 10 years in the future. Considerable uncertainty generally exists concerning the impact of technology, changing product mix, changing demand levels, and organizational designs for the future. Because of the numerous uncertainties that exist, people in the organization tend to "hedge their bets" and provide inflated estimates of space requirements. The facilities planner then has the difficult task of projecting **true space requirements** for the uncertain future.

To further complicate matters, there exists *Parkinson's Law*. Loosely translated, it states that things will expand to fill all available capacity sooner than you plan. Hence, even though the facility might be constructed with sufficient space for the future, when the future arrives there will be no space available for it!

Because of the nature of the problem involving the determination of space requirements, we recommend that it be approached systematically. Specifically, space requirements should be developed "from the ground up."

In determining space requirements for warehousing activities, inventory levels, storage units, storage methods and strategies, equipment requirements, building constraints, and personnel requirements must be considered. Space requirements for storage of materials and supplies will be addressed in more depth in subsequent chapters.

In manufacturing and office environments, space requirements should be determined first for individual workstations; next, departmental requirements should be determined, based on the collection of workstations in the department.

As explained before, modern manufacturing approaches are changing drastically space requirements in production, storage areas, and offices. Specifically, space requirements are being reduced because (1) products are delivered to the points of use in smaller lot and unit load sizes; (2) decentralized storage areas are located at the points of use; (3) less inventories are carried (products are "pulled" from preceding processes using kanbans and internal and external inefficiencies have been eliminated); (4) more efficient layout arrangements (i.e., manufacturing cells) are used; (5) companies are downsizing (focused factories, leaner organizational structures, decentralization of functions, multifunctional employees, high-performance team environments) and (6) offices are shared and telecommuting is used.

The determination of production space requirements is presented in the following sections. A section is also included to address the issue of visual management and space requirements. Personnel and storage space requirements are presented in Chapters 4 and 9, respectively.

Workstation Specification

Recall that a facility was defined in Chapter 1 as including the fixed assets required to accomplish a specific objective. Because a workstation consists of the fixed assets needed to perform specific operations, a workstation can be considered to be a facility. Although it has a rather narrow objective, the workstation is quite important. Productivity of a firm is definitely related to the productivity of each workstation.

A workstation, like all facilities, includes space for equipment, materials, and personnel. The equipment space for a workstation consists of space for:

1. The equipment
2. Machine travel
3. Machine maintenance
4. Plant services

Equipment space requirements should be readily available from machinery data sheets. For machines already in operation, machinery data sheets should be available from either the maintenance department's equipment history records or the accounting department's equipment inventory records. For new machines, machinery data sheets should be available from the equipment supplier. If machinery data sheets are not available, a physical inventory should be performed to determine at least the following:

1. Machine manufacturer and type
2. Machine model and serial number
3. Location of machine safety stops
4. Floor loading requirement
5. Static height at maximum point
6. Maximum vertical travel
7. Static width at maximum point
8. Maximum travel to the left
9. Maximum travel to the right
10. Static depth at maximum point
11. Maximum travel toward the operator
12. Maximum travel away from the operator
13. Maintenance requirements and areas
14. Plant service requirements and areas

Floor area requirements for each machine, including machine travel, can be determined by multiplying total width (static width plus maximum travel to the left and right) by total depth (static depth plus maximum travel toward and away from the operator). To the floor area requirement of the machine add the maintenance and plant service area requirements. The resulting sum represents the total machinery area for a machine. The sum of the machinery areas for all machines within a workstation gives the machinery area requirement for the workstation.

The materials areas for a workstation consist of space for:

1. Receiving and storing inbound materials
2. In-process materials
3. Storing outbound materials and shipping
4. Storing and shipping waste and scrap
5. Tools, fixtures, jigs, dies, and maintenance materials

To determine the area requirements for receiving and storing materials, in-process materials, and storing and shipping materials, the dimensions of the unit loads to be handled and the flow of material through the machine must be known. Sufficient space should be allowed for the number of inbound and outbound unit loads typically stored at the machine. If an inventory holding zone is included within a department for incoming and outgoing materials, one might provide space for only two unit loads ahead of the machine and two unit loads after the machine. Depending on the material handling system, the minimum requirement for space might include that required for one unit load to be worked next, one unit load being worked from, one unit load being worked to, and one unit load that has been completed. Additional space may be needed to allow for in-process materials to be placed into the machine, for material such as bar stock to extend beyond the machine, and for the removal of material from the machine. Space for the removal of waste (chips, trimmings, etc.) and scrap (defective parts) from the machine and storage prior to removal from the workstation must be provided.

Organizations that use kanbans have reported the need for less space for materials. Typically, only one or two containers or pallet loads of materials are kept close to the workstation and the rest of the materials (regulated by the number of kanbans) are located in a decentralized storage area (supermarket) located close by.

The only remaining material requirement to be added to that previously mentioned in determining the total material area requirement is the space required for tools, fixtures, jigs, dies, and maintenance materials. A decision with respect to the storage of tools, fixtures, jigs, dies, and maintenance materials at the workstation or in a central storage location will have a direct bearing on the area requirement. At the very least, space must be provided for the accumulation of tools, fixtures, jigs, dies, and maintenance materials required while altering the machine setup.

As the number of setups for a machine increases, so do the work station area requirements for tools, fixtures, jigs, dies, and maintenance materials. Also, from a security, damage, and space viewpoint, the desirability of a central storage location increases.

Consequently, organizations should allocate space to the workstations according to the concepts to be employed.

The personnel area for a workstation consists of space for:

1. The operator
2. Material handling
3. Operator ingress and egress

Space requirements for the operator and for material handling depend on the method used to perform the operation. The method should be determined using a motion study of the task and an ergonomics study of the operator. The following general guidelines are given to illustrate the factors to be considered.

1.   Workstations should be designed so the operator can pick up and discharge materials without walking or making long or awkward reaches.

2.   Workstations should be designed for efficient and effective utilization of the operator.

3.   Workstations should be designed to minimize the time spent manually handling materials.

4.   Workstations should be designed to maximize operator safety comfort and productivity.

5.   Workstations should be designed to minimize hazards, fatigue, and eye strain.

In addition to the space required for the operator and for material handling, space must be allowed for operator ingress and egress. A minimum of a 30-in. aisle is needed for operator travel past stationary objects. If the operator walks between a stationary object and an operating machine, a minimum of a 36-in. aisle is required. If the operator walks between two operating machines, a minimum of a 42-in. aisle is needed.

Figure 3.35 illustrates the space requirements for a workstation. A sketch like Figure 3.35 should be provided for each workstation in order to visualize the operator's activities. The facilities planner should simulate the operator reporting to the job, performing the task, changing the setup, maintaining the machine, reacting to emergency situations, going to lunch and breaks, cleaning the workstation, evaluating quality, working in teams, responding to feedback boards, and leaving at the end of the shift. Such a simulation will assure the adequacy of the space allocation and may aid in significantly improving the overall operation.

Department Specification

Once the space requirements for individual workstations have been determined, the space requirements for each department can be established. To do this, we need to establish the departmental service requirements. Departmental area requirements are not simply the sum of the areas of the individual workstations included within the department. It is quite possible tools, dies, equipment maintenance, plant services, housekeeping items, storage areas, operators, spare parts, kanban boards, information-communication-recognition boards, problem boards, and andons may be shared to save space and resources (see Figure 3.16). However, care must be taken to ensure that operational interferences are not created by attempting to combine areas needed by individual workstations. Additional space is required within each department for material handling within the department. Aisle space requirements still cannot be determined exactly, as the department configurations, workstation alignments, and material handling system have not been completely defined. However, at this point we can approximate the space requirement for aisles, since the relative sizes of the loads to be handled are known. Table 3.3 provides a guide for use in estimating aisle space requirements.

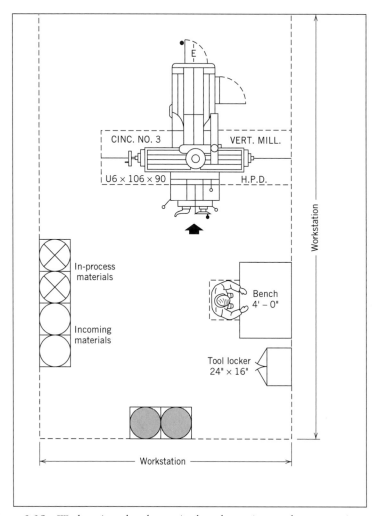

Figure 3.35 Workstation sketch required to determine total area requirements.

Departmental service requirements equal the sum of the service requirements for the individual workstations to be included in a department. These requirements, as well as departmental area requirements, should be recorded on a departmental service and area requirements sheet. Such a sheet is shown in Figure 3.36.

Table 3.3   *Aisle Allowance Estimates*

If the Largest Load Is	Aisle Allowance Percentage Is[a]
Less than 6 ft^2	5-10
Between 6 and 12 ft^2	10-20
Between 12 and 18 ft^2	20-30
Greater than 18 ft^2	30-40

[a]Expressed as a percentage of the net area required for equipment, material, and personnel.

DEPARTMENTAL SERVICE AND AREA REQUIREMENT SHEET

Company ___ A.B.C., Inc. ___ Prepared by ___ J.A. ___ Sheet 1 of 1

Department ___ Turning ___ Date ___

Work Station	Quantity	Service Requirements			Floor Loading	Ceiling Height	Area (square feet)			
		Power	Compressed Air	Other			Equipment	Material	Personnel	Total
Turret lathe	5	440 V AC	10 CFM @ 100 psi		150 PSF	4'	240	100	100	440
Screw machine	6	440 V AC	10 CFM @ 100 psi		190 PSF	4'	280	240	120	640
Chucker	2	440 V AC	10 CFM @ 100 psi		150 PSF	5'	60	100	40	200

Net area required   1280
13% aisle allowance   167
Total area required   1447

Figure 3.36 Department service and area requirements sheet.

Example 3.3

A planning department for the ABC Company consists of 13 machines that perform turning operations. Five turret lathes, six automatic screw machines, and two chuckers are included in the planning department. Bar stock, in 8-ft bundles, is delivered to the machines. The "footprints" for the machines are 4×12 ft for turret lathes, 4×14 ft for screw machines, and 5×6 ft for chuckers. Personnel space footprints of 4×5 ft are used. Material storage requirements are estimated to be 20 ft^2 per turret lathe, 40 ft^2 per screw machine, and 50 ft^2 per chucker. An aisle space allowance of 13% is used. The space calculations are summarized in Figure 3.36. A total of 1447 ft^2 of floor space is required for the planning department. If space is to be provided in the planning department for a supervisor's desk, it must be added to the total for equipment, materials, personnel, and aisles.

Notice that in this example, Company ABC has organized the machines in a process planning department and that provisions for tooling, feedback boards, autonomous maintenance, quick changeovers, quality assurance, and team meetings have not been included in the design. It is also assumed that flexible material handling equipment alternatives are used to move materials, in-process inventory, and finished goods.

Aisle Arrangement

Aisles should be located in a facility to promote effective flow. Aisles may be classified as departmental aisles and main aisles. Consideration of departmental aisles will be deferred until departmental layouts are established. Recall that space was allotted for departmental aisles on the department service and area requirement sheet.

Planning aisles that are too narrow may result in congested facilities having high levels of damage and safety problems. Conversely, planning aisles that are too wide may result in wasted space and poor housekeeping practices. Aisle widths should be determined by considering the type and volume of flow to be handled by the aisle. The type of flow may be specified by considering the people and equipment types using the aisle.

Table 3.4 specifies aisle widths for various types of flow. If the anticipated flow over an aisle indicates that only on rare occasions will flow be taking place at the same time in opposite directions, the aisle widths for main aisles may be

Table 3.4 *Recommended Aisle Widths for Various Types of Flow*

Type of Flow	Aisle Width (feet)
Tractors	12
3-ton Forklift	11
2-ton Forklift	10
1-ton Forklift	9
Narrow aisle truck	6
Manual platform truck	5
Personnel	3
Personnel with doors opening into the aisle from one side	6
Personnel with doors opening into the aisle from two sides	8

obtained from Table 3.4. If, however, the anticipated flow in an aisle indicates that two-way flow will occur frequently, the aisle width should equal the sum of the aisle widths required for the types of flow in each direction.

Curves, jogs, or non-right angle intersections should be avoided in planning for aisles. Aisles should be straight and lead to doors. Aisles along the outside wall of a facility should be avoided unless the aisle is used for entering or leaving the facility. Column spacing should be considered when planning aisle spacing. When column spacing is not considered, the columns will often be located in the aisle. Columns are often used to border aisles, but rarely should be located in an aisle.

Visual Management and Space Requirements

Manufacturing management approaches can impact the way facilities are designed. For example, consider the impact a visual management system (see Figure 3.37) might have. Notice that the example shown has the following features:

A. Identification, housekeeping, and organization (one place for everything and everything in place). Teams need to relate to a place they can identify as their own. A clean environment where they work, meet, review indicators of the status of the work, post information, display team identity symbols and examples of their products, display standard production methods, and identify properly locations for materials, tools, dies, fixtures, and so on. Figure 3.37 illustrates:
    1. identification of the department;
    2. identification of activities, resources, and products;
    3. identification of the team;
    4. markings on the floor (kanban squares, dedicated location for material handling equipment);
    5. markings of tools, racks, fixtures;
    6. technical area;
    7. communication and rest area;
    8. information and instructions; and
    9. housekeeping tools.
B. Visual documentation (tolerances, work instructions, operating instructions for machinery, self-inspection instructions, auditing procedures, plant layout, and floor charts). Figure 3.37 illustrates
    10. manufacturing instructions and technical procedures area.
C. Visual production, maintenance, inventory, and quality control (wall-size schedule charts, kanban boards, autonomous maintenance boards, alarm lamps for machine malfunctions-andons, visual pass/fail templates for quick reading of gauges, charts generated by statistical process control methods, and boards on which problems are recorded). Figure 3.37 illustrates
    11. computer terminal,
    12. production schedule,
    13. maintenance schedule,
    14. identification of inventories and work-in-process,

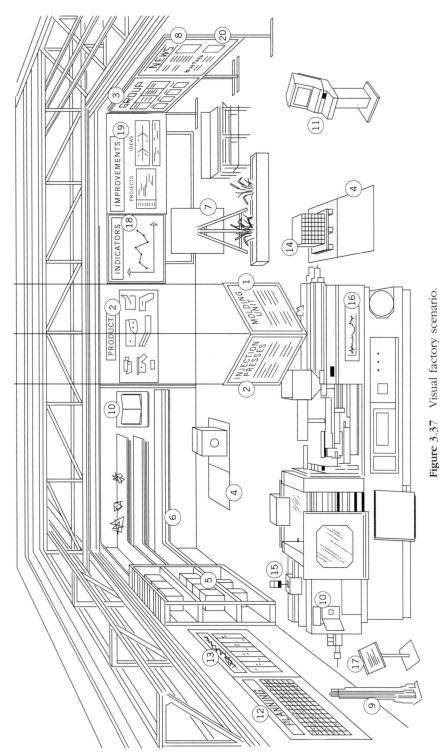

Figure 3.37 Visual factory scenario.

15. monitoring signals for machines,

16. statistical process control, and

17. record of problems.

D.   Performance measurement (objectives, goals, indicators) to show the actual game score. Supervisors, facilitators and workers together look at the situation and examine ways to improve it. Figure 3.37 illustrates

18. objectives, results, and differences.

E.   Progress status (visual mechanisms for tracking and celebrating progress and improvement). Figure 3.37 illustrates

19. improvement activities and

20. company project and mission statement.

It is obvious that a visual management system will make a department look better and will help production and support personnel achieve production and maintenance schedules; control inventories, spare parts, and quality; conform to standards; focus on objectives and goals; and provide follow up to the continuous improvement process. It is also obvious that to use space efficiently, facilities planners need to use walls and aisles to display as much information as possible and need to allow for dedicated areas for materials, dies, housekeeping and maintenance tools, team meetings, and computer terminals.

It is also obvious that a visual management system like that depicted in Figure 3.37 might be difficult to justify economically. If space and budgetary constraints preclude implementing the ideal solution, then you should endeavor to incorporate in the final design as many of the desirable features of the visual management system as can be justified. Sight lines are important, as are good housekeeping and displaying essential information. Having a well-organized working environment will pay dividends in the productivity of the workforce.

3.8   SUMMARY

Activity relationships and space requirements are essential elements of a facilities plan. In this chapter we have emphasized the importance of their determination, as well as indicated that such a determination will not be a simple process. In a sense, the activity relationships and space requirements used in facilities planning provide the foundation for the facility plan. How well the facility achieves its objectives is dependent on the accuracy and completeness of activity relationships and space requirements.

Among the activity relationships considered, flow relationships are of considerable importance to the facilities planner. Included in the consideration of flow relationships are the movement of goods, materials, information, and people. Flow relationships were viewed on a microlevel as the flow within a workstation and on a macrolevel as a logistics system. Flow relationships may be specified by defining the subject, resources, and communications comprising a flow process. Flow relationships may be conceptualized by considering overall flow patterns and may be analyzed by considering general flow principles. Both quantitative and qualitative measures of flow are to be considered. The evaluation of flow relationships is a primary criterion for good facilities plans and serves as the basis for developing most facilities layouts.

We have also emphasized in this chapter the impact that cellular manufacturing, visual management, and many other modern manufacturing approaches have on flow, space, and activity relationships determination. The use of these concepts could change dramatically building size and shape; external and internal flow; and location of production, support, and administrative areas.

BIBLIOGRAPHY

1. Al-Qattan, I., "Designing Flexible Manufacturing Cells Using a Branch and Bound Method," *International Journal of Production Research,* vol. 28, no. 2, 1990, pp. 325–336.
2. Ballakur, A., and Steudel, H. J., "A Within-Cell Utilization Based Heuristic for Designing Cellular Manufacturing Systems," *International Journal of Production Research,* vol. 25, no. 5, 1987, pp. 639–665.
3. Black, J T., "Cellular Manufacturing Systems Reduce Setup Time, Make Small Lot Production Economical," *Industrial Engineering,* November 1983, pp. 36–48.
4. Black, J T., *The Design of the Factory with a Future,* McGraw-Hill, New York, 1991.
5. Burbidge, J. L., "Production Flow Analysis," *The Production Engineer,* April–May, 1971, pp. 139–152.
6. Burbidge, J. L., *The Introduction of Group Technology,* John Wiley, New York, 1975.
7. Chan, H. M., and Milner, D. A., "Direct Clustering Algorithm for Group Formation in Cellular Manufacturing." *Journal of Manufacturing Systems,* vol. 1, no. 1, 1982, pp. 65–74.
8. Deming, W. E., *Out of the Crisis,* MIT Center for Advanced Engineering Study, Cambridge, MA, 1986.
9. Dertouzos, M. L., Lester, R. K., and Solow, R. M., *Made in America: Regaining the Productive Edge,* MIT Press, Cambridge, MA, 1989.
10. Dumolien, W. J., and Santen, W. P., "Cellular Manufacturing Becomes Philosophy of Management at Components Facility," *Industrial Engineering,* November 1983, pp. 72–76.
11. Gabor, A., *The Man Who Discovered Quality,* Penguin Books, New York, 1990.
12. Greif, Michael, *The Visual Factory,* Productivity Press, Cambridge, MA, 1991.
13. Gryna, F. M., *Quality Planning and Analysis: From Product Development Through Use,* 4th ed., McGraw-Hill, New York, 2000.
14. Gupta, T., and Seifoddini, H., "Production Data Based Similarity Coefficient for Machine-Component Cellular Manufacturing System, Grouping Decisions the Design of a Cellular Manufacturing System," *International Journal of Production Research,* vol. 28, no. 7, 1990, pp. 1247–1269.
15. Ham, I., *Introduction to Group Technology,* Pennsylvania State University, University Park, PA, 1975.
16. Han, C., and Ham, I., "Multiobjective Cluster Analysis for Part Family Formations," *Journal of Manufacturing Systems,* vol. 5, no. 4, 1986, pp. 223–229.
17. Harhalakis, G., Nagi, R., and Proth, J. M., "An Efficient Heuristic in Manufacturing Cell Formation for Group Technology Applications," *International Journal of Production Research,* vol. 28, no. 1, 1990, pp. 185–198.
18. Hirano, H., *JIT Factory Revolution: A Pictorial Guide to Factory Design of the Future,* Productivity Press, Cambridge, MA, 1989.
19. Imai, M., *Kaizen: The Key to Japan's Competitive Success,* McGraw-Hill, New York, 1986.

20. Irani, S. A., editor, *Handbook of Cellular Manufacturing Systems,* John Wiley, New York, 1999.
21. Ishikawa, K., *What Is Total Quality Control?,* translated by D. J. Lu, Prentice-Hall, Englewood Cliffs, NJ, 1985.
22. Juran, J. M., *Juran on Leadership for Quality: An Executive Handbook,* The Free Press, New York, 1989.
23. Juran, J. M., *Juran on Quality by Design: The New Steps for Planning Quality into Goods and Services,* Free Press, New York, 1992.
24. Juran, J. M., and Godfrey, A. B., editors, *Quality Handbook,* 5th ed., McGraw-Hill, New York, 1999.
25. Khator, S. K., and Irani, S. A., "Cell Formation in Group Technology: A New Approach," *Computers in Industrial Engineering,* vol. 12, no. 2, 1987, pp. 131–142.
26. King, J. R., "Machine-Component Grouping in Production Flow Analysis: An Approach Using a Rank Order Clustering Algorithm," *International Journal of Production Research,* vol. 18, no. 2, 1980, pp. 213–232.
27. Konz, S., *Facility Design,* 1985, John Wiley, New York, 1985.
28. Kusiak, A., "The Generalized Group Technology Concept," *International Journal of Production Research,* vol. 25, no. 4, 1987, pp. 561–569.
29. Kusiak, A., editor, *Concurrent Engineering: Automation, Tools, and Techniques,* John Wiley, New York, 1993.
30. Logendran, R., and West, T. M., "A Machine-Part Based Grouping Algorithm in Cellular Manufacturing," *Computers in Industrial Engineering,* vol. 19, no. 1–4, 1990, pp. 57–61.
31. McAuley, J., "Machine Grouping for Efficient Production," *The Production Engineer,* February 1972, pp. 53–57.
32. Mizuno, S., *Company-Wide Total Quality Control,* Asian Productivity Organization, Tokyo, Japan, 1988.
33. Morris, W. T., *Analysis for Material Handling Management,* Richard D. Irwin, Homewood, IL, 1962.
34. Muther, R., *Systematic Layout Planning,* Cahners Books, Boston, 1973.
35. Ranson, G. M., *Group Technology,* Pergamon, New York, 1970.
36. Riley, F. J., *Assembly Automation: A Management Handbook,* Industrial Press, New York, 1983.
37. Rother, M. and Harris, R., *Creating Continuous Flow,* Lean Enterprise Institute, Brookline, MA, 2001.
38. Scherkenbach, W. W., *The Deming Route to Quality and Productivity: Road Maps and Roadblocks,* Mercury Press, Rockville, MD, 1991.
39. Schonberger, R. J., "Plant Layout Becomes Product-Oriented with Cellular, Just-In-Time Production Concepts," *Industrial Engineering,* November 1983, pp. 66–71.
40. Schonberger, R. J., *Japanese Manufacturing Techniques: Nine Hidden Lessons in Simplicity,* Free Press, New York, 1982.
41. Schonberger, R. J., *World Class Manufacturing,* Free Press, New York, 1986.
42. Schonberger, R. J., *World Class Manufacturing: The Next Decade,* Free Press, New York, 1996.
43. Schonberger, R. J., *Let's Fix It!: Overcoming the Crisis in Manufacturing,* Free Press, New York, 2001.
44. Seifoddini, H., "A Probabilistic Model for Machine Cell Formation," *Journal of Manufacturing Systems,* vol. 9, no. 1, 1990, pp. 69–75.
45. Shingo, S., *A Study of the Toyota Production System from an Industrial Engineering Viewpoint,* Translated by A. P. Dillon, Productivity Press, Cambridge, MA, 1989.
46. Singh, N. and Rajamani, D., *Cellular Manufacturing Systems: Design, Planning and Control,* Chapman & Hall, New York, 1996.

47. Srinivasan, G., Narendran, T. T., and Mahadevan, B., "An Assignment Model for the Part Families Problem in Group Technology," *International Journal of Production Research,* vol. 28, no. 1, 1990, pp. 215–222.

48. Stoner, D. L., Tice, K. J., and Asthon, J. E., "Simple and Effective Cellular Approach to a Job Shop Machine Shop," *Manufacturing Review,* vol. 2, June 1989, pp. 119–125.

49. Sule, D. R., *Manufacturing Facilities: Location, Planning, and Design,* 2nd ed., PWS Publishing, Boston, MA, 1994.

50. Sule, D. R., *Logistics of Facility Location and Allocation,* Marcel Dekker, New York, 2001.

51. Suzaki, K., *The New Manufacturing Challenge: Techniques for Continuous Improvement,* Free Press, New York, 1987.

52. Tsuchiya, S., *Quality Maintenance: Zero Defects Through Equipment Management,* Productivity Press, Cambridge, MA, 1992.

53. Walton, M., *The Deming Management Method,* Perigee Books, New York, 1986.

54. Walton, M., *Deming Management at Work,* Perigee Books, New York, 1990.

55. Womack, J. P., Jones, D. T., and Roos, D., *The Machine that Changed the World,* Macmillan, New York, 1990.

56. Wantuck, K. A., *Just-In-Time for America,* The Forum, Ltd., Milwaukee, WI, 1989.

57. Wemmerlov, U., and Hyer, N. L., "Cellular Manufacturing in the U.S. Industry: A Survey of Users," *International Journal of Production Research,* vol. 27, no. 9, 1989, pp. 1511–1530.

58. Wu, H. L., "Design of a Cellular Manufacturing System: A Syntactic Pattern Recognition Approach," *Journal of Manufacturing System,* vol. 5, no. 2, 1986, pp. 81–87.

59. *The Idea Book,* edited by Japan Human Relations Association, Productivity Press, Cambridge, MA, 1988.

PROBLEMS

3.1 Explain the meaning of a material management system, a material flow system, and a physical distribution system for a bank.

3.2 When would you recommend a group technology layout?

3.3 The paper flow to be expected through one section of city hall consists of the following:

> 10 medical records per day from records to marriage licenses
>
> 7 certificates per day from printing to marriage licenses
>
> 6 blood samples per day from marriage licenses to lab
>
> 6 blood sample reports per day from lab to marriage licenses
>
> 1 box of medical records per week from marriage licenses to records

The following load-equivalence conversions can be used:

> One medical record is equivalent to one certificate.
>
> One medical record is equivalent to one-half of a blood sample.
>
> One medical record is equivalent to one blood sample report.
>
> One medical record is equivalent to one-tenth of a box of medical records.
>
> Develop a from-to chart for this section of city hall and then develop a relationship chart.

3.4 In designing the layout for an orthopedic hospital, describe how the design might be affected by the most important flow component being designated to be either patients, doctors, nurses, or access to expensive diagnostic equipment such as x-ray, cat-scan, and magnetic resonance imaging equipment.

3.5 When would you recommend a fixed position layout?

3.6 Which layout type is very popular in Just-In-Time facilities? Explain why.

3.7 Mention three limitations of the process layout type.

3.8 Use the direct clustering algorithm to form cells for the machine-part matrix shown below. If conflicts exist, propose alternative approaches for resolving the conflicts.

	Machine #			
Part #	1	2	3	4
1		1	1	
2	1			
3			1	
4	1			1

3.9 For the machine-part matrix shown below, form cells using the direct clustering algorithm and, if conflicts exist, propose alternative approaches for resolving the conflicts.

	Machine #				
Part #	1	2	3	4	5
1	1		1		
2					
3		1		1	1
4	1		1		
5		1			
6				1	1

3.10 Singh and Rajamani [46] provide data for a local wood manufacturer that wants to decrease material handling by changing from a process layout to a GT layout. It is considering installing a conveyor for intracellular movement of parts. It wishes to restrict interacellular movement. The machine-part matrix for the wood manufacturer is shown below. Use DCA to form the cells and, if conflicts exist, propose alternative approaches for resolving the conflicts.

	Machine #					
Part #	1	2	3	4	5	6
1					1	1
2	1			1		
3			1			1
4	1	1				
5	1					
6					1	1
7		1		1		
8			1			1

3.11 For Problem 3.10, suppose the machine-part matrix for the wood manufacturer is as shown below. Use DCA to form the cells and, if conflicts exist, propose alternative approaches for resolving the conflicts.

Machine #

Part #	1	2	3	4	5	6
1			1		1	1
2	1			1		
3			1			1
4	1	1				
5	1		1			
6					1	1
7		1		1		
8			1			1

3.12 Visit a local fast-food restaurant. Describe in detail the arrangement of aisles for the customers (the service line, the booth area, etc.) and the arrangement of aisles for the workers (the area behind the counter) by sketching a layout. Draw the flow of the customers and employees on the layout. Explain how the flow of the workers and customers would change if the layout was different. Is there adequate space for the customers and the employees to operate efficiently? Can any improvement in the layout create a more efficient flow? Describe any visual management approaches being used.

3.13 Choose ten main components of a kitchen, (e.g., oven, sink). Develop a relationship chart of the ten components. Sketch a kitchen containing the ten components and arrange them based upon the relationship chart findings.

3.14 What is the impact of "backtracking" in a manufacturing process? Discuss several methods to prevent "backtracking" from occurring.

3.15 What are the pros and cons of having multiple input/output points (receiving and shipping areas) in a given manufacturing facility (list at least three of each)? What types of considerations should a facilities designer take into account when determining whether or not to use multiple input/output points?

3.16 Which information does the facilities designer need from top management, product designer, process designer, and schedule designer? Describe at least ten modern manufacturing approaches that impact drastically the facilities design process.

3.17 Provide at least 10 benefits of using manufacturing cells. Which modern manufacturing approaches are usually employed in conjunction with manufacturing cells (include at least 10)?

3.18 Develop a part-machine matrix like the one in Figure 3.2 for a situation with ten parts and 15 machines. Use the DCA methodology to identify clusters (cells) of parts and machines. Do you have a bottleneck machine? Recommend at least three different ways to resolve this situation.

3.19 Define a kanban. Which are the different types of kanbans? List at least five benefits of using kanbans. Why are kanbans important in cellular manufacturing?

3.20 Mention at least seven benefits of using U-shape arrangements in manufacturing cells.

3.21 Describe the impact of modern manufacturing approaches on the logistics system.

3.22 Which new functions are being managed by multifunctional operators? Describe the impact of this new roles on activity relationships and flow and space requirements.

3.23 Why is it important for a facilities planner to consider the logistics system?

3.24 Go to a local quick food hamburger restaurant and assess the impact of the order taker's workstation on the overall facility flow. Describe the process and make recommendations for improvement.

3.25 What are the trade-offs involved in parallel, perpendicular, and diagonal flow within parking lots?

3.26 How are the flow principles taken into consideration by you from the time you wake up in the morning to go to school until you go to bed that evening? (Assume you did not cut your classes and you did not stay up all night studying.)

3.27 Given the spatial schematic below, evaluate the flow path lengths for the following components:

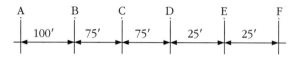

a. Component 1 routing A-B-C-D-E-F.

b. Component 2 routing A-C-B-D-E-F.

c. Component 3 routing A-F-E-D-C-B-A-F.

d. Component 4 routing A-C-E-B-D-F.

3.28 What is the impact on a city having streets that do not meet at right angles? Relate this to aisles.

3.29 A waiter at the famous French restaurant Joe's is interested in obtaining your help in improving the layout of the serving area. After establishing the space needs for the salad serving, beverage serving, dessert serving, soup serving, entree serving, and bill writing areas, you decide your next step is to develop a from-to chart. To do this you need to establish equivalency of loads. Establish these equivalencies and explain your reasoning.

3.30 Given the following from-to chart, recommend an overall flow pattern that will reduce flow.

From \ To	A	B	C	D
A			4	
B				4
C		4		
D	4			

3.31 Develop a relationship chart for the relationships between you, your professor, department chair, college dean, and university president. Develop another relationship chart from your professor's view point for the same people.

3.32 Design a survey that is to be used to determine the relationships among departments in a local bank.

3.33 Develop an activity relationship chart reflecting a student's perspective for the following campus areas: classroom A, classroom B, classroom C, classroom D, professor's office, registrar's office, financial aid office, resident hall room, dining area, parking place for student's automobile, library, computer lab, post office, barber shop or hair salon, and coffee house for students, faculty, and staff. Assume the student takes 2 classes in the morning in alphabetical order in classrooms A and B, plus 2 classes in the afternoon in alphabetical order in classrooms C and D.

3.34 Develop an activity relationship chart that you believe reflects a professor's perspectives for the following campus areas: professor's office, departmental office, academic dean's office, chief academic officer's (provost's) office, university president's office, library, classroom for teaching, laboratory for research, staff assistant's office, graduate assistant's office, professor's parking space, faculty lounge, faculty dining area, restroom, post office, barber shop or hair salon, departmental conference room, and coffee house.

4

PERSONNEL REQUIREMENTS

4.1 INTRODUCTION

The planning of personnel requirements includes planning for employee parking, locker rooms, restrooms, food services, drinking fountains, and health services. This challenge has been compounded by the advent of the 1989 Americans with Disabilities Act (ADA). The facilities planner must integrate barrier-free designs in addressing the personnel requirements of the facility.

Personnel requirements can be among the most difficult to plan because of the number of philosophies relating to personnel. For example:

1. "Our firm is responsible for our employees from the moment they leave their home until they return. We must provide adequate methods of getting to and from work."
2. "Employees should earn their parking locations; all spaces should be assigned to specific individuals."
3. "Employees spend one third of their life within our facility; we must help them enjoy working here."
4. "A happy worker is a productive worker."
5. "A hot lunch makes a worker more productive since it supplies them energy."
6. "Workers who do not feel well are unsafe workers; we should provide medical care to maintain health."
7. "Our company has an obligation to our personnel; we will make all our facilities ADA compliant."
8. "Except for individuals in private offices, no one is allowed to smoke in our building."

9.   "Our employees work hard; the least we can do is provide a place for them to unwind after a hard day's work. Employees who play together will work better together."

10. "Personnel considerations are of little importance in our facility. We pay people to work, not to have a good time."

All of these philosophies are debatable and none of them are universally accepted. Nevertheless, if the management of a facility firmly adopts one of these philosophies, little can be done by the facilities planner other than plan the facility to conform with these philosophies. This chapter presents how to plan personnel requirements into a facility, given the desires of management.

4.2 THE EMPLOYEE–FACILITY INTERFACE

An interface between an employee's work and nonwork activities must be provided. The interface functions as a storage area for personal property of the employee during work hours. Personal property typically includes the automobile and the employee's personal belongings, such as coats, clothes, purses, and lunches.

Employee Parking

Planning employee parking areas is very similar to planning stores or warehouse areas. The procedure to be followed is

1.   Determine the number of automobiles to be parked.
2.   Determine the space required for each automobile.
3.   Determine the available space for parking.
4.   Determine alternative parking layouts for alternative parking patterns.
5.   Select the layout that best utilizes space and maximizes employee convenience.

Care must be used when determining the number of automobiles to be parked. General rules of thumb may not be applicable. For remote sites not being serviced by public transportation, a parking space may be required for every 1.25 employees. At the other extreme, a centralized location served by public transportation may require a parking space for every three employees. The number of parking spaces to be provided must be specifically determined for each facility and must be in accordance with local zoning regulations. Attention should be paid to the requirement for handicapped parking. Although minimum requirements can be as low as two handicapped spaces per 100 parking spaces, five handicapped spaces per 100 parking spaces is not uncommon. The key is to check with your local or state building agency for the required standard. Surveys of similar facilities in the area of the new facility will provide valuable data with respect to the required number of parking spaces. If considerable variability exists between the employee-parking space

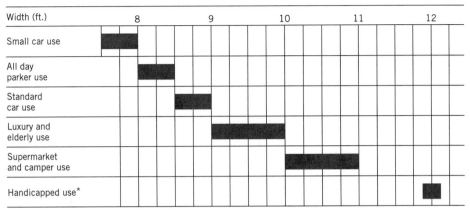

* Minimum requirements = 1 or 2 per 100 stalls or as specified by local, state, or federal law; convenient to destination.

Figure 4.1   Recommended range of stall widths (SW). (*Source:* Ramsey and Sleeper [9].)

requirement ratio for similar facilities the reasons for this variability should be determined. The new facility should be planned in accordance with existing facilities with which it has the greatest similarity.

The size of a parking space for an automobile, which is expressed as the stall width × stall depth, can vary from 7 × 15 ft to 9.5 × 19 ft, depending on the type of automobile and the amount of clearance to be provided. The total area required for a parked automobile depends on the size of the parking space, the parking angle, and the aisle width. Figure 4.1 shows the recommended range of stall widths in feet for various can types and uses.

The factors to be considered in determining the specification for a specific parking lot are

1.   The percentage of automobiles to be parked that are compact automobiles. As a planning guideline, if more specific or current data are not available, 33% of all parking is often allocated to compact automobiles.

2.   Increasing the area provided for parking decreases the amount of time required to park and de-park.

3.   Angular configurations allow quicker turnover; perpendicular parking often yields greater space utilization, although it also requires wider aisles.

4.   As the angle of a parking space increases, so does the required space allocated to aisles.

Automobiles parked in employee parking lots will typically be parked for an entire shift. Therefore, parking and de-parking time is not as important as space utilization. A parking stall width of 7.5 ft for compact automobiles and 8.5 ft for standard-sized automobiles is recommended. In Table 4.1, there are three car groups (G1—small cars, G2—standard cars, and G3—large cars). For a given group, there are corresponding stall width (SW) options. Group I has one SW option (8'-0"). Group II has three (8'-6", 9'-0", and 9'-6") and similarly, Group III has three (9'-0", 9'-6", and 10'-0"). For each stall width option, there are four configurations,

Table 4.1 *Module Width for Each Car Group as a Function of Single and Double Loaded Module Options*

| | | | | | | | θ ANGLE OF PARK | | | | | |
	SW	W	45°	50°	55°	60°	65°	70°	75°	80°	85°	90°
Group I: small cars	8'0"	1	25'9'	26'6"	27'2"	29'4"	31'9"	34'0"	36'2"	38'2"	40'0"	41'9"
		2	40'10"	42'0"	43'1"	45'8"	48'2"	50'6"	52'7"	54'4"	55'11"	57'2"
		3	38'9"	40'2"	41'5"	44'2"	47'0"	49'6"	51'10"	53'10"	55'8"	57'2"
		4	36'8"	38'3"	39'9"	42'9"	45'9"	48'6"	51'1"	53'4"	55'5"	57'2"
Group II: standard cars	8'6"	1	32'0"	32'11"	34'2"	36'2"	38'5"	41'0"	43'6"	45'6"	46'11"	48'0"
		2	49'10"	51'9"	53'10"	56'0"	58'4"	60'2"	62'0"	63'6"	64'9"	66'0"
		3	47'8"	49'4"	51'6"	54'0"	56'6"	59'0"	61'2"	63'0"	64'6"	66'0"
		4	45'3"	46'10'	49'0"	51'8"	54'6"	57'10"	60'0"	62'6"	64'3"	66'0"
	9'0"	1	32'0"	32'9"	34'0"	35'4"	37'6"	39'8"	42'0"	44'4"	46'2"	48'0"
		2	49'4"	51'0"	53'2"	55'6"	57'10"	60'0"	61'10"	63'4"	64'9"	66'0"
		3	46'4"	48'10"	51'4"	53'10"	56'0"	58'8"	61'0"	63'0"	64'6"	66'0"
		4	44'8"	46'6"	49'0"	51'6"	54'0"	57'0"	59'8"	62'0"	64'2"	66'0"
	9'6"	1	32'0"	32'8"	34'0"	35'0"	36'10"	38'10"	41'6"	43'8"	46'0"	48'0"
		2	49'2"	50'6"	51'10"	53'6"	55'4"	58'0"	60'6"	62'8"	64'6"	65'11"
		3	47'0"	48'2"	49'10"	51'6"	53'11"	57'0"	59'8"	62'0"	64'3"	65'11"
		4	44'8"	45'10"	47'6"	49'10"	52'6"	55'9"	58'9"	61'6"	63'10"	65'11"

128

Group III: large cars

Group III: large cars		32'7"	33'0"	34'0"	35'11"	38'3"	40'11"	43'6"	45'5"	46'9"	48'0"
9'0"	1	32'7"	33'0"	34'0"	35'11"	38'3"	40'11"	43'6"	45'5"	46'9"	48'0"
	2	50'2"	51'2"	53'3"	55'4"	58'0"	60'4"	62'9"	64'3"	65'5"	66'0"
	3	47'9"	49'1"	52'3"	53'8"	56'2"	59'2"	61'11"	63'9"	65'2"	66'0"
	4	45'5"	46'11"	49'0"	51'8"	54'9"	58'0"	61'0"	63'2"	64'10"	66'0"
9'6"	1	32'4"	32'8"	33'10"	34'11"	37'2"	39'11"	42'5"	45'0"	46'6"	48'0"
	2	49'11"	50'11"	52'2"	54'0"	56'6"	59'3"	61'9"	63'4"	64'8"	66'0"
	3	47'7"	48'9"	50'2"	52'4"	55'1"	58'4"	60'11"	62'10"	64'6"	66'0"
	4	45'3"	46'8"	48'5"	50'8"	53'8"	57'0"	59'10"	52'2"	64'1"	66'0"
10'0"	1	32'4"	32'8"	33'10"	34'11"	37'2"	39'11"	42'5"	45'0"	46'6"	48'0"
	2	49'11"	50'11"	52'2"	54'0"	56'6"	59'3"	61'9"	63'4"	64'8"	66'0"
	3	57'7"	48'9"	50'2"	52'4"	55'1"	58'4"	60'11"	62'11"	64'6"	66'0"
	4	45'3"	46'8"	48'5"	50'8"	53'8"	57'0"	59'10"	62'2"	64'1"	66'0"

Source: Ramsey and Sleeper [9].

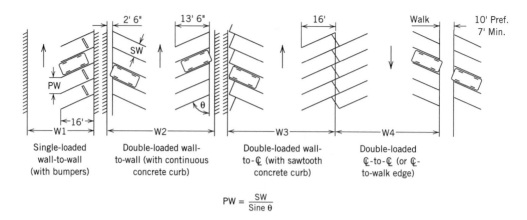

$$PW = \frac{SW}{Sine\ \theta}$$

θ is the parking angle, PW is parking width and SW is the stall width. At an angle of 90° (sine 90° = 1), PW = SW. As the parking angle decreases, PW increases accordingly.

Figure 4.2   Single- and double-loaded module options. (*Source:* Ramsey and Sleeper [9].)

W1, W2, W3, and W4. These configurations are as shown in Figure 4.2. Each configuration for a given SW has 10 corresponding angles of park, θ, and an associated module width. The parking angle is as indicated in Figure 4.2; it is the angle defined by the parking lane and curb, as shown for example in Figure 4.2, W2. Once the car group and associated stall width is determined (let us say GII SW 9'-0")' a parking configuration (W1, W2, W3 or W4) can then be chosen as the initial iteration, depending on the lot size and lot design constraints. Figure 4.2 outlines typical parking configuration options available to the facilities planner.

Using the information from Figures 4.1, 4.2 and Table 4.1, the facilities planner can generate several parking layout alternatives that will optimize the space allocated for parking and maximize employee convenience.

An important issue related to parking lot planning is the location of facility entrances and exits or ingress and egress conditions. Employees should not be required to walk more than 500 ft from their parking place to the entrance of the facility. The entrances should be convenient not only to their parking location, but also to their place of work. If multiple parking lots and entrances are needed in order to accommodate employees, then employees should be assigned to specific lots and entrances. All plant entrances and exits must be carefully planned to meet appropriate insurance and safety codes. Sections 1910.36 and 1910.37 of the Occupational Safety and Health Act (OSHA) standards describe the requirements for entrances and exits.

Example 4.1

A new facility is to have 200 employees. A survey of similar facilities indicates that one parking space must be provided for every two employees and that 40% of all automobiles driven to work are compact automobiles. Five percent of the spaces should be allocated for the handicapped. The available parking lot space is 180 ft and 200 ft deep. Assuming no walls and no walking edge, determine the best parking layout using SW of 8'-6" for standard cars.

If the new facility were to have the same number of parking spaces as similar facilities, 100 spaces would be required. Of these 100 spaces, 40 could be for compact automobiles. However, not all drivers of compact cars will park in a compact space. Therefore, only 30 compact spaces will be provided. Begin the layout of the lot using 90° double-loaded, two-way traffic because of its efficient use of space to determine if the available lot is adequate. From Figure 4.2 W4 is the required module option. Using the W4 module and Table 4.1, we can obtain the following:

Compact cars (8'-0")	Module width
90°, W4	57'-2"
Standard cars (8'-6")	66'-0"
90°, W4	

Check to see if the depth of the lot (200 ft) can accommodate a parking layout consisting of 2 modules of standard cars and 1 compact module,

2(66 ft) + 1(57'-2") = 189'-2"

189-2" < 200 ft, therefore depth requirement OK

Each compact module row will yield a car capacity based on the width of the lot (180 ft) divided by the width requirement per stall (8') times the rows per module (2).

$$\left(\frac{180}{8}\right) \times 2 = 44 \text{ potential number of compact cars}$$

Similarly, each standard module row will yield a car capacity based on the width of the lot (180 ft) divided by the width requirement per stall (8.5') times the number of rows per module (2) times the number of modules (2).

$$\left(\frac{180}{8.5}\right) \times 2 \times 2 = 84 \text{ potential number of standard cars}$$

Total possible = 44 + 84 = 128, which is greater than the required number. Therefore, module configuration (W4) is feasible. A possible alternative of (2 rows/modules × 2 standard modules) + (2 rows/module × 1 compact module) for a total of six rows is a starting point for the layout.

Modifying the layout to account for handicap requirements and circulation reveals the following:

Row 1 will handle all five handicap spaces = 5(12') = 60

The remaining space will be occupied by standard cars

(180 − 60)/8.5 = 14 spaces

Row adjusting for two circulation lanes of 15' each number 2 will handle

(180 − (15 × 2))/8.5 = 17 spaces

Rows 3 and 4 will yield the same number of spaces

Row 5 will have   (180 − 30)/8 = 18 spaces

Row 6 will handle   180/8 = 22 spaces

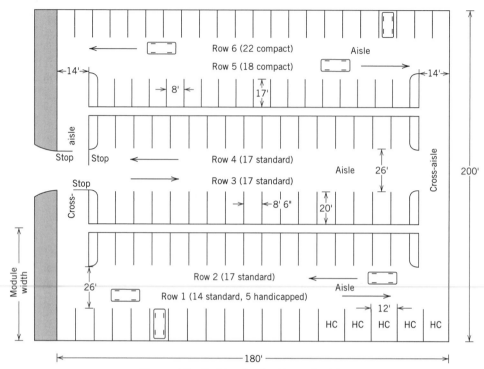

Figure 4.3 Parking lot for Example 4.1.

See Figure 4.3 for the recommended layout. The assignment of compact, standard and handicap spaces is as follows:

Row	Compact	Standard	Handicap
#1		14	5
#2		17	
#3		17	
#4		17	
#5	18		
#6	22		
Total	40	65	5

Note that the *module width* depends on the aisle width and the stall depth (SD), while the *module depth* depends on the stall width (SW) and parking width (PW). Of course, both the module width and module depth depend on the parking angle (θ). In Example 4.1, we assumed the cars are parked at 90 degrees. Therefore, the number of stalls along the module depth depended only on stall width (SW). If the cars are parked at an angle, however, as we show in Example 4.2, we need to explicitly specify the stall depth (SD). (The module widths shown in Table 4.1 implicitly account for the stall depth.)

Example 4.2

Using the same data given for Example 4.1, let us compute the number of standard car stalls that can be placed along the module depth assuming a stall depth of 16 feet (SD = 16′) and a

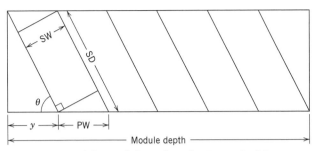

Figure 4.4   Module outline for Example 4.2.

parking angle of 60 degrees ($\theta = 60$). Before we compute the above number, it is instructive to compare the module widths. With 90-degree parking under option W4, earlier we obtained a module width of 66′-0″ from Table 4.1 for standard cars. With 60-degree parking, however, from Table 4.1 we obtain 51′-8″ for the module width for standard cars. Hence, parking at an angle reduced the module width by more than 14 feet, which is a positive outcome (since it may allow us to place more modules in a given lot). However, as we next show, the number of stalls per module decreases when the parking angle is reduced from 90 to 60 degrees.

We first compute the parking width (PW). Recall that PW = SW/sine θ. That is, PW = 8.5/sine 60 = 9.8′. In reference to Figure 4.4, we next compute the value of y, which represents the distance lost due to parking at an angle. By definition, y = SD cosine θ = 16 cosine 60 = 8′. Since the lot width is equal to 180′, we have

Module depth = y + (no of stalls × PW) ≤ 180

which yields number of stalls ≤17.55. That is, we can place 17 standard car stalls along the module depth assuming a parking angle of 60 degrees. Using a 90-degree parking angle in Example 4.1, we had obtained 180/8.5 = 21 cars. Hence, we reduced the module width by about 14′ but lost 4 × 2 = 8 cars per module in the process. Also, parking at angles other than 90 degrees almost always requires the aisles to be one-way. The parking angle that will maximize the number of cars parked, in general, depends on the dimensions of the lot and how the individual modules are arranged within the lot. With increasing demand for parking spaces in virtually every major city and airport around the world—not to mention University campuses—parking lot design and management continues to be a topic of interest.

Storage of Employees' Personal Belongings

A location for storage of employee personal belongings should be provided between the employee entrance and work area. Employees typically store lunches, briefcases, and purses at their place of work. Employees who are not required to change their clothes and who work in an environment where toxic substances do not exist need only to be provided with a coat rack. Employees who either change their clothes or work where toxic substances are present should be provided with lockers. The lockers may be located in a corridor adjacent to the employee entrance if clothes changing does not take place. More commonly, locker rooms are provided for each sex even if clothes changing is not required. Each employee should be assigned a locker. For planning purposes, 6 ft^2 should be allocated for each person using the locker room. If shower facilities are to be provided, they should be located in the locker room. Sinks and mirrors are also typically included there. If toilet facilities are to be included, they

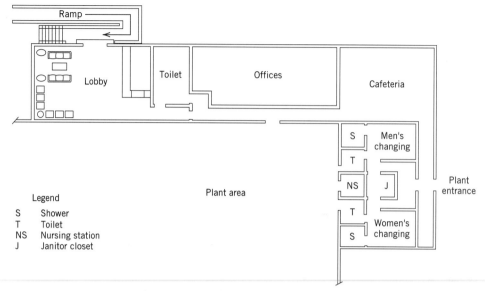

Figure 4.5 Plant entrance and changing room layout.

must be physically separated from the locker room area should lunches be stored in the lockers. Locker rooms are often located along an outside wall adjacent to the employee entrance. This provides excellent ventilation and employee convenience while not interfering with the flow of work within the facility. Figure 4.5 provides an example of a plant entrance and changing room layout.

4.3 RESTROOMS

A restroom should be located within 200 ft of every permanent workstation. Decentralized restrooms often provide greater employee convenience than large, centralized restrooms. Mezzanine restrooms are common in production facilities as they may be located where employees are centered without occupying valuable floor space. However, access to restrooms must be available to handicapped employees. Hence, some restrooms must be at ground level. In any event, the location should comply with local zoning regulations.

Unless restrooms are designed for single occupancy, separate restrooms should be provided for each sex. The recommended minimum number of toilets (i.e., water closets) for the number of employees working within a facility is given in Table 4.2 [9]. In restrooms for males, a urinal may be substituted for a toilet, provided that the number of toilets is not reduced to less than two-thirds the minimum recommended, in Table 4.2. For space planning purposes, 12.5 ft² (2.5' × 5') or 15 ft² (3' × 5') should be allowed for each toilet, and 6 ft² for each urinal. Toilets and urinals must be designed to accommodate wheelchairs for handicapped employees as well.

Table 4.2 *Plumbing Fixture Requirements for Number of Employees*

Business, Mercantile, Industrial Other than Foundry and Storage			
Water Closets	Employees	Lavatories	Employees
1	1–15	1	1–20
2	16–35	2	21–40
3	36–55	3	41–60
4	56–80	4	61–80
5	81–110	5	81–100
6	111–150	6	101–125
7	151–190	7	126–150
		8	151–175
One additional water closet for each 40 in excess of 190		One additional lavatory for each 30 in excess of 175	

Industrial, Foundries, and Storage			
Water Closets	Employees	Lavatories	Employees
1	1–10	1	1–8
2	11–25	2	9–16
3	26–50	3	17–30
4	51–80	4	31–45
5	81–125	5	46–65
One additional water closet for each 45 in excess of 125		One additional lavatory for each 25 in excess of 65	

Assembly, Other than Religious, and Schools					
Water Closets	Occupants	Urinal	Male Occupants	Lavatories	Occupants
1	1–100	1	1–100	1	1–100
2	101–200	2	101–200	2	101–200
3	201–400	3	201–400	3	201–400
4	401–700	4	401–700	4	401–700
5	701–1100	5	701–1100	5	701–1100
One additional water closet for each 600 in excess of 1100		One additional urinal for each 300 in excess of 1100		One additional lavatory for each 1500 in excess of 1100. Such lavatories need not be supplied with hot water.	

Assembly, Religious

One Water Closet and One Lavatory

Assembly, Schools

For pupils use:

1. Water closets for pupils; in elementary schools, one for each 100 males and one for each 75 females; in secondary schools, one for each 100 males and one for each 45 females.
2. One lavatory for each 50 pupils.
3. One urinal for each 30 male pupils.
4. One drinking fountain for each 150 pupils, but at least one on each floor having classrooms.

Where more than five persons are employed, provide fixtures as required for group C1 occupancy.

Table 4.2 *(continued)*

Institutional

(Persons whose movements are not limited.) Within each dwelling unit:

1.   One kitchen sink.
2.   One water closet.
3.   One bathtub or shower.
4.   One lavatory.

Where sleeping accommodations are arranged as individual rooms, provide the following for each six sleeping rooms:

1.   One water closet.
2.   One bathtub and shower.
3.   One lavatory.

Where sleeping accommodations are arranged as a dormitory, provide the following for each 15 persons (separate rooms for each sex except in dwelling units):

1.   One water closet.
2.   One bathtub or shower.
3.   One lavatory.

Institutional, Other Than Hospitals

(Persons whose movements are limited.) On each story:

1.   Water closets: one for each 25 males and one for each 20 females.
2.   One urinal for each 50 male occupants.
3.   One lavatory for each ten occupants.
4.   One shower for each ten occupants.
5.   One drinking fountain for each 50 occupants.

Fixtures for employees the same as required for group C1 occupancy, in separate rooms for each sex.

Institutional, Hospitals

For patients' use:

1.   One water closet and one lavatory for each ten patients.
2.   One shower or bathtub for each 20 patients.
3.   One drinking fountain or equivalent fixture for each 100 patients.

Fixtures for employees the same as required for group C1 occupancy, and separate from those for patients.

Institutional, Mental Hospitals

For patients' use:

1.   One water closet, one lavatory, and one shower or bathtub, for eight patients.
2.   One drinking fountain or equivalent fixture for each 50 patients.

Fixtures for employees the same as required for group C1 occupancy, and separate from those for patients.

Table 4.2 *(continued)*

Institutional, Penal Institutions

For inmate use:

1. One water closet and one lavatory in each cell.

2. One shower on each floor on which cells are located.

3. One water closet and one lavatory for inmate use available in each exercise area.

Lavatories for inmate use need not be supplied with hot water. Fixtures for employees the same as required for group C1 occupancy, and placed in separate rooms from those used by inmates.

Business (C-1 Occupancy)

Building used primarily for the transaction of business with the handling of merchandise being incident to the primary use.

Mercantile (C-2 Occupancy)

Building used primarily for the display of merchandise and its sale to the public.

Additional Requirements

1. One drinking fountain or equivalent fixture for each 75 employees.

2. Urinals may be substituted for not more than one third of the required number of water closets when more than 35 males are employed.

Industrial (C-3 Occupancy)

Building used primarily for the manufacture or processing of products.

Storage (C-4 Occupancy)

Buildings used primarily for the storage of or shelter for merchandise, vehicles, or animals.

Additional Requirements

1. One drinking fountain or equivalent fixture for each 75 employees.

2. Urinals provided where more than 10 males are employed: One for 11–29; two for 30–79; one additional urinal for every 80 in excess of 79.

Assembly (C-5 Occupancy)

Building used primarily for the assembly for athlete educational, religious, social, or similar purposes.

Additional Requirement

One drinking fountain for each 1000 occupants but at least one on each floor.

General Notes

1. Plumbing fixture requirements shown are based on New York State Uniform Fire Prevention and Building Code and can serve only as a guide. Consult codes in force in area of construction and state and federal agencies (Labor Department, General Services Administration, etc.) and comply with their requirements.

2. Plumbing fixture requirements are to be based on the maximum legal occupancy and not on the actual or anticipated occupancy.

3. Proportioning of toilet facilities between men and women is based on a 50–50 distribution. However, in certain cases conditions of occupancy may warrant additional facilities for men or women above the basic 50–50 distribution.

Source: Ramsey and Sleeper [9], Plumbing Fixture Requirements.

Sinks (i.e., lavatories) and worksinks should be provided in accordance with Table 4.2. In no restroom should less than one sink per three toilets be provided. When multiple users may use a sink at a time, 24 linear inches of sink or 20 inches or circular basin may be equated to one sink. For planning purposes, 6 ft^2 should be allowed for each sink.

Entrance doorways into restrooms should be designed such that the interior of the restroom is not visible from the outside when the door is open. A space allowance of 15 ft^2 should be used for the entrance.

The above space requirements, combined with necessary aisles/clearances, can be used to obtain approximate floor space requirements for restrooms. To provide a clearance between the wall and a row of fixtures, the aisle width recommended in [1] ranges from 3'-6" to 4'-0" (for aisles up to 16 feet long), and from 4'-0" to 6'-0" (for longer aisles), if the toilet doors open in. If the toilet doors (2'-2") open out toward the aisle, the recommended aisle width is 4'-6" (for aisles up to 16 feet long), and it ranges from 4'-6" to 6'-6" (for longer aisles). Likewise, to provide a clearance between two rows of fixtures, the aisle width recommended in [1] ranges from 5'-6" to 6'-0" (for aisles up to 16 feet long), and from 6'-0" to 8'-0" (for longer aisles), if the toilet doors open in. If the toilet doors (2'-2") open out toward the aisle, the recommended aisle width between two rows of fixtures is 6'-6" (for aisles up to 16 feet long), and it ranges from 7'-0" to 9'-0" (for longer aisles).

A sample restroom layout showing typical minimum clearances is shown in Figure 4.6 (taken from [1]). In the sample layout there are four 2.5' × 5' toilets, one 3' × 5' shower stall, seven sinks, and three urinals. Using 6 ft^2 per sink and per urinal, and 15 ft^2 for the entrance, the space requirement without aisles and other clearances comes to $(4 \times 12.5) + 15 + 42 + 18 + 15 = 140$ ft^2. The total floor space required for the restroom, however, is approximately $16.5' \times 11.5' \approx 190$ ft^2. Hence, in this particular example, adding aisle space and other clearances increases the floor space requirement by approximately 35%. In other cases, the increase in floor space requirement may be slightly less or considerably greater than 35% depending on the exact layout of the fixtures and the shape of the room. Also, adding other

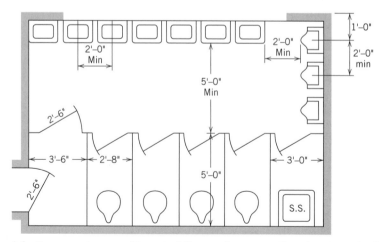

Figure 4.6 Restroom layout with typical fixture clearances. Based on New York State Labor Code. (Taken from [1] with permission of The McGraw-Hill Companies.)

features such as a baby changing station, a janitor's closet, and so on is likely to increase the floor space requirement further.

In some women's restrooms, especially in offices and administration buildings, a coach or bed/cot is provided. If bed(s) are to be provided, the following guidelines are used: (1) the area should be segregated from the restroom by a partition or curtain; (2) if between 100 and 250 women are employed, two beds should be provided; one additional bed should be provided for each additional 250 female employees; and (3) a space allowance of 60 ft^2 should be used for each bed. In any case, both the men's and the women's restrooms must conform to local codes that apply to the type of facility for which the restrooms are being planned.

4.4   FOOD SERVICES

The shifting of new facilities away from central business districts, the shortening of meal breaks, OSHA forbidding consumption of foods where toxic substances exist, and the increased importance of employee fringe benefits have all contributed to making food services more important to the facilities planner. Food service activities may be viewed by a firm as a necessity, a convenience, or a luxury. The viewpoint adopted, as well as a firm's policy on off-premises dining, subsidizing the costs of meals, and the amount of time allowed for meals, has a significant impact on the planning of food service facilities.

Food service facilities should be planned by considering the number of employees who eat in the facilities during peak activity time. Kitchen facilities, on the other hand, should be planned by considering the total number of meals to be served. If employees eat in shifts, the first third of each shift will typically be used by the employee preparing to eat and obtaining the meal. The remainder of the time will be spent at a table eating. Therefore, if a 30-min meal break is planned, dining shifts, as shown in Table 4.3, may begin every 20 min. In a like manner, if a 45-min. break is planned, shifts may begin every 30 min.

Food services requirements may be satisfied by any of the following alternatives:

1. Dining away from the facility
2. Vending machines and cafeteria
3. Serving line and cafeteria
4. Full kitchen and cafeteria

Table 4.3   *Shifting Timing for 30-min Lunch Breaks*

Beginning of Lunch Break	Time Sat Down in Chair	End of Lunch Break
11:30 A.M.	11:40 A.M.	12:00 Noon
11:50 A.M.	12:00 Noon	12:20 P.M.
12:10 P.M.	12:20 P.M.	12:40 P.M.
12:30 P.M.	12:40 P.M.	1:00 P.M.

The first alternative certainly simplifies the task of the facilities planner. However, requiring employees to leave the facility for meals results in the following disadvantages:

1. Meal breaks must be longer.
2. Employee supervision is lost, which may result in
   a. Employees not returning to work.
   b. Horseplay.
   c. Returning to work intoxicated.
   d. Returning to work late.
3. A loss of worker interaction
4. A loss of worker concentration on the tasks to be performed.

In addition, for many facilities off-site restaurants are not sufficient to handle employee meal breaks conveniently. For these reasons it is typically recommended that employees not leave the facility for meals and that space be planned within the facility for dining.

For each of the three remaining food service alternatives, a cafeteria is required. Cafeterias should be designed so that employees can relax and dine conveniently. Functional planning, ease of cleaning, and aesthetic considerations should all be considered, as the cafeteria will be utilized for purposes other than standard employee meals. (Cafeterias are often used as auditoriums.) Also, movable partitions may be used to create conference rooms and more private luncheon meeting rooms.

An integral part of the cafeteria is the food preparation or serving facilities. The option of a serving line or full-service kitchen will be contingent on the number of employees to be served. If a facility employs over 200 people, a serving line is a feasible alternative. Figure 4.7 shows an efficient walkthrough serving line which can be used for sandwiches in combination with catered food service.

Space requirements for cafeterias should be based on the maximum number of employees to eat in the cafeteria at any one time. Table 4.4 gives general area requirement guidelines. The space allocated within the ranges given in Table 4.4 should be determined by the type of tables to be utilized. Popular table sizes are 36-, 42-, and 48-in. square tables and rectangular tables 30 in. wide and 6, 8, and 10 ft long. Square tables require more aisle space than rectangular tables, but result in more attractive cafeterias. Table sizes depend on whether or not employees retain their trays during the meal. A 36-in. square table is adequate for four employees if they do not retain their trays. Standard trays are 14 × 18 in. therefore, a 48-in. square table is most suitable if employees retain their trays. Square tables (42 in. × 42 in.) are quite spacious when trays are not retained or when trays are smaller than standard used. If 36-in. square tables are used, an average space figure from Table 4.4 should be utilized. If 42-in. square tables are used, space requirements between the average and highest values stated in Table 4.4 should be utilized. If 48-in. square tables are used, the highest values stated in Table 4.4 should be utilized for space planning.

Rectangular tables 6, 8, and 10 ft long adequately seat three, four, and five employees, respectively, on each side of the table with no end seats. If individual 6-ft rectangular tables are to be utilized, an average space figure from Table 4.4 should be utilized. Figures between the average and lowest figures given in Table 4.4 should be used for tables placed end-to-end that seat between 6 and 12 people.

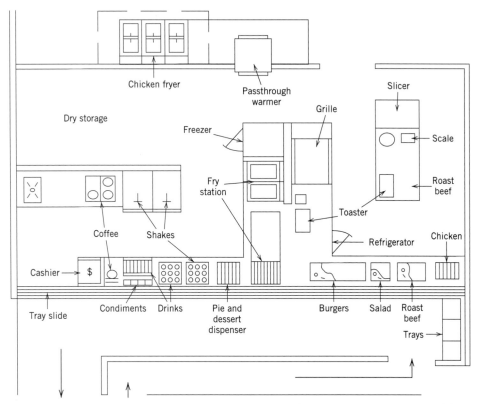

Figure 4.7 Serving line with caterer's preparation area. (*Source:* Ramsey and Sleeper [9].)

If more than 12 people may be seated in a row of tables, the lowest figures given in Table 4.4 may be utilized.

The use of vending machines is the least troublesome way of providing food services for employees. It is also the most flexible of on-site food service alternatives. Employees not wishing to purchase their lunches typically feel more at ease if vending machines are utilized than if serving lines or full kitchens are provided. For space planning purposes 1 ft^2 per person should be allowed for the vending machine area, based on the maximum number of persons eating at one time. Figure 4.8 shows a typical institutional vending area layout.

If a facility employs over 200 people, a serving line is a feasible alternative. When a serving line is utilized, a caterer is frequently contracted to prepare all food off site and to serve the food to employees. The advantage of serving lines is that they offer employees the benefits of a full kitchen but require little effort by the management of

Table 4.4 *Space Requirements for Cafeterias*

Classification	Square Footage Allowance per Person
Commercial	16–18
Industrial	12–15
Banquet	10–11

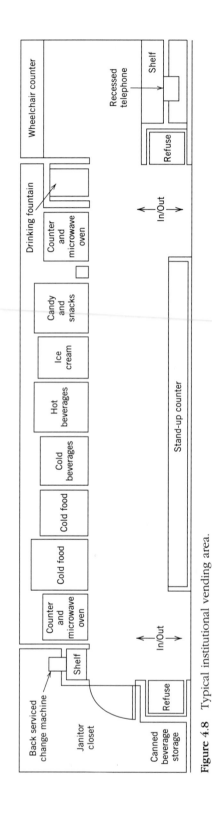

Figure 4.8 Typical institutional vending area.

Note: Refrigerated vending machines and microwave ovens require a rear wall clearance of up to 8 in. to permit cooling. No hot water is required. Some beverage dispensing units require a cold water line with a shut-off valve. Overflow waste disposes into internal bucket or tray. A separate, 115-V electrical circuit for each machine is suggested. Delivery amperages range from approximately 2 to 20 amperes (A) per machine.

(*Source:* Ramsey and Sleeper [9].)

Table 4.5   *Space Required for Full Kitchens*

Number of Meals Served	Area Requirements (ft^2)
100–200	500–1000
200–400	800–1600
400–800	1400–2800
800–1300	2400–3900
1300–2000	3250–5000
2000–3000	4000–6000
3000–5000	5500–9250

Source: Kotschevar and Terrall [5].

the facility. The cost of a meal is often quite competitive with the cost of running a full kitchen for facilities employing less than 400 people. A typical industrial serving line requires 300 ft^2 and can service seven employees per minute [5]. A sufficient number of service lines should be provided so that during peak demand for service employees receive their meals in one-third of the lunch period.

When a full kitchen is used, a serving line (see Figure 4.7) and a kitchen must be included in the facility. A full kitchen can usually be justified economically only if there are over 400 employees within a facility. A kitchen allows the management of a facility to have full control over food service and be able to respond to employee desires. Such control is often a costly and troublesome undertaking; it is one that many firms do not wish to undertake. Space planning for kitchens to include space for food storage, food preparation, and dishwashing should be based on the total number of meals to be prepared. Space estimates for full kitchen are given in Table 4.5

Food services should be located within a facility using the following guidelines.

1. Food services should be located within 1000 ft of all permanent employee workstations. If employees are required to travel more than 1000 ft, decentralized food services should be considered.
2. Food services should be centrally located and positioned so that the distance from the furthest workstation is minimized.
3. Food services should be located to allow delivery of food and trash pickup.
4. Food services should be located to allow employees an outside view while eating.
5. Food services should be located so that adequate ventilation and odors and exhaust do not interfere with other activities within the facility.

Example 4.3

If an industrial facility employs 600 people and they are to eat in three equal 30-min shifts, how much space should be planned for a cafeteria with vending machines, serving lines, or a full kitchen?

If 36″ square tables are to be utilized, Table 4.4 indicates 12 ft^2 is required for each of the 200 employees to eat per shift. Therefore, a 2400-ft^2 cafeteria should be planned.

If a vending machine area is to be used in conjunction with the cafeteria, an area of 200 ft^2 should be allocated for vending machines. Thus, a vending machine food service facility would require 2600 ft^2.

A serving line may serve 70 employees in the first third of each meal shift. Therefore, three serving lines of 300 ft^2 each should be planned. A total of 3300 ft^2 would be required for a food service facility using serving lines.

A full kitchen will require 3300 ft^2 for cafeteria and serving lines plus (from Table 4.5) 2100 ft^2 for the kitchen. Therefore, a total of 5400 ft^2 would be required for a full service food service facility.

Two issues related to food services are drinking fountains and break areas. Drinking fountains should be located within 200 ft of any location where employees are regularly engaged in work. Local building codes should be consulted for determining their exact location. Drinking fountains should be conveniently located but must not be located where employees using drinking fountains may be exposed to a hazard. Drinking fountains are often located near restrooms and locker rooms for the convenience of employees and because plumbing is readily available.

If the cafeteria is within 400 ft of most employees, rest breaks should be confined to the cafeteria. If a cafeteria serving line exists and it may be used to dispense drinks and snacks, it should be utilized for breaks. If a serving line may not be used or if one does not exist, drinks and snacks should be available in the cafeteria from vending machines during break periods.

Locating vending machines throughout a facility may cause supervision, food containment, and trash disposal problems. Nevertheless, if the cafeteria is more than 400 ft from most employees, some vending machines should be located within the facility to reduce employee travel time. When vending machines are used, break areas should be conveniently located for employees while at the same time not interfering with other activities.

4.5 HEALTH SERVICES

It is very difficult to predict facility requirements for health services. Some facilities have little more than a well-supplied first and kit, while other facilities have small hospitals. Therefore, local building codes should be checked in establishing a facility's requirements. The types of health services that may be provided within a facility include

1. Preemployment examinations,
2. first aid treatment,
3. major medical treatment,
4. dental care, and
5. treatment of illnesses.

The facilities planner should check the firm's operating procedure to determine what types of services are to be offered and what health services staff is to be housed within the facility. At the very least, a small first aid room should be included. The minimal requirements for a first aid room are an approved first aid kit, a bed, and two chairs. A minimum of 100 ft^2 is required. If a nurse is to be employed, the first aid room should have two beds and should be expanded to 250 ft^2. In addition, a 75 ft^2 waiting room should be included. For each additional nurse to be employed,

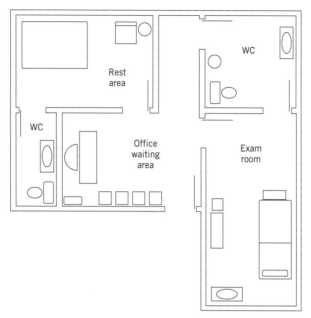

Figure 4.9 Nursing station layout.

250 ft^2 should be added to the space requirements for the first aid room and 25 ft^2 should be added to the space requirements for the waiting room. Figure 4.9 shows a typical nurse/first aid area for a plant facility. If a physician is to be employed on a part-time basis to perform preemployment physicals, a 150 ft^2 examination room should be provided. If physicians are to be employed on a full-time basis the space requirement should be planned in conjunction with a physician, based on the types of services to be offered.

Health services should be located such that examination rooms are adjacent to first aid rooms and close to the most hazardous tasks. Health services should include toilet facilities and should either be soundproofed or located in a quiet area of the plant. They are often located next to restrooms or locker rooms.

4.6 BARRIER-FREE COMPLIANCE

Facilities planners must incorporate the intent of the Americans with Disabilities Act (ADA) [3, 7]. The intent is to ensure that disabled persons shall have the same right as the able-bodied to the full and free use of all facilities that serve the public. To this extent, all barriers that would impede the use of the facility by the disabled person must be removed, thereby making the facility barrier free. As facilities planners, the definition of barriers must be understood. What are considered barriers? A barrier is a physical object that impedes a disabled person's access to the use of a facility, for example, a door that is not wide enough to accommodate a wheel chair or stairs without ramp access to a facility. Note also that communication barriers that are an integral part of the physical structure of the facility are included, for example, permanent signage. The facilities planner must recognized that this applies to all public facility use groups:

- Assembly
- Business
- Educational
- Factories and Industrial
- Institutional

The ADA will fundamentally impact the way industrial engineers approach the design of a facility—from the parking lot to entering and exiting the facility, moving within the interior of the facility, workstations, offices, and restrooms. To remain effective as facilities planners we must account for the handicapped person's space requirements versus that of an able-bodied person. Consider Figure 4.10 and the wheelchair dimension's reach and maneuverability requirements.

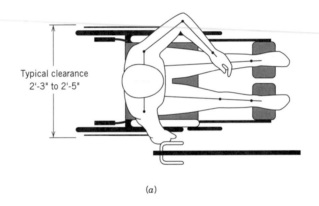

(a)

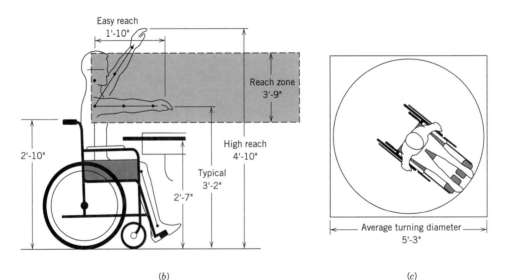

(b)   (c)

Figure 4.10   Wheelchair dimensions and turning radius.

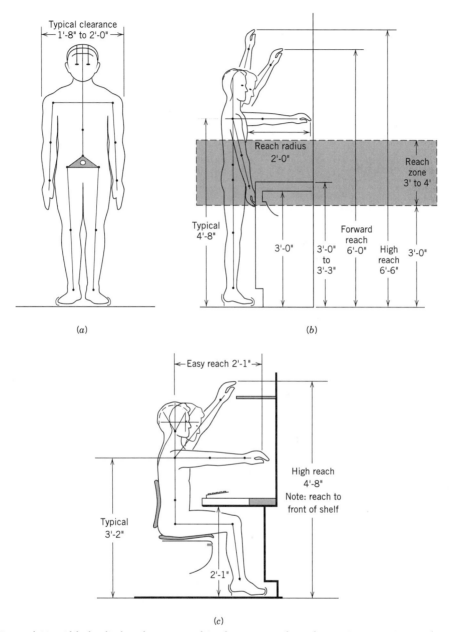

Figure 4.11 Able-bodied anthropomorphic clearance and reach requirements in standing and sitting positions.

Compare these dimensions in Figure 4.10 with the dimension of an able-bodied person's typical clearance and reach requirements as given in Figure 4.11. Although there are significant physical differences between able-bodied and physically disadvantaged individuals, there exists a reach zone where both groups can comfortably access objects placed in this zone (see Figures 4.10b and 4.11b). This zone, as shown, is typically 3 ft to 4 ft above floor level. These upper and lower limits were defined by the reach limits of both groups. The upper limit is dictated by the height that a physically disadvantaged person can easily reach an object, whereas the lower limit is defined by the reach of the average able-bodied person without bending.

From analyzing the clearances for both groups, it is obvious that facility entrances, doors, hallways, and so on must be wide enough to accommodate the wheelchair, typically a 3 ft. minimum. Also, fixed facility elements in laboratories or other work or study areas using workbenches requires a minimum clear width of 3 ft. This width should be increased to a 3 ft-6 in. minimum if the aisle is to accommodate access simultaneously by an able-bodied person and a wheelchaired person, for example in library stalls. Additionally, by mapping the reach requirement of the average person against that of the handicapped, there exists a zone 3 ft. to 4 ft. where both groups can comfortably access items. Thus, the placement of telephones, towel racks, trash, cup disposers, switches, thermostat, fire alarm, emergency call box, door latches, elevator control panel, elevator call button, and so on must be typically a maximum of 40 in. above floor level to effectively accommodate both groups.

4.7 OFFICE FACILITY PLANNING

Offices are among the most challenging facilities to plan. One reason for this is each office employee has an opinion on how space should be planned. Presenting an office layout to an office employee creates the same reactions as giving a group photograph to a person who is not in the photograph. The person immediately focuses on how he or she relates to the others. In the case of office planning, the employee focuses on size, noise, proximity of facilities such as the boss, coffee, cafeteria, elevator, duplicating machines, windows, entrance/exit doors, and so on. The facilities planner beginning an office planning project is well advised to follow a methodical planning process and to fully investigate the activities planned for the facility.

The starting point for planning office facilities is the collection of data concerning the objectives of the facility, and the activities to be performed to accomplish these objectives. The data are collected to determine departmental interrelationships and departmental area requirements. When the data gathering is completed, a series of management interviews should be conducted to verify and refine office requirements.

Approaches to Office Planning

An open office is one in which the office area is free from temporary or permanent partition walls. The space is open and no floor-to-ceiling walls exist. A closed office structure is a structure where floor-to-ceiling temporary or permanent partition walls break up the office facility into smaller rooms.

The motivation for open offices is to increase flexibility for changes in office operations and to easily accommodate expansion. Additional benefits of open offices include:

1. Improved communications
2. Improved supervision
3. Better access to common files and equipment
4. Easier to illuminate, heat, cool, and ventilate
5. Lower maintenance costs
6. Reduced space requirements due to space flexibility

Some objections frequently cited concerning open offices are:

1. Lack of privacy
2. Lack of status recognition
3. Difficulty in controlling noise
4. Easy access for interruptions and interference

The privacy problem is actually a problem of visual and audible privacy. Visual privacy is achieved by using screens or movable partitions, and locating workstations so that lines of sight do not intersect. Figure 4.12 shows several types of office systems. Audible privacy is related to noise control and worker interference and may be partially achieved by providing visual privacy. In open offices, soundproofing the partitions reduces noise. Another method is to introduce electronic noise to mask sound. The objective of electronic noise is not to eliminate noise but to mask the sounds so that they are not discernible.

Closed offices are recommended when the work is confidential, when noise from various sources cannot be masked, and when the tasks to be performed require undisturbed concentration.

Most office structures are a combination of open and closed office structures. Although traditional office layouts consist of private, semiprivate, and general closed offices, more contemporary offices are a combination of closed private and semiprivate, and general open offices.

Area Requirements

Planning for new office facilities requires decisions beyond just the determination of square footage requirements. Nevertheless, the specification of square footage requirements is an essential step in the planning of office facilities. The following data are suggested in [12]:

*   President's office: 250–400 sq ft
*   Vice president's office: 150–250 sq ft
*   Executive office: 100–150 sq ft
*   Partitioned open space—supervisor or manager: 80–110 sq ft

Figure 4.12 Examples of office systems. *(Courtesy of Herman Miller, Inc.)*

Figure 4.12   (continued)

- Open space—clerical or secretary: 60–110 sq. ft
- Conference rooms
  - 15 sq. ft per person (theater style)
  - 20–30 sq. ft per person (conference seating)
- Mail room: 8 to 9 ft wide; length depends on the amount of usage
- Reception area
  - 125–200 sq. ft (receptionist and 2–4 people)
  - 200–300 sq. ft (receptionist and 6–8 people)
- File room: 7 sq. ft per file with a 3–4 ft aisle width

Office Planning in High-Tech, High-Growth Environments

The high-tech, high-growth era beginning in the late 1900s spurred the development of *office campuses*. One particular company's office campus concept is worth noting. In providing office services to its clients, Enfrastructure, Inc. [11] provides full-service infrastructure and facilities support. Their approach is to provide "a completely integrated and supportive work and lifestyle environment for growth companies. The company overcomes typical operating obstacles by combining workspace, technology infrastructure, business support and professional services— leaving companies free to concentrate on their core business." They act as enablers to growth companies. Three things are achieved:

1. *Speed to market.* The campus is "preloaded with all the necessary business and technology infrastructure, enabling companies to accelerate their speed to market."
2. *Capital preservation.* Initial capital investment is reduced since office-related capital investments are not needed. The concept is one of rent rather than ownership.
3. *Scalability.* Growth needs are accommodated for small and large fluctuations in workspace demands. Companies can "utilize as much or as little as they need to succeed."

The two charts in Figure 4.13 compare traditional outsourcing of office products and services versus a pyramid structure. In the traditional method, the company has to do the outsourcing itself. Under the pyramid structure, Enfrastructure provides all the office products and administrative services and the tenant company occupying the campus pays monthly fees to cover the expenses it consumed each month. A cost comparison of these two types of operations is given in Figure 4.14 based on a 23-person firm.

The office suites are completely scalable with modular furnishings including desks, chairs, drawers, and file cabinets. Standard hardware and services included in every office and workstation are: PBX lines, T-1 voice and DSL lines, analog lines, telephone cabling, and firewall/routers. Other services include artwork, shared administrative office staff, receptionist, security system monitoring, mailroom facilities, photocopier, and so forth. All offices are furnished with the same type of ergonomic

Traditional Outsourcing

Outsourcing with Enfrastructure

Figure 4.13   Comparison between traditional and pyramid structure outsourcing. *(Courtesy of Enfrastructure, Inc.)*

chairs, desks, and drawers. They all look alike so that when it is time to move to a bigger office cluster, the movers come in and the next day you are in a different room but everything is the same—chairs, desks, and so on—even the same telephone number. Figure 4.15 shows the floor plan of the office areas.

The conference and meeting rooms are common to all clients with a fixed number of hours per month included in the rent of the office space and extra time charged on a use basis. Conference rooms with dry-erase walls (not boards) are equipped with all the technology that can be assembled, including computer

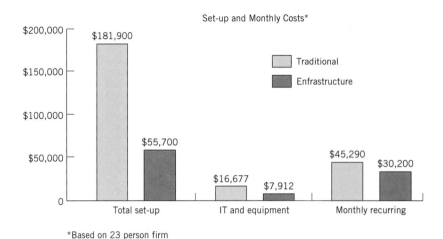

Figure 4.14 Cost comparison based on a 23-person firm. *(Courtesy of Enfrastructure, Inc.)*

hookups, wall-mounted flat-screen interactive monitors, teleconferencing (audio and video) equipment, and so on. Figure 4.16 shows one of the conference rooms.

In addition, an office client is offered other amenities including a copy center, branch bank, concierge for travel and entertainment services, sanctuary facilities where client company staff can "relax, meditate, or sleep in total silence," a training center and a spa, a café and bar for meals, snacks and refreshments, laundry service, and a backyard garden. Figure 4.17 is the floor plan for the ground floor that shows all the above amenities.

A company can literally walk in, sign a contract, and operate the business in a week with all the office infrastructure and services available to large companies. At the end of the month, the company will be invoiced for the use of the facilities.

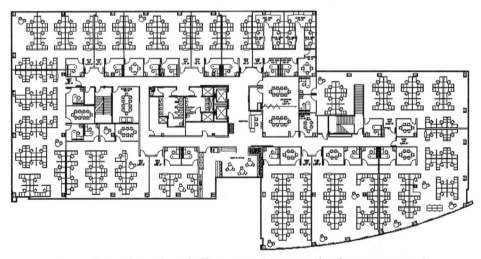

Figure 4.15 Floor plan of office areas. *(Courtesy of Enfrastructure, Inc.)*

Figure 4.16   Technology-equipped conference rooms. *(Courtesy of Enfrastructure, Inc.)*

Figure 4.16   (continued)

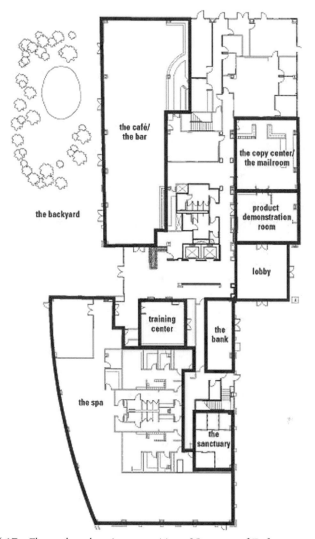

Figure 4.17   Floor plan showing amenities. *(Courtesy of Enfrastructure, Inc.)*

Its targeted clients are incubator-type companies. As B. Dreyer [2] of Enfrastructure stated,

> We look for clients that are high-growth and/or start-ups. Technology, biotech, and other high-growth potential vertical markets are prime candidates for the Enfrastructure model. These companies typically have tighter windows of market opportunity. The clients that come here have to place actual dollar value to the office infrastructure facilities and the soft benefits of the amenities that are offered in-house.
>
> By providing the client a month-to-month agreement (or up to one year), we provide them with maximum flexibility with respect to growth, capital preservation, and scalability. We can get them on campus in about 1–2 weeks and scale them in-house in less than 24 hours to 1 week depending on our occupancy capacity.

As you can see from our discussion, office planning is no longer planning for individual office space requirements. It is now transformed into "office systems" that are modular, flexible, and scalable. It is an "office environment" design problem with the goal of providing growth companies and their employees with the environment that will make them more productive.

4.8   SUMMARY

Guidelines to determine facility requirements for parking lots, locker rooms, restrooms, food services, health services, ADA compliance, and office facilities have been provided in this chapter. Answers to the following questions are *not* the responsibility of the facilities planner:

1.   Should employees be assigned parking locations or should parking be random?
2.   Should locker rooms be provided even though clothes changing does not take place?
3.   Should restrooms be centralized or decentralized?
4.   Should food be provided to employees from vending machines or serving lines?
5.   Should a physician be employed to provide health services or is a nurse adequate?

The answers to these questions are generally the responsibility of the human resources department, industrial relations department, or personnel department. Once these answers are provided, the facilities planner can utilize the guidelines set forth in this chapter to plan personnel services within the facility. However, because the individuals who are responsible for answering such questions might not understand the cost impact of their decisions, the facilities planner should play an active role in the overall facilities planning decision-making process.

Finally, we provided some information on office facilities. We reiterate that while space allocation is a necessary task, the office planning process should address the broader issue of creating an environment where people will be more productive.

BIBLIOGRAPHY

1.   Callendar, J. H., *Time Saver Standards for Architectural Design Data,* 6th ed., McGraw-Hill, New York, 1982.
2.   Dreyer, B., *Personal communication,* Auegst 2001.
3.   Kearney, D. S., *The New ADA: Compliance and Costs,* R. S. Means Company, Inc., Kingston, MAP, 1992.
4.   Konz, S., *Facility Design,* 2nd ed., Publishing Horizons, Inc., Scottsdale, AZ, 1994.
5.   Kotschevar, L. H., and Terrall, M. E., *Food Service Planning: Layout and Equipment,* 2nd ed., Wiley, New York, 1977.
6.   Mariotti, J., "The Landscape Office: New Look in Efficiency," *Industrial Engineering,* pp. 30–35, September 1970.

7. Mumford S., "The ADA: Striving to Comply," *Buildings,* vol. 87, no. 4, bb. 46–48, April 1993.
8. Stokes, J. W., *Food Service in Industry and Institutions,* Brown, Dubuque, IAP, 1973.
9. Ramsey, C. G. and Sleeper, H. R., *Architectural Graphic Standards,* Wiley, New York, 1988.
10. *Boca National Building Code,* 11th ed., Building Officials & Code Administrators International, Inc., Country Club Hills, IL, 1990.
11. Enfrastructure Information Sheet, 2001.
12. "Tenants' Rules of Thumb," OfficeFinder Web site, 1999.

PROBLEMS

4.1 Is it true that a well-fed employee is a happy employee? Is a happy employee a productive employee? Explain.

4.2 Compare the bulkiness of the nonwork belongings that are often brought to a facility by an employee.

4.3 What is the impact on the space required in a parking lot if spaces are assigned or if filled randomly?

4.4 A parking lot is to be 400 ft wide and 370 ft deep. How may standard-sized cars fit in this lot?

4.5 Locate a rectangular or near-rectangular parking lot in your campus or office complex. Using the same parameters as much as possible (i.e., same ratio of compact to standard cars, same fraction of handicapped parking stalls, etc.), determine if you can increase the number of parking spaces by redesigning the lot. If the stall width used in the lot does not appear in Table 4.1, use the closest stall width from Table 4.1 in your calculations.

4.6 Discuss the pros and cons of parking decks (i.e., multilevel parking structures) compared to surface lots.

4.7 A facility is to house 50 female and 50 male employees. Using a 40% allowance for aides and clearances, how much space should be planned for the restrooms?

4.8 The Ajax Manufacturing Company has decided to allow its employees 1 hour for lunch. The lunch breaks are to start at 11:00 A.M. and to end at 1:00 P.M. How many lunch starting times can they accommodate if lunch breaks must being and end on the hour, on the quarter hour, or on the half hour?

4.9 What is the impact on space requirements for a vending machine and cafeteria food service for the following luncheon patterns?

800 employees—one shift

400 employees—two shifts

200 employees—four shifts

4.10 What is the impact on space requirements for a serving line and cafeteria food service for the luncheon patterns given in Problem 4.9?

4.11 What is the impact on space requirements for a full kitchen and cafeteria food service for the luncheon patterns given in Problem 4.9?

4.12 If two nurses and a part-time physician are to be employed in a health services area, how much space should be allowed in the facilities plan?

4.13 Describe the largest personnel service problem on campus. How could this problem be resolved.

4.14   Are personnel services as important in office facilities as they are in production facilities? Describe in detail.

4.15   Is it a good idea to consult with employees who are to utilize personnel services before planning these services? Why or why not?

4.16   Discuss the impact on personnel services of multilevel versus single-level facilities.

4.17   What are the ADA implications on
   a. The building in which this class is taught.
   b. Your school's Football Stadium
   c. Your school's Student Center

4.18   How important is it to involve individual employees in the office planning process?

4.19   Give advantages and disadvantages of open versus closed office structures.

4.20   Compare the traditional office concept with the campus office concept in terms of employees' personal satisfaction.

4.21   Make a list of additional features that an office campus could provide to enhance the experience of its resident companies.

4.22   Make a list of criteria that can be used to evaluate office plans.

Part Two

DEVELOPING ALTERNATIVES: CONCEPTS AND TECHNIQUES

5

MATERIAL HANDLING

5.1 INTRODUCTION

The design of the material handling system is an important component of the overall facilities design. The layout design and the material handling system design are inseparable. It is seldom the case that one can be considered without jointly considering the other case. The integration between these two design functions is particularly critical in the design of a new facility.

Material handling can be observed in one's day-to-day activities—mail delivered in a postal system, parts moved in a manufacturing system, boxes and pallet loads moved in an industrial distribution system, refuse collected in a waste management system, or people moved in a bus or mass transit system. As we will emphasize throughout the discussions in this chapter, material handling is an integral part of the overall facility design process. Material handling problems arise in a wide variety of contexts and a host of alternative solutions usually exists. There is typically more than one "best" solution to a material handling system design problem. This is one area where the material handling engineer must keep a broad perspective and must be cognizant of the "integration effects."

In this chapter, we first discuss the scope and definitions of material handling. We then cover the *principles of material handling*. The overall material handling design process is discussed next. Then, we address the unit load design problem. The unit load is, in our view, one of the most important elements of a material handling system. Finally, we provide a comprehensive listing and discussion of various material handling equipment and alternatives. Discussions on material handling cost estimation and safety considerations are also given in separate sections.

5.2 SCOPE AND DEFINITIONS OF MATERIAL HANDLING

In a typical industrial facility, material handling accounts for 25% of all employees, 55% of all factory space, and 87% of production time (Frazelle [14]). Material handling is estimated to represent between 15% and 70% of the total cost of a manufactured product. Material handling is one activity where many improvements can be achieved, resulting in significant cost savings. The ideal goal is to "totally eliminate" material handling activities, although in most cases reducing the amount of handling is a more appropriate practical goal. Nevertheless, improvements in material handling processes will lead to more efficient manufacturing and distribution flows. A reduction in the number of times a product is handled results in reduced requirements for material handling equipment. Simply handling less, however, is not sufficient. One can view material handling as a means by which *total manufacturing costs are reduced* through more efficient material flow control, lower inventories, and improved safety.

Clearly, simply handling less is not the answer. A well-designed material handling system can be the main backbone of a company's overall production execution strategy. The significance of material handling can be observed in many contemporary manufacturing and distribution operations.

Definitions

Let's take a look at the following definitions of material handling:

1. Material handling is the art and science of moving, storing, protecting, and controlling material.
2. Material handling means providing the right amount of the right material, in the right condition, at the right place, in the right position, in the right sequence, and for the right cost, by the right method(s).

The first definition conveys the fact that the material handling design process is both a science and an art, and that the material handling function involves moving, storing, protecting, and controlling material. It is a science-based discipline involving many areas of engineering, and therefore engineering design methods must be applied. Thus, the material handling design process involves defining the problem, collecting and analyzing data, generating alternative solutions, evaluating alternatives, selecting and implementing preferred alternative(s), and performing periodic reviews. It is an art since material handling systems *cannot* be explicitly designed based solely on scientific formulas or mathematical models. Material handling requires knowledge and appreciation of what is "right and wrong," which is based on significant practical experience in the field.

The second definition captures the essence of the material handling function. We will discuss each element of this definition more thoroughly below.

Right Amount.   The "right amount" refers to the problem of how much inventory is needed. The just-in-time (JIT) philosophy focuses on not having inventories. The right amount is what is needed and not what is anticipated. Thus, a pull-type material flow control structure is advocated. Smaller load sizes are preferred. With significant reductions in setup time, the matching of production lot sizes and transfer batch sizes can result in improved deliveries of the right amount of material.

Right Material.   The two most common errors in manual order picking are picking the wrong amount and picking the wrong material. These errors point to the fact that an accurate identification system is needed. Automatic identification is key to accurate identification. Automatic identification (e.g., a bar-code-based system) cannot be matched by manual approaches. However, improvements such as simplifying the parts numbering system and maintaining the integrity and accuracy of the database system are more fundamental tasks.

Right Condition.   The "right condition" is the state in which the customer desires to receive the material. The customer may specify that the material be delivered packed or unpacked, sorted based on kitting specifications, painted or unpainted, delivered in customer-specified returnable containers, and so on. The goods must also be received without damage.

Right Sequence.   The impact of the "right sequence" of activities performed on the efficiency of a manufacturing or distribution operation is very evident in material handling. Work simplification can help eliminate unnecessary operations or improving those that remain. Combining steps and changing the sequence of operations can also result in more efficient material flow.

Right Orientation.   The "right orientation" means positioning the material for ease of handling. Positioning is particularly critical in automated systems, such as in robot handling operations, where part orientation must be explicitly specified. Often, changing the part design by including "handling tabs" can reduce the handling time. The use of four-way pallets versus two-way pallets can eliminate orientation-induced problems.

Right Place.   The "right place" addresses both transportation and storage. It is desirable to directly transport material to the point of use rather than store the material at some intermediate location. In some situations, materials are left along aisles, causing disruptions in lift-truck operations. The issues associated with centralized versus decentralized storage must be explicitly addressed.

Right Time.   The "right time" means on-time delivery, neither early nor tardy. Reduction in the variance of delivery time is the key to this element of the definition of material handling. A flexible material handling system such as manually operated lift trucks has very wide deviations in transport times, while an automated guided vehicle system has more predictable transport times. The goal is to develop a material handling system that will result in lower production cycle times, and not to lower material handling delivery times. It has been repeated observed in practice that slower average speeds are preferable to faster average speeds if accompanied by reduction in the variation of speeds—variance reduction is the key. The saying "we rush to wait" occurs often in material handling operations.

Right Cost.   The "right cost" is not necessarily the lowest cost. Minimizing cost is the wrong objective in material handling system design. The more appropriate goal is to design the most efficient material handling systems at the most reasonable cost. Put in the right context, material handling is a support function. On-time deliveries often result in increased customer satisfaction that can in turn result in increased demand for the product, thus increasing revenue. Material handling operations should support a company's quest for greater profitability.

Right Method.   There are three aspects of the "right method" that merit a closer look. First, if there are right methods, then there must be wrong methods. Second, it is important to recognize what makes methods right and what makes them wrong. Third, note that it is methods and not method; using more than one method is generally the right thing to do.

   Since the early 1970s, we have advocated requirements-driven material handling systems over solution-driven systems. Solution-driven systems are those in which technologies are chosen without consideration for how the technologies match requirements; instead of defining requirements and matching technology options to requirements, the solution-driven approach force-fits a technology to an application. There have been cases where managers have been enamored by the latest material handling technology only to find out later that more problems arise. The consequences are the tearing down of highly sophisticated, automated material handling systems.

   We also note that equipment selection is the very last step in the process of designing material handling systems. Equipment selection is but the consequence of selecting the best methods from a number of alternatives generated from the examination of the problem of *providing the right amount of the right material, in the right condition, at the right place, at the right time, in the right position, in the right sequence, and for the right cost, using the right methods.*

Scope of Material Handling

As you can see in the discussions above, the scope of material handling is quite broad. Apple [2] identifies three views on the scope of material handling activities—conventional, contemporary, and progressive. The "conventional view" focuses solely on the movement of material from one location to another, usually within the same manufacturing and distribution facility. One question that may be asked is "How do we move material from the receiving dock to the storage area?" Very little attention is given to the interrelationships among the overall handling tasks that may occur within the same facility. The "contemporary view" expands the focus to the overall movement of materials in a factory or warehouse, and an effort is made to develop an integrated material handling plan. The "progressive view" is a total system view. This view looks at material handling as all activities in handling material from all supliers, handling material within the manufacturing and distribution facility, and the distribution of finished goods to customers. The progressive view is what we advocate.

In the next sections, we will discuss the principles and methods for material handling system design.

5.3 MATERIAL HANDLING PRINCIPLES

Material handling principles are important in practice. It is often the case that no mathematical model can provide extensive solutions to the overall material handling problem. These principles provide concise statements of the fundamentals of material handling practice. Condensed from decades of expert material handling experience, they provide guidance and perspective to material handling system designers. However, the use of these principles should not be construed as a substitute for good judgment and experience.

The 20 principles of material handling listed in the 2nd edition of this book have been reduced to 10 principles. The 10 material handling principles and their definitions, recently adopted by the College-Industry Council on Material Handling Education (CIC-MHE), are as follows:

1. *Planning Principle.* A plan is a prescribed course of action that is defined in advance of implementation. In its simplest form a material handling plan defines the material (what) and the moves (when and where); together they define the method (how and who).

2. *Standardization Principle.* Standardization means less variety and customization in the methods and equipment employed.

3. *Work Principle.* The measure of work is material flow (volume, weight, or count per unit of time) multiplied by the distance moved.

4. *Ergonomic Principle.* Ergonomics is the science that seeks to adapt work or working conditions to suit the abilities of the worker.

5. *Unit Load Principle.* A unit load is one that can be stored or moved as a single entity at one time, such as a pallet, container, or tote, regardless of the number of individual items that make up the load.

6. *Space Utilization.* Space in material handling is three-dimensional and therefore is counted as cubic space.

7. *System Principle.* A system is a collection of interacting and/or interdependent entities that form a unified whole.

8. *Automation Principle.* Automation is a technology concerned with the application of electromechanical devices, electronics, and computer-based systems to operate and control production and service activities. It suggests the linking of multiple mechanical operations to create a system that can be controlled by programmed instructions.

9. *Environmental Principle.* Environmental consciousness stems from a desire not to waste natural resources and to predict and eliminate the possible negative effects of our daily actions on the environment.

10. *Life Cycle Cost Principle.* Life cycle costs include all cash flows that will occur from the time the first dollar is spent to plan or procure a new piece of equipment,

Table 5.1 *Material Handling Audit Sheet*

Conditions indicating opportunities for improvement	Condition Observed	Supervisor Attention	Management Attention	Correction Will Require			Other Comments
				Analytical Study	Capital Investment		
1. Production equipment idle due to material shortage							
2. Material piled directly on floor							
3. In-plant containers not standardized							
4. Operators travel excessively for materials and supplies							
5. Excessive demurrage							
6. Misdirected material							
7. Backtracking of material							
8. Automatic data collection system not used							
9. Excessive trash removal							
10. System not capable of expansion and/or change							
11. No pre-kitting of work							
12. Crushed loads in block stacking							

Note. A more complete list is given in Appendix 5.A.

or to put in place a new method, until that method and/or equipment is totally replaced.

These 10 material handling principles are guidelines that will be helpful in solving material handling problems. Obviously, not all principles will apply in every material handling project. These principles can also be used as a checklist but they should become second nature to material handling system designers. Application of these principles to day-to-day activities can enhance material handling solutions.

Material Handling Checklists

Based on these principles, a number of material handling checklists have been developed to facilitate the identification of opportunities to improve existing material handling systems. Checklists can serve a useful purpose in designing new systems as well. They can ensure that nothing falls through the cracks and that everything has been accounted for. There are so many detailed considerations involved in designing material handling systems that it is quite easy to overlook minor issues that can become major problems in the future. An illustration of a checklist for material handling is given in Table 5.1. Appendix 5.A gives a more comprehensive checklist (White [62]). The material handling checklist contains conditions where possible productivity improvement opportunities may exist. Several columns are provided to mark what type of corrective actions may be taken, such as supervisory attention, management attention, analytical study, capital investment, and others.

5.4 DESIGNING MATERIAL HANDLING SYSTEMS

As we emphasize throughout our previous discussions, the material handling systems design process involves the six-step engineering design process. In the context of material handling, these steps are:

1. Define the objectives and scope for the material handling system.
2. Analyze the requirements for moving, storing, protecting, and controlling material.
3. Generate alternative designs for meeting material handling system requirements.
4. Evaluate alternative material handling system designs.
5. Select the preferred design for moving, storing, protecting, and controlling material.
6. Implement the preferred design, including the selection of suppliers, training of personnel, installation, debug and startup of equipment, and periodic audits of system performance.

One item that is worth mentioning again is the periodic audits of system performance. It is unrealistic to expect that the material handling system will operate perfectly the first time around. Adopting a posture of continuous improvement will result in far more efficient operation of a material handling system.

Developing Alternative Material Handling System Designs

To stimulate the development of alternatives, the "ideal systems approach," proposed by Nadler [34], should be considered. This approach consists of four phases:

1.   *Aim* for the theoretical ideal system.
2.   *Conceptualize* the ultimate ideal system.
3.   *Design* the technologically workable ideal system.
4.   *Install* the recommended system.

The **theoretical ideal system** is a perfect system having zero cost, perfect quality, no safety hazards, no wasted space, and no management inefficiencies. The **ultimate ideal system** is a system that would be achievable in the future since the technology exists for its development but its application to a specific material handling application has not been accomplished. The **technologically workable ideal system** is a system for which the required technology is available; however, very high costs or other conditions may prevent some components from being installed now. The **recommended system** is a cost-effective system that will work now without obstacles to its successful implementation.

Following the ideal systems approach allows us to expand our horizon beyond the current state of the technology. One could also see how this approach can expand the search for alternatives beyond what the material system designer knows at present.

The Material Handling System Equation

To help guide the development of alternative material handling system designs, we now look at the value of using the **material handling system equation,** shown in Figure 5.1. Just as a checklist such as the one given in Table 5.1 and Appendix 5.A provides a means to identify opportunities for improvement, the material handling system equation gives us the framework for identifying solutions to material handling problems. The *what* defines the type of materials moved, the *where* and *when*

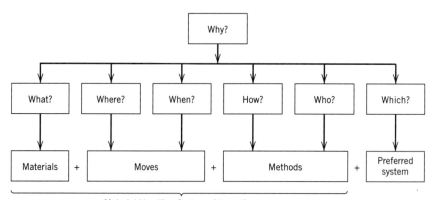

Figure 5.1   Material handling system equation.

identify the place and time requirements, the *how* and *who* point to the material handling methods. These questions all lead us to the recommended system.

The material handling system equation is given by

Materials + Moves + Methods = Recommended System

A detailed listing of the *what, where, when, how, who,* and *which* questions are given below. For each of these questions, we repeatedly ask *why* it is necessary or not.

The *What* Question

1.   What are the types of material to be moved?
2.   What are their characteristics?
3.   What are the amounts moved and stored?

The *Where* Question

1.   Where is the material coming from? Where should it come from?
2.   Where is the material delivered? Where should it be delivered?
3.   Where is the material stored? Where should it be stored?
4.   Where can material handling tasks be eliminated, combined, and simplified?
5.   Where can you apply mechanization or automation?

The *When* Question

1.   When is material needed? When should it be moved?
2.   When is it time to mechanize or automate?
3.   When should we conduct a material handling performance audit?

The *How* Question

1.   How is the material moved or stored? How should material be moved or stored? What are the alternative ways of moving or storing the material?
2.   How much inventory should be maintained?
3.   How is the material tracked? How should the material be tracked?
4.   How should the problem be analyzed?

The *Who* Question

1.   Who should be handling material? What are the required skills to perform the material handling tasks?
2.   Who should be trained to service and maintain the material handling system?
3.   Who should be involved in designing the system?

The *Which* Question

1.   Which material handling operations are necessary?
2.   Which type of material handling equipment, if any, should be considered?
3.   Which material handling system is cost effective?
4.   Which alternative is preferred?

Thus far, our discussion has been at the conceptual level. Discussion at this level gives us the opportunity to view material handling problems from a system perspective. We now look at specific ways to analyze material handling problems. One way to organize data and generate alternatives is to use the *Material Handling Planning Chart.*

Material Handling Planning Chart

A material handling planning chart can be used to gather information pertaining to a specific material handling problem and to provide a preliminary examination of the alternative solutions. The result from analyses using this chart can then be used to further refine solution strategies using methods such as the simulation of alternative solutions.

An illustration of the material handling planning chart is given in Figure 5.2. The first through eighth columns are completed in the same manner as the flow process chart. The flow process chart is an expansion of the operations process in that it includes information on operations, transportations, delays and storages and inspections. Delays are omitted in the material handling planning chart because a delay and storage results in the same facility requirements.

Each operation (O), transport (T), storage (S), and inspection (I) activity should be listed. A transport activity should be recorded before and after most operations, storage, and inspection. All movements, even if performed by a machine operator, and storages, even if only for a short time, should be recorded. Transports should be recorded as "From ___ To ___." These eight columns provide information on the *where* questions.

The ninth through twelfth columns answer the question "What is moved?" A final decision on what is moved usually cannot be made at this point. Unit loads need to been determined (e.g., the type of unit load and the unit load size may not be fully determined at this point). The space provided under method of handling should be utilized to represent alternative unit load designs.

The frequency of move is recorded in the thirteenth column. This number is calculated based on estimated production volumes and the capacity of the unit load. As a practical issue, the frequency of move cannot be determined with certainty. The uncertainty in production forecasts and real-time decisions on moving loads make estimating the frequency of move difficult.

The fourteenth column marks the length of the move. The distance moved cannot be determined prior to the completion of the facility layout design. Here we can see the interdependencies between facility layout design and the material handling system design. Several layouts must be determined from which the length of the move can be calculated. This information can then be used in developing alternative material handling system alternatives. The fifteenth column is where we state the material handling method with the appropriate material handling equipment.

We recommend the use of the material handling planning chart at the rough-cut stage of the material handling system design process. It is an essential part of the data gathering process, and it is here that opportunities for improvement can be seen since the data may show that there are many unnecessary transportations,

MATERIAL HANDLING PLANNING CHART

Company	A.RC. Inc.	Prepared by ____	Date ____	I.A. ____	Layout Alternative __1__ of ____	Sheet __1__ of __8__
Product	Air Speed Control Value					

Step No.	O	T	S	I	Description	Oper. No.	Dept.	Cont. Type	Size	Wt.	Qty. Per Cont.	Freq	Dist	Method of Handling
1			X		Bar stock in Storage (2200)		Stores.							
2		X			Profit Stores to Saw Dept.			LDDSE (FK.TRK)	2.5" × 3.5 × 16"	5 lb	to bars	3 times daily	16 ft	Fork lift
3			X		Store in Saw Department		Saw							
4	X				Cut to length	0101	Saw							
5		X			From Saw to Grinding			TOTE pan	15" × 12" × 7"	30 lb	30	Twice daily	10 ft	Platform hand truck
6			X		Store in Grinding		Grinding							
7	X				Grind to length	0201	Grinding							
8		X						TOTE pan	15" × 12" × 7"	30 lb	30	Twice daily	13 ft	Platform hand truck
9			X		Store in Deburring		Deburring							
10	X				Deburr	0301	Deburring							
11		X			From Deburring to Dr. Prs			TOTE pan	15" × 12" × 7"	30 lb	30	Twice daily	16 ft	Platform hand truck
12			X		Store in Drill Press		Drill Press							
13	X				Dr. CD holes tap. rean,dsk	0401	Drill Press							
14		X			From Dr. Press to Tur. Lathe			TOTE pan	15" × 12" × 7"	30 lb	30	Twice daily	33 ft	Platform hand truck

Figure 5.2 Material handling planning chart for an air flow regulator. Key: Operation—O. transportation—T. storage—S. inspection—I.

storages, and delays inherent in the present method. It also provides the means for generating multiple alterantives. Each of the alternatives generated can then be analyzed in more depth. One such study may involve building a detailed simulation model of each alternative in order to make statistical comparisons of the performance of each system being considered.

5.5   UNIT LOAD DESIGN

Definition of Unit Load Principle

Of the 10 principles of material handling, the **unit load principle** deserves special attention. Bright[1] [2] defines a unit load as

> a number of items, or bulk material, so arranged or restrained that the mass can be picked up and moved as a single object too large for manual handling, and which upon being released will retain its initial arrangement for subsequent movement. It is implied that single objects too large for manual handling are also regarded as unit loads.

Apple [2] makes the following comments on this definition.

> It can be seen that two major criteria are (1) a large number of units and (2) a size too large for manual handling. However, in the overall material handling and physical distribution activity, these two criteria as well as the balance of the above definition, and the "unit size principle" leave some room for misunderstanding, since it is obvious that either a "handful" or a "carload" both bear a relationship to the unit load concept.

Tanchoco[2,3] et al. [59, 57], recognizing the relationship that Apple [2] alluded to, defines a unit load as

> a single item, a number of items or bulk material which is arranged and restrained so that the load can be stored, and picked up and moved between two locations as a single mass.
>
> This definition does not restrict the method of movement to non-manual methods. The definition is restrictive, however, in the sense that a particular type of unit load is defined only for a single move between two locations. This suggests that the nature of the unit load could change each time an item, or a number of items, or bulk material is moved. The "unit" moved is thereby permitted to be a variable quantity and size per move. Philosophically, the authors feel it is important to permit such latitude in the definition of a unit load since, in a given sequence of moves and storages, it may indeed be more economical to handle different types of

---

[1] The historical reference is made to the term *unit load* since the term is still commonly used to mean a large object too heavy for manual handling.
[2] See Tanchoco, Agee, Davis, and Wysk [59].
[3] See Tanchoco [57].

unit loads through a sequence of moves than a single type of unit load through the same move sequence.

By this last definition, it is the "picked up and moved between two locations as a single mass" that defines the unit load. Thus, a single item picked up and moved manually between two locations constitutes one unit load. Two tote pans with identical components picked up moved by a dolly from one machine to another constitute one unit load. One pallet load of nonuniform-size cartons with different products picked up and moved by a lift truck from the packaging area to the shipping dock constitutes one unit load. One full load of products delivered by a truck-trailer from a warehouse to a customer store constitutes one unit load. If the trailer is half full, it is still one unit load. It is the move that defines the unit load.

The size of the unit load can range from a single part carried by a person, to each carton moved through a conveyor system, to a number of cartons on a pallet moved by fork lift trucks, to a number intermodal containers moved by rail across states or by container ships across continents.

The unit load size specification has a major impact on the specification and operation of the material handling system. Large unit loads may require bigger and heavier equipment, wider aisles, and higher floor load capacities. Also, large unit loads increase work-in-process inventory since items have to accumulate to full unit load size before the container or pallet is moved. A major advantage is fewer moves.

Small loads increase the transportation requirements but can potentially reduce work-in-process inventory. Small unit loads often require simple material handling methods such as push carts, and similar devices. Small loads support the concept of just-in-time production. Figure 5.3 illustrates the effect of the unit load size on job completion time. As observed in this figure, the completion time decreases as the unit load size decreases. The material handling time increases. But when the unit load size is 1 piece the completion time is longer. The material handling system is at capacity level and is now the constraining resource. The important conclusion here is that in order to achieve single unit production, the material handling time must be shorter than the unit processing time.

Two important elements in determining the size of the unit load are the "cube" limit and the weight limit (e.g., a single-wall corrugated carton with outside dimension $16'' \times 12'' \times 6''$ and a gross weight limit of 65 lbs). (In specifying box dimensions, the length and the width refer to the long and short side of the box opening. The depth is orthogonal to the length and width.)

The integrity of the unit load can be maintained in a variety of ways. For example, tote boxes, cartons, and pallets can be used to contain the unit load. Likewise, strapping, shrink wrapping, and stretch wrapping can be used to enclose the unit load. Specific considerations should be given on the manner by which the unit load is moved. Apple [2] lists four basic methods.

1.   Lifting under the mass
2.   Inserting the lifting element into the body of the unit load
3.   Squeezing the load between two lifting surfaces
4.   Suspending the load

At the heart of the unit load system design process is the dimensional relationships between the various forms the unit load takes. Figure 5.4 shows several stages in

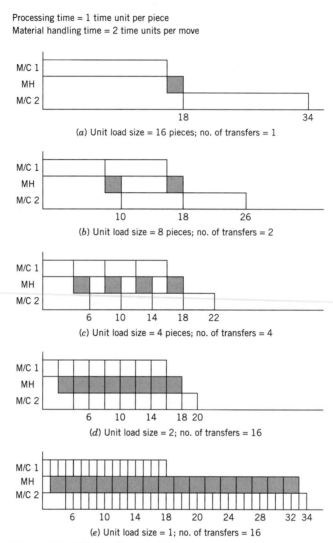

Processing time = 1 time unit per piece
Material handling time = 2 time units per move

(*a*) Unit load size = 16 pieces; no. of transfers = 1

(*b*) Unit load size = 8 pieces; no. of transfers = 2

(*c*) Unit load size = 4 pieces; no. of transfers = 4

(*d*) Unit load size = 2; no. of transfers = 16

(*e*) Unit load size = 1; no. of transfers = 16

Figure 5.3   Effects of unit load size on job completion times.

the material flow process where dimensional relationships play a major role. In this illustration, we assume that the cartons are stacked on pallets and the full pallet loads are loaded directly on trailers or are block-stacked in a warehouse before they are loaded on trailers for shipment to customers.

The use of returnable containers is of particular interest. Containers with good *stacking* and *nesting* features can provide significant reduction in material handling costs. *Stackability* means that a full container can be stacked on top of another full container in the same spatial orientation. Lids or tabs that are integrated in the design of the container are often used to support the container above. *Nestability* means that the shape of the containers permits an empty container to be inserted into another empty container. Figure 5.5 shows why these two features play key roles in moving and storing containers.

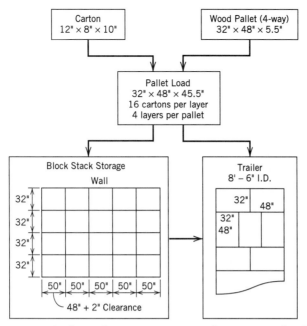

Figure 5.4 Dimensional relationships among various elements in a distribution system.

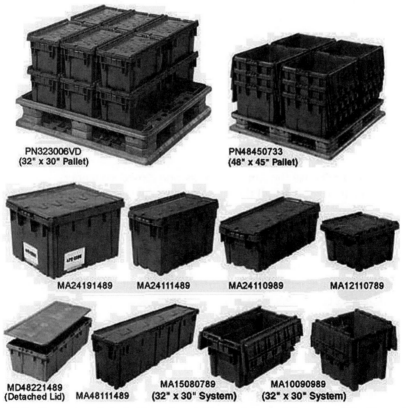

Figure 5.5 Examples of stackable and nestable containers. (*Courtesy of Buckhorn, Inc.*)

Efficiency of Returnable Containers

The discussion below illustrates the importance of selecting the right kind of returnable containers.

Given the following dimensions of a particular type of plastic reusable containers:

Inside dimensions $18'' \times 11'' \times 11''$
Outside dimensions $20'' \times 12'' \times 12''$
Each nested container $20'' \times 12'' \times 12''$

a trailer with inside dimensions of $240'' \times 120'' \times 120''$ is used to transport these containers. The containers are not palletized. Assume that no clearance is needed between containers, and between containers and the walls of the trailer.

Determine the following:

1.   Container space utilization
2.   Storage space efficiency
3.   Container nesting ratio
4.   Trailer space utilization if all containers are stacked vertically in only one orientation
5.   Trailer return ratio

Container space utilization is obtained by dividing the usable cube by the exterior envelope of the container. For this example, the container efficiency is

$$(18'' \times 11'' \times 11'')/(20'' \times 12'' \times 12'') = 0.76 \text{ or } 76\%$$

Storage space efficiency is the ratio of usable cube divided by the storage cube. If the dimension of the storage opening is $24'' \times 16'' \times 14''$, then the storage efficiency is

$$(18'' \times 11'' \times 11'')/(24'' \times 16'' \times 14'') = 0.45 \text{ or } 45\%$$

The *container nesting ratio* is determined by dividing the overall container height by the nested height, that is,

$$12''/2'' = 6; \text{ the ratio is } 6{:}1$$

Six nested containers use the same space as one closed container.

The container takes up all the space in the trailer with $240''/20'' = 12$ containers along the length of the trailer, $120''/12'' = 10$ containers along the width of the trailer, and $120''/12'' = 10$ containers stacked vertically. The total number of containers is $12 \times 10 \times 10 = 1200$. The *trailer space utilization* is

$$(18'' \times 11'' \times 111'')(1200)/(240'' \times 120'' \times 120'') = .76 \text{ or } 76\%$$

One stack of loaded containers has $120''/12'' = 10$ containers. One stack of empty containers has 55 containers, that is,

$$1 + (120'' - 12'')/2'' = 55$$

Thus, the total number of empty containers per trailer is

$$55 \times (240''/20'') \times (120''/12'') = 6600$$

The *trailer return ratio* is

$$6600/1200 = 5.5$$

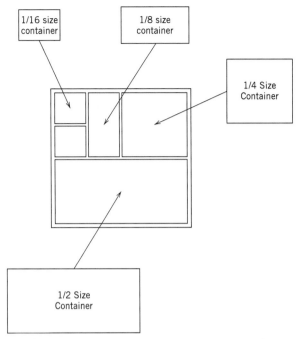

Figure 5.6   Container/pallet system with progressive dimensions.

The impact of trailer return ratio on the overall efficiency of the distribution function cannot be overlooked. Significant cost reductions may be achieved with higher trailer return ratios.

In selecting containers, size progression is one of the important considerations. Figure 5.6 shows a container system with progressive dimensions (e.g., a smaller container is half the size of the larger container). The progression here is $\frac{1}{1}, \frac{1}{2}, \frac{1}{4}, \frac{1}{8}, \frac{1}{16}$, and so on. The use of these types of containers allows the efficient utilization of the load deck of an automated guided vehicle as the vehicle picks up and delivers containers to stations along its route.

Another advantage of using progressive container systems is the simplification of the pallet loading of mixed-sized containers as demonstrated by Pennington and Tanchoco [38] in their study on robotic palletizing of multiple box sizes. Their conclusion is that vertical stacking is simplified, resulting in stacking patterns that can achieve 100% utilization.

Pallets and Pallet Sizes

The use of pallets is another common method of containing the unit load. Pallets can come in a variety of designs that are usually dictated by the application. Figure 5.7 illustrates several types of wood pallets. Among the common pallet sizes are the following:

32 in. × 40 in.	40 in. × 48 in.	48 in. × 40 in.
36 in. × 48 in.	42 in. × 42 in.	48 in. × 48 in.

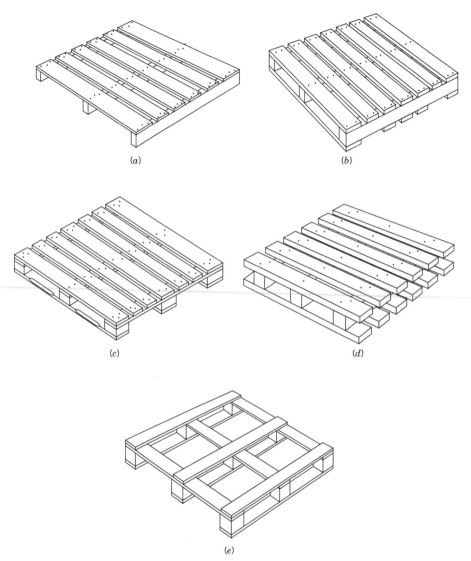

Figure 5.7   Types of wooden pallets.

The first dimension corresponds to the length of the stringer board and the second dimension corresponds to the length of the deckboard of the pallet. Figure 5.8 gives an illustration. Pallets may also be classified as *two-way,* where the fork entry can be only on two opposite sides of the pallet and is parallel to the stringer board, or *four-way,* where the fork entry can be on any side of the pallet.

Nonwooden-type pallets have also become more popular. Table 5.2 shows a comparison of various types of pallets.

The relationship between the container and the pallet, referred to as the *pallet loading problem,* is one that must be addressed explicitly. The objective in the palletloading problem is to maximize the use of space (i.e., the "pallet cube"). Theoretically, for each carton size one could determine from among all combinations

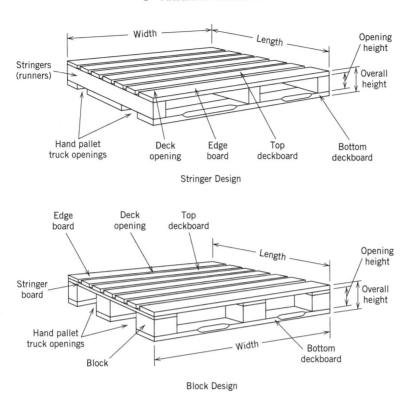

Figure 5.8   Common designs for wooden pallets. (*a*) Stringer design. (*b*) Block design.

the pallet size and the loading pattern that maximizes cube utilization. However, it is common practice to limit the number of alternative pallet sizes to two or three popular sizes. A similar "space utilization" problem exists in the loading of cargo on truck trailers, container ships, and cargo airplanes.

Cube utilization is not the only objective in the pallet loading problem. Quite often, load stability is an equally important consideration for reasons of uniform loading on material handling equipment and safety. Another measure is based on cost (e.g., cubic feet per dollar). Several loading patterns are shown in Figure 5.9.

Unit Load Interactions with Warehouse Components[4]

The purpose of the following discussion is to illustrate the relationship between the unit load size and configuration and certain other system factors, and to emphasize the detailed analysis required in the design process.

The specific system considered consists of the packaging, palletizing, storing, and shipping operations. These postfabrication and -assembly operations can be found in a broad variety of industries. The various interactions involving equipment and facility are examined in detail. A critical factor in these interactions is the

---

[4] The material presented in this section is from Tanchoco and Agee [58].

Table 5.2   *A Comparison of the Different Pallet Types*

Material	Base Weight	Durability	Repairability	Environmental Impact	Typical Applications
Wood	55–112 lb	Medium	High	Material is biodegradable and recyclable	Wide general use, including: grocery; automotive; durable goods; hardware
Pressed Wood Fiber	30–42 lb	Medium	Low	Material is recyclable and can be burned without leaving fuel residues	Bulk bags; order-picking; printing; building materials
Corrugated Fiberboard	8–12 lb	Low	Low	Material is biodegradable and recyclable	Export shipping; one-way shipping applications in grocery; lightweight paper products; industrial parts
Plastic	35–75 lb	High	Medium	Material is recyclable	Captive or closed loop systems; FDA, USDA applications; AS/RS; automotive
Metal	32–100 lb	High	Medium	Material is recyclable	Captive or closed loop systems; FDA, USDA applications; AS/RS; military; heavy equipment; aerospace

specification of the carton used and the pallet size. These two factors directly affect the selection of the material handling equipment and the physical configuration of the storage facility. Further, the utilization of both the warehouse and highway trailer is affected. The focus on unit load design is therefore warranted.

The specific operations included in this example are the following:

- Finished goods are packaged using closed-top cardboard cartons.
- The cartons are transported to a palletizer via a belt conveyor.
- The pallet loads are formed using a mechanized palletizer.
- The full pallet loads are then stored in the finished-goods warehouse using a powered lift truck. This lift truck is used exclusively for warehouse operations.
- Upon receipt of customer orders, full pallet loads are retrieved from the warehouse by a powered lift truck exclusively used for shipping dock operations.
- The retrieved pallet loads are then loaded on highway trailer trucks for delivery to customers.

Figure 5.10 shows a schematic diagram for the system described above.

To further qualify the system under consideration, only one product category is assumed to move through the system. The intent here is to isolate the interactions

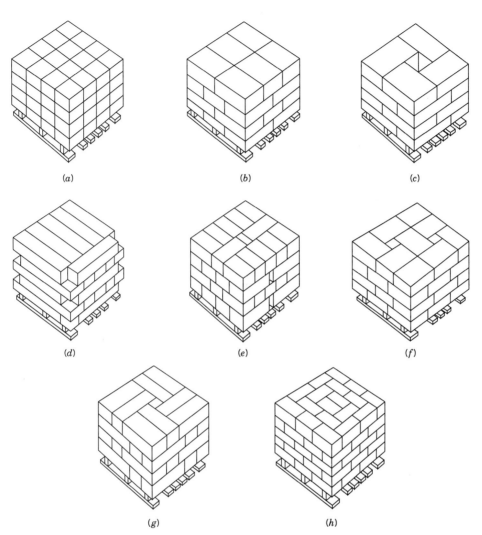

Figure 5.9 Stacking patterns for different pallet sizes. (*a*) Block pattern. (*b*) Row pattern. (*c*) Pinwheel pattern. (*d*) Honeycomb pattern. (*e*) Split-row pattern. (*f*) Split-pinwheel pattern. (*g*) Split-pinwheel pattern for narrow boxes. (*h*) Brick pattern. (*From [7] with permission.*)

among the numerous elements of the system, and to illustrate the importance of the unit load as it relates to these interactions.

The specification of the carton size is perhaps the most critical element in the design of the unit load system. The carton size selected dictates the number of parts contained in each carton and the total number of cartons that may be packaged and transported to the palletizer. Based on the parts flow rate to the packaging stations and the time required to package each carton, the carton flow rate to the palletizer is determined.

The next step is the formation of the pallet load through a palletizer. The type and size of the pallet must be specified and the best pallet loading pattern selected. The rate at which full pallet loads are formed is a function of the capacity of

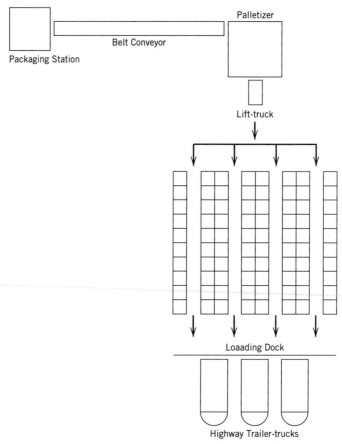

Figure 5.10   Schematic layout of a manufacturing subsystem of packaging, palletization, storage, and shipping.

the palletizer and the two factors previously described, namely the carton size and pallet size.

From the palletizer, the full pallet loads are stored in a warehouse with either a selective-rack or open-rack design. A powered lift truck is used to pick up loads from the palletizer and for the subsequent stacking of these loads in the warehouse as well as the retrieval of the loads from the storage to a pickup/delivery location in the shipping dock area. Hence, the type of material handling equipment specified will dictate the floor space requirement for the warehouse. For instance, the use of a narrow-aisle truck can significantly reduce the floor space requirement. However, the cost savings on building construction must offset the cost of the more expensive narrow-aisle truck. The question of how high the storage building should be is also relevant. For our example, only one building height (20 ft. clear height) is considered.

The next step is the loading of the pallet loads into a highway trailer-truck for delivery to the customers. The number of pallet loads delivered per truckload is constrained by the inside dimensions of the trailer-truck used as well as the dimensions and capability of the dock lift truck to maneuver the load inside the trailer. Here again, the interactions among the pallet loads, dock lift truck, and highway trailer-truck must be examined. Further tradeoffs must be considered between warehouse storage space utilization and the trailer-truck utilization. The utilization of the trailer-truck dictates the number of trips the vehicle must make to complete the delivery of a specified number of items.

The discussion above highlighted the many interactions possible in a simplified system. It is important to note that the objective in the process described is to deliver the parts to the customer and that the specifications of carton sizes and pallet sizes are merely means to this end. A numerical example is given to illustrate some of the interactions. The following alternatives are considered:

1. Carton size
   - 12″L × 10″W × 10″H
   - 10″L × 80″W × 8″H
2. Pallets
   - 40″L × 48″W (2-way)
   - 48″L × 40″W (2-way)
   - 36″L × 36″W (4-way)
3. Palletizer
   - Maximum load height of 70″
4. Lift truck (warehouse operations)
   - Counterbalance lift truck with simplex mast, rated at 3000 lb at 24-in. load center. The maximum fork height is 106 in.
   - Counterbalance lift truck with duplex mast, rated at 3000 lb at 24-in. load center. The maximum fork height is 130 in.
   - Narrow-aisle high-rise lift truck with lifting capacity to 20 ft with a 3000 lb load.
5. Warehouse storage
   - Single-deep pallet-rack design with 20 ft clear height
   - Block stacking design with 20 ft clear height
6. Lift truck (shipping dock operations)
   - Counterbalanced lift truck with simplex mast, rated at 2000 lb at 24-in. load center. The maximum fork height is 106 in.
7. Highway trailer-truck
   - 7′-6″W × 40′L × 11′H (inside dimensions)

There are a total of 36 system configurations possible based on the combinations of the seven components of the system listed above, that is, $(2)(3)(1)(3)(2)(1)(1) = 36$. An instance of a configuration is using the 10″L × 10″W × 8″H cartons, 48″L × 40″W pallets, the palletizer, a narrow-aisle high-rise lift truck, a pallet rack storage design for the warehouse, a counterbalanced lift truck with simplex mast in the shipping area, and the particular highway trailer-truck given.

For each of the 36 possible system configurations, the total area needed in the warehouse is obtained based on a maximum capacity of 500,000 parts. The following measures are used in comparing the 36 configurations:

- Warehouse area (floor space required)
- Warehouse cube utilization

- Highway trailer utilization
- Number of truckloads needed to deliver 500,000 parts

The procedure for obtaining these measures consists of the following steps:

1. Define system component specifications for
   a. Carton
   b. Pallet
   c. Palletizer
   d. Lift-truck (warehouse)
   e. Warehouse storage
      i. Clear height of the building
      ii. Pallet rack vs. block stacking
   f. Lift truck (shipping dock)
   g. Highway trailer-truck
2. Determine the number of carton layers per unit load based on weight and height limits:
   a. Unit load weight
   b. Unit load height
3. Determine the number of unit loads per bay (or stack) subject to
   a. Building height constraint
   b. Maximum height capacity of lift trucks
4. Calculate the area required and cube utilization of the warehouse.
5. Determine the total number of unit loads per trailer based on the following:
   a. Dimensional constraints
   b. Method of loading trailer
      i. One pallet per entry to a trailer
      ii. One pallet stack per entry to a trailer
6. Calculate the utilization of the highway trailer.
7. Repeat steps 1 through 6 for all combinations of system components.

The pallet loading pattern for each carton-pallet pair must be prescribed as shown in Figures 5.11 and 5.12

Table 5.3 shows the results for 18 combinations of system component specifications. The information contained in this table includes:

a. Number of layers per unit load
b. Number of unit loads per stack
c. Total number of unit loads required to handle 500,000 parts
d. Weight of each unit load
e. Height of each unit load
f. Total warehouse area required
g. Warehouse cube utilization
h. Highway trailer utilization
i. Number of truckloads required to deliver the 500,000 parts

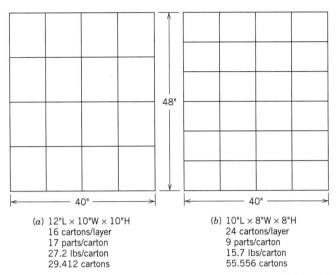

48"

40"

40"

(a) 12"L × 10"W × 10"H
16 cartons/layer
17 parts/carton
27.2 lbs/carton
29.412 cartons

(b) 10"L × 8"W × 8"H
24 cartons/layer
9 parts/carton
15.7 lbs/carton
55.556 cartons

Figure 5.11 Pallet patterns for alternative carton sizes on a 48″ × 40″ pallet.

The material handling system considered is obviously limited both in magnitude (number of parts handled) and scope (only one part type). However, the interactions among the many system components are clearly highlighted. For instance,

a. The combination with the highest warehouse space utilization, 46.1%, requiring the use of 12″L × 10″W × 10″H cartons, 40″L × 48″W pallets, a block-stacking storage design, and a narrow-aisle lift truck, results in 17 highway trailer truckloads.

b. The use of a 48″L × 40″W pallet reduces the warehouse utilization by 1.6% (to 44.5%) but the trailer-truck is utilized more fully at 60.6% (an increase of 23.6%) with the number of truckloads decreased to 11.

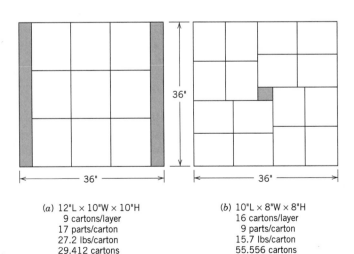

36"

36"

36"

(a) 12"L × 10"W × 10"H
9 cartons/layer
17 parts/carton
27.2 lbs/carton
29.412 cartons

(b) 10"L × 8"W × 8"H
16 cartons/layer
9 parts/carton
15.7 lbs/carton
55.556 cartons

Figure 5.12 Pallet patterns for alternative carton sizes on a 36″ × 36″ pallet.

The specification of more expensive narrow-aisle high-rise equipment in the warehouse is not appropriate for shipping dock operations. The use of less expensive equipment, that is, a counterbalanced lift truck with simplex mast, and 48"L × 40"W pallets results in warehouse and trailer utilizations of 18.8% and 48.5%, respectively. However, the same lift truck can be used in the shipping dock, adding flexibility to overall operations.

There are numerous other tradeoffs affecting the specification of the various components of the system. Some of these tradeoffs can be reduced in economic terms utilizing such measures as minimum investment cost, maximum incremental

Table 5.3   *Results for the 18 Combinations of System Component Specifications*
A. 12"L × 10"W × 10"H carton   17 parts/carton   27.2 lb/carton   29,412 cartons

		Selective-Rack Design			Open-Storage System		
		Simplex	Duplex	Narrow Aisle	Simplex	Duplex	Narrow Aisle
1. 40"L × 48"W	(a)	5	4	5	4	5	5
16/cartons/layer	(b)	2	3	3	3	3	4
	(c)	368	460	368	460	368	368
	(d)	2,201 lb.	1,766	2,201	1,766	2,201	2,201
	(e)	56"	46	56	46	56	56
	(f)	7,856 sq ft	6,575	3,369	5,781	4,618	2,216
	(g)	13.0%	15.5	30.3	17.7	22.1	46.1
	(h)	37.0%	29.6	37.0	29.6	37.0	37.0
	(i)	17	21	17	21	17	17
2. 48"L × 40"W	(a)	5	4	4	4	5	5
16 cartons/layer	(b)	2	3	4	3	3	4
	(c)	368	460	460	460	368	368
	(d)	2,201 lb	1,766	1,766	1,766	2,201	2,201
	(e)	56	46	46	46	56	56
	(f)	7,539 sq ft	6,310	3,334	5,426	4,334	2,293
	(g)	13.5%	16.2	30.6	18.8	23.6	44.5
	(h)	60.6%	48.5	48.5	48.5	60.6	60.6
	(i)	11	13	13	13	11	11
3. 36"L × 36"W	(a)	5	4	3	4	5	5
9 cartons/layer	(b)	2	3	4	3	3	4
	(c)	654	817	1090	817	654	654
	(d)	1,244 lb	999	754	999	1,244	1,244
	(e)	56	46	36	46	56	56
	(f)	10,446 sq ft	8,721	5,581	7,394	5,904	2,842
	(g)	9.8%	11.7	18.3	13.8	17.3	35.9
	(h)	45.5%	36.4	40.9	36.4	45.5	45.5
	(i)	14	18	16	18	14	14

(a) No. of layers/unit load.   (f) Total area required.
(b) No. of unit loads/stack.   (g) Warehouse cube utilization.
(c) Total no. of unit loads.   (h) Highway trailer utilization.
(d) Wt. of one unit load.   (i) No. of truckloads required to deliver 500,000 parts
(e) Ht. of one unit load.

Table 5.3 *continued*
B. 10″L × 8″W × 8″H   9 parts/carton   15.7 lbs./carton   55,556 cartons

		Selective-Rack Design			Open-Storage System		
		Simplex	Duplex	Narrow Aisle	Simplex	Duplex	Narrow Aisle
1. 40″L × 48″W	(a)	6	5	6	5	6	6
24 cartons/layer	(b)	2	3	3	3	3	4
	(c)	386	463	386	463	386	386
	(d)	2289	1911	2289	1911	2289	2289
	(e)	54	46	54	46	54	54
	(f)	8240	6618	3533	5819	4843	2336
	(g)	12.5	15.5	29.1	17.7	21.2	44.0
	(h)	35.6	29.6	35.6	29.6	35.6	35.6
	(i)	18	22	18	22	18	18
2. 48″L × 40″W	(a)	6	5	5	5	6	6
24 cartons/layer	(b)	2	3	4	3	3	4
	(c)	386	463	463	463	386	386
	(d)	2289	1911	1911	1911	2289	2289
	(e)	54	46	46	46	54	54
	(f)	7908	6351	3363	5462	4546	2419
	(g)	13.0	16.2	30.6	18.8	22.6	42.5
	(h)	58.2	48.5	48.5	48.5	58.2	58.2
	(i)	11	13	13	13	11	11
3. 36″L × 36″W	(a)	6	5	6	5	6	6
16 cartons/layer	(b)	2	3	3	3	3	4
	(c)	579	695	579	695	579	579
	(d)	1529	1278	1529	1278	1529	1529
	(e)	54	46	54	46	54	54
	(f)	9264	7411	3946	6283	5227	2513
	(g)	11.1	13.9	26.1	16.4	19.7	40.9
	(h)	51.7	43.1	51.7	43.1	51.7	51.7
	(i)	13	15	13	15	13	13

savings, minimum total unit handling cost, payback period, return on investment, maximum space utilization, and so forth.

Characteristics of a particular system configuration, such as system flexibility, reliability, maintainability, ease of expansion, and other characteristics, may not be as easy to quantify. In the final decision, both quantitative and qualitative factors must be balanced to arrive at the best system specification.

Container and Pallet Pooling

Another alternative to consider in designing unit load systems is to participate in a container/pallet pooling system. Widely accepted in Europe, it is now being adapted by more North American companies. Instead of buying, these companies rent containers and pallets for a fee per day per container or pallet. Whenever you need them, you simply go the nearest depot and get as many as you need. After use, they are returned

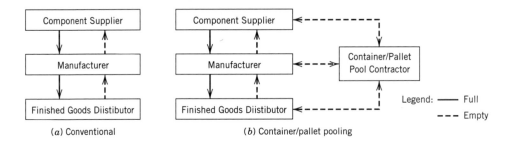

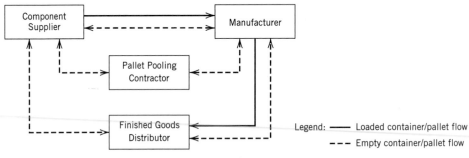

Figure 5.13 Container/Pallet flows in (*a*) conventional system, (*b*) pooling system, and (*c*) Integrated contract logistics and container/pallet pooling.

to the nearest depot or another company in the supply chain can assume possession and they, in turn, assume the daily rent. Figure 5.13 illustrates the container/pallet flow in (a) a conventional system, (b) a container/pallet pooling system, and (c) an integrated logistics and container/pallet pooling system.

The main advantage of a container and pallet pool system is that it minimizes the movement of empty pallets and utilization is increased. Also, there is no need for allocating extra space to store them. Since the operator of the pooling system owns the containers and pallets, they are responsible for maintaining them. Thus, the quality of the containers and pallets tends to be much better, resulting in less product damage and more efficient interfacing with material handling equipment.

An integrated logistics and container/pallet pooling system provides the next level of efficiency in moving products and containers/pallets throughout the entire production and distribution system. Under such a system, empty containers/pallets not requiring repair can be transferred as a regular load to another user facility on the same common carrier delivery.

Remarks

An interesting design question is "Should the material handling system be designed around the unit load or should the unit load system be designed to fit the material handling system?" The obvious answer is "Neither! The unit load system is an integral part of the material handling system and a *simultaneous* determination should be made." While *simultaneous* determination is ideal, it is more common in practice to make *sequential* decisions, particularly in designing large-scale material handling systems. As a result, it is found that either one designs the unit load first or one designs the handling and storage system first. Of these two, the former seems to be the preferred approach among material handling system designers. Many find it difficult to accept the fact that the size of the container and the pallet should be among the first determinations made, rather than among the last.

In cases where an existing material handling system is to be improved, the unit load specifications may be influenced by the physical configuration of an existing building. Door widths, column spacing, aisle widths, turning radii of material handling vehicles, maximum stacking heights of lift trucks, and clear building heights are among the many factors that will influence the design of the unit load system.

From our discussions in this section, it is clear that the unit load system design problem is one that must be done concurrently with the other elements of the overall material handling system design. One key element in this concurrent design process is the specification of progressivesize containers that fit standard pallets. It should be obvious that a reduction in the number of standard containers and pallets can have a significant impact on the overall design and operation of an integrated material handling system.

5.6   MATERIAL HANDLING EQUIPMENT

For many, material handling is synonymous with material handling equipment. That is unfortunate because there is much more to material handling than just equipment specification. This view is a very narrow perspective that is contrary to what we advocate in this text, and that is to view material handling from a systems perspective. There may even be a situation where a specific task within the overall material handling system solution will not require any equipment at all. We will continually emphasize that the focus should be first on the material, second on the move, and third on the method. It is quite easy to encounter a problem situation and immediately think of equipment solutions rather than material handling system solutions. Equipment specification is one of the last steps in the process of determining the preferred material handling system.

Nevertheless, knowledge of equipment alternatives is an essential tool needed by material handling system designer in coming up with alternative designs. New generations of equipment are continually being developed and anyone involved with material handling equipment specification must constantly keep abreast of the most current technologies available.

We classify material handling equipment into the following categories:

1.   Containers and Unitizing Equipment
   - Containers
   - Unitizers

2. Material Transport Equipment
    - Conveyors
    - Industrial Vehicles
    - Monorails, Hoists, and Cranes
3. Storage and Retrieval Equipment
    - Unit Load Storage and Retrieval
        - Unit Load Storage Equipment
        - Unit Load Retrieval Equipment
    - Small Load Storage and Retrieval
4. Automatic Data Collection and Communication Equipment
    - Automatic Identification and Recognition
    - Automatic Paperless Communication

A listing of material handling equipment is given below while a more detailed description of each equipment type is given in Appendix 5.B. Pictures are provided where appropriate. This section is not intended to provide an exhaustive description of material handling equipment. It is intended to be a description of the funcitons, applications, and benefits of the major classifications of material handling equipment.

Material handling cost estimation and safety considerations are discussed in separate sections.

Material Handling Equipment Classifications

I. Containers and Unitizing Equipment
    A. Containers
        1. Pallets
        2. Skids and Skid Boxes
        3. Tote Pans
    B. Unitizers
        1. Stretchwrap
        2. Palletizers
II. Material Transport Equipment
    A. Conveyors
        1. Chute Conveyor
        2. Belt Conveyor
            a. Flat Belt Conveyor
            b. Telescoping Belt Conveyor
            c. Troughed Belt Conveyor
            d. Magnetic Belt Conveyor
        3. Roller Conveyor
        4. Wheel Conveyor

5. Slat Conveyor
6. Chain Conveyor
7. Tow Line Conveyor
8. Trolley Conveyor
9. Power and Free Conveyor
10. Cart-on-Track Conveyor
11. Sorting Conveyor
    a. Deflector
    b. Push Diverter
    c. Rake Puller
    d. Moving Slat Conveyor
    e. Pop-Up Skewed Wheels
    f. Pop-Up Belts and Chains
    g. Pop-Up Rollers
    h. Tilting Slat Conveyor
    i. Tilt Tray Sorter
    j. Cross Belt Sorter
    k. Bombardier Sorter

B. Industrial Vehicles
  1. Walking
    a. Hand Truck and Hand Cart
    b. Pallet Jack
    c. Walkie Stacker
  2. Riding
    a. Pallet Truck
    b. Platform Truck
    c. Tractor Trailer
    d. Counterbalanced Lift Truck
    e. Straddle Carrier
    f. Mobile Yard Crane
  3. Automated
    a. Automated Guided Vehicles
      i. Unit Load Carrier
      ii. Small Load Carrier
      iii. Towing Vehicle
      iv. Assembly Vehicle
      v. Storage/Retrieval Vehicle
    b. Automated Electrified Monorail
    c. Sorting Transfer Vehicle

C. Monorails, Hoists, and Cranes
  1. Monorail

    2. Hoist

    3. Cranes

       a. Jib Crane

       b. Bridge Crane

       c. Gantry Crane

       d. Tower Crane

       e. Stacker Crane

III.   Storage and Retrieval Equipment

    A. Unit Load Storage and Retrieval

       1. Unit Load Storage Equipment

          a. Block Stacking

          b. Pallet Stacking Frame

          c. Single-Deep Selective Rack

          d. Double-Deep Rack

          e. Drive-In Rack

          f. Drive-Thru Rack

          g. Pallet Flow Rack

          h. Push-Back Rack

          i. Mobile Rack

          j. Cantilever Rack

       2. Unit Load Retrieval Equipment

          a. Walkie Stacker

          b. Counterbalance Lift Truck

          c. Narrow Aisle Vehicles

             i. Straddle Truck

             ii. Straddle Reach Truck

             iii. Sideloader Truck

             iv. Turret Truck

             v. Hybrid Truck

          d. Automated Storage/Retrieval Machines

    B. Small Load Storage and Retrieval Equipment

       1. Operator-to-Stock—Storage Equipment

          a. Bin Shelving

          b. Modular Storage Drawers in Cabinets

          c. Carton Flow Rack

          d. Mezzanine

          e. Mobile Storage

       2. Operator-to-Stock—Retrieval Equipment

          a. Picking Cart

          b. Order Picker Truck

          c. Person-aboard Automated Storage/Retrieval Machine

3. Stock-to-Operator Equipment
   a. Carousels
      i. Horizontal Carousel
      ii. Vertical Carousel
      iii. Independent Rotating Rack
   b. Miniload Automated Storage and Retrieval Machine
   c. Vertical Lift Module
   d. Automatic Dispenser

IV. Automatic Identification and Communication Equipment
   A. Automatic Identification and Recognition
      1. Bar Coding
         a. Bar Codes
         b. Bar Code Readers
      2. Optical Character Recognition
      3. Radio Frequency Tag
      4. Magnetic Stripe
      5. Machine Vision
   B. Automatic, Paperless Communication
      1. Radio Frequency Data Terminal
      2. Voice Headset
      3. Light and Computer Aids
      4. Smart Card

Remarks

A variety of material handling equipment is presented in this section and in Appendix 5.B to provide an introduction to the breadth of alternatives available and to emphasize the importance of staying abreast of equipment developments. Just as in the case of methods design, there is no one best way of performing material handling. In fact, there probably exist conditions for which each type of equipment presented is a preferred method of handling. Unfortunately, too many individuals who "design" material handling systems are guilty of using the same solution for many different handling situations. The adage of "slotting the head of a nail in order to use a favorite screwdriver" applies to many material handling designs.

The equipment selection problem, in theory, is simple; namely, reduce the set of alternative approaches to those that are feasible based on the material and move. Unfortunately, in practice there generally exist numerous feasible alternatives as well as an abundant number of manufacturers for each equipment type. Consequently, the material handling system designer must rely on judgment and experience in trimming the set of feasible alternatives to a manageable number. Undoubtedly, the optimal design alternative may be eliminated in the process. However, it is generally the case that no single alternative is superior to all other alternatives by a wide margin. Rather, several good solutions will generally exist and one of them will be selected. The material handling equipment selection process is truly a "satisficing" rather than an "optimizing" process.

Our treatment of material handling equipment is not exhaustive. For example, people-mover systems, front-end loaders, over-the-road trucks, trash and waste disposal equipment, battery chargers, package sealers, labeling machines, elevators, guidance systems, bowl feeders, dock levelers, and conveyor transfers are not described. Additionally, not all types of conveyors, packaging equipment, industrial truck attachments, below-hook lefters, hoists, cranes, monorails, automated storage and retrieval systems, industrial trucks, and containers and supports are treated. Consequently, anyone involved in designing material handling equipment must keep abreast of new developments in the field.

5.7 ESTIMATING MATERIAL HANDLING COSTS

The development of material handling design alternatives covers not just the specification of the "right method of handling." Equally important is that the recommended alternative is at the "right cost." Estimation of the cost of material handling alternatives is not a trivial task. At one end of the spectrum is a "rough-cut" method through the use of standard data and rules of thumb (see Shemesh [51]). As an example of the use of rule-of-thumb data, one can get a reasonably accurate estimate of the purchase cost of specifying several walkie pallet jacks by using the unit cost of each walkie pallet jack. The use of rules of thumb must be approached with caution. The purchase price of material handling equipment can vary significantly from company-to-company, model-to-model, and year-to-year data, and the values contained in such listings can be easily outdated. It is therefore critical that providers of rule-of-thumb data continually update the cost values in their listings. Nevertheless, it is the responsibility of the material handling engineer to verify the accuracy of the information used in any material handling project.

Gerace [21] makes the following comments on the use of rules of thumb.

> A common mistake made by even the most experienced industrial engineers is to fall into the trap of using "rules of thumb" to generate the estimated cost for planned material handling equipment. The deployment of today's material handling equipment is a much more complex undertaking than it was in the past. Rules of thumb, such as $250 per linear foot of installed powered roller conveyor, may have worked in the past, but today will likely create an inaccurate estimate of the true deployment cost. Here is just a short list of the costing factors that affect today's material handling equipment estimating that we didn't have to consider just a few years ago:
>
> • Material handling solutions are no longer a simple case of connecting two locations physically, with sticks of transport conveyor. Almost all solutions today are systems, with logical components for capturing electronic information and making dynamic routing decisions. As a result, solution cost estimates must include controls system hardware, software and integration costs.
>
> • New material handling technologies have dramatically changed the amount of labor required for installation and facility fit-up. A conventional type conveyor may still require approximately 20% of its purchase price to install it, while

newer modular conveyor my only require 10% for installation. In addition, traditional electrical power distribution can be significantly reduced with the application of these new devices as well.

- Material handling equipment is now a global enterprise and as such, depending on global economic pressures, equipment prices can fluctuate by plus or minus 30% each year.

- The equipment marketplace is full of "look-alike" equipment suppliers. Much like the auto industry, we can now purchase two different devices, which perform the same basic functions, where one device is less than one-half the cost of the other. End user technology preference, noise, safety, energy efficiency, modularity, product handling flexibility, reliability and maintainability now drive the choices as to which device is right for any given application.

The correct approach to estimating the deployment cost of a material handling solution is to use a complex pricing model, which includes as many of the pertinent pricing factors as possible. Most experienced systems integrators, who deploy more than a handful of systems each year, would have such a model. The accuracy of these estimating models is directly related to the degree of detail reflected in the elements of the model. The key to maintaining the accuracy of these complex modules is to have a validation process wherein "real and actual data" is continuously factored back into the model, keeping it as current as practicality will allow.

At the other end of the material handling cost estimation spectrum is the use of detailed cost estimation models based on engineering information. An illustration of this approach for estimating conveyor cost can be found in Tanchoco, et al. [59]. A conveyor cost model was presented based on different container sizes and alternative conveyor components such as (1) varying conveyor widths based on the dimensional requirements of a container, (2) the use of slider beds versus roller beds, (3) different types of belt material, (4) different drive-end configurations, and (5) alternative motor sizes based on calculated horsepower requirements. The model presented shows that even for a simple task of moving parts using a conveyor, there exist many possible conveyor equipment component configurations.

For the design of more complex material handling systems, it is inevitable that a more systematic approach be used in order to verify the interactions among the various components of the material handling system design. Approaches such as simulation analysis are needed to perform these verifications. The use of simulation analysis can reveal some internal conflicts among material handling components that can result in blocking delays and deadlock conditions. Finally, we stress that the role of material handling is to facilitate production. Thus, the best material handling system is one that results in *optimal* production operations.

5.8   SAFETY CONSIDERATIONS[5]

Safety should not be an afterthought when designing a material handling solution or any part of a facility. By engineering safety into a design, reliance on process controls or personal protective equipment can be avoided. Many material handling

---

[5] This section is contributed by Albright [1].

Table 5.4 *Recommended Aisle Widths for Facility Design*

Equipment Type	Pick Aisle	Cross Aisle
3-wheel counterbalance	9'–10'	10'
4-wheel counterbalance	10'–12'	12'
Reach truck	8' 6"	10'
Double-deep reach	8' 6"	10'
Order picker truck	5'	10'
Turret truck	5'	12'
Swing-mast truck	5'–6'	12'
Side loader	6'	15'–20'
Fixed-mast truck	5'	20'
Counterbalance w/attachment	12'	14'–20'
Manual pallet jack	6'	8'–10'
Powered pallet jack	7'–8'	8'–10'

equipment suppliers are reputable and provide OSHA-compliant equipment; but having "safe" equipment does not ensure a "safe" work environment. The key to a safe facility is concentrating on the interface between the workforce and the equipment.

Many material handling solutions involve the use of some sort of pallet racking and industrial lift trucks. A common shortfall in these solutions is poor layout of the rack area. Focusing on space efficiency, many planners provide insufficient aisle widths for the type of vehicles being used. As well trained or as safety conscious as the workforce may be, aisles that are too narrow will result in damaged uprights, damaged vehicles, and damaged workers. Table 5.4 shows recommended aisle widths for facility design.

One interface between the workforce and equipment that needs particular attention is the use of aisles by both pedestrians and industrial lift trucks. Each year, there are thousands of injuries as a result of pedestrian and vehicle interactions. In the ideal facility, pedestrian aisles and vehicle travel aisles would be totally separate. Commonly, however, this is not the case. One way to minimize the frequency of interaction is to keep office areas or employee common areas along the perimeter of the building, away from storage or process areas. Entrances into travel aisles should be kept in open areas, away from equipment or building structures that can create "blind spots" for both pedestrians and vehicle operators. Place stanchions or barriers at entrances, so that pedestrians must consciously stop, look, and go around them. If common aisles are to be used, make them an extra three feet wide and mark off a pedestrian walkway either along the left- or right-hand side. Dock areas are especially dangerous, given the high number of vehicles typically in use. Provide lounge areas for incoming and outbound truck drivers, and restrict access to the dock area so that vehicle operators can do their jobs without unnecessary distraction from wandering pedestrians.

Other material handling solutions involve the use of conveyors. Here again, keeping conveyors and industrial lift trucks separate is the best approach. Suspend conveyors from the roof trusses or put them on a mezzanine. If conveyors must be mounted to the floor near travel aisles, use bollards or rails to guard the conveyor supports.

Whether it be a conveyor induction point, a packing bench, or a production process, design of workstations is especially important in the prevention of injuries.

Poor ergonomics, that, is forcing a worker into the job instead of fitting the job to the worker, account for one-third of all workplace injuries. Risk factors to avoid include forcing operators into awkward body postures, repetition, force, contact stress, and vibration. Ideally, workstations should be adjustable to allow for work to be done between the shoulders and the knees, and no further than 18″–24″ away from the worker's body. Requirements to bend, kneel, or squat should also be avoided.

The Americans with Disabilities Act (ADA) imposes specific obligations on employers relative to employment of individuals with disabilities. The ADA requires employers to make reasonable accommodations in regard to which jobs qualified individuals with a disability may perform. A qualified individual with a disability is a person who meets legitimate skill, education, or other requirements and who can perform the essential functions of a position with or without reasonable accommodation. The extent to which an employer must accommodate a "qualified individual" is governed by the ADA. Employers, and their facilities designers, must make judgments based on reliable or objective evidence, if accommodations must be made. Additional information can be found at the U.S. Department of Labor Web site (www.dol.gov). From a safety perspective, no existing or proposed OSHA regulations govern the employment of individuals with disabilities. OSHA's guidelines are to strive for working conditions that will safeguard the safety and health of all workers, including those with special needs and limitations. If employees can perform their job functions in a manner that does not pose a safety hazard to themselves or others, the fact they have a disability is irrelevant.

Facility designers should be familiar with all of the general safety requirements for the type of facility they are designing. OSHA regulations for General Industry (29CFR1910) provide a basis and represent the minimum standards. The regulations are available at the OSHA Web site (www.osha.gov).

5.9 SUMMARY

In this chapter the subject of material handling was introduced. Following a definition of material handling, the principles of material handling were introduced. The use of checklists was discussed. Next, a systematic approach for designing material handling systems was presented. The development of alternatives and solutions were discussed. The unit load principle was covered next. The unit load principle is, in our view, one of the most important elements of a material handling system. In our discussion, we covered the efficiency of returnable containers, standardization of containers and pallets, and the interactions of the unit load with various components of a warehouse and distribution facility. The role of container and pallet pooling systems was introduced. We expect that container and pallet pooling systems will continue to attract the attention of users in their quest for efficiency and cost reduction. A discussion on material handling equipment followed. Finally, the issues of material handling cost estimation and safety were addressed. Throughout this chapter, we emphasized the importance of focusing on the strategic and operational objectives associated with designing material handling systems.

BIBLIOGRAPHY

1. Albright, T., "Safety Considerations in Material Handling Systems Design," Tompkins Associates, Inc., NC, 2001.
2. Apple, J. M., *Material Handling Systems Design,* Ronald Press, New York, 1972.
3. Apple, J. M., *Lesson Guide Outline on Material Handling Education,* Material Handling Institute, Pittsburgh, 1975.
4. Apple, J. M., *Plant Layout and Material Handling, Ronald Press,* 3rd ed., New York, 1977.
5. Apple, J. M. Jr., and Rickles, H. V., "Material Handling and Storage," *Production Handbook,* 4th ed., J. A. White (ed.), Wiley, New York, 1987, pp. 5.15–5.37.
6. Barger, B., "Materials Handling Equipment," *Production Handbook,* 4th ed., J. A. White (ed.), Wiley, New York, 1987, pp. 4.100–4.127.
7. Bolz, H. A., and Hagernan, G. E. (eds.), *Material Handling Handbook,* Ronald Press, New York, 1958.
8. Bonthuis, J. E., "Basic Pointers on Selecting Conveyor Transfers," *Plant Engineering,* vol. 29, no. 15, July 24, 1975, pp. 73–75.
9. Bowlen, R. W., "ABC's of Optical Scanning," *Industrial Engineering,* vol. 5, no. 10, October 1973, pp. 14–22.
10. Chu, H. K., Egbelu, P. J., and Wu, C. T., "Advisor: A Computer-Aided Materials Handling Equipment Selection System," *International Journal of Production Research,* vol. 33, 1995, pp. 3311–3329.
11. Co, C. G., and Tanchoco, J. M. A., "A Review of Research on AGVS Vehicle Management," *Engineering Costs and Production Economics,* vol. 21, no. 1, 1991, pp. 35–42.
12. Egbelu, P. J., "Batch Production with Unit Load Design and Scheduling Considerations," in *Progress in Material Handling and Logistics,* J. A. White and I. W. Pence (ed.), vol. 2, Springer-Verlag, Berlin, 1990.
13. Egbelu, P. J., and Tanchoco, J. M. A. "Characterization of Automatic Guided Vehicle Dispatching Rules," *International Journal of Production Research,* vol. 22, no. 3, 1984, pp. 359–374.
14. Frazelle, E. H., "Material Handling: A Technology for Industrial Competitiveness," *Material Handling Research Center Technical Report,* Georgia Institute of Technology, Atlanta, April 1986.
15. Frazelle, E. H., *Material Handling Systems and Terminology,* Lionhart Publishing, Atlanta, 1992.
16. Frazelle, E. H., "Small Parts Order Picking: Equipment and Strategy," *Material Handling Research Center Technical Report Number 01-88-01,* Georgia Institute of Technology, Atlanta, 30332-0205.
17. Frazelle, E. H., and Apple, J. M. Jr., "Material Handling Technologies," *Logistics Handbook,* W. A. Copacino (ed.), Macmillan, New York, 1993.
18. Frazelle, E. H., and Apple, J. M. Jr., "Warehouse Operations," *Distribution Management Handbook,* J. A. Tompkins (ed.), McGraw-Hill, New York, 1993.
19. Frazelle, E. H., and Ward, R. E., "Material Handling Technologies in Japan," *National Technical Information Service Report #PB93-128197, Washington,* D.C., 1992.
20. Gaskins, R. J., and Tanchoco, J. M. A., "Flow Path Design for Automated Guided Vehicle Systems," *International Journal of Production Research,* vol. 25, no. 5, 1987, pp. 667–676.
21. Gerace, T., "The Dangers of Using Rules of Thumb for Creating Material Handling Cost Estimates," Tompkins Associates, Inc., NC, 2001.
22. Hassan, M. M. D., Hogg, G. L., and Smith, D. R., "A Construction Algorithm for the Selection and Assignment of Materials Handling Equipment," *International Journal of Production Research,* vol 23, 1985, pp. 381–392.

23. Hill, J. M., "Automatic Identification Systems," *Production Handbook,* 4th ed., J. A. White (ed.), Wiley, New York, 1987, pp. 4.128–4.142.

24. Hill, J. M., "Automatic Identification Perspective 1992," *Proceedings of the Material Handling Short Course,* Georgia Institute of Technology, Atlanta, 1992.

25. Hill, J. M., "Automatic Identification from A to Z," *Proceedings of the Material Handling Short Course,* Georgia Institute of Technology, Atlanta, 1993.

26. Horrey, R. J., "Sortation Systems: From Push to High-Speed Fully Automated Applications," *International Conference on Automation in Warehousing Proceedings,* Institute of Industrial Engineers, Atlanta, 1983, pp. 77–83.

27. Kaspi, M. and Tanchoco, J. M. A., "Optimal Flow Path Design of Unidirectional AGV Systems," *International Journal of Production Research,* vol. 28(6), 1023–1030.

28. Kulwiec, R. A., "A Selection and Application Overview: Package and Unit Handling Conveyors," *Plant Engineering,* vol. 28, no. 19, September 19, 1974, pp. 122–128.

29. Kulwiec, R. A., "Material Handling Equipment Guide," *Plant Engineering,* August 21, 1980, pp. 88–99.

30. Kulwiec, R. A. (ed.), *Materials Handling Handbook,* Wiley, New York, 1985.

31. Matson, J. O., Swaminathan, S. R., and J. M. Mellichamp, "Knowledge-Based Materials Handling Equipment Selection," *Proceedings of the 1990 International Industrial Engineering Conference,* Norcross, GA, 1990, pp. 212–227.

32. Miller, J. S., "Some Practical Pointers on Pallets and Containers for Automated Conveyors," *Plant Engineering,* vol. 30, no. 24, pp. 88–90, November 25, 1976.

33. Muther, R., *Systematic Layout Planning,* CBI Publishing Co., Boston, MA, 1973.

34. Nadler, G., "What Systems Really Are," *Modern Materials Handling,* vol. 20, no. 7, July 1965, pp. 41–47.

35. Noble, J. S., and Tanchoco, J. M. A., "Design Justification of Material Handling Systems," in Tanchoco, J. M. A. (ed.), *Material Flow Systems in Manufacturing,* Chapman and Hall, London, U.K., 1994, pp. 54–72.

36. Noble, J. S., and Tanchoco, J. M. A., "Marginal Analysis Guided Design Justification: A Material Handling Example," *International Journal of Production Research,* vol 33, no. 12, 1995, pp. 3439–3454.

37. Noble, J. S., Klein, C. M., and Midha, A., "An Integrated Model of the Material Handling and Unit Load Design Problem," *Journal of Manufacturing Science and Engineering,* vol. 120, 1998, pp. 802–806.

38. Pennington, R. A., and Tanchoco, J. M. A., "Robotic Palletization of Multiple Box Sizes," *International Journal of Production Research,* vol. 26, no. 1, 1988, pp. 95–106.

39. Rembold, B. F. and Tanchoco, J. M. A., "A Modular Framework for the Design of Material Flow Systems," *International Journal of Production Research,* vol. 32, no. 1, 1994, pp. 1–21.

40. Rembold, B. F., and Tanchoco, J. M. A., "Selecting and Sequencing Design Tools in Developing Material Flow Systems Models," International Journal of Production Research, vol. 32, no. 2, 1994, pp. 243–262.

41. Rembold, B. F., and Tanchoco, J. M. A., "Material Flow Model Evaluation and Improvement," *International Journal of Production Research,* vol. 32, no. 11, 1994, pp. 2585–2602.

42. Schultz, G. A., "Selection and Application Guidelines for Belt Conveyors for Unit and Package Loads," *Plant Engineering,* vol. 29, no. 16, August 7, 1975, pp. 71–74.

43. Schultz, G. A., "Selection and Application Guidelines for Belt Conveyors for Semi-bulk Loads," *Plant Engineering,* vol. 29, no. 17, August 21, 1975, pp. 85–88.

44. Schultz, G. A., "Selection and Application Guidelines for Belt Conveyors for Bulk Loads," *Plant Engineering,* vol. 29, no. 18, September 4, 1975, pp. 95–98.

45. Schultz, G. A., "Basic Specification and Application Guidelines for Chain Conveyors and Elevators," *Plant Engineering,* vol. 29, no. 18, September 5, 1975, pp. 95–98.

46. Schultz, G. A., "Chain Conveyors for Unit and Package Handling: A Survey of Equipment Types and Applications," *Plant Engineering,* vol. 30, no. 3, February 5, 1976, pp. 54–57.

47. Schultz, G. A., "Basic Selection and Application Factors for Overhead Trolley Conveyors," *Plant Engineering,* vol. 30, no. 4, February 19, 1976, pp. 117–120.

48. Schultz, G. A., Guidelines for Selecting Chain Conveyors and Elevators for Bulk Materials," *Plant Engineering,* vol. 30, no. 8, April 15, 1976, pp. 73–76.

49. Schultz, G. A., "Factors to Keep in Mind When Selecting Vibrating Conveyors," *Plant Engineering,* vol. 30, no. 17, August 19, 1976, pp. 105–108.

50. Schultz, G. A., "A Practical Procedure for Planning and Budgeting a Package Conveyor System," *Plant Engineering,* vol. 30, no. 18, September 2, 1976, pp. 93–95.

51. Shemesh, L. D., et al., "Gross & Associates Rules of Thumb for Warehousing and Distribution Costs Online," a Gross & Associates publication, Woodbridge, NJ, 2001.

52. Sims, E. R. Jr., "Narrow-Aisle, High-Rise Lift Trucks," *Plant Engineering,* vol. 31, no. 21, October 13, 1977, pp. 186–191.

53. Sinriech, D., Tanchoco, J. M. A. and Herer, Y. T., "The Segmented Bidirectional Single-Loop Topology for Material Flow Systems," *IIE Transactions,* vol. 28, no. 1, 1996, pp. 40–54.

54. Sinriech, D. and Tanchoco, J. M. A., "Solution Procedures and Implementation of the Segmented Flow Topology (SFT) for Discrete Material Flow Systems," *IIE Transactions,* vol. 29, no. 4, 1997, pp. 323–336.

55. Suzuki, J., "Guide to the Installation of Automated Sorters," *Proceedings of the 10th International Conference on Automation in Warehousing,* Institute of Industrial Engineers, Atlanta, October, 1989.

56. Tanchoco, J. M. A. (ed.), *Material Flow Systems in Manufacturing,* Chapman and Hall, London, U.K., 1994.

57. Tanchoco, J. M. A., "Unit Load Design: A Key Element in Integrating Material Movement and Storage," *MHI Advanced Institute for Material Handling Teachers,* Auburn University, Auburn, AL, June 1983, pp. 45–52.

58. Tanchoco, J. M. A. and Agee, M. H., "Plan Unit Loads to Interact with All Components of Warehouse System," *Industrial Engineering,* June 1981, p. 36.

59. Tanchoco, J. M. A., Agee, M. H., Davis, R. P., and R. A. Wysk, "An Analysis of the Interactions between Unit Loads, Handling Equipment, Storage and Shipping," *Unit and Bulk Materials Handling,* ASME Publication, F.J. Loeffler and C.R. Proctor (eds.), pp. 185–190.

60. Tanchoco, J. M. A., Hendricks, G., and Muljono, A., "Tutorial on Automated Guided Vehicle Systems," Unpublished Report, School of Industrial Engineering, Purdue University, West Lafayette, IN, 1995.

61. Tompkins, J. A., "Material Handling Systems," *Encyclopedia of Computers and Technology,* O. Belzer, A. G. Holzman, and A. Kent (eds.), vol. 10, Marcel Dekker, New York, 1978, pp. 254–273.

62. White, J. A., *Yale Management Guide to Productivity,* Yale Industrial Truck Division, Eaton Corp., Philadelphia, 1978.

63. White, J. A., "Determining How Much to Handle," *Modern Materials Handling,* vol. 46, no. 2, February 1991, p. 31.

64. White, J. A., "Providing the Right Materials," *Modern Materials Handling,* vol. 46, no. 3, March 1991, p. 27.

65. White, J. A., "Handle in the Right Condition," *Modern Materials Handling,* vol. 46, no. 5, April 1991, p. 27.

66. White, J. A., "The Right Stuff-in Sequence," *Modern Materials Handling,* vol. 46, no. 6, May 1991, p. 27.

67. White, J. A., "Pay Attention to Orientation," *Modern Materials Handling,* vol. 46, no. 7, June 1991, p. 29.

68. White, J. A., "Put Them in the Right Place!," *Modern Materials Handling,* vol. 46, no. 8, July 1991, p. 27.

69. White, J. A., "Handle at the Right Time!," *Modern Materials Handling,* vol. 46, no. 9, August 1991, p. 25.

70. White, J. A., "The Right Cost of Handling," *Modern Materials Handling,* vol. 46, no. 10, September 1991, p. 27.

71. White, J. A., "Selecting the Right Methods," *Modern Materials Handling,* vol. 46, no. 12, October 1991, p. 27.

72. White, J. A., and Apple, J. M. Jr., "Robots in Material Handling," *Handbook of Industrial Robotics,* S. Y. Nof (ed.), Wiley, New York, 1985, pp. 955–970.

73. White, J. A., and Apple, J. M. Jr., "Material Handling," *International Encyclopedia of Robotics Applications and Automation,* R. Doff (ed.), Wiley, New York, 1988, pp. 873–879.

74. Wilde, J., "Container Design Issues," *Proceedings of the 1992 Material Handling Management Course,* Institute of Industrial Engineers, Atlanta, June 1992.

75. "A Guide for Evaluating and Implementing a Warehouse Bar Code System," Warehousing Education and Research Council, Oak Brook, IL, 1992.

76. *Basics of Material Handling,* Material Handling Institute, Charlotte, NC, 1973.

77. *Automatic Identification Systems for Material Handling and Material Management,* Automatic Identification Manufacturers Product Section, Material Handling Institute, Charlotte, NC, 1976.

78. *Considerations for Planning and Implementing an Automated Storage/Retrieval System,* Material Handling Institute, Charlotte, NC, 1977.

79. *Considerations for Planning and Installing an Automated Storage/Retrieval System,* Automated Storage/Retrieval Systems Product Section, Material Handling Institute, Charlotte, NC, 1977.

80. *Considerations for Planning and Installing Automatic Guided Vehicle Systems,* Automatic Guided Vehicle Systems Section, Material Handling Institute, Charlotte, NC, 1980.

81. "Streamline Your Packaging Operations," *Modern Materials Handling,* May 1996, pp. 24–27.

PROBLEMS

5.1 For each of the following material handling principles, (a) explain what it means, and (b) give a specific example of how it can be applied.
  a. Work principle
  b. Ergonomic principle

5.2 For each of the following material handling principles, (a) explain what it means, and (b) give a specific example of how it can be applied.
  a. Unit load principle
  b. Space utilization principle

5.3 Following the format of Problems 5.1 and 5.2, select two related principles and
  a. Explain what each means
  b. Give a specific example of how each can be applied

5.4 Briefly describe the following types of material handling equipment. (Select three from the list of material handling equipment.)

5.5 Develop a set of attributes for comparing sorting conveyors.

5.6 Develop a set of attributes for comparing automated guided vehicle systems.

5.7   Develop a set of attributes for comparing unit load storage systems.

5.8   Develop a set of attributes for comparing unit load retrieval technologies.

5.9   Develop a set of attributes for comparing small part storage alternatives.

5.10   Develop a set of attributes for comparing automated data collection systems.

5.11   Develop a set of attributes for comparing bar code readers.

5.12   Develop a set of attributes for comparing bar code printers.

5.13   Compare the following:
       a. Pallet truck vs. platform truck
       b. Reach truck vs. turret truck

5.14   Compare the following:
       a. Drive-in rack vs. drive-through rack
       b. Push-back rack vs. mobile rack

5.15   Following the format of Problems 5.13 and 5.14, select two types of material handling equipment and make a comparison.

5.16   Sketch a $40'' \times 48''$ pallet and compare it with a $48'' \times 40''$ pallet.

5.17   Distinguish between a two-way and a four-way pallet.

5.18   Visit a local factory and perform an audit of their containerization system. Make a checklist.

5.19   Visit a local factory and describe the various material handling devices/systems that you observed.

5.20   Develop a case study illustrating unit load interactions with material handling system components.

5.21   The route sheet for a part is as follows:

$$A - F - E - D - C - B - A - F$$

Two thousand pieces will flow through from the first machine A to the final machine F based on the given sequence of operations. A unit load size of 50 is initially specified at the first machine. However, due to lot-sizing decisions the unit load size is doubled after processing on machine D.

If one vehicle (e.g., lift truck) is used to transport the unit loads from machine to machine, determine the total number of trips that the vehicle has to make assuming that the vehicle capacity is one unit load.

APPENDIX 5.A

MATERIAL HANDLING AUDIT CHECKLIST[1]

---

[1] From White [62].

Dept. _____ Building _____ Plant _____
Date _____ Surveyed by _____

Conditions indicating possible productivity improvement opportunities	Condition exists here (✓)	To correct this, we need:			
		Supervisor attention (✓)	Management attention (✓)	Analytical study (✓)	Capital investment (✓)	Other (for comments)
1. Delays in material moving					
2. Excessive material on hand					
3. Production equipment idle for material shortage					
4. Long hauls					
5. Cross traffic					
6. Manual handling					
7. Outmoded handling equipment					
8. Inadequate handling equipment					
9. Insufficient handling equipment					
10. Unbalanced sequence of operations					
11. Idle handling equipment					
12. Obstacles to material flow					
13. Material piled directly on floor					
14. Poor workplace layout for material					
15. Disorderly storage					
16. Cluttered aisles					
17. Cluttered workspace					

18. Crowded dockspace				
19. Motor truck and railroad car tieup				
20. Manual loading techniques				
21. Excessive wasted "cube" in storage				
22. Excessive aisles				
23. Operations unduly scattered				
24. Poor locations of service areas				
25. Lack of in-plant container standardization				
26. Lack of unit load technique				
27. Excessive MH equipment maintenance cost				
28. Rehandling				
29. Handling done by direct labor				
30. Operators traveling for supplies, material				
31. Supplies moved by poor techniques				
32. High indirect payroll				
33. Material waiting for papers				
34. Excessive demurrage				
35. Unexplained delays				
36. Idle labor				
37. Inspection not properly located				
38. Excessive scrap				
39. Hazardous lifting by hand				

MATERIAL HANDLING AUDIT CHECK SHEET*(cont'd)*

Dept. _____ Building _____ Plant _____
Date _____ Surveyed by _____

Conditions indicating possible productivity improvement opportunities	Condition exists here (✓)	To correct this, we need:				
		Supervisor attention (✓)	Management attention (✓)	Analytical study (✓)	Capital investment (✓)	Other (for comments)
40. Misdirected material						
41. Clumsy, dangerous "homemade" handling rigs						
42. Lack of standardization on handling equipment						
43. Long travel distances for material, equipment, and personnel						
44. Backtracking of material						
45. Nonstandard process routing						
46. Opportunity for group technology layout						
47. Opportunity for product layout						
48. Opportunity for process layout						
49. No real-time dispatching of equipment						
50. No modular MH system						
51. No modular work stations						
52. Automatic identification system not used						

208

53. No one-way aisles						
54. MH equipment running empty						
55. Different things treated same						
56. Excessive trash removal						
57. Centralized storage						
58. Decentralized storage						
59. No incentive system for MH labor						
60. Low usage of automated MH equipment						
61. Variable path equipment used for fixed-path handling						
62. System not capable of expansion and/or change						
63. Low usage of industrial robots						
64. No parts preparation performed prior to manufacturing						
65. No prekitting of work						
66. Lack of automated loading/ unloading of trailers						
67. Poor MH at the work station						
68. Lack of industrialized truck attachments and below hook lifters						
69. Equipment capacity not matched to load requirement						
70. Manual palletizing/depalletizing						
71. Lack of equipment for unitizing and stabilizing loads						

Dept. _____ Building _____ Plant _____
Date _____ Surveyed by _____

Conditions indicating possible productivity improvement opportunities	Condition exists here (✓)	To correct this, we need:				
		Supervisor attention (✓)	Management attention (✓)	Analytical study (✓)	Capital investment (✓)	Other (for comments)
72. Lack of a long-range MH plan						
73. No short-interval scheduling of MH equipment						
74. Lack of narrow-aisle and very-narrow-aisle storage equipment						
75. Low bay storage areas						
76. Poor utilization of overhead space						
77. Single-sized pallet rack openings						
78. No palletless handling of unit loads						
79. Storage in part number sequence						
80. Randomized storage						
81. Dedicated storage						
82. Crushed loads in block stacking						
83. No ABC storage classification						
84. Obsolete and inactive material						
85. Floor-stacked material in receiving, QC, and shipping						
86. Aisles and storage locations not clearly marked.						

87. Manual stock locator system						
88. Lack of standardization in part numbers						
89. Cycle counting for physical inventory						
90. No formal audit program in use						
91. No guards to protect racks and columns						
92. Guided aisles without guide rail entry						
93. Loads overhanging pallet						
94. Excessive floor, rack, and structured loading						
95. Equipment operating at excessive speed						
96. Front-to-back rack members not provided						
97. MH equipment does not fit through doors						
98. No sprinklers and smoke detectors						
99. Hazardous and flammable material not segregated and identified						
100. Lack of ventilation in battery charging area						
101. Entrances and exits not secured						
102. Waste and trash containers located near docks						
103. Inadequate number of fire extinguishers						

MATERIAL HANDLING AUDIT CHECK SHEET(contd)

Dept. _____ Building _____ Plant _____
Date _____ Surveyed by _____

Conditions indicating possible productivity improvement opportunities	Condition exists here (✓)	To correct this, we need:				
		Supervisor attention (✓)	Management attention (✓)	Analytical study (✓)	Capital investment (✓)	Other (for comments)
104. No contingency plan for fire loss						
105. Sagging load beams and bent trusses on racks						
106. No formal training for MH equipment operators						
107. No preventive maintenance program						
108. No equipment replacement program						
109. No dock levelers						
110. Unscheduled arrival of outbound and inbound carriers						
111. Decentralized receiving and shipping						
112. Inbound material not unitized						
113. Inadequate number of dock doors						
114. Receiving numbers not preassigned						
115. Picking lists not printed in picking sequence						
116. Orders picked one at a time						
117. Aisle lengths unplanned						
118. Excessive honeycombing in storage						

119. Poor quality pallets, not standardized				
120. Manual sorting in order accumulation				
121. Poor work-in-process control				
122. Energy-inefficient lighting				
123. Lights, heaters, and fans poorly located				
124. No dock enclosures				
125. Excessive heating, ventilation, and air conditioning for material stored				
126. Poorly insulated walls and roof				
127. Poorly designed enclosures for environmentally controlled areas				
128. Lack of scheduled energy use to reduce peak loads				
129. Unclean floors				
130. Battery charging too frequently				
131. Other				

APPENDIX 5.B
MATERIAL HANDLING EQUIPMENT

I. CONTAINERS AND UNITIZING EQUIPMENT

Containers (pallets, skids and skid boxes, and tote pans) and unitizers (stretchwrappers and palletizers) create a convenient unit load to facilitate and economize material handling and storage operations. These devices also protect and secure material.

A. Containers

Containers are frequently used to facilitate the movement and storage of loose items. In grocery shopping we use boxes and bags to facilitate the handling tasks. In industrial applications, loose items are often placed in tote pans or in skid boxes. Additionally, depending on the size and configuration of items to be moved or stored, they might be placed on a pallet or a skid to facilitate their movement and storage using lift trucks or other material handling equipment.

Consider, for example, the need to convey oats unitized in large burlap bags. While the bags might convey well using a belt conveyor, they would not do so using a roller conveyor. However, if the bags were placed on a pallet, then a roller conveyor could be used. Depending on how many degrees of freedom one has in designing the handling system, as well as the distribution and frequency of material handling movements to be performed, one might change the means of conveyance or the means of unitizing the load to be conveyed.

In general, decisions as to how material is to be moved, stored, or controlled are influenced by how the material is contained. The container decision, in essence, becomes the critical decision; it is the cornerstone for the material handling system.

Figure 5.14   Skid boxes.

For this reason, if you are able to choose the container(s) to be used, then you should consider the impact of the decision(s) on subsequent choices of movement, storage, and control technologies.

A.1.   Pallets

Pallets are by far the most common form of unitizing device.

A.2.   Skids and Skid Boxes

Skids and skid boxes are used frequently in manufacturing plants. Often made of metal, they are quite rigid and are well-suited for unitizing a wide variety of items. Generally, skid boxes are too heavy to be lifted manually. (See Figure 5.14.)

A.3.   Tote Pans

Tote pans are used to unitize and protect loose items. Returnable tote pans have become a popular alternative to shipping in cardboard containers. When empty, tote pans should nest or collapse to insure high space utilization. Also, tote dimensions should coordinate with case and pallet dimensions to insure high utilization of material handling vehicles. (See Figure 5.15.)

B.   Unitizers

In addition to the use of containers (skid boxes and tote pans) and platforms (pallets and skids) to unitize a load, special equipment has been designed to facilitate the formation of a unit load. To motivate our consideration of unitizers, suppose you are faced with the need to move pallet loads of cartons. How do you ensure that the cartons placed on the pallet do not shift and become separated from the load? How

Figure 5.15   Plastic reusable tote pans. (*Courtesy of Buckhorn Inc.*)

could you automatically create the pallet load of cartons? How might you automatically remove cartons from the unit load?

To answer the first question, we consider the use of stretchwrap, shrinkwrap, strapping, and banding equipment. To answer the second and third questions, we consider the use of automatic palletizers and de-palletizers.

B.1.   Stretchwrap

Shrinkwrap and stretchwrap equipment, as well as strapping and banding equipment, is used to unitize a load. Strapping can be performed using steel, fiber, and plastic materials; shrinkwrapping and stretchwrapping are performed using plastic film. The

Figure 5.16   Stretchwrap equipment. (*Courtesy of Infra Pak*)

Figure 5.17   Carton palletizer. (*Courtesy of Lambert Material Handling*)

strapping process can be performed manually or mechanically and is best suited for compact loads. Shrinkwrapping is performed by placing a plastic bag over the load and applying heat and suction to enclose the load; shrinkwrapping is similar to the "blister packaging" process used for small articles. Stretchwrapping is performed by wrapping plastic film tightly around a load; multiple layers can be applied to obtain the same degree of weather protection provided by shrinkwrapping. (See Figure 5.16.)

B.2.   Palletizers

Palletizers and de-palletizers are used for case goods handling as well as can and bottle handling. Palletizers receive products and place them on a pallet according to prespecified patterns; de-palletizers receive pallet loads and remove the product from the pallet automatically. A variety of sizes and styles are available for both palletizers and de-palletizers. (See Figure 5.17.)

II.   MATERIAL TRANSPORT EQUIPMENT

Material transport equipment (conveyors, industrial vehicles, monorails, hoists, and cranes) are distinguished from the other categories of material handling equipment by their primary function: material transport. Material transport equipment types are distinguished from one another by their

- Degree of automation (walking, riding, and automated),
- Flow pattern (continuous vs. intermittent, synchronous vs. nonsynchronous),
- Flow path (fixed vs. variable),
- Location (underground, in-floor, floor level, overhead), and
- Throughput capacity.

Well-designed material transport systems achieve an effective match between material move requirements and these material move characteristics.

A. Conveyors

Conveyors are used when material is to be moved frequently between specific points; they are used to move material over a fixed path. Hence, there must exist a sufficient volume of movement to justify dedicating the equipment to the handling task. Depending on the materials to be handled and the move to be performed, a variety of conveyors can be used.

Conveyors can be categorized in several ways. For example, the type of product being handled (bulk or unit) and the location of the conveyor (overhead or floor) have served as bases for classifying conveyors. Interestingly, such classification systems are not mutually exclusive. Specifically, a belt conveyor can be used for bulk and unit materials; likewise, a belt conveyor can be located overhead or on the floor.

Bulk materials, such as soybeans, grain, dry chemicals, and saw dust, might be conveyed using chute, belt, pneumatic, screw, bucket, or vibrating conveyors; unit materials such as castings, machined parts, and materials placed in tote boxes, on pallets, and in cartons might be conveyed using chute, belt, roller, wheel, slat, vibrating, pneumatic, trolley, or two conveyors. Material can be transported on belt, roller, wheel, slat, vibrating, screw, pneumatic, and tow conveyors that are mounted either overhead or at floor level.

A.1. Chute Conveyor

The chute conveyor is one of the most inexpensive methods of conveying material. The chute conveyor is often used to link two powered conveyor lines; it is also used to provide accumulation in shipping areas; a spiral chute can be used to convey items between floors with a minimum amount of space required. While chute conveyors are economical, it is also difficult to control the items being conveyed by chutes; packages may tend to shift and turn so that jams and blockages occur. (See Figure 5.18.)

A.2. Belt Conveyors

There are a wide variety of belt conveyors employed in modern material handling systems. The most popular types are flat belt conveyor, telescoping belt conveyor, trough belt conveyor, and magnetic belt conveyor.

A.2.a. Flat Belt Conveyor. A flat belt conveyor is normally used for transporting light- and medium-weight loads between operations, departments, levels, and buildings. It is especially useful when an incline or decline is included in the conveyor path. Because of the friction between the belt and the load, the belt conveyor provides considerable control over the orientation and placement of the load; however, friction also prevents smooth accumulation, merging, and sorting on the belt. The belt is generally either roller or slider bed supported. If small and irregularly shaped items are being handled, then the slider bed would be used; otherwise, the roller support is usually more economical. (See Figure 5.19.)

A.2.b. Telescoping Belt Conveyor. A telescoping belt conveyor is a flat belt conveyor that operates on telescoping slider beds. They are popular at receiving and shipping docks where the conveyor is extended into inbound/outbound trailers for unloading/loading. (See Figure 5.20.)

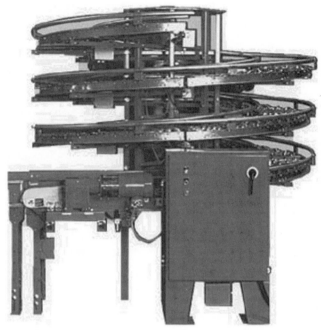

Figure 5.18   Spiral chute conveyor. (*Courtesy of Carter Control Systems, Inc.*)

Figure 5.19   Flat belt conveyor. (*Courtesy of Siemens Dematic Corp.*)

Figure 5.20   Telescoping conveyor. (*Courtesy of Siemens Dematic Corp.*)

A.2.c. Magnetic Belt Conveyor.   A magnetic belt conveyor consists of a steel belt and either a magnetic slider bed or a magnetic pulley. It is used to transport ferrous materials vertically, upside down, and around corners, as well as for the separation of ferrous and nonferrous materials. This type of conveyor can move parts up and over a production line and walkway to save valuable floor space and eliminate the need to relocate equipment. (See Figure 5.21.)

A.3. Roller Conveyor

The roller conveyor is a very popular type of material handling conveyor; it may be powered or nonpowered. The nonpowered roller conveyor is referred to as a "gravity" conveyor, as motion is achieved by inclining the roller section. Powered (or live) roller conveyors are generally either belt or chain driven. However, one manufacturer has employed a revolving drive shaft to power the rollers; rollers are connected individually to the drive shaft by an elastomeric belt. The roller conveyor is well suited for accumulating loads and merging/sorting operations. Because of the roller surface, the materials being transported must have a rigid riding surface. (See Figure 5.22.)

A.4. Wheel Conveyor

The wheel or skate-wheel conveyor is similar to the roller conveyor in design and function; a series of skate wheels are mounted on a shaft. Spacing of the wheels is dependent on the load being transported. Although wheel conveyors are generally more economical than the roller conveyor, they are limited to light-duty applications. (See Figure 5.23.)

Figure 5.21   Magnetic belt conveyor. (*Courtesy of Bunting Magnetics Co.*)

Figure 5.22   Roller conveyor. (*Courtesy of Siemens Dematic Corp.*)

Figure 5.23 Skate wheel conveyor.

A.5. Slat Conveyor

The slat conveyor consists of discretely spaced slats connected to a chain. The slat conveyor functions much like a belt conveyor in that the unit being transported retains its position relative to the conveying surface. Because the conveying surface moves with the product, the orientation and placement of the load are controlled. Heavy loads with abrasive surfaces or loads that might damage a belt are typically conveyed using a slat conveyor.

Additionally, bottling and canning plants use flat chain or slat conveyors because of wet conditions, temperature, and cleanliness requirements. (See Figure 5.24.)

A.6. Chain Conveyor

The chain conveyor consists of one or more endless chains on which loads are carried directly. In transporting bulk materials, a chain is located in the bottom of a trough and pulls the material through the trough. Chain conveyors are often used to transport tote boxes and pallets, as only two or three chains typically are required to provide sufficient contact with the rigid support or container to effect movement. (See Figure 5.25.)

Figure 5.24 Slat conveyor. (*Courtesy of Fredenhagen, Inc. and Acco Babcock*)

A.7. Towline Conveyor

The tow or towline conveyor is used to provide power to wheeled carriers such as trucks, dollies, or carts that move along the floor. Essentially, the tow conveyor provides power for fixed-path travel of carriers that have variable path capability. The towline can be located either overhead, flush with the floor, or in the floor. Towline systems often include selector-pin or pusher-dog arrangements to allow automatic switching between power lines or onto a nonpowered spur line for accumulation. Tow conveyors are generally used when long distances and a high frequency of movement are involved. (See Figure 5.26.)

A.8. Trolley Conveyor

The trolley conveyor consists of a series of trolleys supported from or within an overhead track. They are generally equally spaced in a closed loop path and are suspended from a chain. Specially designed carriers can be used to carry multiple

Figure 5.25 Chain conveyor. (*Courtesy of Jervis B. Webb Co.*)

Figure 5.26 Towline conveyor. (*Courtesy of Jervis B. Webb Co.*)

units of product. They have been used extensively in processing, assembly, packaging, and storage operations. (See Figure 5.27.)

A.9. Power-and-Free Conveyor

The overhead power-and-free conveyor is similar to the trolley conveyor in that discretely spaced carriers are transported by an overhead chain. However, the power-and-free conveyor utilizes two tracks: one powered and the other nonpowered or free. Carriers are suspended from a set of trolleys that run on the free track. Linkage between the power chain and the trolleys on the free track is achieved using a "dog." The dogs on the power chain mate with similar extensions on the carrier trolleys and push the carriers forward on the free track. The advantage of the power-and-free design is that carriers can be disengaged from the power chain and accumulated or switched onto spurs. A variation of the overhead power-and-free conveyor is floor mounted and is called the inverted power-and-free conveyor. This type of conveyor can be found in many car assembly plants. (See Figure 5.28.)

A.10. Cart-on-Track Conveyor

The cart-on-track conveyor is used to transport a cart along a track. Employing the principle of a screw, a cart is transported by a rotating tube. Connected to the cart

Figure 5.27   Trolley conveyor. (*Courtesy of Leas Siegler/Registar Division and Acco Babcock*)

is a drive wheel that rests on the tube; the speed of the cart is controlled by varying the angle of contact between the drive wheel and the tube. The basic elements of the conveyor are the rotating tube, track, and cart. The carts are independently controlled to allow multiple carts to be located on the tube. Carts can accumulate on the tube because they will be stationary when the drive wheel is parallel to the tube. (See Figure 5.29.)

A.11.   Conveyor Sortation Devices

A sorting conveyor is used to assemble material (i.e., cases, items, totes, garments) with a similar characteristic (i.e., destination, customer, store) by correctly identifying the similar merchandise and transporting it to the same location.

A.11.a. Deflector.   A deflector is a stationary or moveable arm which deflects product flow across a belt or roller conveyor to the desired location. A deflector is necessarily in position before the item to be sorted reaches the discharge point.

Figure 5.28   Inverted power-and-free conveyor. (*Courtesy of Jervis B. Webb Co.*)

Stationary arms remain in a fixed position and represent a barrier to items coming in contact with them. With the stationary arm deflector, all items are deflected in the same direction. Movable arm or pivoted paddle deflectors are impacted by the item to be sorted in the same manner as the stationary arm deflector. However, the element of motion has been added. With the movable arm deflector (i.e., the paddle), items are selectively diverted. Pivoted deflectors may be equipped with a belt conveyor flush with the surface of the deflector (a power face) to speed or control the divert. Paddle deflecting systems are sometimes referred to as steel belt sorters since at one time steel belts were used to reduce the friction encountered in diverting products across the conveyor. Deflectors can support medium (1200 to 2000 cartons per hour) throughput for up to 75-lb loads. (See Figure 5.30.)

A.11.b. Push Diverter.   A push diverter is similar to a deflector in that it does not contact the conveying surface but sweeps across to push the product off the opposite side. Push diverters are mounted beside (air or electric powered) or above the conveying surface (paddle pushers) and are able to move items faster and with greater control than a deflector. Overhead push diverters are capable of moving products to either side of the conveying surface whereas side-mounted diverters move conveyed items in one direction only to the side opposite that on which they are mounted. Push diverters have a capacity of 3600 cases per hour for loads up to 100 lb. (See Figure 5.31.)

Figure 5.29   Cart-on-track. (*Courtesy of SI Systems, Inc.*)

A.11.c. Rake Puller.   A rake puller is best applied when the items to be sorted are heavy and durable. Rake puller tines fit into slots between powered or nonpowered roller conveyors. Upon command, a positioning stop device and the tines pop up from beneath the roller conveyor surface to stop the carton. The tines pull the carton across the conveyor, and then drop below the roller surface for a noninterference return to the starting position. During the return stroke, the next carton can be moving into position.

A.11.d. Sliding Shoe Conveyor.   The sliding shoe conveyor is differentiated from the other sorting conveyors by the fact that the divert takes place in-line along the roller conveyor. (See Figure 5.32.)

A.11.e. Pop-Up Skewed Wheels.   Pop-up skewed wheels are capable of sorting flat-bottomed items. The skewed wheel device pops up between the rollers of a powered roller conveyor or between belt conveyor segments and directs sorted items onto a powered takeaway lane. Rates of between 5000 and 6000 cases per hour can be achieved. (See Figure 5.33.)

Figure 5.30   Deflector arm sorter. (*Courtesy of Siemens Dematic Corp.*)

Figure 5.31   Push diverter. (*Courtesy of Ermanco, Inc.*)

Figure 5.32 Sliding shoe sorter. (*Courtesy of Siemens Dematic Corp.*)

Figure 5.33 Pop-up wheel sorter. (*Courtesy of Siemens Dematic Corp.*)

Figure 5.34. Pop-up belt. (*Courtesy of Ermanco, Inc.*)

A.11.f. Pop-Up Belts and Chains. Pop-up belt and chain sort devices are similar to pop-up skewed wheels in that they rise from between the rollers of a powered roller conveyor to alter product flow. Belt and chain sortation devices are capable of handling heavier items than the wheeled devices. (See Figures 5.34 and 5.35.)

A.11.g. Pop-Up Rollers. Pop-up rollers rise up between the chains or rollers of a chain or roller conveyor to alter the flow of product. Pop-up rollers provide relatively inexpensive means for sorting heavy loads at rates of 1000 to 1200 cases per hour. (See Figure 5.36.)

A.11.h. Adjustable Angle Conveyor Sorter. This type of roller conveyor sorter allows smooth flow of cartons with the use of roller diverter by changing the roller angle. Cables are used to pull the rollers to the right angles.

A.11.i. Tilt Tray Sorter. Continuous chains of tilting trays are used to sort a wide variety of lightweight merchandise on a tilt tray sorter. The trays may be fed manually, or by one of the many types of induction devices available. Tilt tray systems can sort to either side of the sorter. Tilt tray sorters do not discriminate for the shape of the product being sorted. Bags, boxes, envelopes, documents, software, etc. can all be accommodated. The tilt tray sorter is not appropriate for long items. The capacity of the tilt tray sorter, expressed in pieces or sorts per hour, is expressed as the ratio of the sorter speed to the pitch or length of an individual tray. Rates of 10,000 to 15,000 items per hour can be achieved. (See Figure 5.37.)

Figure 5.35 Pop-up chain. (*Courtesy of Litton UHS*)

A.11.j. Tilting Slat Conveyor.   In a tilting slat conveyor, the product occupies the number of slats required to contain its length. The sort is executed by tilting the occupied slats. Hence, tilting slat sorters are best applied when a wide variety of product lengths will be handled. The tilting slat is capable of tilting in either direction. Slats may be arranged in a continuous over-and-under configuration.

A.11.k. Cross-Belt Sorter.   A cross-belt sorter is so named because each item rests on a carrier equipped with a separate powered section of belt conveyor which operates orthogonal to the direction of material transport. Hence, the sorting capacity is enhanced and the width of the accumulating chutes can be reduced. (See Figure 5.38.)

Figure 5.36 Pop-up roller sorter. (*Courtesy of Ermanco, Inc.*)

Figure 5.37 Tilt tray sorter. (*Courtesy of Siemens Dematic Corp.*)

Figure 5.38 Cross-belt sorter. (*Courtesy of Siemens Dematic Corp.*)

Figure 5.39 Bombardier sorter. (*Courtesy of SI Systems, Inc.*)

A.11.l. Bombardier Sorter. A bombardier sorter is so-named because items are dropped down through swinging doors much like bombs are dropped through the belly of an airplane. (See Figure 5.39.)

B.   Industrial Vehicles

Industrial vehicles represent a versatile method of performing material handling; they are referred to as variable-path equipment. Industrial vehicles are generally used when movement is either intermittent or over long distances. Apple [2] notes that conveyors are used when the primary function is conveying, cranes and hoists are used when transferring is the primary function, and industrial trucks are used when the primary function is maneuvering or transporting.

Three categories of industrial vehicles are defined for this description—walking, riding, and automated.

B.1.   *Walking Industrial Vehicles*

There are four major classes of walking industrial vehicles:

1.   Hand trucks and hand carts
2.   Pallet jacks
3.   Platform trucks
4.   Walkie stackers

Their popularity stems from their simplicity and low price.

B.1.a. Hand Truck and Hand Cart. The hand truck or cart is one of the simplest and most inexpensive types of material handling equipment. Used for small loads and short distances, the hand truck is a versatile method of moving material manually. A host of carts have been designed to facilitate manual movement of material, as well as the movement of material by conveyor and industrial vehicles.

Figure 5.40   Pallet jack. (*Courtesy of Crown Equipment Corp.*)

B.1.b. Pallet Jack.   The pallet jack is used to lift, maneuver, and transport a pallet load of material short distances. The pallet jack can be either manual or battery powered for both lifting and transporting. The lifting capability is typically from 6 in. to 10 in. (See Figure 5.40.)

B.1.c. Platform Truck.   The platform truck is a version of the industrial truck. Instead of having forks for lifting and supporting the pallets being transported, the platform truck provides a platform for supporting the load. It does not have lifting capability and is used for transporting; consequently, an alternative method of loading and unloading must be provided. (See Figure 5.41.)

B.1.d. Walkie Stacker.   The walkie stacker extends the lifting capability of the pallet jack to allow unit loads to be stacked or placed in storage racks. The walkie stacker can have either a straddle or reach design. (See Figure 5.42.) The straddle type straddles the load with its outriggers; the reach type uses a pantograph or scissors device to allow the load to be retrieved from and placed in storage. (See Figure 5.43.)

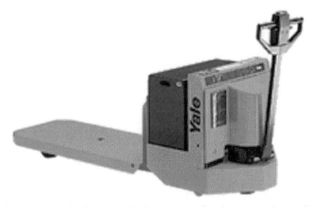

Figure 5.41   Walkie platform truck. (*Courtesy of Yale Material Handling Corp.*)

Figure 5.42 Walkie stacker. (*Courtesy of Crown Equipment Corp.*)

A walkie stacker allows a pallet to be lifted, stacked, and transported short distances. The operator steers from a walking position behind the vehicle. In a situation where there is low throughput, short travel distances, and low vertical storage height, and a low cost solution is desired, the walkie stacker may be appropriate.

B.2. Riding Industrial Vehicles

Riding industrial vehicles allow the vehicle operator to ride to, from, and between locations. Hence, they are typically used for longer moves than walking vehicles. Riding vehicles also offer additional weight and storage height capacity.

Figure 5.43 Walkie reach straddle truck. (*Courtesy of The Raymond Corp.*)

Figure 5.44   Stand-up rider pallet truck. (*Courtesy of Crown Equipment Corp.*)

B.2.a. Pallet Truck.   The pallet truck extends the transporting capability of the pallet jack by allowing the operator either to ride or walk. The pallet truck is used when the distance to be traveled precludes walking. (See Figure 5.44.)

B.2.b. Tractor-Trailer.   The tractor-trailer combination extends the transporting capability of the hand truck by providing a powered, rider-type vehicle to pull a train of connected trailers. A wide variety of tractor-trailers are available, including walkie and rider trucks and remote control alternatives. (See Figure 5.45.)

B.2.c. Counterbalanced Lift Truck.   The workhorse of materials handling is the counterbalanced lift truck (Figure 5.46). Although it is usually referred to as a fork truck, not all designs use forks for lifting, for example, ram attachments and platform designs. As with most of the industrial trucks, a counterbalanced lift truck can be either battery powered (BP) or powered by an internal combustion engine (ICE); the latter can use either gasoline, propane, or diesel fuel. The tires used on the industrial truck can be either cushion (CT) for indoor operation or pneumatic (PT) for outdoor operation. The load-carrying capability of the counterbalanced lift truck can range from 1000 to over 100,000 lb.

Figure 5.45   Stand-up rider two tractor. (*Courtesy of Crown Equipment Corp.*)

Figure 5.46 Sit-down rider counterbalanced lift truck. (*Courtesy of Crown Equipment Corp.*)

As the name implies, counterbalance lift trucks employ a heavy counterbalance over the rear wheels to achieve lift-weight capacities of up to 100,000 lbs and lift-height capacities between 25 and 30 ft. Counterbalance lift trucks may be gas or battery powered. A counterbalanced truck may not be used to store double deep.

As the selective pallet rack is the benchmark pallet storage mode, the counterbalanced truck may be considered the benchmark storage/retrieval vehicle. When it is desirable to use the same vehicle for loading and unloading trucks and storing and retrieving loads, the counterbalanced truck is the logical choice. For use in block stacking, drive-in and drive-thru rack and pallet stacking frames, the operating aisles normally provided are suitable for counterbalanced trucks. Since counterbalance trucks must turn within a storage aisle to retrieve a pallet load, the aisle width required (10–13 ft) to operate is wider than required for some other lift truck alternatives. The relatively low cost and flexibility of counterbalance trucks are their main advantage.

Besides forks, other attachments may be used to lift unique load configurations on a vertical mast. The primary function of an industrial truck attachment is either to save time, conserve space, reduce product damage, save labor, or eliminate equipment.

Industrial truck attachments are available for a wide variety of applications; indeed, the design of industrial truck attachments seems limited only by the ingenuity and creativity of the design engineer. Some of the more popular attachments are:

- The side-shifter, which permits the forks to be shifted from side to side so that the operator can pick up or spot a pallet without having to reposition the truck (hence, very slow and precise positioning of the truck is not necessary)
- The load push/pull attachment, which allows a slipsheet to be substituted for the pallet
- The cinder block attachment, which eliminates the need for a pallet, saves space, and reduces the weight of a load
- The carton clamp
- The paper roll clamp
- The ram attachment

B.2.d. Straddle Carrier.   The straddle carrier carries the load underneath the driver. The truck straddles a load, picks it up, and transports it to the desired location. Used primarily outdoors for long, bulky loads, the straddle carrier can handle up to 120,000 lb and can accommodate intermodal containers.

B.2.e. Mobile Yard Crane.   The mobile crane illustrates the relationship between crane operations and industrial vehicles. Numerous varieties of mobile cranes exist as well as crane-type attachments for industrial vehicles.

B.3. Automated Industrial Vehicles

The category of automated industrial vehicles is distinguished from the other industrial vehicles by the elimination of human intervention from powering and guiding the movement of the vehicle. Instead, vehicles are automatically guided by electrified wires buried in the floor, magnetic tape lined along the floor, rails mounted in the ceiling, cameras mounted on the vehicle, or inertial guidance systems. The types of vehicles included in this category are all types of automated guided vehicles (AGVs), automated electrified monorails (AEMs), and sorting transfer vehicles (STVs).

B.3.a. Automated Guided Vehicles.   *See Appendix 5.C for a more detailed discussion on Automated Guided Vehicle Systems.*

An automated guided vehicle (AGV) is essentially a driverless industrial truck. It is a steerable, wheeled vehicle, driven by electric motors using storage batteries, and it follows a predefined path along an aisle. AGVs may be designed to operate as a tractor, pulling one or more carts, or may be unit load carriers. The unit load AGV is the most common type in manufacturing and distribution.

The path followed by an AGV may be a simple loop or a complex network, and there may be many designated load/unload stations along the path. The vehicle incorporates a path-following system, typically electromagnetic, although some optical systems are in use.

With an electromagnetic path-following system, a guidewire that carries a radio frequency (RF) signal is buried in the floor. The vehicle employs two antennae, so that the guidewire can be bracketed. Changes in the strength of the received signal are used to determine the control signals for the steering motors so that the guidewire is followed accurately. When it is necessary for the vehicle to switch from one guidewire to another (e.g., at an intersection), two different frequencies can be used, with the vehicle being instructed to switch from one frequency to the other. Obviously, a significant level of both analog and digital electronic technology is incorporated into the vehicle itself. In addition, the vehicle routing and dispatching system will employ not only programmable controllers, but minicomputers or even mainframe computers.

The electromagnetic path-following system may also support communication between a host control computer and the individual vehicles. In systems with a number of vehicles, the host computer may be responsible for both the routing and dispatching of the vehicles and collision avoidance. A common method for collision avoidance is zone blocking, in which the path is partitioned into zones and a vehicle is never allowed to enter a zone already occupied by another vehicle. This type of collision avoidance involves a high degree of active host computer–directed control. Optical path-following systems use an emitted light source and track the reflection from a special chemical stripe painted on the floor. In a similar fashion, codes can be painted on the floor to indicate to the vehicle that it should stop for a load/unload station. Simple implementations will require all actions of the vehicle to be preprogrammed (i.e., stop and look for a load, stop and unload, or proceed to the next station). The vehicles typically will not be under the control of a host computer and must therefore employ some method for collision avoidance other than zone blocking. Proximity sensors on the vehicles permit several vehicles to share a loop without colliding.

The state of the art in AGVs is advancing rapidly. There are now "smart" vehicles that can navigate for short distances without an electromagnetic or optical path. Similarly, vehicles are being equipped with sufficient on-board computing capability to manage some of the routing control and dispatching functions. Current developments in the field are leading to path-free, or "autonomous" vehicles, which do not require a fixed path and are capable of "intelligent" behavior. Much of the work currently under way in the AGV field is driven by the availability of small, powerful, but relatively inexpensive microcomputers.

The classifications of automated guided vehicles include (Figures 5.47–5.53):

- Pallet trucks
- Fork trucks
- Unit load vehicles
- Towing vehicles
- Work platform vehicles
- Store/retrieve vehicles

B.3.a.i. Fork Truck AGV

Figure 5.47 Fork truck AGV. (*Courtesy of Jervis B. Webb Co.*)

B.3.a.ii. Unit Load AGV

Figure 5.48   Small load AGVs. (*Courtesy of Jervis B. Webb Co.*)

B.3.a.iii. Towing AGV

Figure 5.49   Medium load AGVs. (*Courtesy of Jervis B. Webb Co.*)

Figure 5.50   Heavy load AGV. (*Courtesy of Meator*)

B.3.a.iv. Work Platform AGV

Figure 5.51   AGV towing vehicle. (*Courtesy of Siemens Dematic Corp.*)

B.3.a.v. Store/Retrieve AGV

Figure 5.52 Work platform AGV. (*Courtesy of Jervis B. Webb Co.*)

B.3.b. Automated Electrified Monorails. Self-powered monorails (SPMs), also referred to as automated electrified monorails (AEMs), resemble overhead power-and-free conveyors in much the same way that AGVs resemble towline conveyors. An SPM moves along an overhead monorail, but rather than being driven by a moving chain it is driven by its own electric motor. The SPM provides many of the benefits of both overhead power-and-free conveyors and AGVs. It is overhead, so it does not create obstructions on the factory floor, and self-powered and programmable, so it has a great deal of route flexibility. In contrast to the AGV, it may use a bus bar to provide the electric current required by the drive motors, rather than carrying a storage battery. (See Figure 5.54.)

Another distinctive feature of SPMs is that they require no sophisticated path-following system, because the path must conform to the monorail structure. On the other hand, path selection (e.g., at an intersection or branch point) requires coordination between a vehicle-tracking function and a physical track-switching function. With an AGV using wire guidance, the path selection may be accomplished by the vehicle itself, simply by switching frequencies.

SPMs have been designed to interface with a wide variety of production equipment. The carriers can be switched from one monorail line to another and can even be raised or lowered between two different levels. As with AGVs, the SPM employs a high degree of computer control.

Figure 5.53   Store/retrieve AGV. (*Courtesy of Jervis B. Webb Co.*)

B.3.c. Sorting Transfer Vehicles.   Sorting transfer vehicles automatically load and unload large and small unit loads at pickup and deposit points located around a fixed path. The vehicle's rapid and precise acceleration and deceleration creates a high-throughput unit load handling system. (See Figure 5.55.)

Figure 5.54   Automated electrified monorail. (*Courtesy of Jervis B. Webb Co.*)

Figure 5.55   Sorting transport vehicle. (*Courtesy of Siemens Dematic Corp.*)

C.   Monorails, Hoists, and Cranes

Monorails and cranes are generally used to transfer material from one point to another in the same general area. Hoists are used to facilitate the positioning, lifting, and transferring of material within a small area. Monorails, hoists, and cranes generally provide more flexibility in the movement path than do conveyors; however, they do not have the degree of flexibility provided by variable-path equipment, such as industrial trucks.

Typically, the loads handled by monorails, hoists, and cranes are much more varied than those handled by a conveyor, also, the movement of materials is generally much more intermittent when using monorails, hoists, and cranes than when using conveyors.

C.1.   Monorail

A monorail consists of an overhead track on which a carrying device rides. The carrier can be either top running or underhung powered or nonpowered. If powered, then the carrying device itself is generally powered electrically or pneumatically. Additionally, intelligent carriers have been developed through the application of microprocessors. The monorail functions like a trolley conveyor, except the carrying devices operate independently and the track need not be a closed loop. (See Figure 5.56.)

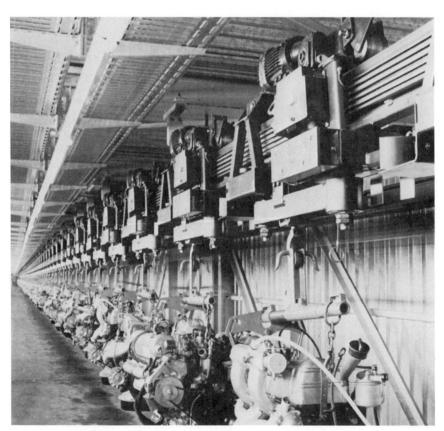

Figure 5.56   Monorails. (*Courtesy of Acco Babcock and Mannesmann Demag*)

Figure 5.57   Hoist. (*Courtesy of Acco Babcock*)

C.2. Hoist

A hoist is a lifting device that is frequently attached to a monorail or crane. The hoist may be manually, electrically, or pneumatically powered. (See Figure 5.57.)

C.3. Cranes

C.3.a. Jib Crane.   A jib crane has the appearance of an arm that extends over a work area. A hoist is attached to the arm to provide lifting capability. The arm may be mounted on a wall or attached to a floor-mounted support. The arm can rotate and the hoist can move along the arm to achieve a wide range of coverage. (See Figure 5.58.)

C.3.b. Bridge Crane.   As the name implies, a bridge crane resembles a bridge that spans a work area. The bridge is mounted on tracks so that a wide area can be covered. The bridge crane and hoist combination can provide three-dimensional coverage of a department. The bridge can be either top riding or underhung. The top-riding crane can accommodate heavier loads. However, the underhung crane is considered to be more versatile than the top-riding crane because of its ability to transfer loads and interface with monorail systems. (See Figure 5.59.)

C.3.c. Gantry Crane.   The gantry crane spans a work area in a manner similar to the bridge crane; however, it is generally floor supported rather than overhead supported on one or both ends of the spanning section. The support can either be fixed in position or travel on runways. (See Figure 5.60.)

Figure 5.58 Jib cranes. (*Courtesy of Acco Babcock*)

Figure 5.59 Bridge crane. (*Courtesy of Mannesmann Demag*)

Figure 5.60 Gantry cranes. (*Courtesy of Harnischfeger Corp.*)

Figure 5.61 Tower crane. (*Courtesy of Potain/America, Inc.*)

C.3.d. Tower Crane.   A tower crane is most often seen on construction sites, but is also usable for ongoing material handling operations. The tower crane consists of a single upright that may be fixed or on a track having a cantilever boom. A hoist operates on the boom, which may be rotated 3600 about the upright. (See Figure 5.61.)

C.3.e. Stacker Crane.   The stacker crane is similar to a bridge crane. Instead of using a hoist, a mast is supported by the bridge; the mast is equipped with forks or a platform, which are used to lift unit loads. The stacker crane is often used for storing and retrieving unit loads in storage racks; however, it can also be used without storage racks when materials are stackable. The stacker crane can be controlled either remotely or with an operator on board in a cab attached to the mast. Stacker cranes can operate in multiple aisles. They are often used in high-rise applications, with storage racks more than 50 ft high. (See Figure 5.62.)

Figure 5.62   Stacker crane. (*Courtesy of Harnischfeger Engineers, Inc.*)

III. STORAGE AND RETRIEVAL EQUIPMENT

Storage and retrieval equipment is distinguished from material transport equipment by its primary function—to house material for staging or building inventory and to retrieve material for use. In some cases, the retrieval equipment is one of the material transport systems described above. In other cases a new equipment type will be introduced.

For discussion purposes, two major classifications of storage and retrieval equipment are defined—unit load and small load systems. Unit load systems typically house large loads such as full pallets, large boxes, and/or large rolls of material. Their applications include housing inventory prior to full unit load shipping, housing reserve storage for replenishment of forward picking areas, and/or housing inventory for partial unit load picking.

Small load storage and retrieval systems typically house small inventory quantities of one or more items in a storage location. The maximum storage capacity for each storage location is typically less than 500 lbs.

A.   Unit Load Storage and Retrieval

For discussion purposes, unit load storage and retrieval systems are further subdivided into unit load storage systems, which house unit loads, and unit load retrieval systems, which allow access to unit loads for retrieval.

A.1.   Unit Load Storage (Racking) Equipment

Unit load storage equipment types are distinguished from one another by their rack configuration, lane depth capacity, stacking capacity, unit load access, and capital expense.

A.1.a. Block Stacking.   Block stacking refers to unit loads that are stacked on top of each other and stored on the floor in storage lanes (blocks) 2 to 10 loads deep. Depending on the weight and stability of the loads, the stacks may range from 2 loads high to a height determined by acceptable safe limits or by the building clear height. Block stacking is particularly effective when there are multiple pallets per SKU (stock-keeping unit) and when inventory is turned in large increments (i.e., several loads of the same SKU are received or withdrawn at one time). (See Figure 5.63.)

A phenomenon referred to as honeycombing occurs with block stacking as loads are removed from a storage lane. Since only one SKU can be effectively stored in a lane, empty pallet spaces are created, which cannot be utilized effectively until an entire lane is emptied. Therefore, in order to maintain high utilization of the available storage positions, the lane depth (number of loads stored from the aisle) must be carefully determined. Because no investment in racks is required, block stacking is easy to implement and allows flexibility with floor space.

Obviously block stacking is not a type of equipment. However, since it is the benchmark for evaluating all other types of unit load storage equipment, it is described here.

Figure 5.63 Block stacking.

A.1.b. Stacking Frame. A stacking frame is portable and enables the user to stack material, usually in pallet-sized loads, on top of one another. They are typically used with unit loads that do not conveniently stack upon themselves. The design issues are the same as those faced in block stacking systems. (See Figure 5.64.)

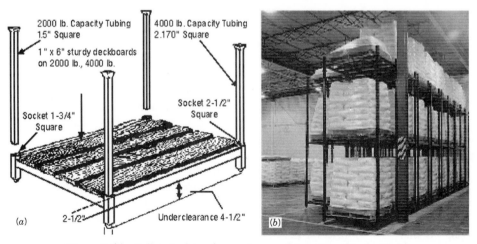

Figure 5.64 Pallet stacking frame storage. (*Courtesy of Jarke Corp.*)

Figure 5.65   Single-deep selective pallet rack. (*Courtesy of Interlake Material Handling, Inc.*)

Portable racks can be either frames that are attached to standard wooden pallets or self-contained steel units made up of decks and posts. When not in use the racks can be disassembled and stored in a minimum of space.

A.1.c. Single-Deep Selective Rack.   A single-deep selective rack is a simple construction of metal uprights and cross-members and provides immediate access to each load stored. Unlike block stacking, when a pallet space is created by the removal of a load, it is immediately available. Loads do not need to be stackable and may be of varying heights and widths. In instances where the load depth is highly variable, it may be necessary to provide load supports or decking. (See Figure 5.65.)

The selective pallet rack might be considered as the benchmark storage mode, against which other systems may be compared for advantages and disadvantages. Most storage systems benefit from the use of at least some selective pallet racks for items whose storage requirement is less than six units.

A.1.d. Double-Deep Rack.   A double-deep rack is merely a single-deep selective rack that is two unit load positions deep. The advantage of the double-deep feature is that fewer aisles are needed, which results in a more efficient use of floor space. In most cases 50% aisle space savings is achieved versus a single-deep selective rack. Double-deep racks are used where the storage requirement for an SKU is six units or greater and when product is received and picked frequently in multiples of two unit loads. Since units loads are stored two deep, a double-reach fork lift is required for storage/retrieval. (See Figure 5.66.)

A.1.e. Drive-In Rack.   A drive-in rack extends the reduction of aisle space begun with double-deep rack. Drive-in racks typically provide for storage lanes from 5 to 10 loads deep. They allow a lift truck to drive into the rack several positions and store or retrieve a unit load. This is possible because the rack consists of upright columns that have horizontal rails to support unit loads at a height above that of the lift truck. This construction permits a second or even a third level of storage,

Figure 5.66 Double-deep selective pallet rack. (*Courtesy of Ridg-U-Rak, Inc.*)

with each level being supported independently of the other. A drawback of the drive-in rack is the reduction of lift truck travel speed needed for safe navigation within the confines of the rack construction. (See Figure 5.67.)

A.1.f. Drive-Thru Rack. A drive-thru rack is merely drive-in rack that is accessible from both sides of the rack. It is used for staging loads in a flow-through fashion where a unit load is loaded at one end and retrieved at the other end. The same design considerations for drive-in racks apply to drive-thru racks. (See Figure 5.68.)

A.1.g. Pallet Flow Rack. Functionally, a pallet flow rack is used like a drive-thru rack, but loads are conveyed on wheels, rollers, or air cushions from one end of a storage lane to the other. As a load is removed from the front of a storage lane the next load advances to the pick face. The main purpose of the flow rack is to simultaneously provide high throughput and storage density. Hence, it is used for those items with high inventory turnover. (See Figure 5.69.)

A.1.h. Push-Back Rack. With a rail-guided carrier provided for each pallet load, a push-back rack provides last-in-first-out deep lanc storage. As a load is placed into storage, its weight and the force of the putaway vehicle pushes the other loads in the lane back into the lane to create room for the additional load. As a load is removed from the front of a storage lane, the weight of the remaining load automatically advances the remaining loads to the rack face. (See Figure 5.70.)

A.1.i. Mobile Rack. A mobile rack is essentially a single-deep selective rack on wheels or tracks. This design permits an entire row of racks to move away from

Figure 5.67   Drive-in rack. (*Courtesy of Paltier LLC*)

adjacent rows. The underlying principle is that aisles are justified only when they are being used. The rest of the time they are occupying valuable space. Access to a particular storage row is achieved by moving the adjacent row and creating an aisle in front of the desired row. Mobile racks are useful when space is scarce and inventory turnover is low. (See Figure 5.71.)

Figure 5.68   Drive-thru rack. (*Courtesy of Ridg-U-Rak, Inc.*)

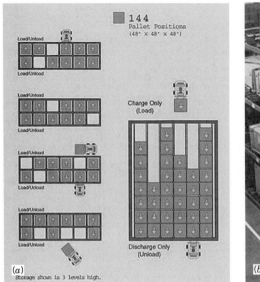

Figure 5.69 Pallet flow rack. (*Courtesy of Paltier LLC*)

Figure 5.70 Push-back rack without loads. (*Courtesy of Paltier LLC*)

Figure 5.71 Mobile racks.

A.1.j. Cantilever Rack. The load-bearing arms of a cantilever rack are supported at one end, as the name implies. The racks consist of a row of single upright columns, spaced several feet apart, with arms extending from one or both sides of the uprights to form supports for storage. The advantage of cantilever racks is that they provide long unobstructed storage shelves with no uprights to restrict the use of horizontal space. The arms can be covered with decking of wood or metal or can be used without decking. They are applicable for long items such as sofas, rugs, rod, bar, pipe, and sheets of metal or wood. (See Figure 5.72.)

A.2. Unit Load Retrieval Equipment

Unit load retrieval equipment types are distinguished from one another by their degree of automation, capital expense, lift height capacity, and aisle width requirements. Equipment offering greater lift height capacity, operating in narrower aisles, and offering greater degrees of automation come with higher prices. Those higher prices may be justified for the associated space and labor savings.

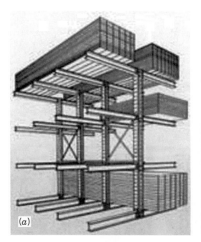

Figure 5.72   Cantilever rack. (*Courtesy of Paltier LLC*)

Two popular devices for retrieving unit loads are walkie stackers and coun-terbalanced lift trucks. Their characteristics were described previously in the section headed "Industrial Vehicles." The other major categories of unit load retrieval equip-ment are narrow-aisle vehicles and automated storage/retrieval (AS/R) machines.

A.2.a. Walkie Stacker.   (See section on "Industrial Vehicles.")

A.2.b. Counterbalance Lift Truck.   (See section on "Industrial Vehicles.")

A.2.c. Narrow-Aisle Vehicles.   Narrow-aisle vehicles are distinguished from walkie stackers and counterbalanced lift trucks by their design to operate in space-efficient storage aisles between 5 and 9 ft wide and 25–60 ft tall. The equipment types typ-ically included in this category are straddle trucks, straddle-reach trucks, sideloader trucks, turret trucks, and hybrid trucks.

A.2.c.i. Straddle Truck   A straddle truck is most often used in warehouses where aisle space is scarce and/or excessively expensive. The principle is to pro-vide load and vehicle stability using outriggers instead of counterbalanced weight, thereby reducing the aisle-width requirement to 7 to 9 feet. To access loads in stor-age, the outriggers are driven into the rack, allowing the forks to come flush with the pallet face. Hence, it is necessary to support the floor-level load on rack beams. (See Figure 5.73.)

A.2.c.ii. Straddle-Reach Truck   The straddle-reach truck was developed from the conventional straddle truck by shortening the outriggers on the straddle truck and providing a reach capability. In so doing, the outriggers do not have to be driv-en under the floor-level load to allow access to the storage positions. Hence, no rack beam is required at the floor level, conserving rack cost and vertical storage requirements. (See Figure 5.74.)

Two basic straddle-reach truck designs are available: mast- and fork-reach trucks. The mast-reach design consists of a set of tracks along the outriggers that

Figure 5.73   Straddle truck. (*Courtesy of Yale Materials Handling Corp.*)

Figure 5.74   Straddle reach truck with scissor mechanism. (*Courtesy of The Raymond Corp.*)

support the mast. The fork-reach design consists of a pantograph or scissors mounted on the mast.

The double-deep-reach truck, a variation of the fork-reach design, allows the forks to be extended to a depth that permits loads to be stored two deep.

A.2.c.iii. Sideloader Truck   The sideloader truck loads and unloads from one side, thus eliminating the need to turn in the aisle to access storage positions. There are two basic sideloader designs. Either the entire mast moves on a set of tracks transversely across the vehicle, or the forks project from a fixed mast on a pantograph. (See Figure 5.75.)

Aisle-width requirements are less than that for straddle trucks and reach trucks. A typical aisle would be 5–7 ft wide, rail or wire guided. Sideloaders can generally access loads up to 30 ft high.

The sideloader truck must enter the correct end of the aisle to access a particular location, which adds an additional burden to routing the truck. A variety of load types can be handled using a sideloader. The vehicle's configuration particularly lends itself to storing long loads in a cantilever rack.

A.2.c.iv. Turret Truck   The swing-mast truck, turret truck, and shuttle truck are members of the modern family of designs that do not require the vehicle to make a turn within the aisle to store or retrieve a pallet. Rather, the load is lifted

Figure 5.75   Sideloader truck. (*Courtesy of The Raymond Corp.*)

Figure 5.76 Swing-mast truck. (*Courtesy of Drexel Industries LLC*)

either by forks that swing on the mast, a mast that swings from the vehicle, or a shuttle fork mechanism. (See Figures 5.76 and 5.77.)

Generally, these types of trucks provide access to load positions at heights up to 40 ft, which provides the opportunity to increase storage density where floor space is limited. They can also run in aisles 5–7 ft wide, further increasing storage density.

Narrow-aisle side-reach trucks generally have good maneuverability outside the aisle, and some of the designs with telescoping masts may be driven into a shipping trailer. Since narrow-aisle side reach trucks do not turn in the aisle, the vehicle may

Figure 5.77 Operator on-board turret truck. (*Courtesy of Yale Materials Handling Corp.*)

be wire guided or the aisles may be rail guided, allowing for greater speed and safety in the aisle and reducing the chances of damage to the vehicle and/or rack.

    A.2.c.v. Hybrid Truck   A hybrid truck (so-called because of its resemblance to an automated storage/retrieval (AS/R) vehicles) is similar to a turret truck, except the operator's cab is lifted with the load. The turret vehicle evolved from the S/R machine used in automated storage/retrieval systems. Unlike the S/R machine, the hybrid truck is not captive to an aisle, but may leave one aisle and enter another. Available models are somewhat clumsy outside the aisle, but operate within the aisle at a high throughput rate. Hybrid trucks operate in aisle widths ranging from 5 to 7 ft, allow rack storage up to 60 ft high in a rack-supported building, and may include an enclosed operator's cab, which may be heated and/or air conditioned. Sophisticated hybrid trucks are able to travel horizontally and vertically simultaneously to a load position. (See Figure 5.78.)

    Excellent floor-space utilization and the ability to transfer vehicles between storage aisles are the major benefits of hybrid trucks. The lack of reconfiguration flexibility, high capital expense, and high dimensional tolerance in the rack are the disadvantages of hybrid trucks.

A.2.d. Automated Storage and Retrieval Machines.   An automated storage and retrieval system (AS/RS) is defined by the AS/RS product section of the Material Handling Institute as a storage system that uses fixed-path storage and retrieval (S/R) machines running on one or more rails between fixed arrays of storage racks. (See Figure 5.79.)

    A unit load AS/RS usually handles loads in excess of 1000 pounds and is used for raw material, work-in-process, and finished goods. The number of systems installed in the United States is in the hundreds, and installations are commonplace in all major industries.

    A typical AS/RS operation involves the S/R machine picking up a load at the front of the system, transporting the load to an empty location, depositing the load in the empty location, and returning empty to the input/output (I/O) point. Such an operation is called a single command (SC) operation. Single commands accomplish either a storage or a retrieval between successive visits to the I/O point. A more efficient operation is a dual command (DC) operation. A DC involves the S/R machine picking up a load at the I/O point, traveling loaded to an empty location (typically the closest empty location to the I/O point), depositing the load, traveling empty to the location of the desired retrieval, picking up the load, traveling loaded to the I/O point, and depositing the load. The key idea is that in a DC, two operations, a storage and a retrieval, are accomplished between successive visits to the I/O point.

    A unique feature of the S/R machine travel is that vertical and horizontal travel occur simultaneously. Consequently, the time to travel to any destination in the rack is the maximum of the horizontal and vertical travel times required to reach the destination from the origin. Horizontal travel speeds are up to 600 ft/min; vertical, 150 ft/min.

    The typical unit load AS/RS configuration, if there is such a thing, would include unit loads stored one deep (i.e., single deep), in long narrow aisles (4 to 5 ft wide), each of which contains a S/R machine. The one I/O point would be located at the lowest level of storage and at one end of this system.

Figure 5.78 Hybrid truck. (*Courtesy of Jervis B. Webb Co.*)

More often than not, however, one of the parameters defining the system is atypical. The possible variations include the depth of storage, the number of S/R machines assigned to an aisle, and the number and location of I/O points. These variations are described in more detail as follows.

When the variety of loads stored in the system is relatively low, throughput requirements are moderate to high, and the number of loads to be stored is high, it is often beneficial to store loads more than one deep in the rack. Alternative configurations include:

Figure 5.79   Unit load AS/RS. (*Courtesy of Siemens Dematic Corp.*)

- *Double-deep storage with single-load width aisles.* Loads of the same stock-keeping unit (SKU) are typically stored in the same location. A modified S/R machine is capable of reaching into the rack for the second load.
- *Double-deep storage with double-load-width aisles.* The S/R machine carries two loads at a time and inserts them simultaneously into the double-deep cubicle.
- *Deep-lane storage with single-load-width aisles.* An S/R machine dedicated to storing will store material into the lanes on either side of the aisle. The lanes may hold up to 10 loads each. On the output side, a dedicated retrieval machine will remove material from the racks. The racks may be dynamic, having gravity or powered conveyor lanes.
- *Rack-entry module (REM) systems* in which a REM moves into the rack system and places/receives loads onto/from special rails in the rack.
- *Twin-shuttle systems* in which the S/R machine is equipped with two shuttle tables and is capable of handling two unit loads per trip.

- *Multiple S/R machines operating within the same aisle.* Though rare, these systems are used to simultaneously achieve high throughput and storage density.

Another variation of the typical configuration is the use of transfer cars to transport S/R machines between aisles. Transfer cars are used when the storage requirement is high relative to the throughput requirement. In such a case, the throughput requirement does not justify the purchase of an S/R machine for each aisle, yet the number of aisles of storage must be sufficient to accommodate the storage requirement.

A third system variation is the number and location of I/O points. Throughput requirements or facility design constraints may mandate multiple I/O points at locations other than the lower-left-hand corner of the rack. Multiple I/O points might be used to separate inbound and outbound loads and/or to provide additional throughput capacity. Alternative I/O locations include the type of system at the end of the rack (some AS/RSs are built underground) and the middle of the rack.

B. Small Load Storage and Retrieval Equipment

Small load storage and retrieval equipment is classified as *operator-to-stock* (OTS, sometimes referred to as *man-to-part* or *in-the-aisle* systems) if the operator travels to the storage location to retrieve material and as *stock-to-operator* (sometimes referred to as *part-to-man* or *end-of-aisle* systems) if the material is mechanically transported to an operator for retrieval. Stock-to-operator equipment types often offer higher productivity, easier supervision, and better item security and protection than operator-to-stock alternatives. At the same time, stock-to-operator options often are more expensive, more difficult to reconfigure, and require more maintenance than operator-to-stock options.

In OTS options, the design and selection of the storage mode may be separated from the design and selection of the retrieval mechanism. Hence, OTS storage equipment and OTS retrieval equipment are described separately.

B.1. Operator-to-Stock Storage Equipment

The three principal operator-to-stock equipment types for housing small loads are bin shelving, modular storage drawers in cabinets, and gravity flow rack. To improve space utilization, each of the storage systems can be incorporated into mezzanine or mobile storage configurations.

B.1.a. Bin Shelving.   Bin shelving is the oldest and still the most popular (in terms of sales volume and number of systems in use) equipment alternative in use for small parts order picking. The low initial cost, easy reconfigurability, easy installation, and low maintenance requirements are at the heart of this popularity. (See Figure 5.80.)

It is important to recall that the lowest initial cost alternative may not be the most cost effective alternative, or the alternative that meets the prioritized needs of an operation. With bin shelving, savings in initial cost and maintenance may be offset by inflated space and labor requirements.

Space is frequently underutilized in bin shelving, since the full inside dimensions of each unit are rarely usable. Also, since people are extracting the items, the

Figure 5.80   Bin shelving. (*Courtesy of Equipto*)

height of shelving units is limited by the reaching height of a human being. As a result, the available building cube may also be underutilized.

The consequences of low space utilization are twofold. First, low space utilization means that a large amount of square footage is required to store the products. The more expensive it is to own and operate the space, the more expensive low space utilization becomes. Second, the greater the square footage, the greater the area that must be traveled by the order pickers, and thus, the greater the labor requirement and costs.

Two additional disadvantages of bin shelving are supervisory problems and item security/protection problems. Supervisory problems arise because it is difficult to supervise people through a maze of bin shelving units. Security and item protection problems arise because bin shelving is one of a class of open systems (i.e., the items are exposed to and accessible from the picking aisles).

As with all of the equipment types, these disadvantages must be evaluated and compared with the advantages of low initial cost and low maintenance requirements in order to make an appropriate equipment selection.

B.1.b. Modular Storage Drawers in Cabinets.   Modular storage drawers/cabinets are called modular because each storage cabinet houses modular storage drawers that are subdivided into modular storage compartments. Drawer heights range from 3 in. to 24 in., and each drawer may hold up to 400 lb worth of material. The storage cabinets can be thought of as shelving units that house storage drawers. (See Figure 5.81.)

Figure 5.81   Modular storage drawers in cabinets. (*Courtesy of Equipto*)

The primary advantage of storage drawers/cabinets over bin shelving is the large number of SKUs that can be stored and presented to the order picker in a small area. The cabinet and drawer suppliers inform us that one drawer can hold from 1 to 100 SKUs (depending on the size, shape, and inventory levels of the items), and that a typical storage cabinet can store the equivalent of 2 to 4 shelving units worth of material. This dense storage stems primarily from the ability to create item housing configurations within a drawer that very closely match the cubic storage requirements of each SKU. Also, since the drawers are pulled out into the aisle for picking, space does not have to be provided above each SKU to provide room for the order picker's hand and forearm. This reach space must be provided in bin shelving storage, otherwise items deep in the unit could not be accessed.

Several benefits accrue from the high-density storage characteristics of storage drawer systems. First, and obviously, the more material that can be packed into a smaller area, the smaller the space requirement. Hence, space costs are reduced. When the value of space is at a true premium, such as on a battleship, on an airplane, or on the manufacturing floor, the reduction in space requirements alone can be enough to justify the use of storage drawers and cabinets. A second benefit resulting from a reduction in square-footage requirements is a subsequent reduction in the travel time and hence labor requirements for order picking.

Additional benefits achieved by the use of storage drawers include improved picking accuracy and protection for the items from the environment. Picking accuracy is improved over that in shelving units because the order picker's sight lines to the items are improved, and the quantity of light falling on the items to be extracted is increased. With bin shelving, the physical extraction of items may occur anywhere from floor level to 7 ft off the ground, with the order picker having to reach into the shelving unit itself to achieve the pick. With storage drawers, the drawer is pulled out into the picking aisle for item extraction. The order picker looks down into the contents of the drawer, which are illuminated by the light source for the picking aisle. (The fact that the order picker must look down in the drawer necessitates that storage cabinets be less than 5 ft in height.) Item security and protection are achieved since the drawers can be closed and locked when items are not being extracted from them.

As one would expect, these benefits are not for free. Storage cabinets equipped with drawers are relatively expensive. Price is primarily a function of the number of drawers and the amount of sheet metal in the cabinet.

B.1.c. Carton Flow Rack. Carton flow rack is another popular OTS equipment alternative. Flow rack is typically used for active items that are stored in fairly uniform-sized and -shaped cartons. The cartons are placed in the back of the rack from the replenishment aisle, and advance/roll toward the pick face as cartons are depleted from the front. This back-to-front movement insures first-in-first-out (FIFO) turnover of the material. (See Figure 5.82.)

Figure 5.82.   Carton flow rack. (*Courtesy of Paltier LLC*)

Essentially, a section of flow rack is a bin shelving unit turned perpendicular to the picking aisle with rollers placed on the shelves. The deeper the sections, the greater the portion of warehouse space that will be devoted to storage as opposed to aisle space. Further gains in space efficiency can be achieved by making use of the cubic space over the flow rack for full pallet storage.

Flow rack costs depend on the length and weigth capacity of the racks. As is the case with bin shelving, flow rack has very low maintenance requirements and is available in a wide variety of standard section and lane sizes from a number of suppliers.

The fact that just one carton of each line item is located on the pick face means that a large number of SKUs are presented to the picker over a small area. Hence, walking and therefore labor requirements can be reduced with an efficient layout.

B.1.d. Mezzanine. Bin shelving, modular storage cabinets, flow rack, and even carousels can be placed on a mezzanine. The advantage of using a mezzanine is that nearly twice as much material can be stored in the original square footage, inexpensively. The major design issues for a mezzanine are the selection of the proper grade of mezzanine for the loading that will be experienced, the design of the material handling system to service the upper levels of the mezzanine, and the utilization of the available clear height. At least 14 ft of clear height should be available for a mezzanine to be considered. (See Figure 5.83.)

B.1.e. Mobile Storage. Bin shelving, modular storage cabinets, and flow rack can all be "mobilized." The most popular method of mobilization is the *train-track* method. Parallel tracks are cut into the floor, and wheels are placed on the bottom

Figure 5.83 Mezzanine. (*Courtesy of Paltier LLC*)

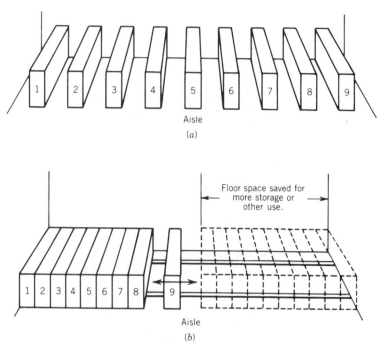

Figure 5.84 Mobile storage or sliding racks.

of the storage equipment to create mobilized equipment. The space savings accrue from the fact that only one aisle is needed between all the rows of storage equipment. The aisle is created by separating two adjacent rows of equipment. As a result, the aisle "floats" in the configuration between adjacent rows of equipment. The storage equipment is moved by simply sliding the equipment along the tracks, by turning a crank located at the end of each storage row, or by invoking electric motors that may provide the motive power. The disadvantage to this approach is the increased time required to access the items. Every time an item must be accessed, the corresponding storage aisle must be created. (See Figure 5.84.)

B.2. Operator-to-Stock Retrieval Equipment

In operator-to-stock alternatives, the operator either walks or rides a vehicle to the pick location. The four retrieval options distinguished here include picking carts, order picker trucks, person-aboard AS/R machines, and robotic retrieval.

B.2.a. Picking Cart. A variety of picking carts are available to facilitate the accumulation, sortation, and/or packing of orders as an order picker makes a picking tour. The carts are designed to allow an order picker to pick multiple orders on a picking tour, thus dramatically improving productivity as opposed to strict single-order picking for small orders. The most conventional vehicles provide dividers for order sortation, a place to hold paperwork and marking instruments, and a step ladder for picking at levels slightly above reaching height. Additional levels of sophistication and cost bring powered carts, light-aided sortation, on-board computer terminals, and on-board weighing. (See Figure 5.85.)

Figure 5.85 Mobile batch pick cart. (*Courtesy of SI Systems, Inc.*)

B.2.b. Order Picker Truck. The order picker truck, sometimes referred to as a cherry picker or stock picker, allows an operator to travel to retrieval locations well above floor level. In so doing the operator's productivity is reduced. However, productivity can be enhanced by minimizing vertical travel through popularity-based storage and/or intelligent pick tour construction. (See Figure 5.86.)

B.2.c. Person-Aboard Automated Storage/Retrieval Machine. The person-aboard AS/R machine, as the name implies, is an automated storage and retrieval machine in which the picker rides aboard a storage/retrieval machine to, from and between retrieval locations. The storage modes may be stacked bin shelving units, stacked storage cabinets, and/or pallet rack. The storage/retrieval machine may be aisle captive or free roaming. (See Figure 5.87.)

Typically, the picker will leave from the front of the system at floor level and visit enough storage locations to fill one or multiple orders, depending on order size. Sortation can take place on board if enough containers are provided on the storage/retrieval (S/R) machine. The person-aboard AS/R machine offers significant square-footage and order picking time reductions over the previously described picker-to-part systems. Square-footage reductions are available because storage heights are no longer limited by the reach height of the order picker. Shelves or storage cabinets can be stacked as high as floor loading, weight capacity, throughput requirements, and/or ceiling heights will permit. Retrieval times are reduced because the motive power for traveling is provided automatically, hence freeing the operator to do productive work while traveling and because search time is reduced since the operator is automatically delivered to the correct location.

Figure 5.86   Order picker truck. (*Courtesy of Raymond Corp.*)

B.3.   Stock-to-Operator Equipment

The major difference between stock-to-operator and operator-to-stock equipment is the answer to the question, Does the operator travel to the stock location, or does the stock travel to the operator? If the stock travels to the operator, the equipment is classified as stock-to-operator equipment.

In stock-to-operator options, the travel time component of total order picking time is shifted from the operator to a device for bringing locations to the picker. Also, the search time component of total order picking time is significantly reduced since the correct location is automatically presented to the operator. In well-designed systems, the result is a large increase in productivity. In poorly designed systems, the improvements may be negligible if the operator is required to wait for the device to present him or her with stock.

The two most popular classes of stock-to-operator equipment are carousels and miniload automated storage and retrieval machines. A more expensive, yet highly productive class is automated dispensers. Each equipment type is described as follows.

B.3.a. Carousels.   Carousels, as the name implies, are mechanical devices that house and rotate items for storage and retrieval. Three classes of carousels are

Figure 5.87   Person-aboard AS/RS. (*Courtesy of Demag*)

currently available for small load storage and retrieval applications—horizontal carousels, vertical carousels, and independently rotating racks.

B.3.a.i. Horizontal Carousel   Horizontal carousel is a linked series of rotating bins of adjustable shelves driven on the top or on the bottom by a drive motor unit. Rotation takes place about an axis perpendicular to the floor at between 80 and 100 ft/min. (See Figure 5.88.)

Items are extracted from the carousel by order pickers who occupy fixed positions in front of the carousel(s). Operators may be responsible for controlling the rotation of the carousel. Manual control is achieved via a keypad, which tells the carousel which bin location to rotate forward, and a foot pedal, which releases the carousel to rotate. Carousels may also be computer controlled, in which case the sequence of pick locations is stored in the computer and brought forward automatically.

A popular design enhancement is the installation of a light tree in front of a pair of carousels. The light tree has light displays for each level of the carousel. The correct retrieval location and quantity are displayed on the light. These displays significantly improve the productivity and accuracy of the storage and retrieval transactions.

A management option with carousels is the flexible scheduling of operators to carousels. If an operator is assigned to one carousel unit, he or she must wait for

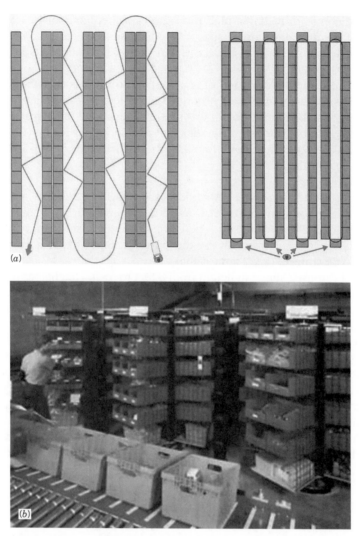

Figure 5.88 Horizontal carousel. (*Courtesy of White Systems, Inc.*)

the carousel to rotate to the correct location between picks. If an operator is assigned to two or more carousels, he or she may pick from one carousel while the other is rotating to the next pick location. Remember, the objective of stock-to-operator alternatives is to keep the operators busy extracting stock. Humans are excellent extractors of items; the flexibility of our limbs and muscles provides us with this capability. Humans are not efficient searchers, walkers, or waiters.

Horizontal carousels vary in length from 15 ft to 100 ft, and in height from 6 ft to 25 ft. The length and height of the units are dictated by the pick rate requirements and building restrictions. The longer the carousel, the more time required, on average, to rotate the carousel to the desired location. Also, the taller the carousel, the more time required to access the items. Heights over 6 ft necessitate the use of ladders or robot arms on vertical masts to access the items.

In addition to providing a high pick rate capacity, horizontal carousels make good use of the available storage space. Very little space is required between adjacent carousels, and the only lost space is that between parallel sections of bins on the same carousel unit.

One important disadvantage of horizontal carousels is that the shelves and bins are open. Consequently, item security and protection can be a problem.

A "twin-bin" horizontal carousel was recently introduced to the material handling market. In the twin-bin carousel, the traditional carousel carrier is split vertically in half and rotated 90 degrees. This allows for shallower carriers, thus improving the storage density for small parts.

B.3.a.ii. Vertical Carousel A vertical carousel is a horizontal carousel turned on its end and enclosed in sheet metal. As with horizontal carousels, an order picker operates one or multiple carousels. The carousels are indexed either automatically via computer control, or manually by the order picker working a keypad on the carousel's work surface. (See Figure 5.89.)

Vertical carousels range in height from 8 feet to 35 ft. Heights (as lengths were for horizontal carousels) are dictated by throughput requirements and building restrictions. The taller the system, the longer it will take, on average, to rotate the desired bin location to the pick station.

Retrieval times for vertical carousels are theoretically less than those for horizontal carousels. The decrease results from the items always being presented at the order picker's waist level. This eliminates the stooping and reaching that goes on with horizontal carousels, further reduces search time, and promotes more accurate picking. (Some of the gains in item extract time are negated by the slower rotation speed of the vertical carousel. Recall that the direction of rotation is against gravity.)

Additional benefits provided by the vertical carousel include excellent item protection and security. In the vertical carousel, only one shelf of items is exposed at one

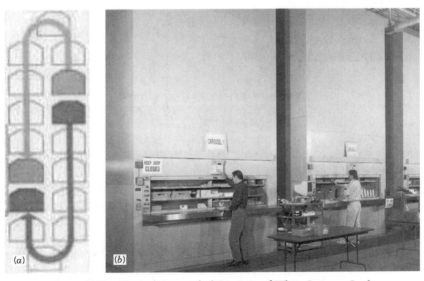

Figure 5.89 Vertical carousel. (*Courtesy of White Systems, Inc.*)

time, and the entire contents of the carousel can be locked up. On a per-cubic-foot-of-storage basis, vertical carousels are typically 400/(T-70%) more expensive than horizontal carousels. The additional cost of vertical carousels over horizontal carousels is attributed to the sheet metal enclosure, and the extra power required to rotate against the force of gravity.

B.3.a.iii. Independent Rotating Rack An independent rotating rack (IRR) carousel is like multiple one-level horizontal carousels stacked on top of one another. As the name implies, each level rotates independently. As a result, several storage locations are ready to be accessed by an operator at all times. Consequently, the operator is continuously picking.

Remote order picking, assembly, and/or order or kit staging are more common applications of independent rotating racks. In those applications, a robot is positioned at the end of the IRR to store and retrieve inbound and outbound unit loads. In that way, the IRR is used to simultaneously achieve high throughput and storage density.

Clearly, for each level to operate independently, each level must have its own power and communication link. These requirements force the price of independent rotating rack well beyond that of vertical or horizontal carousels.

*B.3.b. **Miniload Automated Storage and Retrieval Machine.*** In the miniload automated storage and retrieval system, a miniload storage/retrieval (S/R) machine travels horizontally and vertically simultaneously in a storage aisle, transporting storage containers to and from an order picking station located at one end of the system. The order picking station typically has two pick positions. As the order picker is picking from the container in the left pick position, the S/R machine is taking the container from the right pick position back to its location in the rack and returning with the next container. The result is the order picker rotating between the left and right pick positions. (A system enhancement used to improve the throughput of miniload operations is to assign multiple S/R devices per aisle.) (See Figure 5.90.)

The sequence of containers to be processed is determined manually by the order picker keying in the desired line item numbers or rack locations on a keypad, or the sequence is generated and processed automatically by computer control.

Miniloads vary in height from 8 to 50 ft, and in length from 40 to 200 ft. As in the case with carousels, the height and length of the system are dictated by the throughput requirements and building restrictions. The longer and taller the system, the longer the time required to access the containers. However, the longer and taller the system, the fewer the aisles and S/R machines that will have to be purchased.

The transaction rate capacity of the miniload is governed by the ability of the S/R machine (which travels approximately 500 ft/min horizontally and 120 ft/min vertically) to continuously present the order picker with unprocessed storage containers. This ability, coupled with the human factors benefits of presenting the containers to the picker at waist height in a well-lit area, can produce impressive pick rates.

Square-footage requirements are reduced for the miniload due to the ability to store material up to 50 ft high, the ability to size and shape the storage containers and the subdivisions of those containers to very closely match the storage volume requirements of each SKU, and an aisle width that is determined solely by the width of the storage containers.

Figure 5.90   Miniload AS/RS. (*Courtesy of Siemens Dematic Corp.*)

The disadvantages of the miniload system are probably already apparent. As the most sophisticated of the system alternatives described thus far, it should come as no surprise that the miniload carries the highest price tag of any of the order picking system alternatives. Another result of its sophistication is the significant engineering and design effort that accompanies each system. Finally, greater sophistication leads to greater maintenance requirements. It is only through a disciplined maintenance program that miniload suppliers are able to advertise uptime percentages between 97% and 99.9%.

Recently, preengineered modular miniload system designs have been introduced to the U.S. material handling system market. Preengineered systems offer the same range and degree of benefits as conventional miniload system designs, are less expensive, and are delivered and installed sooner.

B.3.c. High-Density Vertical Storage.   High-density vertical storage equipment is a relatively new addition to the family of small load storage and retrieval equipment. Simply put, this type of storage is a small miniload AS/RS turned on its side. Systems range in height from 15 to 35 ft and can be interfaced with automated material

Figure 5.91   High-density vertical storage. (*Courtesy of Richards-Wilcox, Inc.*)

transport systems. High-density vertical storage offers many of the benefits of miniload systems at a reduced price. (See Figure 5.91.)

B.3.d. Automatic Dispenser.   An automatic dispenser works much like vending machines for small items of uniform size and shape. Each item is allocated a vertical dispenser ranging from 2 to 6 inches wide and from 3 to 5 ft tall. (The width of each dispenser is easily adjusted to accommodate variable product sizes.) The dispensing mechanism acts to kick the unit of product at the bottom of the dispenser out onto a conveyor running between two rows of dispensers configured as an A-frame over a belt conveyor. A tiny vacuum conveyor or small finger on a chain conveyor is used to dispense the items. (See Figure 5.92.)

Virtual order zones begin at one end of the conveyor and pass by each dispenser. If an item is required in the order zone it is dispensed onto the conveyor. Merchandise is accumulated at the end of the belt conveyor into a tote pan or carton. A single dispenser can dispense at a rate of up to 6 units per second. Automatic item pickers are popular in industries with high throughput for small items of uniform size and shape. Cosmetics, wholesale drugs, compact discs, videos, publications, and polybagged garments are some examples.

Replenishment is performed manually from the back of the system. This manual replenishment operation significantly cuts into the savings in picking labor requirements associated with pick rates on the order of 1500 picks per hour per pick head.

One new design for automated dispensing machines is an inverted A-frame that streamlines the replenishment of automated dispensers and increases the storage density along the picking line. Another new design allows automated dispensing for polybagged garments.

Figure 5.92   Automatic dispenser. (*Courtesy of SI Systems, Inc.*)

IV. AUTOMATED DATA COLLECTION AND COMMUNICATION EQUIPMENT

Automated status control of material requires that real-time awareness of the location, amount, origin, destination, and schedule of material be achieved automatically. This objective is in fact the function of automatic identification and communication technologies—technologies that permit real-time, nearly flawless data collection and communication. Examples of automatic identification and communication technologies at work include

- A vision system reading and interpreting labels to identify the proper destination for a carton traveling on a sortation conveyor
- A laser scanner to relay the inventory levels of a small parts warehouse to a computer via radio frequency communication
- A voice recognition system to identify parts received at the receiving dock.
- A radio frequency (RF) or surface acoustical wave (SAW) tag used to permanently identify a tote pan or pallet
- A card with a magnetic stripe that travels with a unit load to identify the load through the distribution channels

For discussion purposes we distinguish automatic identification systems from automatic communication systems. Automatic identification systems allow machines to identify material and capture key data concerning the status of the material. Automatic communication systems allow paperless communication of the captured data.

A. Automatic Identification and Recognition

The list of automatic identification and recognition technologies is expanding and includes bar coding, optical character recognition, radio frequency tags, magnetic stripes, and machine vision.

A.1. Bar Coding

A bar code system includes the bar code itself, bar code reader(s), and bar code printer(s). Each system component is described below. More comprehensive descriptions are provided by [23, 77].

A.1.a. Bar Codes. A bar code consists of a number of printed bars and intervening spaces. The structure of unique bar/space patterns represents various alphanumeric characters. Examples appear on virtually every consumer item allowing checkout clerks in retail and grocery stores to scan the code on the item to automatically record its identification and price. (See Figure 5.93.)

The same bar/space pattern may represent different alphanumeric characters in different codes. Primary codes or symbologies for which standards have been developed include

- *Code 39:* An alphanumeric code adopted by a wide number of industry and government organizations for both individual product identification and shipping package/container identification.

Figure 5.93 Sample automatic identification product codes.

- *Interleaved 2 of 5 Code:* A compact, numeric-only code still used in a number of applications where alphanumeric encoding is not required.

- *Codabar:* One of the earlier symbols developed, this symbol permits encoding of the numeric-character set, six unique control characters, and four unique stop/start characters that can be used to distinguish different item classifications. It is primarily used in nongrocery retail point-of-sale applications, blood banks, and libraries.

- *Code 93:* Accommodating all 128 ASCII characters plus 43 alphanumeric characters and four control characters, Code 93 offers the highest alphanumeric data density of the six standard symbologies. In addition to allowing for positive switching between ASCII and alphanumeric, the code uses two check characters to ensure data integrity.

- *Code 128:* Provides the architecture for high-density encoding of the full 128-character ASCII set, variable-length fields, and elaborate character-by-character and full symbol integrity checking. Provides the highest numeric-only data density. Adopted in 1989 by the Uniform Code Council (U.S.) and the International Article Number Association (EAN) for shipping container identification.

- *UPC/EAN:* The numeric-only symbols developed for grocery supermarket point-of-sale applications and now widely used in a variety of other retailing environments. Fixed-length code suitable for unique manufacturer and item identification only.

- *Stacked Symbologies:* Although a consensus standard has not yet emerged, the health and electronics industries have initiated programs to evaluate the feasibility of using Code 16K or Code 49, two microsymbologies that offer significant potential for small item encoding. Packing data in from 2 to 16 stacked rows, Code 16K accommodates the full 128-character ASCII set and permits the encoding of up to 77 characters in an area of less than 0.5 in.[2] Comparable in terms of data density, Code 49 also handles the full ASCII character set. It encodes data in from two to eight rows and has a capacity of up to 49 alphanumeric characters per symbol. Two-dimensional bar codes are a recent development in bar code symbology. Two-dimensional codes yield dramatic improvements in data density, the amount of information encoded per square inch.

A.1.b. Bar Code Readers. Bar codes are read by both contract and noncontact scanners. Contact scanners can be portable or stationary and use a wand or light pen. As the wand/pen is manually passed across the bar code, the scanner emits either white or infrared light from the wand/pen tip and reads the light pattern that is reflected from the bar code. This information is stored in solid-state memory for subsequent transmission to a computer. More sophisticated (and expensive) hand-held units include fixed- and moving-beam scanners employing light-emitting diodes (LEDs), low-power helium-neon lasers, or the latest development, laser diodes. These devices can scan bar code labels from distances up to 6 ft.

A.1.b.i. Contact Reader A contact reader is an excellent substitute for keyboard or manual data entry. Alphanumeric information is processed at a rate of up

Figure 5.94   Contact reader. (*Courtesy of LXE*)

to 50 in./min, and the error rate for a basic scanner connected to its decode is 1 in 1,000,000 reads. Light pens and wand scanners are examples of contact readers. (See Figure 5.94.)

A.1.b.ii. Noncontact Reader   A noncontact reader is usually stationary and includes fixed-beam, moving-beam scanners, and charged couple device (CCD) scanners. These scanners employ fixed-beam, moving-beam, video camera, or raster scanning technology to take from one to several hundred looks at the code as it passes. Most bar code scanners read codes bidirectionally by virtue of sophisticated decoding electronics, which distinguish the unique start/stop codes peculiar to each symbology and decipher them accordingly. Further, the majority of scanner suppliers now provide equipment with an autodiscrimination feature that permits recognition, reading, and verification of multiple symbol formats with no internal or external adjustments. Finally, at least two suppliers have introduced omnidirectional scanners for industrial applications that are capable of reading a code at high speed throughout a large field of view regardless of its orientation. (See Figure 5.95.)

Fixed-beam readers use a stationary light source to scan a bar code. They depend on the motion of the object to be scanned to move past the beam. Fixed-beam readers rely on consistent, accurate code placement on the moving object.

Moving-beam scanners employ a moving light source to search for codes on moving objects. Because the beam of light is moving, the placement of the code on the object is not critical. In fact, some scanners can read codes accurately on items moving at a speed of 1000 ft/min.

CCD scanners have been only recently used to read bar codes. A CCD scanner is more like a camera than a bar code scanner. By changing camera lenses and focal lengths, the scanners read various bar codes at various distances. Like any

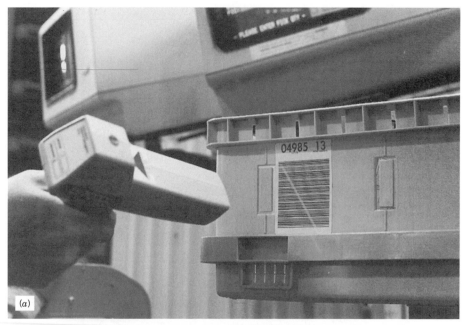

Figure 5.95(*a*)   Handheld bar code scanner.

Figure 5.95(*b*)   Stationery bar code scanner.

Figure 5.96   Omnidirectional bar code reader.

camera system, the quality of the picture (the accuracy of the read) depends on the available light. When installed correctly, the CCD scanner is 99% accurate.

Omnidirectional scanners are a recent development in bar-code-reading technology. Omnidirectional scanners can successfully read and interpret a bar code regardless of its orientation. (See Figure 5.96.)

A.1.c. Bar Code Printers.   Bar code printers are the bar code system components that typically receive the least attention. However, the design and selection of the printer to a great extent determines the ultimate success of the bar code system since the quality of the printed code is critical to the acceptance and capacity of the system. There are five main classes of bar code printers—laser, thermal transfer, serial, impact, and ink jet.

A.2.   Optical Character Recognition

Optical characters are human and machine readable. An example is the account number printed along the bottom of most bank checks. Optical character recognition (OCR) systems read and interpret alphanumeric data so that people as well as computers can interpret the information contained in the data.

OCR labels are read with handheld scanners, much like bar codes. Optical character recognition systems operate at slower read rates than bar code systems and are priced about the same. OCR systems are attractive when both human- and machine-readable capabilities are required.

Until recently, the commercial applications of optical character recognition have been confined to document reading and limited use for merchandise tag reading at the retail point of sale. Without tight control of character printing and the reading environment, OCR's performance has not met the criteria established by other automatic identification techniques. A single printing anomaly, such as an inkspot or void can easily obscure or transpose an OCR character, rendering the label unreadable or liable to misreading. On the other hand, where encoding space is at a premium and the environment is relatively contaminant-free, OCR may be a viable alternative.

A.3. Radio Frequency Tag

Both the radio frequency (RF) tag and the surface accoustical wave (SAW) tag encode data on a chip that is encased in a tag. When a tag is within range of a special antenna, the chip is decoded by a tag reader. Radio frequency tags can be programmable or permanently coded and can be read from up to 30 ft away. SAW tags are permanently coded and can be read only within a 6-ft range. A common example is the tag now placed in the windshield of automobiles that permits high-speed auto recognition and account debiting at high-speed toll booths.

Radio frequency and SAW tags are typically used for permanent identification of a container, where advantage can be taken of the tag's durability. RF and SAW technologies are also attractive in harsh environments where printed codes may deteriorate and become illegible.

A.4. Magnetic Stripe

A magnetic stripe is used to store a large quantity of information in a small space. The stripe appears in common use on the back of credit cards and bank cards. The stripe is readable through dirt or grease, and the data contained in the stripe can be changed. The stripe must be read by contact, thus eliminating high-speed sortation applications. Magnetic stripe systems are generally more expensive than bar code systems.

A.5. Machine Vision

Machine vision relies on cameras to take pictures of objects and codes and send the pictures to a computer for interpretation. Machine vision systems "read" at moderate speeds, with excellent accuracy at least for limited environments. Obviously, these systems do not require contact with the object or code. However, the accuracy of a read is strictly dependent on the quality of light. Machine vision is becoming less costly but is still relatively expensive.

B. Automatic, Paperless Communication

Devices for automatically communicating with material handling operators include radio frequency data terminals, voice headsets, light and computer aids, and smart cards.

B.1. Radio Frequency Data Terminal

A handheld, arm-mounted, or lift-truck-mounted radio data terminal (RDT) is a reliable device for inventory and vehicle/driver management. The RDT incorporates a multicharacter display, full keyboard, and special function keys. It communicates and receives messages on a prescribed frequency via strategically located antennae and a host computer interface unit. Beyond the basic thrust toward tighter control of inventory, improved resource utilization is most often cited in justification of these devices. Further, the increasing availability of software packages that permit ROT linkage to existing plant or warehouse control systems greatly simplifies their implementation. The majority of ROTs installed in the plant environment use hand-held wands or scanners for data entry, product identification, and location verification. This marriage of technologies provides higher levels of speed, accuracy, and productivity than could be achieved by either technique alone. A recent advance places small radio frequency terminals and noncontact bar code readers on the wrists of warehouse operators. This allows handsfree operation of the technology. More detailed descriptions of the technology and its applications are provided in [23, 77]. (See Figure 5.97.)

Figure 5.97(a) Handheld RF terminal. (*Courtesy of Telxon Corp.*)

Figure 5.97(*b*) Wrist-mounted RF Terminal. (*Courtesy of Telxon Corp.*)

Figure 5.97(*c*) Truck-mounted RF Terminal. (*Courtesy of LXE.*)

Appendix 5.B MATERIAL HANDLING EQUIPMENT

Figure 5.98 Voice I/O system. (*Courtesy of Vocollect, Inc.*)

B.2. Voice Headset

Synthesized voice communication and human voice recognition (VR) systems are computer-based systems that translate computer data into synthesized speech and translate human speech into computer data. A headset with earphones and an attached microphone is used to interact with the computer system. Radio frequencies are used to transmit communications to mobile operators. These systems are attractive when an operator's hands and eyes must be freed up for productive operations. Though VR systems are in their infancy, some systems recognize up to 1000 words and are 99.5% accurate. Systems are in use today for receiving inspection, lift truck storage and retrieval operations, and order picking. (See Figure 5.98.)

B.3. Light and Computer Aids

The objectives of light or computer aids are to reduce the search time, extract time, and documentation time portions of total order picking time and to improve order picking accuracy. Search time is reduced by a computer automatically illuminating a lighted display at the correct storage or retrieval location. Extract time is reduced since the lighted display also indicates the correct quantity to pick. Documentation time is reduced by allowing the order picker to push a button at the pick location

Figure 5.99   Light-aided order picking. (*Courtesy of Kingston-Warren Corp.*)

to inform the host computer that the pick has been completed. The result is accurate order picking at a rate of up to 600 picks per man hour. Light displays are available for bin shelving, flow rack, and carousel systems. (See Figure 5.99.)

Computer aids provide a computer display over each pick station. On the display is a picture of the configuration of the storage container in that pick station. A light shines on the compartment within the container from which the pick is to be performed; the quantity to be picked is displayed on the computer screen. Systems of this type are also available for use with vertical carousels, miniload AS/RS, and vertical lift modules.

The basic solution approach of light- and computer-aided systems is to take the thinking out of order picking. Consequently, an unusual human factors challenge is introduced. The job may become too easy, and hence boring and even mentally degrading to some operators.

B.4. Smart Card

A smart card (essentially a credit card with a magnetic stripe) is now used to capture information ranging from employee identification, to the contents of a trailer

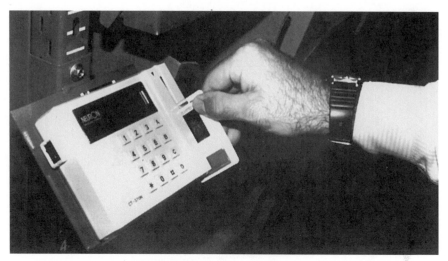

Figure 5.100 Smart card. (*Courtesy of Kingston-Warren Corp.*)

load of material, to the composition of an order picking tour. For example, at a large cosmetics distribution center, order picking tours are downloaded onto smart cards. The smart cards are in turn inserted into a smart card reader on each order picking cart. In so doing, the picking tour is illuminated on an electronic map of the warehouse appearing on the front of the cart. (See Figure 5.100.)

APPENDIX 5.C

AUTOMATED GUIDED VEHICLE SYSTEMS[1]

OVERVIEW

Automated guided vehicles (AGVs) are considered one of the most flexible types of material handling systems. These vehicles, which range from mail deliverers to 125-ton transporters, are equipped with a variety of functions from guidance and routing systems to traffic management and load transfers.

AGVs allow manufacturing systems to individually track items. While operating with a manufacturing or service environment the AGVs require continuous on-board or centralized computer control to coordinate their movements with other material handling devices or other AGVs. Extensive planning and strict adherence to pickup and dropoff points must be maintained. The precise movements and distances that the AGV follows require system discipline. Manual or automatic load transfer may be achieved once an AGV has reached its destination. An operator can board the AGV and position the vehicle to be loaded or unloaded. Otherwise, the AGV can transfer loads automatically using roller conveyors, clamping devices, lift tops, and so on, depending on the the material requirements.

Several methods of guidance can be implemented that allow the AGV to follow a fixed path or a free path. The determination of fixed or free path will depend on cost, flexibility requirements, and future expansion alternatives. While fixed-path systems may be less expensive, the installation may disrupt production, causing subsequent costs. Free-path guidance requires a software-programmable path and is more flexible to layout changes and capacity requirements than a fixed-path layout. Routing system and traffic management are also advantages of an AGV system.

---

[1]The material in this section is based on Tanchoco, Hendricks, and Muljono [60].

The routing system selects the vehicle with the optimum path to the required destination. Additionally, the AGVs must be able to avoid collisions within the facility with other vehicles, equipment, and operators.

Additional specifications and examples of applications in industry can be found in the following pages. Also, many of the pictures can be viewed larger. To determine if this is possible, place the cursor over the picture and if a hand appears you can select the picture to be viewed with greater detail.

TYPES OF AGVS

1. Pallet Trucks

Pallet trucks are used to transport palletized loads in distribution applications. They can be selfguided or operator-guided.

In order to load a pallet, there must be clearance below the load for it to be able to be lifted. The pallet truck may be boarded by an operator to be loaded or may auto-reverse into the pallet. If the truck is operating independently, the load must be accurately positioned. The auto-reverse capability will add considerable expense to the pallet truck.

The pallet truck then goes to a specified destination in the storage area to unload the pallet in a spur and automatically returns to be loaded.

2. Fork Trucks

High system discipline must be agreed on before designing the intricate path layout for a fork truck system. It is often the most expensive AGV; however, no humans are necessary for daily operation. The fork truck is capable of picking up and discharging a load automatically, stacking loads, and servicing palletized loads.

They are applied in total automation applications. The system requires auto-pickup/dropoff of loads from the floor or a stand. This AGV has greater flexibility in integrating other subsystems.

3. Unit Load Vehicles

For applications involving high-volume movement short to moderate distances, single or multiple unit load vehicles may be employed. As a means of efficient horizontal transportation, these vehicles are integrated into warehousing and distribution systems and storage/retrieval systems. They are capable of interfacing with conveyors, workstations, machine tools, load-stands, and automated systems.

There are several types of unit load vehicles equipped with decks to transfer loads differently. They include: lift and lower, powered or nonpowered roller, chain or belt decks, and custom decks with multiple components.

A tight turn and sweep radius, bidirectional and/or omnidirectional travel, and a speed of 2 mph contribute to high maneuverability in tight manufacturing settings. Their ability to operate independently requires accurate positioning for automatic transport and transfer.

3a.   Light Load Vehicles

Smaller in size than other AGVs, light load vehicles have capacity of up to 500 lb. Similar to unit load vehicles, they have high maneuverability in small, tight areas. These vehicles are often used to transfer small parts in trays and/or baskets from part storage to the workstation in the manufacturing environment. Electronics fabrication, small assembly manufacturing, and parts-kitting are a few of its applications. In some cases, the light load vehicle is driven by demand. Another possible application for light load vehicles is within the office environment for delivering mail or other documents.

3b.   Medium Load Vehicles

These vehicles range in carrying capacity from 500 lb to several tons. These vehicles are typically used in transporting drums, pallets, and bins. They are also not typically custom made like the heavy load vehicles.

3c.   Heavy Load Vehicles

These unit load vehicles are almost always custom made. They have lifting and carrying capacity of hundreds of tons. Their applications usually involve steel coils and other material traditionally transported by cranes.

4.   Towing Vehicles

Towing vehicles may eliminate or become an alternative to fork-lift trucks, manual trains, and even operators. They have capacity to tow 8000 to 50,000 lb. These AGVs can be applied in a variety of uses, including bulk movement into and out of the warehouse, distribution, and on the shop floor.

Several trailers can be attached to form a train that moves as a chain. There are two types of trailers that may be attached to the lead: closed and open unit. Closed unit trailers will not detach from the lead. Open unit trailers can be coupled and uncoupled manually or automatically. Automatic open unit trailers can eliminate operators.

5.   Work Platforms

Work platforms are used in two main manufacturing environments: assembly line and automotive assembly. The work platform allows the operator to work at an individual pace. There is also an ergonomic benefit to this vehicle. It is possible to design this vehicle to tip or tilt so that the operator will have better access to the subassembly being worked on.

Small light loads or subassemblies such as motors and transmissions in a serial assembly process may be produced on work platforms. The AGV moves down, the operator takes parts out of the tray on the AGV and assembles them onto the sub-assembly, then the platform moves to the next parts staging area and the next operator.

6.   Store/Retrieve Vehicles

Store/retrieve vehicles carry masts with a shuttle that moves vertically to store/retrieve loads in storage racks. This type of vehicle is preferred when the frequency of required moves is low and shuttles dedicated to each lane in an AS/RS system are not cost effective.

GUIDANCE METHODS

Guidance allows the vehicle to follow a predetermined route that is optimized for specific applications. In general, guidance can be classified into two major areas: wire guidance and nonwire guidance.

1.   Wire Guidance

Wire buried underground, which carries electrical current at predetermined frequencies, functions as a guide-path for AGVs. Different frequencies are used for defining different paths. In case a central computer is used to control the vehicles, communication between the central computer and individual vehicles is established by means of separate control wires laid in the same grooves as the guidance wires with messages transmitted on a separate carrier frequency and multiplexed to each vehicle in turn.

There are two advantages of an embedded-wire AGV system. First, it can handle dirty environments and heavy traffic in the workplace. Second, the system is free of interference from outside sources because it does not use electromagnetic signals or light in its operation. The disadvantages of the wire system are that it has a low degree of flexibility, the guide-path cannot be easily altered, and it is costly to repair when the wire is damaged.

The cost of creating a new path for a wire-guided vehicle is higher compared to most non-wire guidance. As long as routes are well defined and do not need to be changed frequently, wire guidance offers reliable material movement at relatively low cost. The system is not restricted in range, can be used in any type of environment, indoors and outdoors, and can run conveniently through tunnels, over bridges and into lifts.

2.   Non-Wire Guidance

There are situations where it is desirable that the vehicle should not be restricted to a fixed path. For example, the routes may need to be changed frequently, or exact load/unload locations may vary at short notice. For this reason, non-wire guidance is preferred to wire guidance. Based on industrial statistics, in 1994 non-wire guidance outnumbered wire guidance in North America.

Free-ranging vehicles incorporate some method for navigating their own routes to preassigned destinations. Sensing and correction are carried out continuously and virtually instantaneously using the vehicle's on-board computer. Another attraction of the free-ranging vehicle is that it is capable of avoiding obstacles in its path.

There are two general types of non-wire guidance method:

2a. Non-Wire Floor-Path Guidance

Non-wire floor-path guidance may take the form of optical guidance, or painted or chemical stripe. Advantages of non-wire floor-path guidance over traditional wire guidance are flexibility and easy setting.

i. Optical Guidance. For optical guidance, any type of light that comes from underneath of the vehicle is reflected. The reflected light signals are sensed by photosensors under the vehicle. Spot marks on the floor can be used in optical guidance. The mark is retroreflective glass bead approximately 10 mm in diameter. The vehicle is equipped with on-board sensors of a set of halogen lamps and a camera. The light from the halogen lamps is reflected by the spot mark and captured by the camera. Different spot mark configurations at intersections indicate the decision to be made by the vehicle. For example, a vehicle can be programmed to turn left once it hits a triangular spot mark at the intersection.

ii. Painted or Chemical Stripe. The floor is painted with a fluorescent color or chemical stripe that forms a guide-path for the AGV. The vehicles are equipped with sensors underneath to follow the guide-path. Applications include the use of ultraviolet light on the vehicle that stimulates chemical particles in the guide-path, generating visible light. The sensor on the AGV follows the path and reads various information codes along the guide-path. The information codes instruct the AGV to perform various operations and to identify stop locations.

2b. Non-Wire Non-Floor-Path Guidance

i. Dead Reckoning. Dead reckoning, usually using odometer, is a technique where the relative position of a vehicle is tracked by using an optical encoder to measure the precise rotation of the drive wheel and steer angle. The vehicle is equipped with a computer that calculates the position of the vehicle based on the wheel rotations referenced to a starting point of motion. This guidance system is frequently used in short-distance material handling. A major source of error of dead reckoning comes from wheel slippage. Wheel slippage can cause distance to be underestimated. Load variations can also distort the odometer wheels and introduce additional errors.

ii. Laser Beam. Several laser scanners scan laser beams to introduce a path. These scanners are mounted along the guide-path. The AGV is equipped with a photosensing system to capture the laser beam from the laser scanners. Using the signal related to the scanner's axis as steering information, the vehicle can be reliably guided off-line.

The laser beam guidance system is preferred to other guidance systems, such as the reflecting-tape system, in a superclean room application that provides more flexibility. It is also used in places that are not able to use ordinary pilot-line-guided vehicles. The laser beam can also be generated by the AGV to measure its position based on corner cubes that are located at fixed positions along a guide-path. The vehicle detects the corner cube by the intensity of the laser reflected back. For safety purposes, the energy density of the laser should not be hazardous to the human eye.

iii. Combination of Dead Reckoning and Laser Beam. A combination of dead reckoning and laser beam guidance system can also be implemented to cover the weaknesses of each guidance system. Triangularization using laser beam covers the error introduced by wheel slippage in dead reckoning method. The position information obtained by dead reckoning is modified when the laser beam system detects a corner.

iv. Beacon System. The system uses beacons that are fixed at reference locations in the plant. The vehicle's location can be tracked by a tracking device attached to the vehicle that measures its distance and direction from any one beacon. One type of beacon guidance system is laser beacon. The laser beacon is installed at a predetermined position and transmits angular data to the vehicle for positioning.

v. Inertial Guidance. The intertial guidance system is based on the use of an on-board gyroscope. The solid-state electronic gyroscope serves as the platform for the vehicle's guidance. In operation, the gyroscope provides heading information for the AGV's on-board control computer and increases the accuracy of autonomous travel. The biggest advance made by this guidance system is a proprietary positioning system that uses position codes placed along the guide-path. Each of these position codes emits a unique vehicle location address that the control computer uses to recalibrate its position. Two gyros (the free gyro and the rate gyro) are used in practice. In the free gyro system, a gyroscope along with an odometer is used to determine the vehicle's direction and to measure distance relative to a specific point. In the rate gyro system, a gas rate gyroscope and odometer are used to determine a vehicle's position and direction.

ROUTING

A routing system is used to select the vehicle that is positioned with the optimum path. A network controller gives the destination, while the on-board controller navigates the vehicle. If inductive guidance is used, incremental or absolute digital routing requires fixed positions within the floor. This requires the system to determine its own position and can be used only over short ranges. One disadvantage is that the magnetic devices for the floor positions can be very expensive. The incremental system counts the number of fixed floor magnets to determine routing. Errors occur when counting becomes frequent. The absolute system is superior to the incremental system.

TRAFFIC MANAGEMENT

In order to successfully manage the interaction of an AGV within any facility, the vehicle must have the ability to avoid collisions with other vehicles. This can be accomplished through programming in the guidance system. Additionally, the vehicle flow should be maximized in order to enhance the material handling system. Traffic tie-ups can cause unnecessary delays in production as the system may be awaiting the arrival of a vehicle for pickup or delivery. The system can be broken down into zones, blocks, and regions to alleviate the problem of traffic management.

LOAD TRANSFER

Picking up, delivering, and transferring loads from an AGV system may be accomplished manually or automatically. An operator can manually couple and uncouple a train of towed vehicles, remove a pallet from a fork truck, or load a unit load vehicle. All of these functions can be performed automatically as well.

Manual coupling of towing vehicles requires the operator to lead the AGV to the workstation or loading area. A manually controlled fork truck will be boarded or driven by the operator to position the load accurately. Fork Trucks may also be used to load/unload other AGVs. Rollers, belts, or chain conveyors can be employed on unit load vehicles to allow the operator to push or pull the load onto or off of the workstation.

Towing vehicles can be controlled by a central computer that directs the AGV to spurs where coupling should occur. The vehicle will then proceed to the next spur determined by the central computer to be uncoupled. Towing vehicles and unit load vehicles can be equipped with automatic conveyors that will transfer the load using a belt, chain, or roller device when the vehicle has been properly aligned with the workstation.

Fork trucks, unit load vehicles, and pallet trucks have the capability of power lift-and-lower load transfer. Upon reaching the assigned destination, these vehicles will raise or lower their load to deposit, or raise or lower their load decks to retrieve a load.

COMPONENTS

There are several hardware components that may be optional pieces of equipment on an AGV. Some typical hardware components include a battery with an on-board charger, a combination steering/drive motor, and speed-reducers. Additional fixtures may be installed providing safety features. They are bumpers, flashing lights and audio signal, directional signals, emergency stops, and a halt mechanism.

DESIGN DECISIONS

The following is a checklist of items to consider when designing an AGV system:

- Number of vehicles required
- Type of load required
- Height and width of aisles and doorways

- Appropriate vehicle for the determined load
- Integration with existing material handling system
- Method of guidance
- Floors/ramps/elevators

(This list is by no means complete but rather the beginning of a customized checklist.)

OPERATIONAL ISSUES

There are three key factors that should be explicitly determined prior to a decision to implement an AGV system.

1. Vehicle management
   - How are the vehicles going to be maintained?
   - Will there be a tracking system pinpointing specific locations at all times?
2. Throughput requirement
   - Will the proposed system meet or exceed the current throughput requirements?
3. Material flow
   - How will this system interact within the current flow?
   - Will other changes in the material flow have to be made?

APPLICATIONS

AGV applications can be found in many industries including:

- Aerospace
- Automotive
- Container ports
- Electronics
- Hospitals
- Metal refineries
- Mobile robots
- Paper mills
- Steel mills
- Warehousing

6

LAYOUT PLANNING MODELS AND DESIGN ALGORITHMS

6.1 INTRODUCTION

The generation of layout alternatives is a critical step in the facilities planning process, since the layout selected will serve to establish the physical relationships between activities. Recognizing that the layout ultimately selected will be either chosen from or based on one of the alternatives generated, it is important for the facilities planner to be both creative and comprehensive in generating layout alternatives.

In this chapter, we focus on developing layout alternatives. More specifically, we are concerned with developing a *block layout* (which shows the relative locations and sizes of the planning departments) as opposed to a *detailed layout* (which shows the exact location of all equipment, work benches, and storage areas *within* each department). We previously addressed the determination of requirements. In Chapter 1 we treated the strategic relationships between facilities planning and manufacturing, distribution, and marketing. From that discussion we recognized the importance of taking a long-range viewpoint and coordinating the facilities plan with the plans of other organizational units. A facilities layout strategy should emerge from the overall strategic plan. Product, manufacturing, marketing, distribution, management, and human resource plans will be impacted by and will impact on the facilities layout.

The facilities requirements resulting from product design, process design, and schedule design decisions were examined in Chapter 2. The impact of personnel requirements on space, proximities, and special facility features was treated in Chapter 4.

Chapter 3 provided a comprehensive treatment of activity relationships and space requirements as they affect facilities planning. From that discussion, as well as the emphasis in Chapter 1 on the establishment of activity relationships and determination of space requirements, it is clear that both are critically important in the design of a layout for a facility.

298

Before proceeding further, it seems appropriate to ask the following question, Which comes first, the material handling system or the facility layout? Many appear to believe the layout should be designed first and then the material handling system should be developed. Yet, material handling decisions can have a significant impact on the effectiveness of a layout. For example, the following decisions will affect the layout:

1. Centralized versus decentralized storage of work-in-process (WIP), tooling, and supplies
2. Fixed-path versus variable-path handling
3. The handling unit (unit load) planned for the systems
4. The degree of automation used in handling
5. The type of level of inventory control, physical control, and computer control of materials

Each of these considerations affects the requirements for space, equipment, and personnel, as well as the degree of proximity required between functions.

Why do people tend to focus first on layout? Perhaps one reason is an over emphasis on the manufacturing process. For example, it seems perfectly logical to place department B next to department A, if process B occurs immediately after process A. In such a situation, the handling problem is reduced to, What is the best way to move materials from A to B? Conventional wisdom suggests that the handling problem can be addressed after the layout is finalized. However, if materials cannot flow directly from a machine in A to a machine in B, then WIP storage is required either in A, B, or elsewhere. Depending on the storage and control requirements, a centralized WIP storage area might be used, so that materials flow from A to S (storage), then from S to B. With such a centralized WIP storage system, materials do not flow from A to B and B no longer needs to be placed next to A. Furthermore, the centralized system provides added flexibility when the process sequence changes.

Another reason for the "layout first" syndrome could be a misapplication of the "handling less is best" adage. When the total system is considered including storage and control, then handling more might be better than handling less, if more and less relate to distance. Our experience indicates "handling less is best" in terms of the number of times materials are handled, but that handling distance is not always the major concern. So, which comes first, the material handling system or the facility layout? Our answer is, "Both!" The layout and the handling system should be designed simultaneously. However, the complexity of the design problem generally requires that a sequential process be used. For this reason, we recommend that a number of alternative layout plans be developed and the appropriate handling system be designed for each. The preferred layout will be that which results from a consideration of the total system [67].

6.2 BASIC LAYOUT TYPES

In Chapter 3 we identified four types of planning departments:

1. Fixed material location departments
2. Production line departments

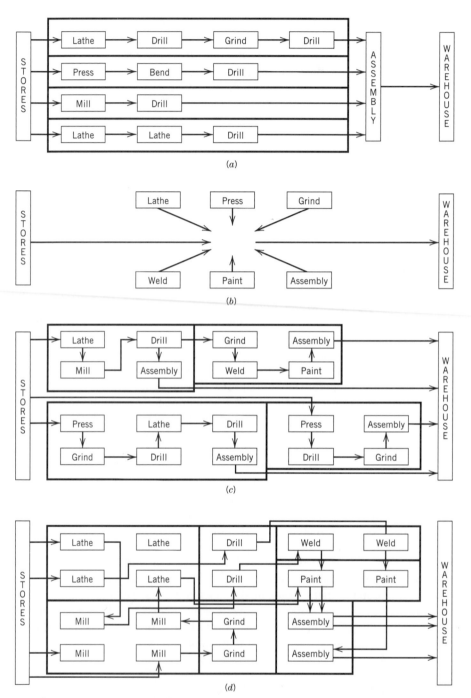

Figure 6.1   Alternative types of layouts. (*a*) Production line product layout. (*b*) Fixed product layout. (*c*) Product family layout. (*d*) Process layout.

3. Product family departments
4. Process departments

Figure 3.9 illustrated the material flows for each type of department. Also provided in Figure 3.9 were the associated layouts common to each type planning department. This figure is repeated here as Figure 6.1 with the departments emphasized.

The layout for a fixed material location department differs in concept from the other three. With the other layouts, material is brought to the workstation; in the case of the fixed material location departments the workstations are brought to the material. It is used in aircraft assembly, shipbuilding, and most construction projects. The layout of the fixed material location department involves the sequencing and placement of workstations around the material or product. Although fixed material location departments are generally associated with very large, bulky products, they certainly are not limited in their application. For example, in assembling computer systems it frequently occurs that the materials, subassemblies, housings, peripherals, and components are brought to a systems integration and test workstation and the finished product is "built" or assembled and tested at that single location. Such layouts will be referred to as fixed product layouts.

The layout for a production line department is based on the processing sequence for the part(s) being produced on the line. Materials typically flow from one workstation directly to the next adjacent one. Nice, well-planned flow paths generally result in this high-volume, low-variety environment. Such layouts will be referred to as product layouts.

The layout for a product family department is based on the grouping of parts to form product families. Nonidentical parts may be grouped into families based on common processing sequences, shapes, material composition, tooling requirements, handling/storage/control requirements, and so on. The product family is treated as a pseudoproduct and a pseudoproduct layout is developed. The processing equipment required for the pseudoproduct is grouped together and placed in a manufacturing cell. The resulting layout typically has a high degree of intradepartmental flow and little interdepartmental flow; it is variously referred to as a group layout and a product family layout.

The layout for a process department is obtained by grouping like processes together and placing individual process departments relative to one another based on flow between departments. Typically, there exists a high degree of interdepartmental flow and little intradepartmental flow. Such a layout is referred to as a process layout and is used when the volume of activity for individual parts or groups of parts is not sufficient to justify a product layout or group layout. Fixed product, product, group, and process layouts are compared in Table 6.1. Typically, one will find that a particular situation has some products that fit each of the layout types. Hence, a *hybrid* or *combination* layout will often result in practice.

6.3 LAYOUT PROCEDURES

A number of different procedures have been developed to aid the facilities planner in designing layouts. These procedures can be classified into two main

Table 6.1   *Advantages and Limitations of Fixed Product Layout, Product Layout, Group Layout, and Process Layout*

Fixed Product Layout	
Advantages	Limitations
1. Material movement is reduced.	1. Personal and equipment movement is increased.
2. When a team approach is used, continuity of operations and responsibility results.	2. May result in duplicate equipment.
3. Provides job enrichment opportunities.	3. Requires greater skill for personnel.
4. Promotes pride and quality because an individual can complete the "whole job."	4. Requires general supervision.
5. Highly flexible; can accommodate changes in product design, product mix, and production volume.	5. May result in increased space and greater work-in-process.
	6. Requires close control and coordination in scheduling production.

Product Layout	
Advantages	Limitations
1. Smooth, simple, logical, and direct flow lines result.	1. Machine stoppage stops the line.
2. Small work-in-process inventories should result.	2. Product design changes cause the layout to become obsolete.
3. Total production time per unit is short.	3. Slowest station paces the line.
4. Material handling requirements are reduced.	4. General supervision is required.
5. Less skill is required for personnel.	5. Higher equipment investment usually results.
6. Simple production control is possible.	
7. Special-purpose equipment can be used.	

Group Layout	
Advantages	Limitations
1. By grouping products, higher machine utilization can result.	1. General supervision required.
2. Smoother flow lines and shorter travel distances are expected than for process layouts.	2. Greater labor skills required for team members to be skilled on all operations.
3. Team atmosphere and job enlargement benefits often result.	3. Critically dependent on production control balancing the flows through the individual cells.
4. Has some of the benefits of product layouts and process layouts; it is a compromise between the two.	4. If flow is not balanced in each cell, buffers and work-in-process storage are required in the cell to eliminate the need for added material handling to and from the cell.
5. Encourages consideration of general-purpose equipment.	5. Has some of the disadvantages of product layouts and process layouts; it is a compromise between the two.
	6. Decreases the opportunity to use special-purpose equipment.

Table 6.1 *(continued)*

Process Layout	
Advantages	Limitations
1. Increased machine utilization.	1. Increased material handling requirements.
2. General-purpose equipment can be used.	2. More complicated production control required.
3. Highly flexible in allocating personnel and equipment.	3. Increased work-in-process.
4. Diversity of tasks for personnel.	4. Longer production lines.
5. Specialized supervision is possible.	5. Higher skills required to accommodate diversity of tasks required.

categories: construction type and improvement type. *Construction-type* layout methods basically involve developing a new layout "from scratch." *Improvement* procedures generate layout alternatives based on an existing layout.

Although many papers in the literature on facility layout concentrate on the development of construction-type procedures, most layout work still involves some form of improving the layout of existing facilities. As Immer [27, p. 32] observed as early as the 1950s,

> Much of (the) work will consist of making minor changes in existing layout, locating new machines, revising a section of the plant, and making occasional studies for material handling. The plans for a complete new production line or a new factory may make the headlines, but except for a war or a new expansion, the average layout (planner) will very seldom have to consider such a problem.

However, the passage on material handling can be a challenge since as we stated in our previous discussion that the layout and material handling system should be designed simultaneously and the process of revising a facility layout should be viewed as an opportunity to also review the effectiveness of existing material handling methods.

We begin our discussion of layout procedures by discussing some of the original approaches to the layout problem. The concepts in these approaches continue to serve as the foundation of many of the methodologies proposed today.

Apple's Plant Layout Procedure

Apple [1] proposed the following detailed sequence of steps in producing a plant layout.

1. Procure the basic data.
2. Analyze the basic data.
3. Design the productive process.
4. Plan the material flow pattern.
5. Consider the general material handling plan.
6. Calculate equipment requirements.
7. Plan individual workstations.
8. Select specific material handling equipment.

9. Coordinate groups of related operations.
10. Design activity interrelationships.
11. Determine storage requirements.
12. Plan service and auxiliary activities.
13. Determine space requirements.
14. Allocate activities to total space.
15. Consider building types.
16. Construct master layout.
17. Evaluate, adjust, and check the layout with the appropriate persons.
18. Obtain approvals.
19. Install the layout.
20. Follow up on implementation of the layout.

Apple noted that these steps are not necessarily performed in the sequence given. As he put it,

> Since no two layout design projects are the same, neither are the procedures for designing them. And, there will always be a considerable amount of jumping around among the steps, before it is possible to complete an earlier one under consideration. Likewise, there will be some backtracking going back to a step already done—to re-check or possibly re-do a portion, because of a development not foreseen. [1, p. 14]

Reed's Plant Layout Procedure

Reed [51] recommended the following "systematic plan of attack" as required steps in "planning for and preparing the layout."

1. Analyze the product or products to be produced.
2. Determine the process required to manufacture the product.
3. Prepare layout planning charts.
4. Determine workstations.
5. Analyze storage area requirements.
6. Establish minimum aisle widths.
7. Establish office requirements.
8. Consider personnel facilities and services.
9. Survey plant services.
10. Provide for future expansion.

Reed calls the layout planning chart "the most important single phase of the entire layout process" [51, p. 10]. It incorporates the following:

1. Flow process, including operations, transportation, storage, and inspections.
2. Standard times for each operation.
3. Machine selection and balance.
4. Manpower selection and balance.
5. Material handling requirements.

An example of a layout planning chart is given in Figure 6.2.

LAYOUT PLANNING CHART

PART NO. 1 — PART NAME PLASTIC CONTACT PAD — PCS/ASSY 1 — PCS REQ/HR 70.4 — SHEET 1 OF 1

ASSY NO. — — ASSY NAME — — ASSY/PRODUCT 2 — PRODUCTION HRS/DAY 6.0 — PREPARED BY J.G.D. — DATE 1-4-60

MATERIAL PLASTIC — SIZE 11/2" OD × 3/8" ID (from 4' × 8' × 1/4" SHEETS) — PCS/DAY 422 — LOT SIZE 1 — APPROVED BY ____ — DATE ____

ST NO.	F M S I	DECRIPTION	OPER NO.	DEPT NO.	TIME PER PIECE	MACHINE OR EQUIPMENT	MACHINES REQD			OPER PER MACH	TOTAL MANPOWER				HANDLING REQUIREMENTS				REMARKS
							MACH FRAC	COMB WITH	MACH REQD		CREQ FRAC	MAN FRAC	COMB WITH	MEN REQD	HOW MOVED	CONT TYPE	LOAD SIZE	DIST MOVED	
1		FROM MATERIALS STORAGE													FORK LIFT	PALLET	4 SHEETS	100'	
2		ON PALLET BY SAW		2															
3		TO SAW TABLE																	
4		SAW INTO STRIPS 2 1/2" × 8'	10	2	.02	TABLE SAW	.028	9 – 10 16 – 10	1	2	.028	.056	9 – 10 16 – 10	2					
5		TO RACK BY SAW											1 – 20 14 – 30						
6		IN RACK		2									17 – 20 3 – 20						
7		TO HEATER											6 – 30 13 – 20						
8		IN RACK BY HEATER		2									14 – 20						
9		FEED INTO HEATER																	
10		HEAT	10	2	.04	HEATER	.055	9 – 20 16 – 20	1	1	.055	.055	6 – 20 6 – 10	1					
													3 – 10 2 – 10						
													9 – 20 16 – 20						
													13 – 10 14 – 10						
													17 – 10						
11		FEED TO PUNCH PRESS																	
12		PUNCH TO SHAPE	10	2	.04	PUNCH PRESS	.055	4 – 10 4 – 20	1	1	.055	.055	4 – 10 4 – 20	1					
13		TP BIN BY PUNCH PRESS						9 – 30 16 – 30					9 – 30 16 – 30						
14		IN BIN		2				19A – 10 19B – 10					19A – 10 19B – 10						
15		TO PARTS STORAGE											19A – 20 5A – 10		4 WHEEL HAND TRUCK	TOTE BOX	400 PER BOX	150'	
16		IN PARTS STORAGE		5															
17		TO ASSEMBLY													4 WHEEL HAND TRUCK	TOTE BOX	400 PER BOX	60'	
18		IN BIN IN ASSEMBLY		6															
19		TO TABLE																	

Figure 6.2 Layout planning chart.

Muther's Systematic Layout Planning (SLP) Procedure

Muther [47] developed a layout procedure he named systematic layout planning or SLP. The framework for SLP is given in Figure 6.3. It uses as its foundation the activity relationship chart described in Chapter 3 and illustrated in Figure 6.4.

Based on the input data and an understanding of the roles and relationships between activities, a material flow analysis (from-to chart) and an activity relationship analysis (activity relationship chart) are performed. From the analyses performed, a relationship diagram is developed (Figure 6.5).

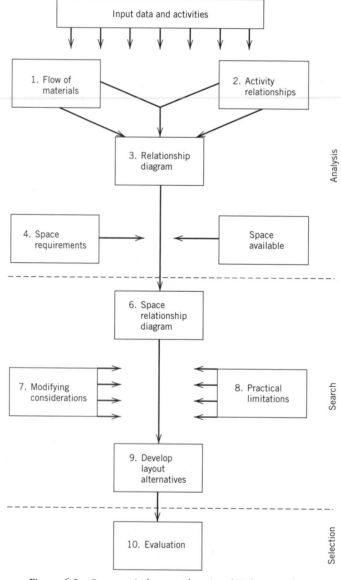

Figure 6.3   Systematic layout planning (SLP) procedure.

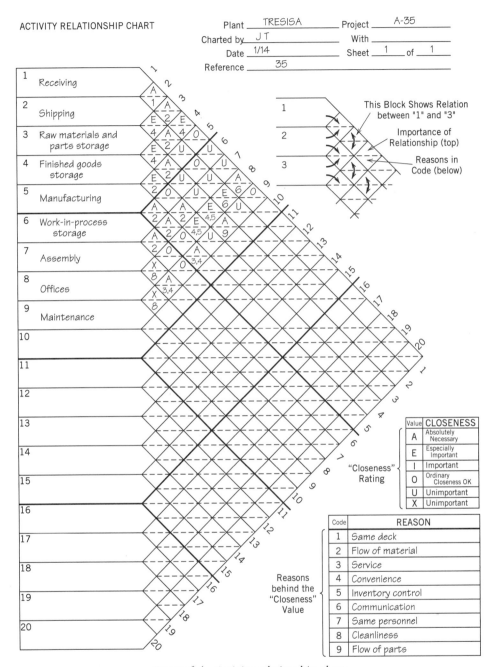

Figure 6.4 Activity relationship chart.

The relationship diagram positions activities spatially. Proximities are typically used to reflect the relationship between pairs of activities. Although the relationship diagram is normally two dimensional, there have been instances in which three dimensional diagrams have been developed when multistory buildings, mezzanines, and/or overhead space were being considered.

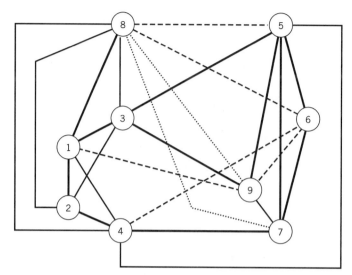

Figure 6.5 Relationship diagram.

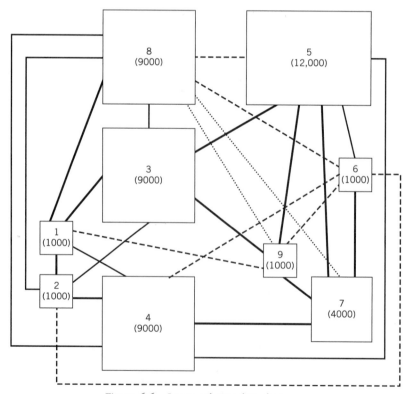

Figure 6.6 Space relationship diagram.

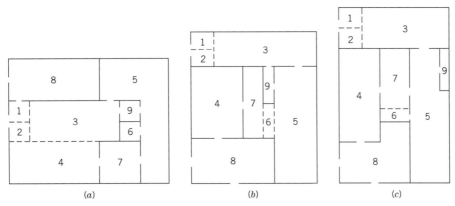

Figure 6.7 Alternative block layouts.

The next two steps involve the determination of the amount of space to be assigned to each activity. From Chapter 3, departmental service and area requirement sheets would be completed for each planning department. Once the space assignments have been made, space templates are developed for each planning department and the space is "hung on the relationship diagram" to obtain the space relationship diagram (Figure 6.6).

Based on modifying considerations and practical limitations, a number of layout alternatives are developed (Figure 6.7) and evaluated. The preferred alternative is then identified and recommended.

While the process involved in performing SLP is relatively straightforward, it does not necessarily follow that difficulties do not arise in its application. We addressed in Chapter 3 such issues as the a priori assignment of activity relationships and the use of proximity as a criterion for measuring the degree of satisfaction of activity relationships. In addition to those concerns, it should be noted that alternative relationship diagrams can often be developed, with apparent equivalent satisfaction of activity relationships. Likewise, the shapes of the individual space templates used in constructing the space relationship diagram can influence the generation of alternatives. Finally, the conversion of a space relationship diagram into several feasible layout alternatives is not a mechanical process; intuition, judgment, and experience are important ingredients in the process.

The SLP procedure can be used sequentially to develop first a block layout and then a detailed layout for each planning department. In the latter application, relationships between machines, workstations, storage locations and entrances to and exits from the department are used to determine the relative location of activities within each department.

6.4 ALGORITHMIC APPROACHES

The relative placement of departments on the basis of their "closeness ratings" or "material flow intensities" is one that can be reduced to an algorithmic process. The layout procedures we presented in the previous section provide an excellent framework and overall process to construct or improve a layout but they do not provide a formal procedure or algorithm for some of the critical steps associated with layout design and evaluation. The models and algorithms we present in this section are formal procedures/algorithms that can help the layout analyst develop or improve a layout, and at the same time provide him or her with objective criteria to facilitate the evaluation of various layout alternatives that emerge in the process.

The layout algorithms we present, at least in theory, can be executed by hand. However, for most practical, real-world problems, the algorithms we show here are intended or best-suited for computer implementation. Currently available computer-based layout algorithms cannot replace human judgment and experience, and they generally do not capture the qualitative characteristics of a layout. However, computerized layout algorithms can significantly enhance the productivity of the layout planner and the quality of the final solution by generating and numerically evaluating a large number of layout alternatives in a very short time. Computerized layout algorithms are also very effective in rapidly performing "what if" analyses based on varying the input data or the layout itself.

The algorithms presented in this section, for the most part, represent the outgrowth of university research. As such, commercial versions of the algorithms we present either do not exist or they must be obtained through the original authors. Also, due to limited space, we are only able to present what we hope is a diverse and representative subset of many layout algorithms that have been developed to date.

A number of commercial packages are available for facility layout. However, with some exceptions (see Section 6.8), such packages are either intended only for presentation purposes (that is, they are electronic drafting tools which facilitate the drawing or maintenance of a given layout) or they are designed primarily as a layout evaluation tool (that is, they can evaluate a layout provided that one has been supplied by the layout planner). While such tools can also significantly enhance the productivity of the layout planner, in this section we will focus on those algorithms that can actually improve a given layout or develop a new layout from scratch.

Algorithm Classification

Most layout algorithms can be classified according to the type of input data they require. Some algorithms accept only qualitative "flow" data (such as a relationship chart) while others work with a (quantitative) flow matrix expressed as a from-to chart. Some algorithms (such as BLOCPLAN) accept both a relationship chart and a from-to chart; however, the charts are used only one at a time when evaluating a layout. The current trend appears to be toward algorithms that use a from-to chart, which generally requires more time and effort to prepare but provides more information on parts-flow (or material handling trips) when completed. *Given that flow values can be converted to relationship ratings and vice versa,* most algorithms can be used with either type of data. Of course, if a relationship chart is converted to a from-to chart (by assigning numerical values to the closeness ratings), the "flow" values picked by the layout planner represent only an ordinal scale.

With respect to data input, the time and effort required to compile a relationship chart or a from-to chart increases rapidly with the number of departments.

Since the relationship chart is based on user-assigned closeness ratings, its construction often requires input from multiple individuals. After obtaining such input, the analyst needs to identify and resolve possible inconsistencies. For example, for the same pair of departments, one person may assign an "A" while another assigns an "I." Rather than attempt to decide who is "right" and who is "wrong," the analyst needs to meet with both individuals to understand the reasons behind their ratings, and with agreement from all concerned parties, determine the final closeness rating to be used for the pair. Given the significant increase in the number of all possible department pairs as a function of the number of departments, relationship charts are often not practical for problems with 20 or more departments.

The same increase in the number of department pairs also applies to from-to charts. For medium- to large-sized problems (say, 20–30 or more departments), filling out each entry in the from-to chart would not be practical. In such cases, however, the from-to chart can be constructed in reasonable time by using the production route data for each product (or product family). For example, if product type A is processed through departments 1-2-5-7, and is moved at a rate of, say, 20 loads/hr, then we set $f_{12} = f_{25} = f_{57} = 20$ in the from-to chart. Repeating this process for each product (or product type) completes the from-to chart, which is often a "sparse matrix" (i.e., it contains many blank entries). In fact, in many cases it is useful to first construct a separate and complete from-to chart for each product (or product type) so product-level flow data remain available to the analyst at all times. Subsequently, individual product-level from-to charts can be combined into one cumulative chart, using appropriate weights for individual product types if necessary. Of course, the construction of the from-to chart in the above manner is done by computers in most cases. Also, if the unit of flow for the product changes as it moves from one process to the next, appropriate multipliers can be inserted into the routing data to scale the flow intensity up or down based on the unit of flow.

Layout algorithms can also be classified according to their objective functions. There are two basic objectives: One aims at minimizing the sum of flows times distances while the other aims at maximizing an adjacency score. Generally speaking, the former, that is, the "distance-based" objective—which is similar to the classical Quadratic Assignment Problem (QAP) objective—is more suitable when the input data is expressed as a from-to chart, and the latter, that is, the "adjacency-based" objective, is more suitable for a relationship chart.

Consider first the distance-based objective. Let m denote the number of departments, f_{ij} denote the flow from department i to department j (expressed in number of unit loads moved per unit time), and c_{ij} denote the cost of moving a unit load one distance unit from department i to department j. The objective is to minimize the cost per unit time for movement among the departments. Expressed mathematically, the objective can be written as

$$\min z = \sum_{i=1}^{m} \sum_{j=1}^{m} f_{ij} c_{ij} d_{ij}, \tag{6.1}$$

where d_{ij} is the distance from department i to j. In many layout algorithms d_{ij} is measured rectinearly between department centroids; however, it can also be measured according to a particular aisle structure (if one is specified).

Note that the c_{ij} values in Equation 6.1 are implicitly assumed to be independent of the utilization of the handling equipment, and they are linearly related to the length of the move. In those cases where the c_{ij} values do not satisfy the above assumptions, one may set $c_{ij} = 1$ for all i and j and focus only on *total unit load travel* in the facility, i.e., the product of the f_{ij} and the d_{ij} values. In some cases, it may also be possible to use the c_{ij} values as relative "weights" (based on unit load attributes such as size, weight, bulkiness, etc.) and minimize the weighted sum of unit load travel in the facility.

Consider next the adjacency-based objective where the adjacency score is computed as the sum of all the flow values (or relationship values) between those departments that are adjacent in the layout. Letting $x_{ij} = 1$ if departments i and j are adjacent (that is, they share a border) in the layout, and 0 otherwise, the objective is to maximize the adjacency score; that is,

$$\max z = \sum_{i=1}^{m} \sum_{j=1}^{m} f_{ij} x_{ij}. \tag{6.2}$$

Although the adjacency score obtained from Equation 6.2 is helpful in comparing two or more alternate layouts, it is often desirable to evaluate the *relative* efficiency of a particular layout with respect to a certain lower or upper bound. For this purpose, the layout planner may use the following "normalized" adjacency score:

$$z = \frac{\sum_{i=1}^{m} \sum_{j=1}^{m} f_{ij} x_{ij}}{\sum_{i=1}^{m} \sum_{j=1}^{m} f_{ij}}. \tag{6.3}$$

Note that the normalized adjacency score (which is also known as the *efficiency rating*) is obtained simply by dividing the adjacency score obtained from Equation 6.2 by the total flow in the facility. As a result, the normalized adjacency score is always between zero and 1. If the normalized adjacency score is equal to 1, it implies that all department pairs with positive flow between them are adjacent in the layout.

In some cases, the layout planner may represent an X relationship between departments i and j by assigning a negative value to f_{ij}. The exact negative value to be used should be determined with respect to the "real" (i.e., positive) flow values in the from-to chart. If such "negative flow" values are used, the normalized adjacency score must be modified as follows:

$$z = \frac{\sum_{(i,j) \in F} f_{ij} x_{ij} - \sum_{(i,j) \in \bar{F}} f_{ij} (1 - x_{ij})}{\sum_{(i,j) \in F} f_{ij} - \sum_{(i,j) \in \bar{F}} f_{ij}} \tag{6.4}$$

where F and $\bar{F}$ represent the set of department pairs with positive and negative flow values, respectively.

The adjacency-based objective has been used in a number of algorithms; see, for example, [13], [15], and [43], among others. Although such an objective is easy to use and intuitive (i.e., department pairs with high closeness ratings or large flows need to be adjacent in the layout), it is generally not a complete measure of layout efficiency since it disregards the distance or separation between non-adjacent departments. Therefore, as remarked in [5], it is possible to construct two layouts that have identical or similar adjacency scores but different travel distances in terms of parts flow.

Layout algorithms can further be classified according to the format they use for layout representation. Most layout algorithms use the discrete representation (shown in Figure 6.8*a*) which allows the computer to store and manipulate the

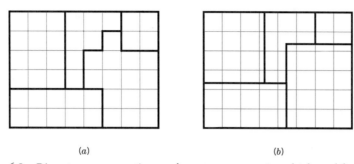

(a)   (b)

Figure 6.8 Discrete versus continuous layout representation. (*Adapted from* [16])

layout as a matrix. With such a representation, the area of each department is rounded off to the nearest integer number of grids. If the grid size is too large (too small), small (large) departments may have too few (too many) grids. Also, the grid size determines the overall "resolution" of the layout; a smaller grid size yields a finer resolution which allows more flexibility in department shapes. However, since the total area is fixed, a smaller grid size results in a larger number of grids (or a larger matrix) which can considerably increase the computational burden. Hence, in algorithms that use the discrete representation, selecting the appropriate grid size is an important decision that must be made early in the planning process.

The alternate representation (see Figure 6.8*b*) is the continuous representation where there is no underlying grid structure. (The grid in Figure 6.8*b* was retained only for comparison purposes.) Although such a representation is theoretically more flexible than its discrete counterpart, it is also more difficult to implement on a computer. In fact, except for one case (where the shapes of certain departments are "adjusted" to accommodate a nonrectangular building), present computerized layout algorithms that use the continuous representation are restricted to a rectangular building and rectangular department shapes.

While it is possible to model nonrectangular buildings by using fixed "dummy" departments (see the section on CRAFT), generally speaking, defining L-shaped, U-shaped, and other arbitrary, nonrectangular departments with the continuous representation is not straightforward. If a department is rectangular and we know the area it requires, then we need to know only the x,y-coordinate of its centroid and the length of its side in the north-south direction to specify its exact location and shape. (How would you specify the exact location and shape of an L-shaped department or a T-shaped department with a known area? Among alternative specifications, which one requires the minimum amount of data? Which one makes it easier to identify and avoid overlapping departments?)

Department shapes play an important role in computerized layout algorithms. Although we discuss specific shape measures in Section 6.5, first we need to present some basics. Since, by definition, a department represents the smallest indivisible entity in layout planning, a layout algorithm should not "split" a department into two or more pieces. If some departments are "too large," the layout planner can go back and reconsider how the departments were defined and change some of them as necessary. In the process, one large department may be redefined as two smaller departments. Once all the departments have been defined, however, a layout algorithm cannot change them.

Although the human eye is adept at judging shapes and readily identifying split departments, for the computer to "recognize" a split department we need to develop formal measures that can be incorporated into an algorithm. Consider, for example, the discrete representation, where a department is represented as a collection of grids. Suppose there is a "dot" that can move *only from one grid to an adjacent grid.* (Two grids are adjacent only if they share a border of positive length; two grids that "touch" each other at the corners are not considered adjacent.) We say that department i is not split if the above dot can start at any grid assigned to department i and travel to any other grid assigned to department i *without* visiting any grid that has not been assigned to department i. In other words, given the restrictions imposed on the movements of the dot, any grid assigned to department i must be "reachable" from any other such grid.

For example the department shown in Figure 6.9a and 6.9b is split, while those shown in Figures 6.9c and 6.9d are not. However, according to the above definition, the department shown in Figure 6.9e is also not a split department. Departments such as the one shown in Figure 6.9e are said to contain an "enclosed void" [16] and as a rule-of-thumb are not considered practical or reasonable for facility layout purposes. (Of course, one possible exception to this rule is an "atrium." However, an atrium should generally be modeled by placing a fixed "dummy department" inside the building; refer to the section on CRAFT for dummy departments.) The shape measures we present in Section 6.5 are devised to generally avoid department shapes such as those shown in Figures 6.9d and 6.9e. In the above discussion of department shapes we focused on the discrete representation. Although it has been used effectively only with rectangular departments, our comments apply to the continuous representation as well.

Lastly, as we remarked earlier, layout algorithms can be classified according to their primary function; that is, layout *improvement* versus layout *construction*.

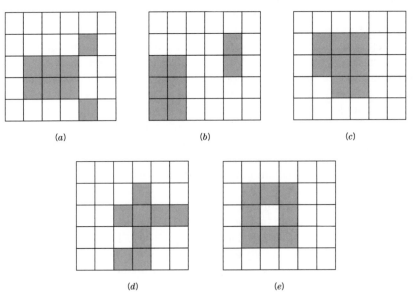

Figure 6.9 Examples of split and unsplit departments.

Improvement-type algorithms generally start with an initial layout supplied by the analyst and seek to improve the objective function through "incremental" changes in the layout. Construction-type layout algorithms generally develop a layout "from scratch." They can be further divided into two categories: those that assume the building dimensions are given and those that assume they are not. The first type of construction algorithm is suitable when an operation is being moved into an existing, vacant building. The second type of construction algorithm is suitable for "green field" applications.

Even with "green field" applications, however, there is usually a *site plan,* which shows the property, surrounding roads, etc. Given the constraints imposed by the site plan, it is often necessary to construct the new building within a certain "envelope." If a construction algorithm of the second type is used, it is often difficult to ensure that the resulting building will properly fit into such an envelope. With construction algorithms of the first type, on the other hand, one may model the above envelope as an "existing building" to obtain a proper fit between the actual building and the envelope. Primarily for the above reason, the construction algorithms we present in this chapter are of the first type (i.e., they assume the building dimensions are given).

In the following sections, we describe the overall modeling techniques and/or methods used in various layout algorithms; namely, the pairwise exchange method, a graph-based method, CRAFT, BLOCPLAN, MIP, LOGIC, and MULTIPLE. (For ease of reference, in those cases where the original authors did not use an acronym, we created our own.) Detailed information such as input data format or output variables are not presented since such information is relevant and available only if the reader obtains a copy of the computer program. However, for each algorithm we present sufficient detail so that not only the fundamental concepts are covered but interested readers may develop their own basic implementation of the algorithm. The algorithms we present are relatively recent ones with a few exceptions such as CRAFT, which we consider because it is one of the first layout algorithms developed and it provides a good platform for the others. Furthermore, all the algorithms we present require a basic understanding of heuristic search techniques (such as the "steepest descent" procedure) and the difference between "locally" and "globally" optimal solutions. We selected these methods based on their conceptual contributions to layout planning methods and/or their unique approach to layout construction/improvement.

Pairwise Exchange Method

In an earlier discussion, we suggested that majority of layout problems involves the redesign of an existing facility, which is typically triggered by the addition of new machines, changes in product mixes, decisions related to the contraction and expansion of storage areas, or a simple realization that the old layout is no longer adequate for its current needs. Thus, we are given an existing layout, and the problem is to come up with an improved layout.

The pairwise exchange method ([51] and [8]) is an improvement-type layout algorithm. Although it can be used with both an adjacency-based and distance-based objective, it is often used with the latter. We will illustrate the pairwise exchange method below through an example based on equal-area departments for simplicity. Its implementation with unequal-area departments (which is the case in practice) will be shown later via CRAFT, MULTIPLE, and others.

Table 6.2 *Material Flow Matrix*

		To Department			
		1	2	3	4
From	1	—	10	15	20
Department	2		—	10	5
	3			—	5
	4				—

Consider four departments of equal sizes. The material flows between departments are given in Table 6.2. The existing layout is shown in Figure 6.10a. A distance matrix can be obtained based on the existing layout as given in Table 6.3.

The objective function value (or "total cost") for the existing layout is computed as follows:

$$TC_{1234} = 10(1) + 15(2) + 20(3) + 10(1) + 5(2) + 5(1) = 125$$

The subscript notation indicates the order of the departments in the initial layout.

The pairwise exchange method simply states that for each iteration, all feasible exchanges in the locations of department pairs are evaluated and the pair that results in the *largest reduction* in total cost is selected. (Moving in the direction of largest cost reduction is also known as "steepest descent" in optimization.) Since all departments areas are assumed to be of equal size, the feasible exchanges are 1-2, 1-3, 1-4, 2-3, 2-4, and 3-4. The distance matrix is recomputed each time an exchange is performed. The total costs resulting from these exchanges are

$$TC_{2134}(1\text{-}2) = 10(1) + 15(1) + 20(2) + 10(2) + 5(3) + 5(1) = 105$$
$$TC_{3214}(1\text{-}3) = 10(1) + 15(2) + 20(1) + 10(1) + 5(2) + 5(3) = 95$$
$$TC_{4231}(1\text{-}4) = 10(2) + 15(1) + 20(3) + 10(1) + 5(1) + 5(2) = 120$$
$$TC_{1324}(2\text{-}3) = 10(2) + 15(1) + 20(3) + 10(1) + 5(1) + 5(2) = 120$$
$$TC_{1432}(2\text{-}4) = 10(3) + 15(2) + 20(1) + 10(1) + 5(2) + 5(1) = 105$$
$$TC_{1243}(3\text{-}4) = 10(1) + 15(3) + 20(2) + 10(2) + 5(1) + 5(1) = 125$$

Thus, we select the pair 1-3 and perform the exchange in the layout. This is shown in Figure 6.10*b*.

For the next iteration, we consider all feasible exchanges which consist of the same set as in iteration 1. The resulting total costs are

$$TC_{3124}(1\text{-}2) = 10(1) + 15(1) + 20(2) + 10(1) + 5(1) + 5(3) = 95$$
$$TC_{1234}(1\text{-}3) = 10(1) + 15(2) + 20(3) + 10(1) + 5(2) + 5(1) = 125$$
$$TC_{3241}(1\text{-}4) = 10(2) + 15(3) + 20(1) + 10(1) + 5(1) + 5(2) = 110$$
$$TC_{2314}(2\text{-}3) = 10(2) + 15(1) + 20(1) + 10(1) + 5(3) + 5(2) = 90$$

(a) Iteration 0 | 1 | 2 | 3 | 4 |

(b) Iteration 1 | 3 | 2 | 1 | 4 |

(c) Iteration 2 | 2 | 3 | 1 | 4 |

Figure 6.10 Layouts corresponding to each iteration.

Table 6.3 *Distance Matrix Based on Existing Layout*

		To Department			
		1	2	3	4
From	1	—	1	2	3
Department	2		—	1	2
	3			—	1
	4				—

$$TC_{3412}(2\text{-}4) = 10(1) + 15(2) + 20(1) + 10(3) + 5(2) + 5(2) = 105$$
$$TC_{4213}(3\text{-}4) = 10(1) + 15(1) + 20(2) + 10(2) + 5(1) + 5(3) = 105$$

The pair 2-3 is selected with a total cost value of 90. Figure 6.10c shows the resulting layout after two iterations. Continuing on, the third iteration calculations are

$$TC_{3214}(1\text{-}2) = 10(1) + 15(2) + 20(1) + 10(1) + 5(2) + 5(3) = 95$$
$$TC_{1324}(1\text{-}3) = 10(2) + 15(1) + 20(3) + 10(1) + 5(1) + 5(2) = 120$$
$$TC_{3421}(1\text{-}4) = 10(1) + 15(3) + 20(2) + 10(2) + 5(1) + 5(1) = 125$$
$$TC_{2134}(2\text{-}3) = 10(1) + 15(1) + 20(2) + 10(2) + 5(3) + 5(1) = 105$$
$$TC_{3142}(2\text{-}4) = 10(2) + 15(1) + 20(1) + 10(3) + 5(1) + 5(2) = 100$$
$$TC_{4123}(3\text{-}4) = 10(1) + 15(2) + 20(1) + 10(1) + 5(2) + 5(3) = 95$$

Since the lowest total cost for this iteration, 95, is worse than the total cost value of 90 in the second iteration, the procedure is terminated. The final layout arrangement is 2-3-1-4 as shown in Figure 6.10c. The final layout is also known as a 2-opt layout since there are no two-way exchanges that can further reduce the layout cost.

Remarks.   The pairwise exchange procedure described above is not guaranteed to yield the optimal layout solution because the final outcome is dependent on the initial layout, that is, a different initial layout can result in another solution. Thus, we can only claim local optimality. Also, you may have observed that it is possible to cycle back to one of the alternative layout arrangements from a previous iteration. For instance, the layout arrangement 1-2-3-4 is what we started with and we see the same arrangement in the second iteration when departments 1 and 3 were exchanged based on the solution from iteration one, that is, layout arrangement 3-2-1-4. Additionally, symmetric layout arrangements may also occur, for example, 4-3-2-1 in iteration 3 is identical to 1-2-3-4.

Pairwise exchange can be easily accomplished only if the pair of departments considered are of equal size (as we assumed in this example). Otherwise, we would have to expend some effort in rearranging the two departments being exchanged and possibly other departments in the layout. We will discuss the exchange of unequal-area departments in CRAFT and subsequent algorithms.

Graph-based Method

The graph-based method is a construction-type layout algorithm; it has its roots in graph theory. It is often used with an adjacency-based objective. The recognition of

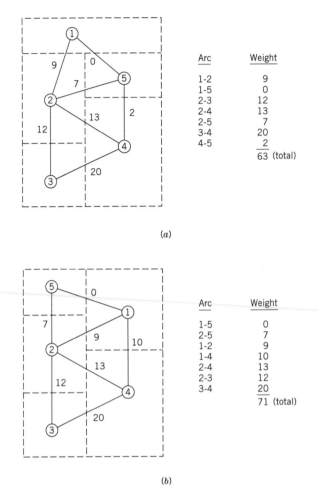

Arc	Weight
1-2	9
1-5	0
2-3	12
2-4	13
2-5	7
3-4	20
4-5	2
	63 (total)

(a)

Arc	Weight
1-5	0
2-5	7
1-2	9
1-4	10
2-4	13
2-3	12
3-4	20
	71 (total)

(b)

Figure 6.11   Adjacency graphs for alternative block layouts.

the usefulness of graph theory as a mathematical tool in the solution of facilities planning problems dates back to the late 1960s (Krejcirik [33]) and early 1970s (Seppanen and Moore [55]). The use of graph theory methods has strong similarities with the SLP method developed by Muther [47].

Consider the block layout shown in Figure 6.11*a*. We first construct an adjacency graph where each node represents a department and a connecting arc between two nodes indicate that two departments share a common border. A similar graph is constructed for the alternative block plan layout shown in Figure 6.11*b*. We observe that the two graphs shown in Figure 6.11 are subgraphs of the graph shown in Figure 6.12*b*, which is derived from the relationship chart in Figure 6.12*a*. The relationship chart displays numerical "weights" rather than alphabetic closeness ratings.

Given the adjacency-based objective, block layout (b) is better than block layout (a) with scores of 71 and 63, respectively. Thus, finding a maximally weighted block layout is equivalent to obtaining an adjacency graph with the maximum sum of arc weights.

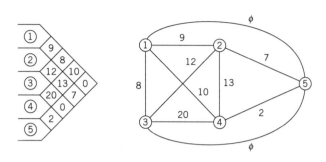

(a) Relationship Chart   (b) Relationship Diagram

Figure 6.12   Relationship chart and relationship diagram for graph-based example.

Before we describe a method for determining adjacency graphs, we first make the following observations:

a.   The adjacency score does not account for distance, nor does it account for relationships other than those between adjacent departments.
b.   Dimensional specifications of departments are not considered; the length of common boundaries between adjacent departments are also not considered.
c.   The arcs do not intersect; this property of graphs is called planarity. We note that the graph obtained from the relationship diagram is usually a nonplanar graph.
d.   The score is very sensitive to the assignment of numerical weights in the relationship chart.

There are two strategies we can follow in developing a maximally weighted planar adjacency graph. One way is to start with the graph from the relationship diagram and selectively prune connecting arcs while making sure that the final graph is planar. A second approach is to iteratively construct an adjacency graph via a node insertion algorithm while retaining planarity at all times. A heuristic procedure based on the second approach is described below.

Procedure:
Step 1.   From the relationship chart in Figure 6.12a, select the department pair with the largest weight. Ties, if any, are broken arbitrarily. Thus, departments 3 and 4 are selected to enter the graph.
Step 2.   Next, select the third department to enter. The third department is selected based on the sum of of the weights with respect to departments 3 and 4. From Figure 6.13a department 2 is chosen with a value of 25. The columns in this figure correspond to the departments already in the adjacency graph and the rows correspond to departments not yet selected. The last column gives the sum of the weights for each unassigned department.
Step 3.   We then pick the fourth department to enter by evaluating the value of adding one of the unassigned departments represented by a node on a face of the graph. A face of a graph is a bounded region of a graph. For instance, a triangular face is the region bounded by arcs 2-3, 3-4, and 4-2 in Figure 6.13a. We will denote this face as 2-3-4. The outside region is

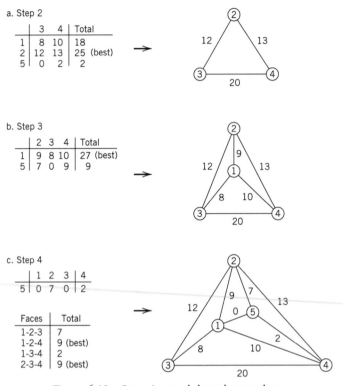

a. Step 2

	3	4	Total
1	8	10	18
2	12	13	25 (best)
5	0	2	2

b. Step 3

	2	3	4	Total
1	9	8	10	27 (best)
5	7	0	9	9

c. Step 4

	1	2	3	4
5	0	7	0	2

Faces	Total
1-2-3	7
1-2-4	9 (best)
1-3-4	2
2-3-4	9 (best)

Figure 6.13   Steps in graph-based procedure.

referred to as the external face. For our example, the value of adding departments 1 and 5 are 27 and 9, respectively. Department 1 is selected and placed inside the region 2-3-4, as shown in Figure 6.13*b*.

Step 4.   The remaining task is to determine on which face to insert department 5. For this step, department 5 can be inserted on faces 1-2-3, 1-2-4, 1-3-4, and 2-3-4. Inserting 5 on faces 1-2-4 and 2-3-4 yields identical values of 9. We arbitrarily select 1-2-4. The final adjacency graph is given in Figure 6.13*c*. This solution is optimal with a total sum of arc weights equal to 81.

Step 5.   Having determined an adjacency graph, the final step is to construct a corresponding block layout. A block layout based on the final adjacency graph is shown in Figure 6.14. The manner by which we constructed the block layout is analogous to the SLP method. We should note that in constructing the block layout, the original department shapes had to be altered significantly in order to satisfy the requirements of the adjacency graph. In practice, we may not have as much latitude in making such alterations since department shapes are generally derived from the geometry of the individual machines within the department and the internal layout configuration. We will discuss department shapes and their control later in this chapter (see Section 6.5). Finally, we should point out that there are algorithmic methods for performing this step as demonstrated by Giffin, et al. [18] and Hassan and Hogg [21].

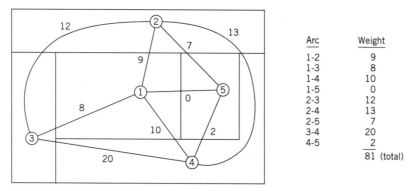

Arc	Weight
1-2	9
1-3	8
1-4	10
1-5	0
2-3	12
2-4	13
2-5	7
3-4	20
4-5	2
	81 (total)

Figure 6.14   Block layout from the final adjacency graph.

CRAFT

Introduced in 1963 by Armour, Buffa, and Vollman (see [2] and [8]), CRAFT (*Computerized Relative Allocation of Facilities Technique*) is one of the earliest layout algorithms presented in the literature. It uses a from-to chart as input data for the flow. Layout "cost" is measured by the distance-based objective function shown in Equation 6.1. Departments are not restricted to rectangular shapes and the layout is represented in a discrete fashion.

Since CRAFT is an improvement-type layout algorithm, it starts with an initial layout, which typically represents the actual layout of an existing facility but may also represent a prospective layout developed by another algorithm. CRAFT begins by determining the centroids of the departments in the initial layout. It then calculates the rectilinear distance between pairs of department centroids and stores the values in a distance matrix. The initial layout cost is determined by multiplying each entry in the from-to chart with the corresponding entries in the unit cost matrix (i.e., the c_{ij} values) and the distance matrix.

CRAFT next considers all-possible two-way (pairwise) or three-way department exchanges and identifies the best exchange, that is, the one that yields the largest reduction in the layout cost. (No department can be split as a result of a two-way or three-way exchange.) Once the best exchange is identified, CRAFT updates the layout according to the best exchange, and computes the new department centroids as well as the new layout cost to complete the first iteration. The next iteration starts with CRAFT once again identifying the best exchange by considering all-possible two-way or three-way exchanges in the (updated) layout. The process continues until no further reduction in layout cost can be obtained. The final layout obtained in such a manner is also known as a two-opt (three-opt) layout since no two-way (three-way) exchanges can further reduce the layout cost.

Since computers were relatively "slow" in the 1960s, the original implementation of CRAFT deviated slightly from the above description. When the program considered exchanging the locations of departments i and j, instead of actually exchanging the department locations to compute their new centroids and the actual layout cost, it computed an *estimated* layout cost simply by temporaily treating the centroid of department i in the current layout as the centroid of department j and vice versa; that is, it simply swapped the *centroids* of departments i and j.

The error incurred in estimating the layout cost as described above depends on the relative size of the two departments being exchanged. If the departments differ in size, the estimated centroids may deviate significantly from their correct locations. (Of course, if the departments are equal in area, no error will be incurred.) As a result, the actual reduction in the layout cost may be overestimated or underestimated. Although it does not fully address the above error, once the best exchange *based on the estimated layout cost* is identified, CRAFT exchanges the locations of the departments and computes their new centroids (and the actual layout cost) before continuing to the next iteration.

A more important refinement we need to make in the above description of CRAFT is concerned with the exchange procedure. Although at first it may seem "too detailed," this refinement is important from a conceptual point of view since it demonstrates an intricate aspect of using computers for layout purposes. When CRAFT considers exchanging two departments, instead of examining all-possible exchanges as we stated above, it actually considers exchanging only those departments that are either adjacent (i.e., share a border) or equal in area. Such a restriction is not arbitrary. Given that departments cannot be split, it would be impossible to exchange two departments without "shifting" the location of the other departments in the layout, *unless the two departments are either adjacent or equal in area.* (Why?) Since CRAFT is not capable of "automatically" shifting departments in such a manner, it considers exchanging only those that are adjacent or equal in area.

Obviously, two departments with equal areas, whether they are adjacent or not, can always be exchanged without shifting the other departments in the layout. However, if two departments are not equal in area, then adjacency is a *necessary but not sufficient* condition for being able to exchange them without shifting the other departments. That is, in certain cases, even if two (unequal-area) departments are adjacent, it may not be possible to exchange them without shifting the other departments. We will later present an example for such a case.

We also need to stress that, while searching for a better solution, CRAFT picks only the best (estimated) exchange at each iteration, which makes it a "steepest descent" procedure. It also does not "look back" or "look forward" during the above search. Therefore, CRAFT will terminate at the first two-opt or three-opt solution that it encounters during the search. Such a solution is very likely to be only locally optimal. Furthermore, with such a search procedure, the termination point (or the final layout) will be strongly influenced by the starting point (or the initial layout). Consequently, CRAFT is a highly "path-dependent" heuristic and to use it effectively we generally recommend trying different initial solutions (if possible) or trying different exchange options (two-way vs. three-way) at each iteration.

CRAFT is generally flexible with respect to department shapes. As long as the department is not split, it can accommodate virtually any department shape. Theoretically, due to the centroid-to-centroid distance measure, the optimum layout (which has an objective function value of zero) consists of concentric rectangles! Of course, the above problem stems from the fact that the centroid of some O-shaped, U-shaped, and L-shaped departments may lie outside the department itself. Unless the initial layout contains concentric departments, CRAFT will not construct such a layout. However, some department shapes may be irregular, and the objective function value may be underestimated due to the centroid-to-centroid measure.

CRAFT is normally restricted to rectangular buildings. However, through "dummy" departments, it can be used with nonrectangular buildings as well. Dummy departments have no flows or interaction with other departments but require a certain area specified by the layout planner. In general, dummy departments may be used to:

1. Fill building irregularities.
2. Represent obstacles or unusable areas in the facility (such as stairways, elevators, plant services, and so on).
3. Represent extra space in the facility.
4. Aid in evaluating aisle locations in the final layout.

Note that, when a dummy department is used to represent an obstacle, its location must be fixed. Fortunately, CRAFT allows the user to fix the location of any department (dummy or otherwise). Such a feature is especially helpful in modeling obstacles as well as departments such as receiving and shipping in an existing facility.

One of CRAFT's strengths is that it can capture the initial layout with reasonable accuracy. This strength stems primarily from CRAFT's ability to accommodate nonrectangular departments or obstacles located anywhere in a possibly nonrectangular building. However, in addition to being highly path dependent, one of CRAFT's weaknesses is that it will rarely generate department shapes that result in straight, uninterrupted aisles as is desired in the final layout. Fixing some departments to specific locations, and in some cases placing dummy departments in the layout to represent main aisles, may lead to more reasonable department shapes. Nevertheless, as is the case with virtually all computerized layout algorithms, the final layout generated by the computer should not be presented to the decision maker before the layout planner "molds" or "massages" it into a practical layout.

As we demonstrate in the following example, molding or massaging a layout involves relatively minor adjustments made to the department shapes and/or department areas in the final layout. The fact that such adjustments are almost always necessary does not imply that computerized layout algorithms are of limited use. To the contrary, by considering a large number of alternatives in a very short time, a computerized layout algorithm narrows down the solution space for the layout planner who can concentrate on further evaluating and massaging a few "promising" solutions identified by the computer.

Example 6.1

Consider a manufacturing facility with seven departments. The department names, their areas, and the from-to chart are shown in Table 6.4. We assume that all the c_{ij} values are set equal to one. The building and the current layout (which we supply as the initial layout to CRAFT) are shown in Figure 6.15, where each grid is assumed to measure $20' \times 20'$. Since the total available space (72,000 ft^2) exceeds the total required space (70,000 ft^2), we generate a single dummy department (H) with an area of 2000 ft^2. (In most cases, depending on the amount of excess space available, it is highly desirable to use two or more dummy departments with more or less evenly distributed space requirements. Note that the allocation of excess space in the facility plays a significant role in determining future expansion options for many departments.) For practical reasons, we assume that the locations of the receiving (A) and shipping (G) departments are fixed.

Table 6.4 *Departmental Data and From-To Chart for Example 6.1*

Department Name	Area (ft^2)	No. of Grids	FLOW							
			A	B	C	D	E	F	G	H
1. A: Receiving	12,000	30	0	45	15	25	10	5	0	0
2. B: Milling	8,000	20	0	0	0	30	25	15	0	0
3. C: Press	6,000	15	0	0	0	0	5	10	0	0
4. D: Screw m/c	12,000	30	0	20	0	0	35	0	0	0
5. E: Assembly	8,000	20	0	0	0	0	0	65	35	0
6. F: Plating	12,000	30	0	5	0	0	25	0	65	0
7. G: Shipping	12,000	30	0	0	0	0	0	0	0	0
8. H: Dummy	2,000	5	0	0	0	0	0	0	0	0

CRAFT first computes the centroid of each department which are shown in Figure 6.15. For each department pair, it then computes the rectilinear distance between their centroids and multiplies it by the corresponding entry in the from-to chart. For example, the rectilinear distance between the centroids of departments A and B is equal to six grids. CRAFT multiplies 6 by 45 and adds the result to the objective function. Repeating the above calculation for all department pairs with nonzero flow yields an initial layout cost of 2974 units. (We caution the reader that CRAFT computes the distances in grids, not in feet. Therefore, the actual layout cost is equal to 2974 × 20 = 59,480 units.)

Subsequently, CRAFT performs the first iteration and exchanges departments E and F to obtain the layout shown in Figure 6.16. Departments E and F are not equal in area; however, since they are adjacent, one can draw a single "box" around E and F such that it contains both departments E and F but no other departments. (Note that, if E and F were not adjacent, drawing such a box would not be possible.) Since the above box contains no other departments, CRAFT will exchange departments E and F without shifting any other department. (Naturally, CRAFT does not "draw boxes." We introduced the box analogy only to clarify our description of CRAFT.)

There are several ways in which departments E and F may be exchanged within the above box. Comparing the locations of departments E and F in Figures 6.15 and 6.16, it is evident that CRAFT started with the left-most column of department F (the larger of the two departments) and labeled the first 20 grids of department F as department E. To complete the exchange, all the grids originally labeled with an E have been converted to an F. Of course, one may implement the above scheme starting from the rightmost column (or, say, the top row) of department F. Regardless of its exact implementation, however,

	1	2	3	4	5	6	7	8	9	10	11	12	13	14	15	16	17	18
1	A	A	A	A	A	A	A	A	A	A	G	G	G	G	G	G	G	G
2	A			•						A	G				•			G
3	A	A	A	A	A	A	A	A	A	A	G	G	G					G
4	B	B	B	B	B	C	C	C	C	C	E	E	G	G	G	G	G	G
5	B		•		B	C		•		C	E	E	E	E	E	E	E	E
6	B				B	C	C	C	C	C	E	E	E	E	E	E	E	E
7	B	B	B	B	B	D	D	D	D	F	F	F	F	F	F	F	E	E
8	D	D	D	D	D	D		D	F				•		F	F	F	
9	D			•			D	D	F	F	F	F	F			F		
10	D	D	D	D	D	D	D	D	H	H	H	H	H	H	F	F	F	F

Figure 6.15   Initial CRAFT layout and department centroids for Example 6.1 (*z* = 2974 × 20 = 59,480 units).

	1	2	3	4	5	6	7	8	9	10	11	12	13	14	15	16	17	18
1	A	A	A	A	A	A	A	A	A	A	G	G	G	G	G	G	G	G
2	A									A	G							G
3	A	A	A	A	A	A	A	A	A	A	G	G	G					G
4	B	B	B	B	B	C	C	C	C	C	F	F	G	G	G	G	G	G
5	B				B	C				C	F	F	F	F	F	F	F	F
6	B				B	C	C	C	C	C	F	F	F	F	F	F		F
7	B	B	B	B	B	D	D	D	D	E	E	E	E	E	E	E	F	F
8	D	D	D	D	D	D			D	E						E	F	F
9	D							D	D	E	E	E	E	E	E	E	F	F
10	D	D	D	D	D	D	D	D	H	H	H	H	H	E	E	F	F	F

Figure 6.16 Intermediate CRAFT layout obtained after exchanging departments E and F ($z = 2953 \times 20 = 59{,}060$ units).

such an exchange scheme generally leads to poor department shapes. In fact, the department shapes in CRAFT often have a tendency to deteriorate with the number of iterations even if all the departments in the initial layout are rectangular.

The *estimated* reduction in the layout cost obtained by exchanging (the centroids of) departments E and F is equal to 202 units. Upon exchanging departments E and F and computing the new department centroids, CRAFT computes the actual cost of the layout shown in Figure 6.16 as 2953 units. Hence, the actual reduction in the layout cost is 21 units as opposed to 202 units. The reader may verify that the above significant deviation is largely due to the fact that the new centroid of department F (after the exchange) deviates substantially from its estimated location.

In the next iteration, based on an estimated reduction of 95 units in the layout cost, CRAFT exchanges departments B and C to obtain the layout shown in Figure 6.17. The layout cost is equal to 2833.50 units, which represents an actual reduction of 119.50 units. This clearly illustrates that the error in estimation may be in either direction. Using estimated costs, CRAFT determines that no other (equal-area or adjacent) two-way or three-way exchange can further reduce the cost of the layout and it terminates with the final solution shown in Figure 6.17.

Recall that a computer-generated layout should not be presented to the decision maker before the analyst molds or massages it into a practical layout. In massaging a layout, the analyst may generally disregard the grids and use a continuous representation. In so doing, he or she may smooth the department borders and slightly change their areas or orientation, if necessary. After massaging the layout shown in Figure 6.17, we obtained the layout shown

	1	2	3	4	5	6	7	8	9	10	11	12	13	14	15	16	17	18
1	A	A	A	A	A	A	A	A	A	A	G	G	G	G	G	G	G	G
2	A									A	G							G
3	A	A	A	A	A	A	A	A	A	A	G	G	G					G
4	C	C	C	B	B	B	B	B	B	B	F	F	G	G	G	G	G	G
5	C			C	C	B				B	F	F	F	F	F	F	F	F
6	C			C	B	B	B	B	B	B	F	F	F	F	F	F		F
7	C	C	C	C	B	D	D	D	D	E	E	E	E	E	E	E	F	F
8	D	D	D	D	D	D			D	E						E	F	F
9	D							D	D	E	E	E	E	E	E	E	F	F
10	D	D	D	D	D	D	D	D	H	H	H	H	H	E	E	F	F	F

Figure 6.17 Final CRAFT layout ($z = 2833.50 \times 20 = 56{,}670$ units).

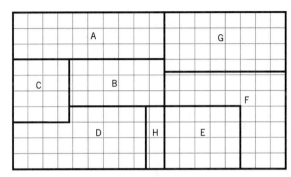

Figure 6.18   Final "massaged" layout obtained with CRAFT.

in Figure 6.18. Note that, other than department H (which is a dummy department), we made no significant changes to any of the departments; yet, the layout shown in Figure 6.18 is more reasonable and perhaps more practical than the one shown in Figure 6.17.

Once it is massaged in the above manner, a layout cannot be generally reevaluated via a computer-based layout algorithm unless one is willing to redefine the grid size and repeat the process using the new grid size. Also, we need to stress that in many real-world problems, layout massaging usually goes beyond adjusting department shapes or areas. In massaging a layout, the analyst must often take into account certain qualitative factors or constraints that may not have been considered by the algorithm.

Earlier we remarked that if two departments are *not equal* in area, then adjacency is a *necessary but not sufficient* condition for being able to exchange them without shifting the other departments. Obviously, adjacency is necessary since it is otherwise physically impossible to exchange two unequal-area departments without shifting other departments. (Recall that extra space is also modeled as a "department.") The fact that adjacency is not sufficient, on the other hand, can be shown through the following example [38].

Consider a 7×5 layout with six departments as shown in Figure 6.19. Note that departments 2 and 4 are not equal in area but they are adjacent, which implies that we can draw a single box around them. Yet, one cannot exchange departments 2 and 4 without splitting department 2. In fact, if we give the above layout to CRAFT, and fix the locations of all the departments except 2 and 4, CRAFT does not exchange departments 2 and 4 even if $f_{1,4}$ is set equal to a large value [38]. The above example is, of course, carefully constructed to show that adjacency is not sufficient. In most cases, two unequal-area departments that are adjacent can be exchanged without splitting either one.

6	6	6	5	5
6	6	6	5	5
6	6	6	5	4
6	6	6	4	4
2	2	2	2	2
1	1	2	3	3
1	1	2	3	3

Figure 6.19   Example to show that CRAFT may not be able to exchange two adjacent departments that are not equal in area.

Specifying the conditions required for a three-way exchange is somewhat more complicated. Suppose departments *i*, *j*, and *k* are considered for a three-way exchange such that department *i* "replaces" *j*, department *j* "replaces" *k*, and department *k* "replaces" *i*. If a single "box" can be drawn around departments *i*, *j*, and *k*, and this "box" does not contain any other department, then (except for cases such as the one shown above for departments 2 and 4) we can perform a three-way exchange without shifting any of the other departments. Note that, in order to draw such a box, each of the three departments need not share a border with the other two; one may still draw the above box if department *i* is adjacent to department *j* (but not *k*), and department *k* is adjacent to department *j* (but not *i*).

If it is not possible to draw such a box, equal-area departments may still permit some three-way exchanges. Suppose departments *i* and *j* are adjacent but department *k* is separated from both. For the above exchange involving departments *i*, *j*, and *k*, for example, one may perform the exchange without shifting any of the other departments if departments *j* and *k* are equal in area. Of course, other combinations (including the case where all three departments are nonadjacent but equal in area) are also possible. Computer implementation of three-way exchanges (i.e., deciding which department to move first and how to reassign the grids) is not straightforward. Furthermore, the number of possible three-way exchanges increases quite rapidly with the number of departments (which may result in long execution times). Since more recent (single-floor) facility layout algorithms focus on two-way exchanges only, and since the adjacency or equal-area requirement has been relaxed with various layout formation techniques, we will not present in detail three-way exchanges performed by CRAFT.

A personal-computer implementation of CRAFT by Hosni, Whitehouse, and Atkins [26] was distributed by the Institute of Industrial Engineers (IIE) under the name MICRO-CRAFT. (Since IIE discontinued software distribution, the reader may refer to the original authors to inquire about the code.) MICRO-CRAFT (or MCRAFT) is similar to CRAFT except that the above constraint is relaxed (i.e., MCRAFT can exchange any two departments whether they are adjacent or not). Such an improvement is obtained by using a layout formation technique which "automatically" shifts other departments when two unequal-area, nonadjacent departments are exchanged. Instead of assigning each grid to a particular department, MCRAFT first divides the facility into "bands" and the grids in each band are then assigned to one or more departments. The number of bands in the layout is specified by the user. MCRAFT's layout formation technique—which was originally used in an earlier algorithm, namely, ALDEP (*Automated Layout DEsign Program*) [54]—is described through the following example.

Example 6.2

Consider the same data given for Example 6.1. Unlike CRAFT (where the user selects the grid size), MCRAFT asks for the length and width of the building and the number of bands. The program then computes the appropriate grid size and the resulting number of rows and columns in the layout. Thus, setting the building length (width) equal to 360 ft (200 ft), and the number of bands equal to, say, three, we obtain the initial layout shown in Figure 6.20, where each band is six rows wide. MCRAFT forms a layout by starting at

```
11111111111111111111111111111111111117777777777777777777777777777777777
11111111111111111111111111111111111117777777777777777777777777777777777
11111111111111111111111111111111111117777777777777777777777777777777777
11111111111111111111111111111111111117777777777777777777777777777777777
11111111111111111111111111111111111117777777777777777777777777777777777
11111111111111111111111111111111111117777777777777777777777777777777777
44444222222222222222222222223333333333333333555555555555555555555555
44444222222222222222222222223333333333333333555555555555555555555555
44444222222222222222222222223333333333333333555555555555555555555555
44444222222222222222222222223333333333333333555555555555555555555555
44444222222222222222222222223333333333333333555555555555555555555555
44444222222222222222222222223333333333333333555555555555555555555555
44444444444444444444444444448888886666666666666666666666666666666666
44444444444444444444444444448888886666666666666666666666666666666666
44444444444444444444444444448888886666666666666666666666666666666666
44444444444444444444444444448888886666666666666666666666666666666666
44444444444444444444444444448888886666666666666666666666666666666666
44444444444444444444444444448888886666666666666666666666666666666666
```

Figure 6.20 Initial MCRAFT layout for Example 6.2 ($z = 60,661.11$ units).

the upper-left-hand corner of the building and "sweeping" the bands in a serpentine fashion. In doing so, it follows a particular sequence of department numbers which we will refer to as the *layout vector* or the *fill sequence*. The layout shown in Figure 6.20 was obtained from the layout vector 1-7-5-3-2-4-8-6, which is supplied by the user as the initial layout vector. (Note that MCRAFT designates the departments by numbers instead of letters; department 1 represents department A, department 2 represents department B, etc.)

MCRAFT performs four iterations (i.e., four two-way exchanges) before terminating with the two-opt layout. The departments exchanged at each iteration and the corresponding layout cost are given as follows: first iteration—departments C and E (59,611.11 units); second iteration—departments C and H (58,083.34 units); third iteration—departments C and D (57,483.34 units); and fourth iteration—departments B and C (57,333,34 units). The resulting three-band layout is shown in Figure 6.21, for which the layout vector is given by 1-7-8-5-3-2-4-6. Except for department 2, the departments in Figure 6.21 appear to have reasonable shapes. Unlike CRAFT, the department shapes obtained from the sweep method tend to be reasonable if an appropriate number of bands is selected. Of course, alternative initial and final layouts may be generated by varying the number of bands and the initial layout vector.

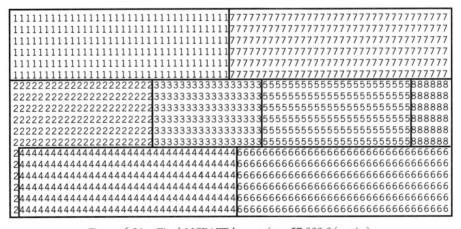

Figure 6.21 Final MCRAFT layout ($z = 57,333.34$ units).

Due to the above layout formation technique, MCRAFT will not be able to capture the initial layout accurately unless the departments are already arranged in bands. As a result, one may have to massage the initial layout to make it compatible with MCRAFT. (The reader may compare the original initial layout shown in Figure 6.15 with MCRAFT's initial layout shown in Figure 6.20.) The actual cost of the initial layout as computed by CRAFT is equal to 2974 × 20 = 59,480 units, whereas MCRAFT's initial layout cost is equal to 60,661.11 units—a relatively small difference considering the fact that the "actual" layout used by CRAFT is itself an approximation of reality. Based on the number of departments and the actual initial layout, however, the above difference may be significant for some problems.

Another limitation of the sweep method used in MCRAFT is that the band width is assumed to be the same for all the bands. As a result, MCRAFT is generally not as effective as CRAFT in treating obstacles and fixed departments. If there is an obstacle in the building, the analyst must ensure that its width does not exceed the band width. Otherwise, the obstacle must be divided into two or more pieces which is likely to further complicate matters. In addition, while it is straightforward in MCRAFT to fix the location of any department, a fixed department may still "shift" or "float" when certain non-equal-area departments are exchanged.

In the above example, the two fixed departments (1 and 7) remained at their current locations since they are the first two departments in the initial layout vector 1-7-5-3-2-4-8-6. However, if we fix, say, department 2, and the algorithm exchanges departments 3 and 4, the location of department 2 will shift when the layout is formed with the new layout vector. (Why?) In general, when two unequal-area departments are exchanged, the location of a fixed department will shift if the two departments fall on either side of the fixed department in the layout vector. Considering that obstacles are also modeled as fixed departments (with zero flow), the above limitation implies that obstacles may shift as well. It is instructive to note that MCRAFT's primary strength (i.e., being able to "automatically" shift other departments as necessary) is also its primary weakness (i.e., fixed departments and obstacles may also be shifted).

BLOCPLAN

BLOCPLAN, which was developed by Donaghey and Pire ([11] and [49]), is similar to MCRAFT in that departments are arranged in bands. However, there are certain differences between the two algorithms. BLOCPLAN uses a relationship chart as well as a from-to chart as input data for the "flow." Layout "cost" can be measured either by the distance-based objective (see Equation 6.1) or the adjacency-based objective (see Equation 6.2). Furthermore, in BLOCPLAN, the number of bands is determined by the program and limited to two or three bands. However, the band widths are allowed to vary. Also, in BLOCPLAN since each department occupies exactly one band, all the departments are rectangular in shape. Lastly, unlike MCRAFT, BLOCPLAN uses the continuous representation.

BLOCPLAN may be used both as a construction algorithm and an improvement algorithm. In the latter case, as with MCRAFT, it may not be possible to capture the initial layout accurately. Nevertheless, improvements in the layout are sought through (two-way) department exchanges. Although the program accepts both a relationship chart and a from-to chart as input, the two charts can be used only one at a time when evaluating a layout. That is, a layout is not evaluated according to some "combination" of the two charts.

BLOCPLAN first assigns each department to one of the two (or three) bands. Given all the departments assigned to a particular band, BLOCPLAN computes the appropriate band width by dividing the total area of the departments in that band by the building length. The complete layout is formed by computing the appropriate width for each band as described above and arranging the departments in each band according to a particular sequence.

The above layout formation procedure and the layout scores computed by BLOCPLAN are explained through the following example.

Example 6.3

Consider the same data given for Example 6.1. As in MCRAFT, BLOCPLAN will not be able to capture the initial layout accurately unless the departments are already arranged in bands. As a result, the analyst may have to massage the initial layout to make it compatible with BLOCPLAN. (The reader may compare the actual initial layout shown in Figure 6.15 with BLOCPLAN's initial layout shown in Figure 6.22.) The actual cost of the initial layout as computed by CRAFT is equal to 59,480 units, whereas BLOCPLAN's initial layout cost is equal to 61,061.70 units—a relatively small difference of less than 3%.

BLOCPLAN offers the analyst a variety of options for improving the layout. The analyst may try some two-way exchanges simply by typing the department indices to be exchanged, or he or she may select the "automatic search" option to have the algorithm generate a prespecified number of layouts. (According to [11], the automatic search option is based on the "procedures that an experienced BLOCPLAN user used in obtaining a 'good' layout"; no details are provided on what these procedures are.) Using BLOCPLAN's improvement algorithm, which interactively considers all possible two-way exchanges, we obtain the layout shown in Figure 6.23, where departments C and H have been exchanged. Since no other two-way exchange leads to a reduction in layout cost, BLOCPLAN was terminated with the final layout as shown in Figure 6.23.

The "cost" of the final layout, as measured by the distance-based objective and the from-to chart given for the example problem—is equal to 58,133.34 units. BLOCPLAN

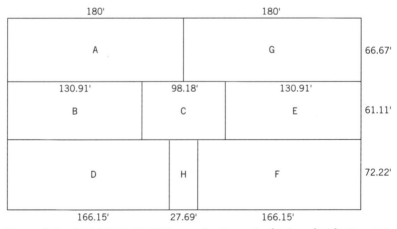

Figure 6.22   Initial BLOCPLAN layout for Example 6.3 ($z = 61,061.70$ units).

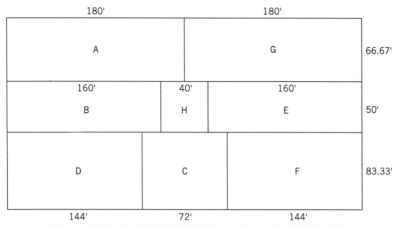

Figure 6.23   Final BLOCPLAN layout ($z = 58{,}133.34$ units).

implicitly assumes that all the c_{ij}'s are equal to 1.0; that is, the user cannot enter a cost matrix. (Recall that in the example problem all the c_{ij}'s are assumed to be equal to 1.0.) The final layout shown in Figure 6.23 can also be evaluated with respect to the adjacency-based objective. Using Equation 6.2, we compute $z = 235$ units, which is obtained by adding the f_{ij} values between all department pairs that are adjacent in Figure 6.23. The normalized adjacency score (or the efficiency rating) given by Equation 6.3 is equal to $235/(235 + 200) = 0.54$.

   If the input data is given as a relationship chart instead of a from-to chart, the adjacency score may still be computed provided that a numerical value is assigned to each closeness rating. More specifically, in place of f_{ij} shown in Equation 6.2, one may use the numerical value of the closeness rating assigned to departments i and j. The following default values are used in BLOCPLAN: $A = 10$, $E = 5$, $I = 2$, $O = 1$, $U = 0$, and $X = -10$. The user may of course specify different numerical values. Although the values $A = 6$, $E = 5$, $I = 4$, $O = 3$, $U = 2$, and $X = 1$ have been used in some algorithms or texts, for the purposes of computing the adjacency score, using a nonzero value for U and a nonnegative value for X would not be appropriate. Department pairs with an U relationship should not affect the score whether they are adjacent or not, while those pairs with an X relationship should adversely affect the score if they are adjacent.

   Even if a from-to chart is provided, BLOCPLAN computes the adjacency score based on the relationship chart. This is accomplished by converting the from-to chart into a relationship chart and using the above numerical values assigned to the closeness ratings. Although one may develop alternate schemes for such a conversion, the one used in BLOCPLAN is straightforward; it is explained through the following example.

Example 6.4

Consider the from-to chart given earlier for Example 6.1 (see Table 6.4). Since BLOCPLAN implicitly assumes that all the unit costs are equal to 1.0, and all the distances are symmetric by definition (i.e., $d_{ij} = d_{ji}$ for all i and j), it first converts the from-to chart into a

Table 6.5   *Flow Between Chart and Relationship Chart for Example 6.4*

	A	B	C	D	E	F	G	H		A	B	C	D	E	F	G	H
A	0	45	15	25	10	5	0	0	A	—	I	U	O	U	U	U	U
B		0	0	50	25	20	0	0	B		—	U	I	O	O	U	U
C			0	0	5	10	0	0	C			—	U	U	U	U	U
D				0	35	0	0	0	D				—	O	U	U	U
E					0	90	35	0	E					—	A	O	U
F						0	65	0	F						—	E	U
G							0	0	G							—	U
H								0	H								—
			(*a*) Flow-between chart									(*b*) Relationship chart					

flow-between chart by adding f_{ij} to f_{ji} for all department pairs. That is, the from-to chart shown in Table 6.4 is converted into the flow-between chart shown in Table 6.5*a*. A flow-between chart is symmetric, by definition.

The maximum flow value in the flow-between chart is equal to 90 units. Dividing the above maximum by 5 we obtain 18, which leads to the following intervals and corresponding closeness ratings: 73 to 90 units (A), 55 to 72 units (E), 37 to 54 units (I), 19 to 36 units (O), and 0 to 18 units (U). That is, BLOCPLAN divides the flow values into five intervals of 18 units each. Any flow value between 73 and 90 units is assigned an A relationship, and so on. Applying this conversion scheme to the above flow-between chart yields the relationship chart shown in Table 6.5*b*.

Given the default numerical values assigned to the closeness ratings (i.e., $A = 10$, $E = 5$, $I = 2$, $O = 1$, $U = 0$, $X = -10$) and the relationship chart shown in Table 6.5*b*, we compute a normalized adjacency score of $30/48 = 0.63$ for the final layout obtained in Example 6.3 (see Figure 6.23). It is instructive to note that the normalized adjacency score for the initial layout in Example 6.3 is also equal to 0.63, while the distance-based objective for the initial and final layouts in Example 6.3 are equal to 61,061.70 and 58,133.34 units, respectively. The above result illustrates the concern we expressed earlier for the adjacency score; that is, it is possible for two layouts to have virtually the same (normalized) adjacency score but different travel distances for parts flow. BLOCPLAN reports both the normalized adjacency score and the distance-based objective for each layout it generates.

In addition to the distance-based layout cost (using the flow-between chart) and the normalized adjacency score (based on the relationship chart), BLOCPLAN computes a "REL-DIST" score, which is a distance-based layout cost that uses the numerical closeness ratings instead of the flow values. That is, the numerical values of the closeness ratings are multiplied by the rectilinear distances between department centroids. For example, for the initial (final) layout shown in Figure 6.22 (Figure 6.23), the REL-DIST score is equal to 2887.01 (2708.33) units. The REL-DIST score is useful when a from-to chart is not available and the layout must be evaluated with respect to a relationship chart only.

BLOCPLAN also computes a normalized REL-DIST score. However, the upper and lower bounds used in normalizing the REL-DIST score depend on the particular layout being evaluated (we refer the reader to [11]). As a result, caution must be exercised when comparing the normalized REL-DIST scores of two competing layouts.

We remarked earlier that BLOCPLAN implicitly assumes that all the unit costs are equal to 1.0. This assumption is not as restrictive as it may first seem. If the unit costs are not equal to 1.0, then provided that they are symmetric, that is, $c_{ij} = c_{ji}$ for all i and j, the user may first multiply each flow value with the corresponding unit cost and then enter the result as the "flow." Note that unit costs must be symmetric since BLOCPLAN works with a flow-between chart as opposed to a from-to chart as shown earlier in Example 6.4.

BLOCPLAN may also be used as a construction algorithm. In doing so, the layout planner may indicate the location of certain departments a priori. As we remarked earlier, whether one inputs the entire initial layout or a few departments that must be located prior to layout construction, it may not be possible to accurately locate all the departments. BLOCPLAN uses the following scheme to specify the location of one or more departments: the entire building is divided into nine cells (labeled A through I) that are arranged in three bands. The top band contains cells A, B, and C (from left to right), while the middle and bottom bands contain cells D, E, F and G, H, I (from left to right), respectively. Each cell is further divided into two halves; a right half and a left half. Hence, there are a total of 18 possible locations, which is also equal to the maximum number of departments that BLOCPLAN will accept.

The location of a department is designated by indicating the appropriate side of a cell. For example, to locate a department on the northeast corner of the building, one would use C-R to indicate the right side of cell C. To locate a department, say, on the west side of the building, one would use D-L, and so on. Of course, regardless of the cell they are assigned to, all departments will be arranged in two or three bands when BLOCPLAN constructs a layout. Each department may be assigned to only one of the 18 cells.

MIP[1]

Using the continuous representation, the facility layout problem may be formulated as a mixed integer programming (MIP) problem if all the departments are assumed to be rectangular. Recall that, with rectangular departments, just the centroid and the length (or width) of a department fully define its location and shape. Although an alternate objective may be used, the model we show here assumes a distance-based objective (given by Equation 6.1) with d_{ij} defined as the rectilinear distance between the centroids of departments i and j. Generally speaking, models based on mathematical programming are regarded as *construction-type* layout models since there is no need to enter an initial layout. However, as we discuss later, such models may also be used to improve a given layout.

The model we show here is based on the one presented by Montreuil [44]. A similar model is also presented by Heragu and Kusiak [23], where the department dimensions are treated as parameters with known values instead of decision variables. Assuming that all department dimensions are given and fixed is appropriate in "machine layout" problems, where each "department" represents the rectangular "footprint" of a machine. Note that, with such an approach, the layout model is actually used as a two-dimensional "packing" algorithm to determine the locations of rectangular objects with known shapes. Aside from possible differences in the objective function, such a packing problem is also known as the "two-dimensional bin packing" problem [7].

---

[1]The model presented in this section requires a basic knowledge of linear and integer programming.

Treating the department dimensions as decision variables, the facility layout problem may be formulated as follows. Consider first the problem parameters. Let:

B_x be the building length (measured along the x-coordinate),

B_y be the building width (measured along the y-coordinate),

A_i be the area of department i,

L_i^ℓ be the lower limit on the length of department i,

L_i^u be the upper limit on the length of department i,

W_i^ℓ be the lower limit on the width of department i,

W_i^u be the upper limit on the width of department i,

M be a large number.

Consider next the decision variables. Let:

α_i be the x-coordinate of the centroid of department i,

β_i be the y-coordinate of the centroid of department i,

x_i' be the x-coordinate of the left (or west) side of department i,

x_i'' be the x-coordinate of the right (or east) side of department i,

y_i' be the y-coordinate of the bottom (or south side) of department i,

y_i'' be the y-coordinate of the top (or north side) of department i,

z_{ij}^x be equal to 1 if department i strictly to the east of department j, and 0 otherwise,

z_{ij}^y be equal to 1 if department i is strictly to the north of department j, and 0 otherwise.

Note that department i would be strictly to the east of department j if and only if $x_j'' \le x_i'$. Likewise, department i would be strictly to the north of department j if and only if $y_j'' \le y_i'$. Two departments are guaranteed not to overlap if they are "separated" along either the x-coordinate (i.e., one of the departments is strictly to the east of the other) or the y-coordinate (i.e., one of the departments is strictly to the north of the other). Of course, it is possible that two departments are separated along both the x and y-coordinates.

The above parameter and variable definitions lead to the following model:

$$\text{Minimize } z = \sum_i \sum_j f_{ij} c_{ij} \left(|\alpha_i - \alpha_j| + |\beta_i - \beta_j| \right) \tag{6.5}$$

$$\text{Subject to: } L_i^\ell \le (x_i'' - x_i') \le L_i^u \qquad \text{for all } i \tag{6.6}$$

$$W_i^\ell \le (y_i'' - y_i') \le W_i^u \qquad \text{for all } i \tag{6.7}$$

$$(x_i'' - x_i')(y_i'' - y_i') = A_i \qquad \text{for all } i \tag{6.8}$$

$$0 \le x_i' \le x_i'' \le B_x \qquad \text{for all } i \tag{6.9}$$

$$0 \le y_i' \le y_i'' \le B_y \qquad \text{for all } i \tag{6.10}$$

$$\alpha_i = 0.5x_i' + 0.5x_i'' \qquad \text{for all } i \tag{6.11}$$

$$\beta_i = 0.5y_i' + 0.5y_i'' \qquad \text{for all } i \tag{6.12}$$

$$x_j'' \le x_i' + M(1 - z_{ij}^x) \qquad \text{for all } i \text{ and } j, \ i \ne j \tag{6.13}$$

$$y_j'' \le y_i' + M(1 - z_{ij}^y) \qquad \text{for all } i \text{ and } j, \ i \ne j \tag{6.14}$$

$$z_{ij}^x + z_{ji}^x + z_{ij}^y + z_{ji}^y \geq 1 \qquad \text{for all } i \text{ and } j, \ i < j \qquad (6.15)$$

$$\alpha_i, \beta_i \geq 0 \qquad \text{for all } i \qquad (6.16)$$

$$x_i', x_i'', y_i', y_i'' \geq 0 \qquad \text{for all } i \qquad (6.17)$$

$$z_{ij}^x, z_{ij}^y \ 0/1 \text{ integer} \qquad \text{for all } i \text{ and } j, \ i \neq j \qquad (6.18)$$

The objective function given by Equation 6.5 is the distance-based objective shown earlier as Equation 6.1. Constraint sets 6.6 and 6.7, respectively, ensure that the length and width of each department are within the specified bounds. The area requirement of each department is expressed through constraint set 6.8, which are the only non-linear constraints in the model. Constraint sets 6.9 and 6.10 ensure that the department sides are defined properly and that each department is located within the building in the x and y directions, respectively. Constraint sets 6.11 and 6.12, respectively, define the x and y coordinates of the centroid of each department.

Constraint set 6.13 ensures that $x_j'' \leq x_i'$ (i.e., department i is strictly to the east of department j) if $z_{ij}^x = 1$. Note that, if $z_{ij}^x = 0$, constraint set 6.13 is satisfied whether x_j'' is less than or equal to x_i' or not. In other words, constraint set 6.13 becomes "active" only when $z_{ij}^x = 1$. Constraint set 6.14 serves the same purpose as constraint set 6.13 but in the y direction. Constraint set 6.15 ensures that no two departments overlap by forcing a separation at least in the east–west or north–south direction.[2] Lastly, constraint sets 6.16 and 6.17 represent the nonnegativity constraints while constraint set 6.18 designates the binary variables.

Due to constraint set 6.8, the above model is nonlinearly constrained. Furthermore, the objective function constains the absolute value operator. There are alternative schemes one may employ to approximate constraint 6.8 via a linear function. In [44], the department area is controlled through its perimeter (which is a linear function of the length and width of the department) and constraint sets 6.6 and 6.7.

Provided that $f_{ij} \geq 0$ for all i and j, the absolute values in the objective function may be removed by introducing a "positive part" and a "negative part" for each term [46]. That is, if we set

$$\alpha_i - \alpha_j = \alpha_{ij}^+ - \alpha_{ji}^- \quad \text{and} \quad \beta_i - \beta_j = \beta_{ij}^+ - \beta_{ij}^-, \qquad (6.19)$$

$$\text{then} \quad |\alpha_i - \alpha_j| = \alpha_{ij}^+ + \alpha_{ij}^- \quad \text{and} \quad |\beta_i - \beta_j| = \beta_{ij}^+ + \beta_{ij}^-, \qquad (6.20)$$

where α_{ij}^+, α_{ij}^-, β_{ij}^+, and β_{ij}^- are all nonnegative variables.

Hence, the nonlinear model given by Equations 6.5 through 6.18 can be transformed into the following linear MIP problem:

$$\text{Minimize } z = \sum_i \sum_j f_{ij} c_{ij} (\alpha_{ij}^+ + \alpha_{ij}^- + \beta_{ij}^+ + \beta_{ij}^-) \qquad (6.21)$$

$$\text{Subject to:} \quad L_i^\ell \leq (x_i'' - x_i') \leq L_i^u \qquad \text{for all } i \qquad (6.22)$$

$$W_i^\ell \leq (y_i'' - y_i') \leq W_i^u \qquad \text{for all } i \qquad (6.23)$$

$$P_i^\ell \leq 2(x_i'' - x_i' + y_i'' - y_i') \leq P_i^u \qquad \text{for all } i \qquad (6.24)$$

$$0 \leq x_i' \leq x_i'' \leq B_x \qquad \text{for all } i \qquad (6.25)$$

---

[2]Actually, given the nature of constraints 6.13 and 6.14, one may rewrite constraint 6.15 as $z_{ij}^x + z_{ji}^x + z_{ij}^y + z_{ji}^y = 1$. Such constraints are known as "multiple-choice" constraints in integer programming; i.e., only one of the binary variables in the constraint may be set equal to one. Some branch-and-bound algorithms aim at reducing the computational effort by taking advantage of multiple-choice constraints.

$$0 \leq y_i' \leq y_i'' \leq B_y \qquad \text{for all } i \qquad (6.26)$$

$$\alpha_i = 0.5x_i' + 0.5x_i'' \qquad \text{for all } i \qquad (6.27)$$

$$\beta_i = 0.5y_i' + 0.5y_i'' \qquad \text{for all } i \qquad (6.28)$$

$$\alpha_i - \alpha_j = \alpha_{ij}^+ - \alpha_{ij}^- \qquad \text{for all } i \text{ and } j, i \neq j \qquad (6.29)$$

$$\beta_i - \beta_j = \beta_{ij}^+ - \beta_{ij}^- \qquad \text{for all } i \text{ and } j, i \neq j \qquad (6.30)$$

$$x_j'' \leq x_i' + M(1 - z_{ij}^x) \qquad \text{for all } i \text{ and } j, i \neq j \qquad (6.31)$$

$$y_j'' \leq y_i' + M(1 - z_{ij}^y) \qquad \text{for all } i \text{ and } j, i \neq j \qquad (6.32)$$

$$z_{ij}^x + z_{ji}^x + z_{ij}^y + z_{ji}^y \geq 1 \qquad \text{for all } i \text{ and } j, i < j \qquad (6.33)$$

$$\alpha_i, \beta_i, \geq 0 \qquad \text{for all } i \qquad (6.34)$$

$$x_i', x_i'', y_i', y_i'' \geq 0 \qquad \text{for all } i \qquad (6.35)$$

$$\alpha_{ij}^+, \alpha_{ij}^-, \beta_{ij}^+, \beta_{ij}^- \geq 0 \qquad \text{for all } i \text{ and } j, i \neq j \qquad (6.36)$$

$$z_{ij}^x, z_{ij}^y \; 0/1 \text{ integer} \qquad \text{for all } i \text{ and } j, i \neq j \qquad (6.37)$$

where P_i^ℓ and P_i^u that appear in constraint set 6.24 represent the lower and upper limits imposed on the perimeter of department i, respectively.

In the above formulation, department shapes as well as their areas are indirectly controlled through constraint sets 6.22 through 6.24. In some cases, one may wish to explicitly control the shape of a department by placing an upper limit on the ratio of its longer side to its shorter side. If we designate the upper limit by R_i (≥ 1.0), such a ratio may be controlled simply by adding the following (linear) constraints to the model for department i:

$$(x_i'' - x_i') \leq R_i(y_i'' - y_i') \qquad (6.38)$$

$$(y_i'' - y_i') \leq R_i(x_i'' - x_i') \qquad (6.39)$$

Note that, depending on which side is the longer one, at most one of the above two constraints will be "binding" or hold as an equality for non-square departments.

We remarked earlier that the layout planner may represent an X relationship between departments i and j by assigning a negative value to f_{ij}. However, the transformation we used earlier to linearize the model (see Equation 6.19) requires that $f_{ij} \geq 0$ for all i and j. While this requirement rules out the use of "negative flow" values, it does not imply that the above model cannot effectively capture an X relationship between two departments. One may insert a minimum required distance between two departments i and j by modifying constraints 6.31 and 6.32, as follows:

$$x_j'' + \Delta_{ij}^x \leq x_i' + M(1 - z_{ij}^x) \qquad \text{for all } i \text{ and } j, i \neq j \qquad (6.40)$$

$$y_j'' + \Delta_{ij}^y \leq y_i' + M(1 - z_{ij}^y) \qquad \text{for all } i \text{ and } j, i \neq j \qquad (6.41)$$

where Δ_{ij}^x and Δ_{ij}^y denote the minimum required distance (or clearance) between departments i and j along the x- and y-coordinates, respectively. Such clearances should be used sparingly; otherwise, the number of feasible solutions can be very limited. Also, note that if departments i and j are separated along, say, only the x-coordinate, then the resulting clearance between them along the y-coordinate may be less than Δ_{ij}^y.

Hence, provided that the departmental area requirements need not be satisfied with precision, the optimum layout with rectangular departments can be, in theory, obtained by solving the MIP model given by Equations 6.21 through 6.37. In practice, however, obtaining an exact solution to the above problem is not straightforward due to the large number of binary variables involved. At present, using "generic" optimization codes, the largest-size problem that one can solve exactly and with "reasonable" computational effort is approximately a seven- or eight-department problem in general; see, for example, [41] for details.

Consequently, to use the MIP model in practice (where the number of departments can easily exceed 15 or 18 departments), one may aim for a heuristic solution rather than the optimal solution. One obvious heuristic approach is to terminate the branch-and-bound search when the difference between the least lower bound and the incumbent solution is less than, say, 5% or 10%. The advantage of such an approach is that the resulting solution is guaranteed to be within 5% or 10% of the optimum; the primary drawback is that it may still take a substantial amount of computer time to identify such a solution.

Another possible heuristic approach [45] is to determine the north–south and east–west relationships between the departments a priori by constructing a "design skeleton." With such an approach, first the z_{ij}^x and the z_{ij}^y values are determined heuristically, and the MIP model shown by Equations 6.21 through 6.37 can then be solved as a linear programming problem to "pack" the departments within the building.

Example 6.5

Consider the same data given for Example 6.1. Assuming that departments A and G are fixed on the north side of the building (see the initial layout shown in Figure 6.15), we use the MIP model to construct the optimal layout. The solution is shown in Figure 6.24. The layout cost is equal to 53,501.17 units. Although some department areas in Figure 6.24 are not exactly equal to the values specified in Table 6.4, the reader may verify that the maximum deviation is only 0.25% (due to department D). Of course, some shape constraints were imposed on the departments to avoid long-and-narrow departments.

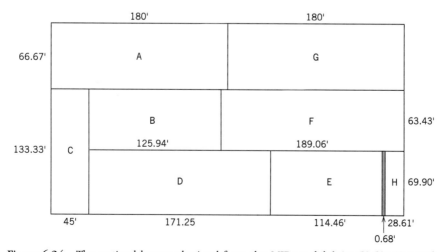

Figure 6.24 The optimal layout obtained from the MIP model ($z^* = 53,501.17$ units).

There are many real-world applications where the ideal shape of a department is a rectangle. However, there are also cases where L-shaped or U-shaped departments may be more appropriate or necessary. Furthermore, if there are fixed departments or obstacles in the building, enforcing rectangular department shapes with exact area values may make it impossible to find a feasible solution (even if the continuous representation is used and the fixed departments and obstacles are rectangular).

For example, consider a 10×10 building and four departments with the following area requirements: $A_1 = 18$, $A_2 = 20$, $A_3 = 27$, and $A_4 = 33$. Note that the total area requirement is actually less than 100 units. If none of the departments are fixed, then the above problem obviously has many feasible solutions. However, if the first department is fixed, say, at the northeast corner of the building (see Figure 6.25a), then there is no feasible solution to the problem. In contrast, if the area of, say, department 3 or 4 can be slightly adjusted, then three feasible solutions can be constructed as shown in Figures 6.25a, 6.25b, and 6.25c.

Thus, retaining some flexibility in departmental area requirements is necessary for the effective use of the MIP model. In fact, approximating the departmental

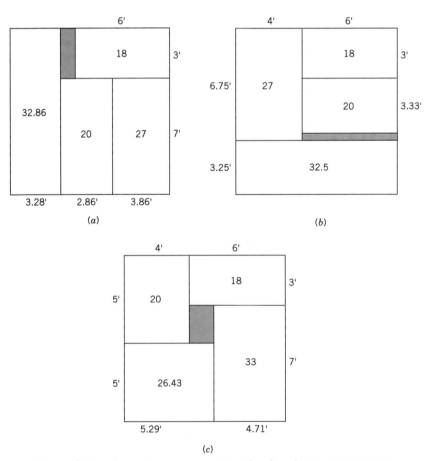

Figure 6.25   Alternative arrangements with relaxed area requirements.

areas through some linear relaxation of constraint 6.8 serves a dual purpose: It allows one to remove a nonlinear constraint and at the same time it significantly reduces the likelihood of having no feasible solutions when obstacles and/or fixed departments are present. Of course, generally speaking, it also increases the number of feasible solutions.

The MIP model may also be used to improve a given layout after specifying the dimensions and location of fixed departments in the initial layout. This is accomplished simply by setting α_i, β_i, x'_i, x''_i, y'_i, and y''_i equal to their appropriate values for all the fixed departments and/or obstacles. (Nonrectangular obstacles may be represented as a collection of rectangular dummy departments.) Note that, with such an approach, the initial layout is "improved" not by exchanging department locations but rather determining the new locations of all the (nonfixed) departments at once. Since such an approach does not explicitly consider "incremental" improvements to the layout and the resulting relocation costs versus material handling savings, it may not be a practical approach.

If relocation costs are significant, the layout planner may systematically fix and unfix certain subsets of the (nonfixed) departments and solve the MIP model several times to affect incremental changes in the layout. Alternately, the layout planner may first determine what the north–south/east–west relationships are going to be for two departments after their exchange, and subsequently use these relationships and the above MIP model to "repack" the departments with two of them exchanged.

LOGIC

LOGIC (*Layout Optimization with Guillotine Induced Cuts*) was developed by Tam [58]. In describing LOGIC, we assume that a from-to chart is given as input data for the flow. We also assume that layout "cost" is measured by the distance-based objective function shown in Equation 6.1. The departments generated by LOGIC are rectangular provided that the building is rectangular. The layout is represented in a continuous fashion.

Although LOGIC can be used as a layout improvement algorithm, we will first present it as a construction algorithm. LOGIC is based on dividing the building into smaller and smaller portions by executing successive "guillotine" cuts; that is, straight lines that run from one end of the building to the other. Each cut is either a vertical cut or a horizontal cut. If a cut is vertical, a department is assigned either to the east side of the cut or to the west. Given the building width (or a portion of it), and the total area of all the departments assigned, say, to the east side of a vertical cut, one may compute the exact x-coordinate of the cut. Likewise, given building length (or a portion of it), and the total area of all the departments assigned, say, to the north side of a horizontal cut, one may compute the exact y-coordinate of the cut.

LOGIC executes a series of horizontal and vertical cuts. With each cut, an appropriate subset of the departments are assigned to the east–west or north–south side of the cut. In order to systematically execute the cuts and keep track of the departments, LOGIC constructs a tree as described in the following example, where we assume that the cuts and the department assignments are made randomly.

Example 6.6

Consider the same data given for Example 6.1, except that we assume none of the departments (including A and G) are fixed. That is, we assume a vacant building which measures 360 ft × 200 ft in length and width, respectively. Suppose the first cut is a vertical cut and departments D, F, and G are assigned to its east while the remaining departments are assigned to its west (see Figure 6.26*a*). Since the total area required by departments D, F, and G is equal to 36,000 ft^2 and the building width is equal to 200 ft, the above vertical cut divides the building into two pieces that are 36,000/200 = 180 ft long each. The first cut is also shown in Figure 6.27, where each node is labeled with a *v* for a vertical cut and a *h* for a horizontal cut.

LOGIC next treats each portion of the building as a "building" by itself and repeats the above procedure until each "building" contains only one department. Consider first the "building" which contains departments A, B, C, E, and H. Suppose the next cut is a horizontal cut and that departments A and B are assigned to the north of the cut while C, E, and H are assigned to its south. Since the total area required by departments A and B is equal to 20,000 ft^2 and the "building" length is equal to 180 ft, the width of the "building" which contains departments A and B is equal to 20,000/180 = 111.11 ft (see Figure 6.26*b*).

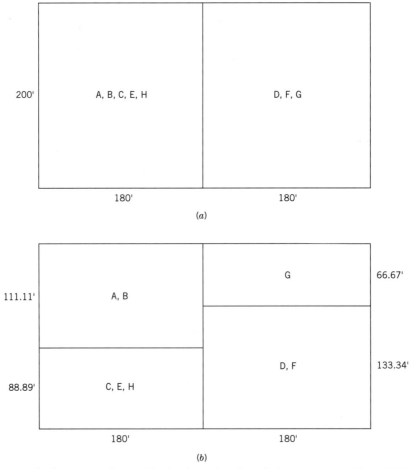

Figure 6.26 Layout obtained by horizontal and vertical cuts executed by LOGIC.

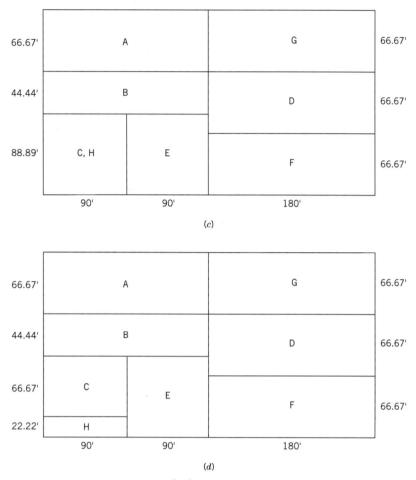

Figure 6.26 (*continued*)

Consider next the "building" which contains departments D, F, and G. Suppose the third cut is again a horizontal cut and that department G assigned to the north of the cut while D and F are assigned to its south. Since department G requires an area of 12,000 ft^2 and the "building" length is equal to 180 ft, the width of department G is set equal to 12,000/180 = 66.67 ft as shown in Figure 6.26*b*. The second and third cuts as described above are also shown in Figure 6.27.

Suppose the fourth cut horizontally divides departments A and B, while the fifth cut vertically divides departments {C, H} and E. Further suppose the sixth cut horizontally divides departments D and F. The layout that results from the above cuts is shown in Figure 6.26*c*. Assuming that the seventh and final cut horizontally divides departments C and H, we obtain the final layout shown in Figure 6.26*d*. (The above cuts are also shown in Figure 6.27.)

LOGIC may also be used as an improvement algorithm in a variety of ways. Here we will show how it can be used to exchange two departments *given that the cut-tree remains the same*. Consider the layout shown in Figure 6.26*d*. Suppose we would like to exchange departments D and E, which are not equal in area. If the cuts remain the same (see Figure 6.27), we simply replace all the D's in the tree

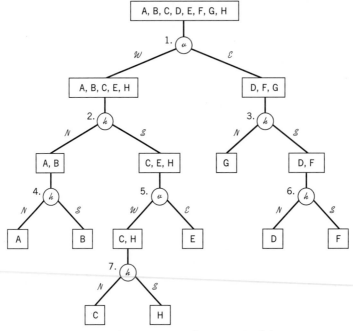

Figure 6.27   Cut-tree for Example 6.6.

with E's and vice versa, and compute the new x- and y-coordinates of the cuts. The resulting layout is shown in Figure 6.28.

It is instructive to note that, since departments D and E are adjacent, CRAFT would have taken a fundamentally different approach to exchange them and the resulting shapes of the two departments would have been different from those shown in Figure 6.28. As the above example illustrates, LOGIC can exchange two unequal-area departments (whether they are adjacent or not). Naturally, other departments have to shift to accommodate such an exchange (We encourage the

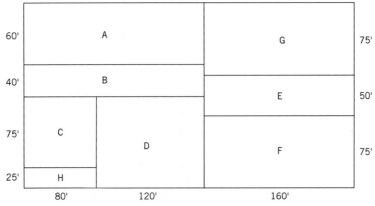

Figure 6.28   LOGIC layout obtained after exchanging departments D and E in Figure 6.26*d*.

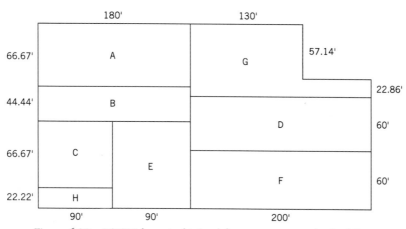

Figure 6.29 LOGIC layout obtained for a nonrectangular building.

reader to compare the layouts shown in Figures 6.26d and 6.28.) Hence, like MCRAFT, LOGIC can "automatically" shift other departments, when necessary. However, again like MCRAFT, this may pose a problem if a fixed department is shifted in the process. One may try excluding all fixed departments from the tree (to retain their current positions). With such an approach, if a cut goes through one or more fixed departments or obstacles, it will complicate the calculation of its x or y coordinate. With LOGIC, it is generally not straightforward to model fixed departments or obstacles relative to CRAFT or MIP. We refer the reader to [58] for details.

LOGIC can be applied in nonrectangular buildings provided that the building shape is "reasonable." If a cut intersects a portion of the building where the length or width changes, then LOGIC uses a simple search strategy to compute the exact location of the cut. For example, consider the layout shown in Figure 6.29. The layout and the building are identical to the one shown earlier in Figure 6.26d except for the change we made to the east side of the building. The third cut shown earlier in Figure 6.27 (i.e., the cut that horizontally divided departments G and {D, F}) has to be computed differently since the "building" length increases from 130 ft to 200 ft within department G.

In presenting LOGIC, we assumed that the above cuts—and the assignment of departments to either side of each cut—are made "randomly." Actually, in order to use LOGIC in a practical application, such decisions are made within the framework of a pseudorandom search strategy such as "simulated annealing," which we discuss in Section 6.6.

We note that layouts obtained by LOGIC are supersets of layouts obtained by BLOCPLAN and similar algorithms that use "bands" for layout formation. That is, any layout developed by BLOCPLAN can be expressed as a cut-tree in LOGIC but not all cut-trees obtained with LOGIC will yield a layout that is composed of bands. Therefore, BLOCPLAN's solution space (and the solution space of other algorithms that use "bands") is a subset of LOGIC's solution space, and consequently one would expect to obtain lower-cost layouts with LOGIC in general. We also note that with both LOGIC and BLOCPLAN it is difficult to treat *nonfixed* departments that may have *prescribed or fixed shapes*. Since the final shape of a department depends on a number of factors that are not known until the cut-tree (or the bands) is constructed, there is no easy way to control the final length and width of a nonfixed department. With few exceptions (such as MIP), many layout algorithms are not very effective in tackling fixed or prescribed department shapes. We will further discuss department shapes in Section 6.5.

MULTIPLE

MULTIPLE (*MULTI-floor Plant Layout Evaluation*) was developed by Bozer, Meller, and Erlebacher [6]. As the name suggests, MULTIPLE was originally developed for multiple-floor facilities, which we address later in this chapter. However, it can also be used in single-floor facilities simply by setting the number of floors equal to one and disregarding all the data requirements associated with the lifts.

Except for the exchange procedure and layout formation, MULTIPLE is similar to CRAFT. It uses a from-to chart as input data for the flow, and the objective function is identical to that of CRAFT (i.e., a distance-based objective with distances measured rectilinearly between department centroids). Departments are not restricted to rectangular shapes, and the layout is represented in a discrete fashion. Also, like CRAFT, MULTIPLE is an improvement-type layout algorithm that starts with an initial layout specified by the layout planner. Improvements to the layout are sought through two-way exchanges, and at each iteration the exchange that leads to the largest reduction in layout cost is selected; that is, MULTIPLE is a steepest-descent procedure.

The fundamental difference between CRAFT and MULTIPLE is that MULTIPLE can exchange any two departments whether they are adjacent or not. Recall that MCRAFT and BLOCPLAN can also exchange any two departments; however, since both algorithms are based on "bands," fixed departments may either shift or change shape. Also, the band approach imposes certain restrictions on the initial layout as well as fixed departments and obstacles as discussed earlier. In essence, MULTIPLE retains the flexibility of CRAFT while relaxing CRAFT's constraint imposed on department exchanges.

MULTIPLE achieves the above task through the use of "spacefilling curves," (SFCs) which were originally developed by the Italian mathematician Peano. Although SFCs (which were considered "mathematical oddities" at the time of their introduction) initially had nothing to do with optimization and industrial engineering, they have been used to construct a heuristic procedure for routing and partitioning problems [3] and for determining efficient locations for items in a storage rack [4]. In MULTIPLE, SFCs are used to reconstruct a new layout when any two departments are exchanged.

MULTIPLE's use of SFCs for the above purpose can perhaps be best described through an example. Consider the SFC shown in Figure 6.30*a*, which is known as the Hilbert curve [24]. (The procedure to generate such a curve is shown in [6]. The interested reader may also refer to [3].) Note that the curve connects each grid such that a "dot" traveling along the curve will always visit a grid that is adjacent to its current grid. Also note that each grid is visited exactly once. Suppose the following

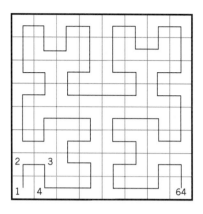

(*a*) Hilbert curve [19]

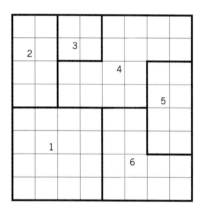

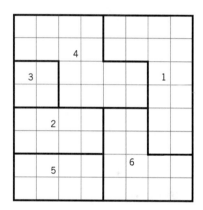

(*b*) Layout vector: 1-2-3-4-5-6   (*c*) Layout vector: 5-2-3-4-1-6

Figure 6.30   MULTIPLE's use of spacefilling curves in layout formation; departments 1 and 5 have been exchanged.

area values (expressed in grids) are given for six departments: $A_1 = 16$, $A_2 = 8$, $A_3 = 4$, $A_4 = 16$, $A_5 = 8$, and $A_6 = 12$. If the *layout vector* or the *fill sequence* is given by 1-2-3-4-5-6, we obtain the layout shown in Figure 6.30*b* by starting from grid 1 and assigning the first 16 grids (along the SFC) to department 1, the next eight grids (along the SFC) to department 2, and so on. In other words, the SFC allows MULTIPLE to map a unidimensional vector (i.e., the layout vector) into a two-dimensional layout. The above mapping can be performed rapidly since the grids are sorted a priori according to their sequence on the SFC.

Given such a mapping, it is now straightforward to exchange any two departments by exchanging their locations in the layout vector. For example, to exchange departments 1 and 5, which are neither adjacent nor equal in area, we first switch their positions in the layout vector to obtain 5-2-3-4-1-6 and then simply reassign the grids (following the SFC). The resulting layout is shown in Figure 6.30*c*. Note that all the departments, except for department 6, have "shifted" to accommodate the above exchange. (The shift is fairly significant since department 5 is only half as large as department 1.) In general, whenever two unequal-area departments are exchanged, all the departments that fall between the two departments on the layout vector will be shifted; the other departments will remain in their original locations.

In order to avoid shifting fixed departments or obstacles, the SFC "bypasses" all the grids assigned to such departments. Also, in those cases where the building shape is irregular or there are numerous obstacles (including walls), MULTIPLE can be used with a "hand-generated" curve, which may start at any grid, and end at any grid, but must visit all the grids exactly once by taking only horizontal or vertical steps (from one grid to an adjacent grid). Diagonal steps are allowed but generally not recommended since they may split a department. (Recall that two grids that "touch" each other only at the corners are not considered adjacent for facility layout purposes.) Unless a fixed department or wall physically separates the building into two disjoint segments, it is always possible to construct such a curve by reducing the grid size. (Why?)

Of course, such hand-generated curves are mathematically no longer SFCs, but they serve the same function. In [6], the authors report that while using MULTIPLE in a "large, four-floor production facility," they opted for hand-generated curves to capture the "exact building shape, the current layout, and all the obstacles." In the following example, we illustrate how a hand-generated curve can be used to capture the current layout. Such curves are also referred to as "conforming curves" since they fully conform to the current layout.

Example 6.7

Consider the same data given for Example 6.1, where departments A and G are assumed to be fixed. The initial layout (which has a cost of 59,480 units) was shown earlier in Figure 6.15. A hand-generated, conforming curve for the initial layout is shown in Figure 6.31a. Note that the curve does not visit departments A and G. Also note that the curve visits all the grids assigned to a particular department before visiting any other department. The initial layout vector is given by C-B-D-H-F-E.

Given the above initial layout vector and the curve, in the first iteration MULTIPLE exchanges departments C and D, which reduces the layout cost to 54,260 units. In the second and last iteration, MULTIPLE exchanges departments C and H to obtain the layout vector D-B-H-C-F-E and the final layout shown in Figure 6.31b. The cost of the final layout is equal to 54,200 units, which is less than the final layout cost obtained by CRAFT (2833.50 × 20 = 56,670 units). In general, MULTIPLE is very likely to obtain lower-cost solutions than CRAFT because it considers a larger set of possible exchanges at each iteration. However, even if both algorithms are started from the same initial layout, MULTIPLE is not guaranteed to find a lower-cost layout than CRAFT. (Why?)

As we indicated for CRAFT, the final layout generated by MULTIPLE may also require massaging to smooth the department borders. Retaining the relative locations of the departments and slightly adjusting some department areas, we obtain the final layout shown in Figure 6.31c. Also, note that the particular curve used not only determines the final layout cost, but it also determines the department shapes. Hence, alternative layouts can be generated by trying different curves.

6 LAYOUT PLANNING MODELS AND DESIGN ALGORITHMS

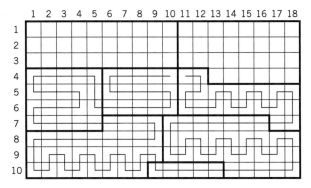

(a) Conforming "hand-generated" curve

	1	2	3	4	5	6	7	8	9	10	11	12	13	14	15	16	17	18
1	A	A	A	A	A	A	A	A	A	A	G	G	G	G	G	G	G	G
2	A									A	G							G
3	A	A	A	A	A	A	A	A	A	A	G	G	G					G
4	D	D	D	D	D	D	D	D	D	D	E	E	G	G	G	G	G	G
5	D									D	E	E	E	E	E	E	E	E
6	D	D	D	D	D	D	D	D	D	D	E	E	E	E	E	E	E	E
7	B	B	B	B	B	B	B	B	B	F	F	F	F	F	F	F	E	E
8	B	B	B	B	B	B	B	B	B	F	F	F	F	F	F	F	F	F
9	B	H	H	C	C	C	C	C	C	F	F	F	F	F	F	F	F	F
10	B	H	H	H	C	C	C	C	C	C	C	C	C	F	F	F	F	F

(b) Final MULTIPLE layout (z = 54,200 units).

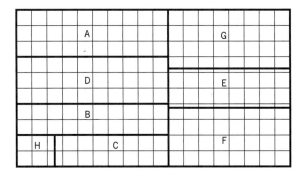

(c) Final "massaged" layout obtained with MULTIPLE.

Figure 6.31 Layouts obtained with MULTIPLE for Example 6.7.

MULTIPLE may also be used as a construction procedure. In such cases, the layout planner may use any SFC or hand-generated curve that best conforms to the (vacant) building and possible obstacles; of course, there is no initial layout that the curve needs to conform to. Generally speaking, curves such as the one shown in Figure 6.30*a* seem to generate more reasonable department shapes than those which "run" in straight lines from one end of the building to the other. Any layout vector may be used as the "initial" layout. It is generally recommended to try alternative layout vectors as the starting point.

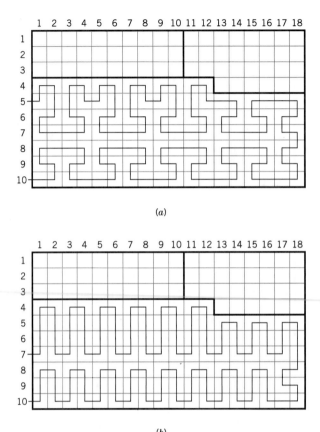

Figure 6.32 Alternative curves for MULTIPLE.

It is also instructive to note that, by considering all possible layout vectors for small problems, one may obtain the "optimal" layout *with respect to a given curve*. For example, for the curve we used in Example 6.7, the layout vector D-B-H-C-F-E is optimal and the final layout shown in Figure 6.31*b* (at 54,200 units) is the "optimal" layout. Two alternative curves are shown in Figures 6.32*a* and *b*, for which the optimal layout vector is given by D-E-F-B-C-H (at 54,920 units) and D-E-F-H-B-C (at 54,540 units), respectively. The above results suggest that, while alternative layouts may be obtained from alternative curves, the cost of these layouts may not be very sensitive to the curve, provided that the curves are not dramatically different. As an exercise, the reader may construct the optimal layouts by using the above two layout vectors and the corresponding curves shown in Figures 6.32*a* and *b*.

6.5 DEPARTMENT SHAPES AND MAIN AISLES

In developing alternative block layouts, the main aisles are typically not represented explicitly until the block layout is finalized. As the layout planner "massages" the final block layout, he or she attempts to correct irregular department shapes and aims for smooth department borders primarily for two reasons: first, a poor department shape may make it virtually impossible to develop an efficient and effective detailed layout for that department; second, since main aisles connect all the departments, by definition, irregular department shapes would lead to irregular main aisles. Generally speaking, for efficient material handling and other reasons (including safety, unobstructed travel, evacuation in an emergency), main aisles should connect all the departments in a facility with minimum travel, minimum number of turns, and minimum "jog overs" (i.e., they should run in straight lines as much as possible). Hence, attaining "good" department shapes is an important consideration in finalizing a block layout.

As we showed in Section 6.4 (see Equations 6.38 and 6.39), controlling department shapes is relatively straightforward for rectangular departments (such as those obtained with BLOCPLAN and MIP). This is due to the fact that it is straightforward to define and measure the shape of a rectangle: it is simply the ratio of its longer side to its shorter side (or vice versa). However, if the layout planner needs or wishes to consider nonrectangular department shapes (such as those obtained with CRAFT and MULTIPLE), then shape measurement and control is not straightforward. In fact, given two alternative but "similar" shapes for the same department, one of the alternatives may be regarded as acceptable while the other one is regarded as poor. Although humans are good at making such (subjective) judgments with respect to department shapes, computer-based algorithms require formal and objective measures.

A few alternative measures have been suggested in the literature to control the shape of nonrectangular departments. As we shall see, such measures are intended only to detect and avoid irregular department shapes; they cannot be used to guarantee or prescribe specific department shapes. Although the shape measures we describe below are more suitable for the discrete layout representation, they can be implemented with the continuous representation as well.

Two measures to control department shapes are presented in [36]. Both measures are based on first identifying the smallest rectangle that fully encloses the department. The first measure is obtained by dividing the area of the smallest-enclosing-rectangle (SER) by the area of the department. Given that the department area is fixed, one would expect the above ratio to increase as the department shape becomes more irregular since a larger rectangle would be required to enclose the department. The second measure is obtained by dividing the longer side of the SER by its shorter side. As before, as the department shape becomes more irregular, one would expect the above ratio to increase.

A third measure, presented in [6], originally appeared in papers concerned with geometric modeling (see [17], among others). It is based on the observation that, given an object with a fixed area, the perimeter of the object generally increases as its shape becomes more irregular. Hence, one may measure the shape of a department by dividing its perimeter by its area. However, unless the layout planner generates alternative shapes for each department "by hand" and computes the above ratio for each shape a priori, it is difficult to predict reasonable values for it. To address this difficulty, in [6] the above ratio is normalized as follows: If the "ideal" shape for a department is a square, then the "ideal" shape factor, say, S^*, is equal to $(P/A)^* = (4\sqrt{A})/A = 4/\sqrt{A}$, where P denotes the perimeter and A denotes the area of the department. The normalized shape factor, say, F, is equal to $S/S^* = (P/A)/(4/\sqrt{A}) = P/(4\sqrt{A})$. Hence, if a department is square shaped, we

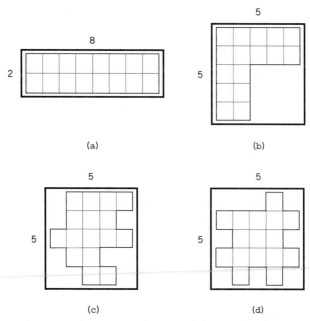

Figure 6.33 Alternative department shapes and the smallest-enclosing-rectangle.

obtain $F = 1.0$; otherwise, we obtain $F > 1.0$. Generally speaking, reasonable shapes are obtained if $1.0 \le F \le 1.4$. If a square is not the "ideal" shape for a department, the analyst may impose a lower bound greater than 1.0 on F.

In [6] the above three shape factors are compared via an example; we present it here with a minor change. Suppose the department area is equal to 16 units. Four alternative shapes, including the SERs, are shown in Figure 6.33. For Figure 6.33*a*, the first measure is equal to 1.0, and it increases to $25/16 = 1.5625$ for Figure 6.33*b*. However, it remains at 1.5625 for Figures 6.33*c* and 6.33*d*. The second measure, on the other hand, is equal to 4.0 for Figure 6.33*a*, and it decreases to 1.0 for the remaining three department shapes. The third measure, F, is equal to 1.25, 1.25, 1.50, and 1.625 for Figures 6.33*a* through 6.33*d*, respectively. (The reader may verify that P is equal to 20, 20, 24, and 26 for Figures 6.33*a* through *d*, respectively.)

As the above example illustrates, no shape measure is perfect, although F seems to generate more accurate results. Furthermore, as we remarked earlier, none of the shape measures can be used to prescribe a particular shape for a department. Rather, they can be used to avoid irregular department shapes by checking the shape factor for each department following an exchange. Since a typical computer run considers many possible department exchanges, any shape measure must be straightforward to compute. The three shape measures we showed above are straightforward to implement within a computer-based layout algorithm that uses the discrete representation; the computational burden they generate is minimal.

For example, consider the third measure. To compute the perimeter of department i, the computer first examines each grid assigned to department i, one at a time, and simply counts the number of adjacent grids which have *not* been assigned to department i. (Unless it is located along the building perimeter, each grid has exactly four adjacent grids. Grids that are located along the perimeter of the building must be treated slightly differently.) Once the perimeter of each department is computed, the computer determines the normalized shape factor for each department. As a result of exchanging two departments, if any department violates a shape constraint (i.e., the normalized shape factor for the department falls outside a user-defined range, which may be department-specific), then the computer simply "rejects" the exchange.

Note that, with the above scheme to compute department perimeters, those with "enclosed voids" will be correctly identified as possibly irregular in shape. For example, for the departments shown in Figures 6.9c, d and e, the normalized shape factor is given by 1.061, 1.591, and 1.414, respectively. Had we not included the "inside perimeter" in Figure 6.9e, the shape factor would have been 1.061, which is a misleading figure.

Although shape measures are useful in avoiding irregular department shapes, they should be used with care for three reasons: First, if strict shape constraints are imposed, many (if not all) possible department exchanges may be rejected regardless of how much they reduce the layout cost. Second, some departments may take irregular shapes only temporarily; that is, a department which is irregularly shaped in the current iteration may assume a reasonable shape in the next iteration—note that this may occur both with CRAFT and MULTIPLE. Third, the analyst has the option to correct the department shapes by massaging the final layout generated by the computer. If strict shape constraints are imposed, the computer will "automatically" discard lower-cost solutions with irregular department shapes even if the analyst could have corrected the problem through massaging. Hence, we generally recommend imposing no shape constraints until the layout planner makes a few runs and obtains preliminary results. In subsequent runs, a shape constraint may be imposed only on those departments that have unreasonable shapes that cannot be corrected through massaging.

6.6 SIMULATED ANNEALING AND GENETIC ALGORITHMS[3]

Simulated Annealing (SA) and Genetic Algorithms (GAs) represent relatively new concepts in optimization. Although both SA and GAs can be used for layout construction, we will limit their presentation to *layout improvement*. Since a formal and thorough treatment of either subject is beyond the scope of this book, we will present only some of the basic concepts of SA and GAs, and show only one SA-based layout algorithm.

The most significant drawback of the steepest descent approach (which we described in Section 6.4) is that it forces the algorithm to terminate the search at the first two-opt or three-opt solution it encounters. As we remarked earlier, such a solution is very likely to be only locally optimal. (Note that, when they consider

---

[3]This section presents research results, rather than detailed algorithms and examples. Intended for first-year graduate students and advanced undergraduate students, it provides some insight into the use of "metaheuristics" in layout design.

two-way or three-way department exchanges, CRAFT, BLOCPLAN, and MULTIPLE are examining the local "neighborhood" of the current solution.) As a result, any algorithm that uses the steepest descent technique becomes highly "path dependent" in that the initial solution and the specific department exchanges made by the algorithm play a significant role in determining the cost of the final solution. (The impact of the initial solution is also known as the *initial layout bias.*) Ideally, a heuristic procedure should consistently identify a low-cost, that is, an optimal or near optimal, solution regardless of the starting point.

One of the primary strengths of SA is that, while trying to improve a layout, it may "occasionally" accept nonimproving solutions to allow, in effect, the algorithm to explore other regions of the solution space (instead of stopping at the first "seemingly good" solution it encounters). In fact, a SA-based procedure may accept non-improving solutions several times during the search in order to "push" the algorithm out of a solution which may be only locally optimal. As a result, the objective function value may actually increase more than once. However, the amount of increase in the objective function that the algorithm will "tolerate" is carefully controlled throughout the search. Also, the algorithm always "remembers" the best solution; that is, the best solution identified since the start of the search is never discarded even as new regions of the solution space are explored.

The fundamental concepts behind SA are based on an interesting analogy between statistical mechanics and combinatorial optimization problems [32]. Statistical mechanics is the "central discipline of condensed matter physics, a body of methods for analyzing aggregate properties of the larger numbers of atoms to be found in samples of liquid or solid matter" [32]. One of the key issues in statistical mechanics is the state of the matter (or the arrangement of its atoms) as its temperature is gradually reduced until it reaches the "ground state" (which is also referred to as the "lowest energy state" or the "freezing point").

According to [68], "In practice, experiments designed to find the (lowest energy) states are performed by careful *annealing,* that is, by first melting the (material) at a high temperature, then lowering the temperature slowly (according to an *annealing schedule*), finally spending a long time at temperatures in the vicinity of freezing, or solidification, point. The amount of time spent at each temperature during the annealing process must be sufficiently long to allow the system to reach thermal equilibrium (steady state). If care is not taken in adhering to the annealing schedule (the combination of a set of temperatures and length of time to maintain the system at each temperature), undesirable random fluctuations may be frozen into the material thereby making the attainment of the ground state impossible."

Hence, the analogy between combinatorial optimization problems and statistical mechanics is that each solution in the former corresponds to a particular arrangement of the atoms in the latter. The objective function value is viewed as the energy of the material, which implies that finding a low-cost solution is analogous to achieving the lowest energy state through an effective annealing schedule. Also, since we are working with mathematical optimization models rather than "condensed matters," the annealing schedule applied to a combinatorial optimization problem is not "real" annealing, it is "simulated" annealing. As we shall see shortly, the annealing schedule, that is, the set of temperatures used (including the initial temperature) and the time spent at each temperature, plays an important role in algorithm development and the cost of the final solution.

The concept of "occasionally" accepting nonimproving solutions goes back to a simple Monte Carlo experiment developed in [42]: Given a current arrangement of the atoms (or "elements"), we randomly make incremental changes to the current arrangement to obtain a new arrangement, and we measure the *decrease* in energy, say, ΔE. If $\Delta E > 0$ (i.e., the energy decreases), we accept the new arrangement as the current one and use it to make subsequent changes. However, if $\Delta E \leq 0$, the new arrangement is accepted with probability $P(\Delta E) = \exp(\Delta E/k_b T)$, where T is the temperature and k_b is Boltzmann's constant. (If we do not accept the new arrangement, we generate another one.)

To apply the above procedure to optimization problems we simply treat the "current arrangement" as the *current solution,* the "new arrangement" as the *candidate solution,* and the "energy" as the *objective function value.* A random incremental change is made, for example, by exchanging two randomly picked departments in the current layout. Last, we set $k_b = 1$ since it has no known significance in optimization problems. Note that, as the increase in the objective function gets larger, the probability of accepting the candidate solution gets smaller. For example, if $T = 150$ and the increase in the objective function is equal to 250 units (i.e., $\Delta E = -250$), the probability of accepting the candidate solution is equal to 0.1889. At $T = 150$, if the increase in the objective function is equal to 400 units, however, the above probability decreases to 0.0695. Also note that the probability of accepting a nonimproving solution decreases as the temperature decreases. For example, for $\Delta E = -250$, if we decreased the temperature from 150 to 100 units, the above probability would decrease from 0.1889 to 0.0821. In other words, we are more likely to accept nonimproving solutions early in the annealing process (due to the relatively high temperatures).

It is instructive to further note that the probability in question ultimately depends on the change in the objective function *relative to* the temperature. Therefore, in a SA-based algorithm, we are not only concerned with how fast we "cool" the system but also with the initial temperature we select to start the annealing process. One possible approach is to set the initial temperature according to the objective function value of the starting solution (see, for example, [30]). In the algorithm that follows, the initial temperature is set equal to $z_0/40$, where z_0 is the cost of the initial layout. Of course, the layout planner should experiment with different initial temperature settings.

The above process of generating candidate solutions and appropriately updating the current solution continues until the system reaches steady state *at the current temperature.* As shown in the following algorithm, the percent change in the *mean* objective function value is used to determine whether steady state has been reached. According to [68], "at each temperature, the annealing schedule *must* allow the simulation to proceed long enough for the system to reach steady state." Once steady state is reached, the temperature is reduced according to a predetermined temperature reduction factor and we continue to generate and evaluate candidate solutions with the new temperature setting.

Typically, the search is terminated either when a user-specified final temperature is reached (which can also be expressed as the maximum number of temperature reductions to be considered) or a user-specified number of *successive* temperature reductions does not produce an improvement in the best solution identified since the start of the annealing process (i.e, the "current best" solution). The algorithm we present below uses both the latter stopping criterion and one that is based on the maximum number of "epochs." An epoch corresponds to a particular set of candidate solutions accepted by the algorithm. The epoch length is expressed as the number of candidate solutions in the set.

To develop a SA-based layout algorithm, we will apply an annealing schedule to MULTIPLE, which was described in Section 6.4. Recall that, in MULTIPLE, given a particular spacefilling curve and the departmental area requirements, the layout vector (i.e., a sequence of department numbers) fully defines a layout. Therefore, any solution is represented only as a layout vector. Except for the technique used in generating the candidate layout vectors, the algorithm we show here is identical to SABLE (*Simulated Annealing-Based Layout Evaluation*), which was developed by Meller and Bozer [39]. (SABLE uses a more general technique for generating candidate layout vectors.) When appropriate, variables were used in a manner similar to their use in [68]. Let

s^0 be the initial layout vector,

s^* be the "current best" layout vector which corresponds to the lowest cost layout identified by the algorithm,

s be the current layout vector,

s' be the candidate layout vector,

α be the temperature reduction factor (which controls how fast the system is "cooled down"),

T be a set of annealing schedule temperatures $\{t_1, t_2, t_3, \ldots \}$, where $t_i = t_0(\alpha)^i$ for all $i > 1$,

t_0 be the initial temperature,

e be the (fixed) epoch length,

$f_j(s)$ be the objective value of the *j*th accepted candidate layout vector, s, in an epoch.

$\bar{f}_e$ be the mean objective function value of an epoch, i.e., $\bar{f}_e = [\sum_{j=1}^{e} f_j(s)]/e$,

$\bar{f}_e'$ be the overall mean objective function value of all the layout vectors accepted during the epochs previous to the current one (for a given temperature),

ϵ_i be a threshold value used to determine whether the system is in equilibrium at temperature i,

M be the maximum total number of epochs to be considered (across all temperatures),

I be a counter to record the last temperature setting that produced the "current best" layout vector, s^*,

N be the maximum number of successive temperature reductions that will be performed with no improvement in s^*.

The initial layout vector s^0 as well as the values of α, t_0, e, ϵ, M, and N are specified by the user a priori. Using the above notation (except for the subscript j, which we will omit for brevity), a simple version of SABLE [39] is presented as follows:

Step 1. Set $s = s^0$, $I = 1$, and $i = 1$. Compute the initial layout cost, z_0; set $t_0 = (z_0)/40$ and $t_1 = \alpha t_0$.

Step 2. For the current layout vector, compute the layout cost $f(s)$.

Step 3a. Randomly pick two departments in s, exchange their locations and store the resulting layout vector (i.e., the candidate layout vector) in s'.

Step 3b. Compute the decrease in the layout cost, i.e., set $\Delta f = f(s) - f(s')$. If $\Delta f > 0$, go to step 3d; otherwise, go to step 3c.

Step 3c. Sample a random variable $x \sim U(0, 1)$. If $x < \exp(\Delta f / t_i)$, go to step 3d; otherwise, go to step 3a.

Step 3d. Accept the candidate sequence; i.e., set $s = s'$ and $f(s) = f(s')$. If $f(s) < f(s^*)$, then update the "current best" solution; i.e., set $s^* = s$, $f(s^*) = f(s)$, and $I = i$. If e candidate sequences have been accepted, go to step 4; otherwise, go to step 3a.

Step 4. If equilibrium has not been reached at temperature t_i, that is, if $|\bar{f}_e - \bar{f}'_e|/\bar{f}'_e \geq \epsilon_i$, reset the counter for accepted candidate solutions and go to step 3a; otherwise, set $i = i + 1$ and $t_i = t_0(\alpha)^i$. If $(i - I) < N$, go to step 5; otherwise, **STOP**—the maximum number of successive nonimproving temperatures has been reached.

Step 5. If the total number of epochs is less than M, go to step 3a; otherwise, **STOP.**

The parameter values selected by the user are very likely to have a more-than-minor impact on the cost of the final solution obtained by the above algorithm as well as its execution time. Setting the initial temperature with respect to the initial layout cost and ensuring that the system is not "cooled down" too rapidly generally improves the performance of the algorithm. In [39], for example, the following parameter values were observed empirically to substantially reduce the initial layout bias and yield generally "good" solutions for all the test problems evaluated: $\alpha = 0.80$, $t_0 = (z_0)/40$, $e = 30$, $\epsilon_1 = 0.25$, $\epsilon_i = 0.05$ for all $i \geq 2$, $M = 37$, and $N = 5$. We caution the reader that the above parameter settings are shown only as an example. (Also, recall that in [39] a more general candidate solution generation procedure is used.) Once a SA-based algorithm is developed, it is generally good practice to experiment with a variety of parameter settings.

The above algorithm is certainly not the only possible application of SA to facility layout problems. In fact, as we remarked in Section 6.4, LOGIC [58] is actually a SA-based layout algorithm. Recall that LOGIC develops a layout by executing a series of horizontal and vertical cuts, and by assigning (an appropriate subset of) the departments to one or the other side of each cut. Given a current solution, which is represented by a cut-tree (see Section 6.4), a candidate solution may be obtained, for example, by randomly changing the orientation of one (or more) of the cuts in the tree and/or by randomly changing the partition of a set of departments over one (or more) of the cuts. (We refer the reader to [58] for further details.) Candidate solutions obtained in the above manner can be evaluated within the framework of an annealing schedule as we showed in the above algorithm. Another application of SA in facility layout is presented in [28]. The algorithm in [28] is intended primarily for the special case where all the departments have equal area requirements; using it with general departmental area requirements is possible but it may lead to numerical problems or split departments.

The "biased sampling technique" (BST) [48], which was applied to CRAFT, has certain similarities to SA. With the BST, at each iteration, given a set of exchanges (that are estimated to reduce the layout cost), instead of always selecting the exchange which is estimated to reduce the layout cost the most, the algorithm assigns a nonzero probability to each exchange in the set and randomly selects one of them. By solving the same problem several times with different random number streams and different starting points, the BST was shown to reduce the initial layout bias [48]. However, the BST is not as "formal" a concept as SA and, more importantly, the BST will still terminate at the first locally optimal solution it encounters (i.e., the BST does not accept an exchange that may increase the objective function value).

It is instructive to note that developing effective and efficient SA-based layout algorithms is not only a matter of finding a "good" annealing schedule but also a matter of finding a "good" representation that makes it possible to *rapidly generate and evaluate* a variety of candidate solutions. Note that we are referring to "solution" representation and not "layout" representation (as in discrete versus continuous layout representation). Naturally, the two are not independent. For example, in MULTIPLE [6] and SABLE [39], the solution is represented as a layout vector and the layout is constructed through the use of spacefilling curves (which work well with the discrete layout representation). In LOGIC [58] the solution is represented as a cut-tree and the layout is constructed by applying vertical and horizontal cuts (which work well with the continuous layout representation).

The application of GAs, on the other hand, to facility layout problems (and other optimization problems) is relatively recent, but it has been gaining momentum. The basic concept behind GAs was developed by Holland [25] who observed that the "survival of the fittest" (SOF) principle in nature may be used in solving decision-making, optimization, and machine-learning problems. Although the SOF principle may first appear to have no relation to optimization problems, the "relationship" between the two is as fascinating as the "relationship" we described between statistical mechanics and optimization problems.

Algorithms based on SA "occasionally" accept nonimproving solutions; however, they still work with only one solution at a time. That is, there is only one current solution, from which we generate only one candidate solution. In contrast, GAs work with a family of solutions (known as the "current population") from which we obtain the "next generation" of solutions. When the algorithm is used properly, we obtain progressively better solutions from one generation to the next. That is, good solutions (or actually "parts of good solutions") propagate from one generation to the next and lead to better solutions as we produce more generations.

The basic GA, at least on the surface, is admirably simple. Given a current population of, say, N solutions, we generate the next population as follows: We randomly pick two "parents" (i.e., two solutions) from the current population and "randomly cross over" the two parents to obtain two "offsprings" (i.e., two "new" solutions) for the next population. We repeat the above process until we have a new generation with N solutions. (Note that, unlike most natural systems, the population size is fixed at N.) Each time we generate two offsprings, their two parents are picked randomly; however, the probability of picking a particular solution as a parent is proportional to the "fitness" of the parent. For minimization problems, the above implies that the probability of picking a solution as a parent is inversely proportional to the objective function value of that parent (i.e., a "fit" parent with a small objective function value is more likely to be selected). The first generation (i.e., the starting generation) is usually created randomly. Typically, the process of creating new generations continues until a prespecified number of generations have been produced or no noticeable reduction in the average population objective function value is detected for a number of successive generations.

In most GAs, the above "basic" algorithm is modified in a number of ways. One common addition is "mutation," which essentially means altering (in some arbitrary fashion) one or more solutions picked at random within a given population. Another addition is the concept of "elitist reproduction," which basically implies that when a new population is created, the best 10% or 20% of the solutions in the current population are "automatically" copied over to the next population. The remaining 90% or 80% of the solutions in the next population are generated by the "two parents-two offsprings" method we described above. Solutions that were copied over are still eligible to act as parents for creating the remainder of the next population.

The population size, the mutation rate, the rate of elitist reproduction, and the number of generations to create are all user-specified parameters that affect the performance and execution time of a GA. Also, there are alternative cross-over operators that specify how two offsprings are "randomly" generated from the two parents. In fact, the type of solution representation used for the problem is critical in terms of defining appropriate cross-over operators (so that we do not create "infeasible offsprings" from "feasible parents"). That is, even though the two parents represent feasible solutions, if we do not use or develop an appropriate cross-over operator, their offsprings may not. A discussion of alternative cross-over operators and appropriate values to assign to the above parameters are beyond the scope of this book. The reader may refer to [19] for an excellent introduction to basic and some advanced concepts in GAs and their use in optimization problems and machine learning. Also, [9] is a good reference book for GAs; it includes comprehensive examples for GA applications. Furthermore, the reader will find GAs applied to the Quadratic Assignment Problem and the facility layout problem in [59], [61], and [62], among others.

In concluding this section, we note that, generally speaking, if the computational effort required to evaluate a single solution is high, SA-based algorithms are likely to outperform GAs. This stems primarily from the fact that a GA is more likely to evaluate a larger number of solutions than a SA-based algorithm. It may also be the case that, in early generations, GAs spend computer time evaluating many poor solutions before they are removed from the "gene pool." On the other hand, there is a "natural fit" between GAs and parallel computers, where each processor in the computer can independently and concurrently generate and evaluate a pair of offsprings for the new population. This way, a large number of new population members can be rapidly generated and evaluated in parallel. Nevertheless, with or without parallel computers, it is too early to make conclusive statements about SA versus GAs in facility layout; further research is needed to fully develop both techniques and compare their performance on various facility layout problems.

6.7 MULTI-FLOOR FACILITY LAYOUT

With some exceptions such as offices, it is perhaps safe to say that most industrial facilities are single-floor facilities. This is especially true for manufacturing, warehousing, and distribution facilities. However, many such facilities also use mezzanines, which can be modeled as a "partial second floor" for layout purposes. More specifically, a mezzanine is modeled as a full second floor with an obstacle. Recall that an obstacle is essentially a fixed department with zero flow. The obstacle is placed on the portion of the second floor that falls outside the mezzanine.

Furthermore, there are still many industrial facilities in use with two or more floors. Some of these multi-floor facilities are older buildings in the United States, while a majority of them are two-story or three-story facilities being used in Asian countries and other countries where land is either limited and/or very expensive, especially as one gets closer to the industrialized zones of the country. Although a multi-floor facility presents certain challenges (see, for example, Schonberger [52], pp. 120–121), many manufacturers in such regions prefer a multi-floor facility because it saves them considerable capital, and they depend on the infrastructure (energy, highways, railways, etc.) that is often lacking in rural areas where land may be readily available.

Generally speaking, a multi-floor facility layout problem is more challenging than a comparable single-floor layout problem. The incremental complexity is mostly due to vertical travel between floors. While advances in material handling technology (to handle material between floors) provide the layout analyst with a number of options (see, for example, [69] and [70]), the number and location of vertical handling devices to use, the congestion and delays they may create, and the possible lack of coordination between departments on separate floors are important factors that will impact the overall quality and effectiveness of any layout developed for a multi-floor facility. Moreover, space constraints in multi-floor facilities may be more limiting. For example, due to floor-loading capacities, floor-to-ceiling clear height, heat generation, chemical processes involved, and so forth, certain departments may be restricted to certain floors.

Also, even if the total usable floor space available in the building exceeds the sum of the floor space requirements of the departments, not all layouts will be feasible. Consider, for example, a three-floor facility with 9 unit squares of available floor space per floor. Suppose 7 departments (numbered 1 through 7) have the following area requirements: 4 unit squares each for departments 1 through 5, and 3 unit squares each for departments 6 and 7, for a total of 26 (<27) unit squares. Although there is excess space in the building, there is no feasible assignment of departments to floors. Therefore, unless the departments are redefined, there is no feasible layout. Recall that splitting a department, within or across floors, is not allowed.

Of course, the above small example is a special case, but the main point is that department-to-floor assignments play an important role in defining the layout alternatives and eventually the layout cost. Although a number of layout algorithms have been developed for the multi-floor facility layout problem (see, for example, BLOCPLAN [11], MSLP [31], SPACECRAFT [29], and SPS [36]), they have certain shortcomings such as splitting departments across two or more floors, not considering any lifts or allowing only a single lift (or single bank of lifts) for vertical handling, and/or assuming all equal-area departments. Two algorithms that work with multiple floors, multiple lift locations, and unequal-area departments (without splitting them across floors) are MULTIPLE [6] and SABLE [39], which we described earlier. Since SABLE was shown to be more effective than MULTIPLE, here we will briefly describe SABLE's implementation in multi-floor facilities.

Recall that SABLE is a simulated annealing-based layout-improvement algorithm. We presented a simplified, single-floor version of SABLE in Section 6.6, where a candidate layout vector is generated by exchanging the locations of two departments in the current layout vector. Given a candidate layout vector, SABLE starts with the first department in the vector and places it in the first floor. The second department in the vector is also placed in the first floor and so on, until the placement of the next department in the vector would create a space overflow in the first floor. At that point, the next department is placed in the second floor and the procedure continues until either all departments in the vector are successfully placed or the candidate layout vector is declared infeasible (in which case a new candidate layout vector is generated from the current vector again by exchanging the locations of two departments in the current layout vector). Each department is placed according to a spacefilling curve defined for that floor; that is, each floor has its own spacefilling curve. The number and location of the lifts are specified by the analyst.

Once a feasible candidate layout vector is generated and all the departments are placed, the resulting multi-floor layout is evaluated by the following expression:

$$\min z = \sum_i \sum_j (c_{ij}^H d_{ij}^H + c_{ij}^V d_{ij}^V) f_{ij}, \tag{6.42}$$

where f_{ij} is the flow from department i to department j (expressed in number of unit loads moved per unit time), $c_{ij}^H (c_{ij}^V)$ is the cost of, or the relative "weight" associated with, moving a unit load one horizontal (one vertical) distance unit from department i to department j, and $d_{ij}^H (d_{ij}^V)$ is the horizontal (vertical) distance from department i to department j. Note that Equation 6.42 is very similar to the single-floor, distance-based layout objective. The horizontal distance from department i to j, d_{ij}^H, is assumed to go through the lift that minimizes the total horizontal rectilinear distance between the two department centroids. Of course, the vertical distance is measured between the floors. Lift congestion and the throughout capacity of the lift are not taken into account.

Based on the value of the objective function as determined by Equation 6.42, the annealing procedure proceeds as described earlier (see step 3b and subsequent steps in Section 6.6) and the algorithm stops when one of the stopping criteria is encountered.

A somewhat different approach to multi-floor layout is to solve the problem in two stages. In the first stage, each department is permanently assigned to one of the floors. In the second stage, the layout of each floor is improved using simulated annealing. Of course, in the first stage, the objective is to minimize the vertical component of the objective function given by Equation 6.42, and in the second stage, the objective is to minimize the horizontal component. Generally speaking, solving a problem in two stages is not as effective as solving it as a single-stage problem even if a good or optimum solution is found for each of the two stages. However, the above two-stage approach is appealing because minimizing vertical travel (by placing two departments with high interdepartmental flow within the same floor as much as possible) has managerial appeal, and it also reduces the solution space we need to explore when we tackle the second stage.

The above multi-floor layout algorithm, STAGES, is presented in [40], where the first stage is solved via a linear mixed integer programming (MIP) model assuming that the inter-floor distances are equal (i.e., the distance between floors 1 and

2 is equal to the distance between floors 2 and 3, and so on). Otherwise, the first stage results in a Quadratic Assignment Problem (QAP), which has a nonlinear objective. (The QAP is described in Chapter 10.) Although there are many heuristic procedures developed for the QAP, it is generally a difficult problem to solve. Therefore, even if the inter-floor distances may not be exactly equal, using the MIP model presented below would probably be the preferable approach in most cases to obtain a solution to stage 1.

Let d_{ij}^{gh} denote the vertical distance from department i to department j if department i is assigned to floor g, and department j is assigned to floor h. Assuming that the inter-floor distance between *any two* adjacent floors is equal to δ, we have $d_{ij}^{gh} = \delta |g - h|$, where $i, j = 1, \ldots, N$ and $g, h = 1, \ldots, G$. Suppose our (binary) decision variable is denoted by x_{ig} such that x_{ig} is equal to 1 if department i is assigned to floor g, and 0 otherwise. If we let y_i denote the floor number of department i, that is, $y_i = \sum_{g=1}^{G} g x_{ig}$, then we obtain $d_{ij}^{gh} = \delta |y_i - y_j|$. Note that the variable y_i is an integer variable but we do not declare it as an integer variable because it "automatically" assumes integer values as long as the x_{ig} values are restricted to 0 or 1.

Suppose the set of positive flows is denoted by $F = \{f_{ij}\}$. That is, $f_{ij} > 0$ for all $i, j \in F$. Let $|F| = M$; that is, let M denote the number of positive flows in the flow matrix. Also, let f_{ij}^m denote the mth positive flow ($m = 1, \ldots, M$). The linear MIP model to assign departments to floors with the objective of minimizing total vertical travel (that is, the vertical component in Equation 6.42) is given as follows:

$$\text{Minimize } z = \sum_{m=1}^{M} V_m \tag{6.43}$$

$$\text{Subject to: } y_i = \sum_{g=1}^{G} g x_{ig} \qquad \text{for } i = 1, \ldots, N \tag{6.44}$$

$$V_m \geq \delta (c_{ij}^V f_{ij}^m)(y_i - y_j) \qquad \text{for } m = 1, \ldots, M \tag{6.45}$$

$$V_m \geq \delta (c_{ij}^V f_{ij}^m)(y_j - y_i) \qquad \text{for } m = 1, \ldots, M \tag{6.46}$$

$$\sum_{g=1}^{G} x_{ig} = 1 \qquad \text{for } i = 1, \ldots, N \tag{6.47}$$

$$\sum_{i=1}^{N} a_i x_{ig} \leq A_g, \qquad \text{for } g = 1, \ldots, G \tag{6.48}$$

where a_i is the floor space required by department i, and A_g is the available floor space on floor g. Constraint set 6.44 determines the floor number of department i based on the value of the binary variables. Constraint sets 6.45 and 6.46 together determine the flow times vertical distance for the mth flow. Constraint set 6.47 ensures that each department is assigned to exactly one floor. Constraint set 6.48 ensures that the total available floor space on each floor is not exceeded.

Once the department-to-floor assignments are determined by solving the above linear MIP model, the floor assignment for each department is fixed and the second stage of the algorithm is executed. In STAGES, a modified version of SABLE is used to tackle the second stage. That is, the current layout vector is perturbed to generate a candidate layout vector *without changing the floor assignments of the departments,* and a spacefilling curve (entered for each floor) is used to re-layout the affected floors. For the simplified version of SABLE we presented in Section 6.6, the second stage of STAGES can be implemented simply by randomly picking a floor and exchanging the location of two randomly picked departments in the layout vector of that floor. (A slightly different and more general approach is used in [40] to perturb the current layout vector.)

According to the numerical results presented in [40], the layout costs obtained from STAGES are, in general, better than those obtained from SABLE and better than another annealing-based algorithm where the department-to-floor assignments obtained in the first stage are allowed to change as the second stage is solved. We refer the interested reader to [40] for further details.

6.8 COMMERCIAL FACILITY LAYOUT PACKAGES

There are a number of commercial packages available for facilities planning. However, a majority of these packages are CAD-based documentation or drawing tools; they typically do not offer layout construction or improvement algorithms such as the ones we described in this chapter. One package, however, that appears to have gained more recognition in industry is VisFactory (or e-Factory), which includes multiple modules, namely, FactoryCAD, FactoryFLOW, FactoryPLAN/OPT, and FactoryVIEW. The original vendor for this package was CIMTechnologies/EAI. More recently, however, VisFactory is offered by Unigraphics Solutions Inc. (UGS). The UGS Web address for product information is www.ugs.com/products/colsol/visfactory/.

In the second edition of the book, we had mentioned LayOPT (by Production Modeling Corporation, PMC) and FactoryModeler (by Systemes Espace Temps, Inc., which is now SET Technologies, Inc.). Unfortunately, both products have been pulled out of the market, although SET Technologies, Inc. reportedly plans to introduce a facility layout module into a larger package they offer, namely Nikan, for manufacturing-oriented supply chain design, planning, and management.

A commercial package for workplace/layout design and material flow simulation, eM-Workplace, is offered by Tecnomatix Technologies Ltd. (www.tecnomatix.com). Two other commercial packages, PLANOPT and PLANET, are offered by Engineering Optimization Software (www.enggoptsoftware.bizland.com). PLANOPT is a construction-type layout algorithm that works with rectangular department shapes.

In the second edition of the book, we had also stated that newer packages would be "based on powerful graphical displays, possibly with multimedia and three-dimensional 'rendering,'" and that they may offer "integrated design, analysis, simulation, and animation environments." The above packages (especially VisFactory and eM-Workplace) and others available in the market support this prediction; we expect this trend to continue. Changes in packages, their features, and the vendors that offer them all point to the dynamic nature of the software business. We therefore encourage the reader to use the Web to keep abreast of new developments. Trade/professional publications (such as *IIE Solutions*, solutions.iienet.org) periodically publish lists of software packages for facilities planning/design, which are also an excellent source of information for the layout analyst/engineer.

6.9   THE IMPACT OF CHANGE

The need for a facility layout study can arise under a variety of circumstances. For example, some of the more common situations that arise in the context of plant layout include the following:

1.   Changes in the design of existing product, the elimination of products from the product line, and the introduction of new products.

2.   Changes in the processing sequences for existing products, replacements of existing processing equipment, and changes in the use of general-purpose and special-purpose equipment.

3.   Changes in production quantities and associated production schedules, resulting in the need for capacity changes,

4.   Changes in the organizational structure as well as changes in management philosophies concerning production strategies such as the adoption of just-in-time concepts, total quality management, etc.

If requirements change frequently, then it is desirable to plan for change and to develop a flexible layout—one that can be easily modified, expanded, or contracted. Robert L. Propst [50, p. 16], in addressing the impact of change on facilities planning for electronics manufacturing, noted:

> Our ancestors could deal with changes as an evolutionary factor. Change could be digested in small steps or ignored for a lifetime. Today, it is the dominating reality; our new natural state of affairs.
>
> Curiously, it is the lack of flexibility in our physical facilities that is proving to be the bottleneck in electronics production. A great many irritations stem from services and facilities that respond too slowly, or not at all, to our buildings, furnishings, and service that have to be revitalized and revisualized.

Flexibility can be achieved by utilizing modular office equipment, workstations, and material handling equipment; installing general-purpose production equipment; utilizing a grid-based utilities and services system; and using modular construction. Additionally, the design of the facility can have a significant impact on the ease and cost of expansion. We will discuss some of these issues in more detail below.

Adapting to Change and Planning for Facility Reorganization

Before getting too specific about how to plan for change, it would be worthwhile to step back and observe that in many manufacturing organizations there are cycles of expansion and decline due to the very nature of the business environment. In other words, the manufacturing environment is very dynamic. The facility layout should also be treated as dynamic. In as much as businesses should have long-term business strategies, we must also have a multiyear master plan for facility layout. This master plan should be consistent with the company's business plan and it should attempt to anticipate future requirements and make provisions for adapting to changes in facility requirements. Oftentimes, prior decisions impose severe constraints which make it very difficult to institute innovative changes in the facility layout. A few examples include placements of shipping and receiving docks, locations of heavy machineries, clear building heights, floors with low load-bearing capacities, load-bearing walls, location of utilities, and so on.

The master layout plan must also provide the means for a facility to react quickly to change, adding capacity in a short period of time or be able to operate efficiently at scaled-down operating levels. The facility design must be flexible in order to provide this high level of responsiveness. We note, however, that variations in production requirements should not be interpreted as signaling a need to change the facility layout. Often, opportunities for adjustments can be made by seeking more efficient machine schedules, better maintenance of equipment, smoother material flows, and closer coordination with customers and suppliers in identifying the criticality of delivery dates. When all else fails, it may be time to do a new layout of the production facility. Or in some cases, it may be the easiest alternative to implement particularly if the cost of change is marginal. And when these occasions arise, the layout must be flexible enough to quickly accommodate these changes.

How do we develop such a flexible layout? Harmon and Peterson [20] suggest the use of the following objectives:

1. Reorganize factory subplants to achieve superior manufacturing status.
2. Provide maximum perimeter access for receiving and shipping materials, components, and products as close to each subplant as practical.
3. Cluster all subplants dedicated to a product or product family around the final process subplant to minimize inventories, shortages, and improve communication.
4. Locate supplier subplants of common component subplants in a central location to minimize component travel distances.
5. Minimize the factory size to avoid wasted time and motion of workers.
6. Eliminate centralized storage of purchased materials, components, and assemblies and move storage to focused subplants.
7. Minimize the amount of factory reorganization that will be made necessary by future growth and change.
8. Avoid locating offices and support services on factory perimeters.
9. Minimize the ratio of aisle space to production process space.

The idea of breaking up a factory into smaller enterpreneurial units is not a new concept, see Skinner [57]. Each subunit, or a subplant, can be organized much more efficiently. Alternative layout configurations can be designed for each subplant to take advantage of their specific product and process requirements. An illustration of a factory, which is organized along the subplant concept, is given in Figure 6.34. The illustration shows a major flow structure revolving around a spine flow and material flows within each subplant organized along U-shaped, I-shaped and S-shaped flow structures,[4] see Figure 6.34a. You will note that facility expansions and contractions are easily achieved with this modular configuration by simply extending outward or contracting inward within each sub-unit as shown in Figure 6.34b. A critical consideration, however, is the central flow structure, which could end up as the bottleneck when the total material handling requirements between subplants exceeds the

---

[4]For a comprehensive discussion on alternative material flow structures, see Tanchoco [60].

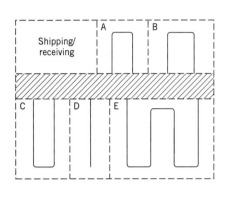

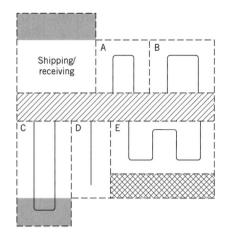

(a) Spine Layout with U, I and S Flows
Within Departments

(b) Layout After Expanding and Contracting
Some Departments

Figure 6.34 Illustration of a flexible layout.

capacity limits of the spine flow server. Sufficient capacity must be incorporated in the initial design in anticipation of future material flow requirements.

Volvo Facilities

Volvo Skovdeverken's factory for the manufacture of four-cylinder engines is designed on the basis of the modular concept. According to Volvo, the factory

> . . . . is characterized by a new concept of layout, environment, technology and work organization. The company's own technicians have worked together with representatives of the employees and with outside experts during the planning and execution. This has resulted in a further development of the striving for a better working environment worker satisfaction and well-being which has been typical for Volvo's activities in various places for many years.
>
> Since practical considerations require that the machining, assembly and test departments shall adjoin each other the factory layout is of basic importance. The objective was to create the atmosphere of a small workshop while retaining the advantages of rational production and a flexible flow of materials that a large factory makes possible. The result is a layout quite different from the traditional pattern and giving many practical and environmental advantages.
>
> The new factory consists of a body containing the assembly and test departments with four arms at right angles which contain the machining departments. These latter are separated from one another by generous garden areas. The method of assembly of the engine is one of the most interesting features, both from the technical and the work organizations points of view. The conventional assembly line principle has been replaced by an extremely flexible system of group assemblies. Electrically driven assembly trucks (AGVS) which are controlled by the assembly personnel, together with other technically advanced solutions, have given the work of building the engine a new interest. The different teams, in the machining departments as well as the assembly, have cooperated in the design of the workplaces and are now taking part in the development of new forms of work organization [66]

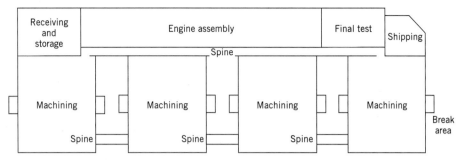

Figure 6.35 Block layout of the Volvo Skovdeverken factory.

As shown in Figure 6.35, the four manufacturing modules are connected by a material flow/personnel flow spine at each end of the module. The plant is approximately 400,000 ft^2 in size. Windows in each manufacturing module and break area overlook landscaped gardens.

Volvo Kalmarverken's automobile assembly plant provides another example of a modular design. As shown in Figure 6.36, the assembly plant consists of four equal-sized hexagonal modules, with three two-story assembly modules and one one-story preparation and finishing module. Additionally, a smaller two-story hexagonal module housing administrative and engineering support offices is located in front of and connected to the assembly building.

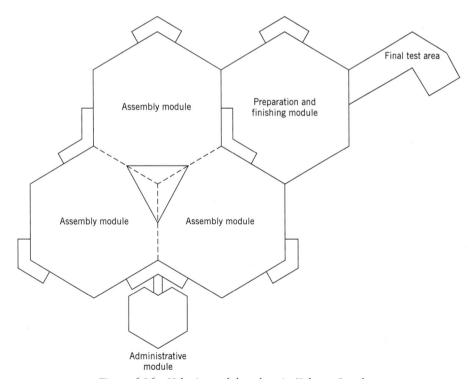

Figure 6.36 Volvo's modular plant in Kalmar, Sweden.

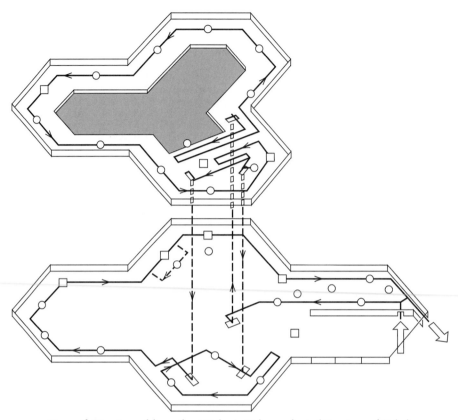

Figure 6.37   Assembly path at Volvo's Kalmar plant. (*Courtesy of Volvo*)

The assembly operations are performed adjacent to the outer walls of the three assembly modules. The assembly paths are indicated in Figure 6.37. From Figure 6.38, notice the location of the material storage area in the center of the plant. Lift trucks are used to store and retrieve materials in the storage racks, as well as to transport materials between the storage area and pickup-and-deposit stations located on each floor. The lift truck operates on the first floor and lifts/lowers materials to/from the second floor.

The Volvo Kalmar plant is recognized internationally for its pioneering work on job enlargement and the team approach to automotive assembly. When the planning for the Kalmar plant began in the early 1970s, Pehr G. Gyllenhammar, managing director of the Volvo Group, gave the following general directive:

> It has to be possible to create a working place which meets the need of the modern human being for motivation and satisfaction in his daily work. It must be possible to accomplish this objective without reducing efficiency. [66]

The team concept was one of the basic objectives established at the beginning of the facilities planning process. The organization was to be built around the team concept. The team members were to collaborate on a common set of tasks and work within an established production framework. They were to be allowed to

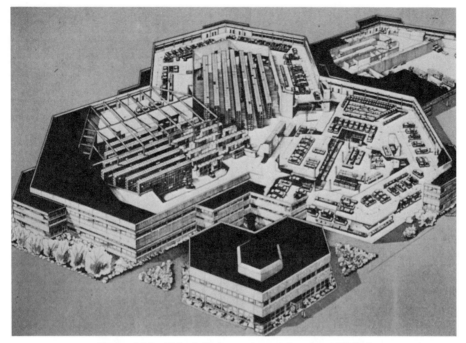

Figure 6.38   Volvo's Kalmar plant. (*Courtesy of Volvo*)

switch jobs among themselves, vary their work pace, carry joint responsibility for quality, and have the possibility to influence their working environment. It was felt by Volvo's management that the larger number of work tasks, combined with team membership, would provide greater meaning and satisfaction for each employee.

As a result of the planning objectives, each work team was provided its own entrance, its own changing room and saunas, its own break area, and its own assembly area. Each assembly area is approximately 10,000 ft^2 in size and is viewed as the team's own small workshop.

Assembly is performed by 20 different teams. Each team completes one system in the car, for example, the electrical system, instruments, and safety equipment. Assembly is carried out on special wire-guided, battery-powered, computer-controlled carriers (automated guided vehicle system, AGVS).

Two assembly approaches are used. As illustrated in Figure 6.39, straight-line assembly and dock assembly methods are used. With straight-line assembly the work to be performed by the team is divided among four or five workstations. The workers work in pairs and follow a car from station to station carrying out the entire work assignment belonging to their team. When a pair of workers complete work on one car, they walk back to the beginning station in their area and repeat the process. Typically, the two-person team members trade work assignments to provide additional variety to their work.

The dock assembly approach is used when the entire assembly task for an area is carried out at one of the four assembly stations by a team of two or three workers. With the dock assembly approach, the AGVS brings a car to an assembly dock where the entire work cycle is performed. The content and quality of the work performed is no different than that for the straight-line assembly approach.

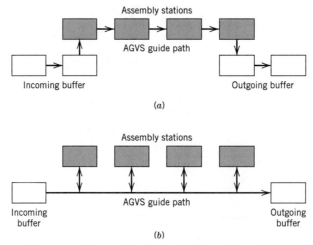

Figure 6.39   Assembly approaches used at Volvo's Kalmar plant. (*a*) Straight-line assembly. (*b*) Dock assembly.

6.10   DEVELOPING LAYOUT ALTERNATIVES

Because the final layout will result from the generation of alternative facility layouts, it is important that the number, quality, and variety of alternative designs be as large as feasible. If one agrees that facilities planning involves both art and science, then there would surely be agreement that "the art" comes in generating layout alternatives. It is at this stage that one's creativity is truly tested.

We have previously emphasized the need to divorce one's thinking from the present method. The ideal system approach was intended to facilitate the creative generation of alternatives. There exist a number of other aids to improve one's ability to generate more and better design alternatives. The following are some suggested approaches that have proven to be beneficial.

1. Exert the necessary effort.
2. Set a time limit.
3. Seek many alternatives.
4. Establish a goal.
5. Make liberal use of the questioning attitude.
6. Don't get bogged down in details too soon.
7. Don't "fail to see the forest for the trees."
8. Think big, then think little.
9. Don't be conservative.
10. Avoid premature rejection.
11. Avoid premature acceptance.
12. Refer to analogous problems of others.
13. Consult the literature.

14. Consult peers in other organizations.
15. Use the brainstorming technique.
16. Divorce your thinking from the existing solution.
17. Spread the effort out over time.
18. Involve operating people.
19. Involve management.
20. Involve experienced people.
21. Involve inexperienced people.
22. Involve those who oppose change.
23. Involve those who promote change.
24. Be aware of what the competition is doing.
25. Recognize your own limitations.
26. Look for trends.
27. Do your homework first.
28. Understand the requirements.
29. Don't overlook an improved present method.
30. Think long range.

The number of possible facility layouts for any reasonably sized problem can be quite large. Furthermore, the specification of constraints and the establishment of an accurate objective function are tasks that can be accomplished "fuzzily" at best. For this reason, the design process does not ask for the selection of the best design from among all possible feasible designs. Rather, it asks for a selection to be made from among several reasonable designs—it asks for the designer to "satisfice" not "optimize."

The problem of designing a facility layout can be distinguished from a number of other problem solving activities. As observed by Simon, "In ordinary language . . . we apply the term 'design' only to problem solving that aims at synthesizing new objects. If the problem is simply to choose among a given set of alternatives, e.g., to choose the location or site for a plant, we do not usually call it a design problem, even if the set of available alternatives is quite large, or possibly infinite" [56, p. 295].

Simon gives two reasons for calling the layout problem, but not, a math programming problem, a design problem. The negative reason he gives is that no simple finite algorithm exists for obtaining directly a solution to the layout problem. The positive reason given by Simon is that the process used for attacking the layout problem involves synthesizing the solution from intermediate, or component, decisions that are "*selective, cumulative,* and *tentative*" [56].

Because of the immense number of possible layout designs available to the facilities planner, the search process becomes a search over a space of design components and partially completed designs, rather than a search over a space of designs. In following the design process, very few *complete* designs are generated, compared, and evaluated. Instead, one generates design components, for example, receiving and shipping component, assembly component, and manufacturing component. Even at the component level, the comparisons and evaluations are typically performed at a macrodesign level. Quite often, only one complete design is produced. The sequential process of generating and evaluating component designs is sufficiently effective as a filter that the complete design is the final design.

6.11 SUMMARY

The objectives of this chapter were fourfold. First, we felt it important to cover the "tried-and-true" material concerning the basic types of layout and alternative layout planning procedures. Second, we wanted to present more contemporary layout algorithms and discuss their relative strengths and weaknesses. Third, we wanted to emphasize the importance of planning for change. Change is the only thing that is certain to occur in the future and planning for change will help guarantee that an organization is able to adapt and remain competitive. Fourth, we wished to encourage you to "enter the world of the architect."

Aspects of form and style as they relate to facilities planning should not be dismissed out of hand. The architect typically approaches the layout planning problem from a different perspective than does the facilities planning engineer or layout analyst. It is important for the engineer to understand "where the architect is coming from." Furthermore, it has been our experience that the success of the layout analyst in performing facilities planning frequently depends on his or her ability to work together with the architect.

Among the layout algorithms we covered in this chapter, CRAFT, MCRAFT and MULTIPLE use the discrete layout representation, and with the possible exception of MCRAFT, they place no restrictions on department shapes. Also, CRAFT and MULTIPLE can capture the initial layout, the building shape, and fixed departments/obstacles with fairly high accuracy. However, with CRAFT and MULTIPLE it is difficult to generate departments with prescribed shapes (such as rectangular or L-shaped departments). Although the spacefilling curve in MULTIPLE may be revised to correct some department shapes, both algorithms are likely to generate layouts that require considerable massaging.

BLOCPLAN, MIP, and LOGIC, on the other hand, use the continuous layout representation and they work only with rectangular department shapes. (LOGIC may generate one or more nonrectangular departments if the building is nonrectangular or there are obstacles present.) Rectangular departments considerably increase our control over department shapes. Also, in practice, a rectangular shape is very likely to be acceptable (if not desirable) for many departments in a facility. However, if fixed departments or obstacles are present, maintaining rectangular departments may increase the cost of the layout. Furthermore, capturing the initial layout, which may contain one or more nonrectangular departments, may not be possible or straightforward (especially with BLOCPLAN and LOGIC).

We also demonstrated the fundamental "idea" behind each algorithm. For example, MCRAFT relies on a sweep technique, while BLOCPLAN uses two or three (horizontal) bands. LOGIC, on the other hand, divides the building progressively into smaller portions by performing vertical and horizontal cuts and by assigning one or more departments to each portion of the building. MULTIPLE uses a spacefilling curve (or a hand-generated curve), which "maps" the two-dimensional layout into a single dimension (i.e., the layout vector). Last, MIP is based on mathematical programming and it uses 0/1 integer variables to ensure that departments do not overlap.

As far as the objective function is concerned, all the algorithms we showed use either the distance-based objective or the adjacency-based objective, or both. In reality, almost any layout algorithm can be adapted to accept one objective function or another. For example, the objective functions of CRAFT, LOGIC, and MULTIPLE can be changed from distance based to adjacency based with no fundamental changes to the algorithm itself. In fact, except for MIP, any one of the above algorithms can be used to generate a layout and evaluate it with respect to a user-specified objective function. One can also use an alternate objective function with MIP; however, care must be exercised to maintain a "reasonable" objective function (or at least a linear objective function) since the objective function, unlike the other algorithms, is an inherent part of MIP.

For those readers who might, at least from a theoretical standpoint, be concerned with layout optimality, we note that only two of the algorithms we presented can be used to identify the optimal layout: MIP will find the optimal layout only if all the departments are required to be rectangular; MULTIPLE will find the optimal layout *only with respect to a particular spacefilling curve* by enumerating over all-possible department sequences in the layout vector. LOGIC may also find the "optimal" layout (by enumerating over all-possible cut-trees); however, if all the departments are rectangular, the cost of any optimal solution obtained by LOGIC will be greater than or equal to the cost of an optimal solution obtained by MIP. (Why? Hint: consider the impact of the "guillotine cuts" in LOGIC.)

There are a number of early layout algorithms that we did not describe in this chapter. Such algorithms include ALDEP [54], COFAD [64], CORELAP [35], and PLANET [10]. Since their introduction, these algorithms have served as the cornerstones of facility layout algorithms, and they fueled interest in computer-based facility layout.[5] We did not present a number of other layout algorithms; see, for example, DISCON [12], FLAC [53], and SHAPE [22] and [65]. The reader may refer to [34] and [37] for a survey and discussion of those and other single-floor and multi-floor layout algorithms.

Some readers, from a theoretical or practical standpoint, might wonder which layout algorithm is the best. From a theoretical standpoint, it is valid to compare certain layout algorithms. For example, since they use the same objective function and the same representation, one may compare the performance of CRAFT, MCRAFT, and MULTIPLE. In fact, in [6], numerical results are shown to compare CRAFT with MULTIPLE. Likewise, as long as all the departments are rectangular, one may perform a meaningful comparison between BLOCPLAN, LOGIC, and MIP (assuming that MIP is solved through a heuristic procedure such as the one shown in [45]). In contrast, comparing, say, MULTIPLE with MIP, requires a more careful approach; the two algorithms use different representations and the resulting department shapes are likely to be different. Hence, any comparison must include the department shapes. However, if the purpose of the comparison is to simply measure the layout cost impact of allowing nonrectangular departments (at the risk of obtaining some irregularly shaped departments), then it may make sense to compare just the layout costs of MULTIPLE and MIP.

From a practical standpoint, we hope that the reader is now in a position to appreciate that each layout algorithm has certain strengths and weaknesses. Particular constraints imposed by a problem (such as the number and location of fixed departments and/or obstacles, department shapes, the building shape, moving

---

[5]Due to their historical significance, ALDEP and CORELAP are described in the appendix to the chapter.

into a vacant building versus improving a given layout), coupled with our description of each algorithm, should help the reader (or the layout planner) select not the "best" algorithm but the "most appropriate" one. Although "dummy or negative flows" can be used in a from-to chart, and some qualitative interactions among the departments may be captured in a relationship chart, we also need to remind the practitioner that no computer-based layout algorithm will capture all the significant aspects of a facility layout problem.

In fact, we do not know if computers will ever be able to fully capture and use human experience and judgment, which play a critical role in almost any facility layout problem. In that sense, a computer-based layout algorithm should not be viewed as an "automated design tool," which can functionally replace the layout planner. To the contrary, we believe the (human) layout planner will continue to play a key role in developing and evaluating the facility layout. The computer-based layout algorithms presented in this chapter (and other such algorithms in the literature) are intended as "design aids" that, when used properly, significantly enhance the productivity of the layout planner. Therefore, for nontrivial problems, we believe that, given two layout planners with comparable experience and skills, the one "armed" with an appropriate computer-based layout algorithm is far more likely to generate a demonstrably "better" solution in a shorter time.

BIBLIOGRAPHY

1. Apple, J. M., *Plant Layout and Material Handling*, 3rd ed., John Wiley, New York, 1977.
2. Armour, G. C., and Buffa, E. S., "A Heuristic Algorithm and Simulation Approach to Relative Location of Facilities," *Management Science*, vol. 9, no. 2, 1963, pp. 294–309.
3. Bartholdi, J. J., and Platzman, L. K., "Heuristics Based on Spacefilling Curves for Combinatorial Problems in Euclidean Space," *Management Science*, vol. 34, no. 3, 1988, pp. 291–305.
4. Bartholdi, J. J., and Platzman, L. K., "Design of Efficient Bin-Numbering Schemes for Warehouses," *Material Flow*, vol. 4, 1988, pp. 247–254.
5. Bozer, Y. A. and Meller, R. D., "A Reexamination of the Distance-based Facility Layout Problem," *IIE Transactions*, vol. 29, no. 7, pp. 549–560, 1997.
6. Bozer, Y. A., Meller, R. D., and Erlebacher, S. J., "An Improvement-Type Layout Algorithm for Single and Multiple Floor Facilities," *Management Science*, vol. 40, no. 7, 1994, pp. 918–932.
7. Brown, A. R., *Optimum Packing and Depletion*, Macdonald and Co., American Elsevier Publishing Co., New York, 1971.
8. Buffa, E. S., Armour, G. C., and Vollman, T. E., "Allocating Facilities with CRAFT," *Harvard Business Review*, vol. 42, no. 2, pp. 136–159, March/April 1964.
9. Davis, L. (ed.), *Handbook of Genetic Algorithms*, Van Nostrand Reinhold, New York, 1991.
10. Deisenroth, M. P., and Apple, J. M., "A Computerized Plant Layout Analysis and Evaluation Technique (PLANET)," *Technical Papers 1962*, Amrican Institute of Industrial Engineers, Norcross, GA, 1972.
11. Donaghey, C. E., and Pire, V. F., "Solving the Facility Layout Problem with BLOCPLAN," Industrial Engineering Department, University of Houston, TX, 1990.
12. Drezner, Z., "DISCON: A New Method for the Layout Problem," *Operations Research*, vol. 20, 1980, pp. 1375–1384.

13. Foulds, L. R., "Techniques for Facilities Layout: Deciding Which Pairs of Activities Should be Adjacent," *Management Science,* vol. 29, no. 12, 1983, pp. 1414–1426.

14. Foulds, L. R., Gibbons, P. B., and Giffin, J. W., "Facilities Layout Adjacency Determination: An Experimental Comparison of Three Graph Theoretic Heuristics," *Operations Research,* vol. 33, no. 5, pp. 1091–1116, 1985.

15. Foulds, L. R., and Robinson, D. F., "Graph Theoretic Heuristics for Plant Layout Problem," *International Journal of Production Research,* vol. 16, 1978, pp. 27–37.

16. Francis, R. L., McGinnis, L. F., and White, J. A., *Facility Layout and Location: An Analytical Approach,* Prentice Hall, Englewood Cliff, NJ, 1992.

17. Freeman, H., "Computer Processing of Line-Drawing Images," *Computing Surveys,* vol. 6, 1974, pp. 57–97.

18. Giffin, J. W., Foulds, L. R., and Cameron, D. C., "Drawing a Block Plan From a Relationship Chart with Graph Theory and Microcomputer," *Computers and Industrial Engineering,* vol. 10, pp. 109–116, 1986.

19. Goldberg, D. E., *Genetic Algorithms in Search, Optimization, and Machine Learning,* Addison-Wesley, Boston 1989.

20. Harmon, R. L., and Peterson, L. D., *Reinventing the Factory: Productivity Breakthroughs in Manufacturing Today,* Free Press, New York, 1990.

21. Hassan, M. M. D., and Hogg, G. L., "On Constructing a Block Layout by Graph Theory," *International Journal of Production Research,* vol. 29, no. 6, pp. 1263–1278, 1991.

22. Hassan, M. M. D., Hogg, G. L., and Smith, D. R., "SHAPE: A Construction Algorithm for Area Placement Evaluation," *International Journal of Production Research,* vol. 24, 1986, pp. 1283–1295.

23. Heragu, S. S., and Kusiak, A., "Efficient Models for the Facility Layout Problem," *European Journal of Operational Research,* vol. 53, 1991, pp. 1–13.

24. Hobson, E. W., *The Theory of Functions of a Real Variable and the Theory of Fourier's Series, Volume I,* 3rd ed., Cambridge University Press and Harren Press, Washington D.C., 1950.

25. Holland, J. H., *Adaptation in Natural and Artificial Systems,* University of Michigan Press, Ann Arbor, MI, 1975.

26. Hosni, Y. A., Whitehouse, G. E., and Atkins, T. S., *MICRO-CRAFT Program Documentation,* Institute of Industrial Engineers, Norcross, GA.

27. Immer, J. R., *Layout Planning Techniques,* McGraw-Hill, New York, 1950.

28. Jajodia, S., Minis, I., Harhalakis, G., and Proth, J., "CLASS: Computerized LAyout Solutions using Simulated annealing," *International Journal of Production Research,* vol. 30, no. 1, 1992, pp. 95–108.

29. Johnson, R. V., "SPACECRAFT for Multi-Floor Layout Planning," *Management Science,* vol. 28, no. 4, 1982, pp. 407–417.

30. Johnson, D. S., Aragon, C. R., McGeoch, L. A., and Schevon, C., "Optimization by Simulated Annealing: An Experimental Evaluation; Part I, Graph Partitioning," *Operations Research,* vol. 37, 1989, pp. 865–892.

31. Kaku, K., Thompson, G. L., and Baybars, I., "A Heuristic Method for the Multi-Story Layout Problem," *European Journal of Operational Research,* vol. 37, 1988, pp. 384–397.

32. Kirkpatrick, S., Gelatt, C. D., and Vecchi, M. P., "Optimization by Simulated Annealing," *Science,* vol. 220, 1983, pp. 671–680.

33. Krejcirik, M., *Computer Aided Plant Layout,* Computer Aided Design, vol. 2, pp. 7–19, 1969.

34. Kusiak, A., and Heragu, S. S., "The Facility Layout Problem," *European Journal of Operational Research,* vol. 29, 1987, pp. 229–251.

35. Lee, R. C., and Moore, J. M., "CORELAP—Computerized Relationship Layout Planning," *Journal of Industrial Engineering,* vol. 18, no. 3, 1967, pp. 194–200.

36. Liggett, R. S., and Mitchell, W. J., "Optimal Space Planning in Practice," *Computer Aided Design,* vol. 13, 1981, pp. 277–288.

37. Meller, R. D., "Layout Algorithms for Single and Multiple Floor Facilities," Ph.D. Dissertation, Department of Industrial and Operations Engineering, University of Michigan, 1992.

38. Meller, R. D., personal communication, 1993.

39. Meller, R. D., and Bozer, Y. A., "A New Simulated Annealing Algorithm for the Facility Layout Problem," *International Journal of Production Research,* vol. 34, no. 6, 1996, pp. 1675–1692.

40. Meller, R. D., and Bozer, Y. A., "Alternative Approaches to Solve the Multi-Floor Facility Layout Problem," *Journal of Manufacturing Systems,* vol. 16, no. 3, pp. 192–203, 1997.

41. Meller, R. D., Narayanan, V., and Vance, P. H., "Optimal Facility Layout Design," *Operations Research Letters,* vol. 23, no. 3–5, 1998, pp. 117–127.

42. Metropolis, N., Rosenbluth, A. W., Rosenbluth, M. N., Teller, A. H., "Equation of State Calculation by Fast Computing Machines," *Journal of Chemical Physics,* vol. 21, 1953, pp. 1087–1092.

43. Montreuil, B., Ratliff, H. D., and Goetschalckx, M., "Matching Based Interactive Facility Layout," *IIE Transactions,* vol. 19, no. 3, 1987, pp. 271–279.

44. Montreuil, B., "A Modeling Framework for Integrating Layout Design and Flow Network Design," *Proceedings of the Material Handling Research Colloquium,* Hebron, KY, 1990, pp. 43–58.

45. Montreuil, B., and Ratliff, H. D., "Utilizing Cut Trees as Design Skeletons for Facility Layout," *IIE Transactions,* vol. 21, no. 2, 1989, pp. 136–143.

46. Murty, K., *Linear and Combinatorial Programming,* Wiley, New York, 1976.

47. Muther, R., *Systematic Layout Planning,* 2nd ed., Cahners Books, Boston, 1973.

48. Nugent, C. E., Vollman, R. E., and Ruml, J., "An Experimental Comparison of Techniques for the Assignment of Facilities to Locations," *Operations Research,* vol. 16, 1968, pp. 150–173.

49. Pire, V. F., "Automated Multistory Layout System," unpublished Master's thesis, Industrial Engineering Department, University of Houston, Houston, TX, 1987.

50. Propst, R. L., *The Action Factory System: An Integrated Facility for Electronics Manufacturing,* Herman Miller Corp., Zeeland, MI, 1981.

51. Reed, R. Jr., *Plant Layout: Factors, Principles and Techniques,* Richard D. Irwin, IL, 1961.

52. Schonberger, R. J., *World Class Manufacturing, The Lessons of Simplicity Applied,* Free Press, 1986.

53. Scriabin, M., and Vergin, R. C., "A Cluster-Analytic Approach to Facility Layout," *Management Science,* vol. 31, no. 1, 1985, pp. 33–49.

54. Seehof, J. M., and Evans, W. O., "Automated Layout Design Program," *Journal of Industrial Engineering,* vol. 18, 1967, pp. 690–695.

55. Seppannen, J., and Moore, J. M., "Facilities Planning with Graph Theory," *Management Science,* vol. 17, no. 4, pp. 242–253, December 1970.

56. Simon, H. A., "Style in Design," in C. M. Eastman (ed.), *Spatial Synthesis in Computer-Aided Building Design,* Wiley, New York, 1975.

57. Skinner, W., "The Focused Factory," *Harvard Business Review,* pp. 113–121, May–June 1974.

58. Tam, K. Y., "A Simulated Annealing Algorithm for Allocating Space to Manufacturing Cells," *International Journal of Production Research,* vol. 30, 1991, pp. 63–87.

59. Tanaka, H., and Yoshimoto, K., "Genetic Algorithm Applied to the Facility Layout Problem," Department of Industrial Engineering and Management, Waseda University, Tokyo, 1993.

60. Tanchoco, J. M. A. (ed) *Material Flow Systems in Manufacturing,* Chapman and Hall, London, U.K., 1994.

61. Tate, D. M., and Smith, A. E., "A Genetic Approach to the Quadratic Assignment Problem," *Computers and Operations Research,* vol. 22, no. 1, 1995, pp. 73–83.

62. Tate, D. M. and Smith, A. E., "Unequal-Area Facility Layout by Genetic Search," *IIE Transactions,* vol. 27, no. 4, pp. 465–472, 1995.

63. Tompkins, J. A. "Modularity and Flexibility: Dealing with Future Shock in Facilities Design," *Industrial Engineering,* vol. 12, no. 9, pp. 78–81, September 1980.

64. Tompkins, J. A., and Reed, R. Jr., "An Applied Model for the Facilities Design Problem," *International Journal of Production Research,* vol. 14, no. 5, 1976, pp. 583–595.

65. Tretheway, S. J., and Foote, B. L., "Automatic Computation and Drawing of Facility Layouts with Logical Aisle Structures," *International Journal of Production Research,* vol. 32, no. 7, 1994, pp. 1545–1555.

66. *Volvo: The Engine Factory That is Different,* a Volvo publication describing the Volvo Skovde plant, Gothenberg, Sweden.

67. White, J. A., "Layout: The Chicken or the Egg?," *Modern Materials Handling,* vol. 35, no. 9, p. 39, September 1980.

68. Wilhelm, M. R., and Ward, T. L., "Solving Quadratic Assignment Problems by Simulated Annealing," *IIE Transactions,* vol. 19, 1987, pp. 107–119.

69. *Material Handling Engineering,* "Vertical Reciprocating Conveyors: Flexible Handling Devices," vol. 45, no. 9, 1990, pp. 81–85.

70. *Modern Materials Handling,* "Vertical Conveyors Increase Plant's Efficiency 33%," vol. 40, Casebook Directory, 1985, p. 113.

APPENDIX 6.A
ALDEP AND CORELAP

6.A1   INTRODUCTION

Because of their historical significance, in the appendix we consider briefly two additional computer-aided layout algorithms, ALDEP [54] and CORELAP [35]. Following the development of CRAFT in 1963, ALDEP and CORELAP were developed in 1967. For two decades, the three computer-aided layout algorithms served as the benchmarks for all other such algorithms.

Developed for main frame computers, we are not aware of microcomputer versions of ALDEP and CORELAP. For this reason, we did not choose to include them in the main body of the chapter. However, for those who want to understand the evolution of computer-aided layout algorithms to the present, it is useful to consider the earliest such algorithms.

In the appendix, we consider first CORELAP, since ALDEP is but a slight variation of CORELAP. Our reason for including both CORELAP and ALDEP is the randomness aspect of ALDEP and its multifloor capability. After learning about CORELAP and ALDEP, we suspect you will recognize the hereditary roots of several algorithms presented in the chapter.

6.A2   CORELAP

CORELAP, an acronym representing **computerized relationship layout planning,** constructs a layout for a facility by calculating the total closeness rating (TCR) for each department where the TCR is the sum of the numerical values assigned to the

```
              2 2
1 1 1        1 1 1          Placing
1 1 1 7 7 7  1 1 1 7 7 7    rating = 64
    (a)          (b)


1 1 1    2 2  Placing     1 1 1 2 2   Placing
1 1 1 7 7 7   rating = 16  1 1 1 7 7 7  rating = 80
    (c)                        (d)
```

Figure 6.A1   Illustration of CORELAP placing rating.

closeness relationships (A = 6, E = 5, I = 4, O = 3, U = 2, X = 1) between a department and all other departments [35]. The department having the highest TCR is then placed in the center of the layout. If there is a tie for the highest TCR, the following tie-breaking rule is applied: the department having the largest area and then the department having the lowest department number. Next, the relationship chart is scanned and if a department is found having an "A" relationship with the *selected department,* it is brought into the layout. If none exists, the relationship chart is scanned for an "E" relationship, then an "I", and so on. If two or more departments are found having the same relationship with the selected department, the department having the highest TCR is selected; if a tie still exists, the tie-breaking rule is utilized. The third department to enter the layout is determined by scanning the relationship chart to see if an unassigned department exists that has an "A" relationship with the first department selected. If so, this department is brought into the layout. If a tie exists, the TCR and then the tie-breaking rule are utilized. If no unassigned department exists that has an "A" relationship with the second department, the procedure is repeated considering "E" relationships, then "I" relationships, and so on. If a tie occurs, the TCR and then the tie-breaking hierarchy are utilized. The same procedure is repeated for the fourth department to enter the layout, except the three departments previously selected are included in the search. The procedure continues until all departments have been selected to enter the layout.

Once a department is selected to enter the layout, a placement decision must be made. The placement decision is made by calculating the placing rating for the available locations of the department, where the placing rating is the sum of the weighted closeness ratings between the department to enter the layout and its neighbors. For example, consider the layout consisting of departments 1 and 7 in Figure 6.A1a. Suppose department 2 is next to enter the layout and it has an "A" relationship with department 1 and an "E" relationship with department 7. If an "A" relationship is weighted 64 and an "E" is weighted 16 by the user, the placing ratings are as shown in 6.A1*b*, 6.A1*c*, and 6.A1*d*.

If a tie occurs for the placing rating, the boundary lengths of the tied locations are compared. The boundary length is the number of unit square sides that the department to enter the layout has in common with its neighbors. In Figures 6.A1*b* and 6.A1*c*, the boundary lengths are 2, and in 6.A1*d* the boundary length is 3.

After the final CORELAP layout has been prepared, CORELAP evaluates the layout by calculating the layout score. The layout score is defined as

$$\text{Layout score} = \sum_{\text{all departments}} \begin{matrix}\text{numerical}\\\text{closeness}\\\text{rating}\end{matrix} \times \begin{matrix}\text{length}\\\text{of shortest}\\\text{path}\end{matrix} \qquad (6.A1)$$

It should be noted that CORELAP utilizes the shortest rectilinear path between departments as opposed to the rectilinear distance between department centroids as in CRAFT. The shortest rectilinear path is used as it is assumed that each department will have a dispatch area and a receiving area on the side of its layout nearest its neighbor.

CORELAP-generated layouts often result in irregular building shapes and will need manual adjustment. Care must also be exhibited with the interpretation of the layout scores, as the shortest rectilinear path between departments may not always be a realistic measure.

Example 6.A1

The data requirements for CORELAP are the department areas and the activity relationship chart. Suppose the activity relationship chart is as given in Figure 6.A2 and the space requirements are as given in Table 6.A1. CORELAP begins by calculating the TCR for each department. Table 6.A2 illustrates these calculations. The first department selected by CORELAP to enter the layout is department 5, as it has the highest TCR. Scanning the relationship chart indicates that only department 6 has an "A" relationship with department 5, so it will be the second department selected to enter the layout. The selection of the third department to enter the layout begins by scanning the relationship chart and determining that there are no additional "A" relationships with department 5 and that there are no unselected departments having an "A" relationship with department 6. "E" relationships with department 5 also result in no selection. An "E" relationship does exist between departments 6 and 7; therefore, department 7 will be the third department to enter the layout. In an effort to select the fourth department, it is noted that no additional departments have "A" or "E" relationships with department 5, 6, or 7. Department 5 has an "I" relationship with departments 2 and 4. By checking the TCR, we see department 2 has the highest value and will be the fourth department to enter the layout. After determining that no "A" relationships exist with the selected departments, "E" relationships are scanned; it may be seen that departments 1 and 4 have an "E" relationship with department 2. Both departments 1 and 4 have a TCR of 19 and an area of 12,000 ft^2, so that lowest numbered department, department 1, will be the fifth department to enter the layout. Department 4 will enter the layout next, followed by department 3. The overall order of departments entering the layout is 5-6-7-2-1-4-3.

The weighted values for placing departments for this example are as assigned in Table 6.A3. Department 5 is the first department to enter the layout and is placed in the center.

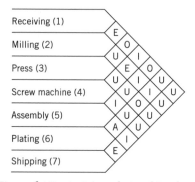

Figure 6.A2   Activity relationship chart.

Table 6.A1   *Department Areas and Number of Unit Area Templates for Example 6.A1*

Code	Function	Area (Square feet)	Number of Unit Area Templates
1	Receiving	12,000	6
2	Milling	8000	4
3	Press	6000	3
4	Screw machine	12,000	6
5	Assembly	8000	4
6	Plating	12,000	6
7	Shipping	12,000	6

Table 6.A2   *Calculation of the Total Closeness Ratings (TCRs) for Example 6.A1*

Department	Department Number	Relationships	TCR
Receiving	1	E, O, I, O, U, U	19
Milling	2	E, U, E, I, I, U	22
Press	3	O, U, U, U, U, U	14
Screw machine	4	I, E, U, I, U, U	19
Assembly	5	O, I, U, I, A, I	23
Plating	6	U, I, O, U, A, E	22
Shipping	7	U, U, U, U, I, E	17

Table 6.A3   *Weights Assigned to Relationships for Example 6.A1*

Relationship	Weight
A	243
E	81
I	27
O	9
U	1

```
0 0 0 0  0  0 0 0        0 0 0  0  0  0 0 0
0 0 0 16 16 0 0 0        0 0 0  16 16 0 0 0
0 0 0 0  15 0 0 0        0 0 17 17 15 0 0 0
0 0 0 0  0  0 0 0        0 0 0  0  0  0 0 0
0 0 0 0  0  0 0 0        0 0 0  0  0  0 0 0
        (a)                      (b)


0 0 0  0  0  0 0 0        0 14 11 11 13 0 0 0
0 0 12 16 16 0 0 0        0 14 12 16 16 0 0 0
0 0 17 17 15 0 0 0        0 0  17 17 15 0 0 0
0 0 0  0  0  0 0 0        0 0  0  0  0  0 0 0
0 0 0  0  0  0 0 0        0 0  0  0  0  0 0 0
        (c)                      (d)
```

Figure 6.A3   CORELAP layout construction for the example problem. (*a*) The first two adjacent departments. (*b*) Department 7 added to departments 5 and 6. (*c*) The layout consisting of departments 5, 6, 7, and 2. (*d*) The final layout.

Table 6.A4   *Calculation of the Layout Score*

Relationship	Preset Value	From	To	Distance	Product of Preset Value and Distance
A	6	15	16	0	0
E	5	11	12	0	0
E	5	12	14	0	0
E	5	16	17	0	0
I	4	11	14	0	0
I	4	12	15	2	8
I	4	12	16	0	0
I	4	14	15	3	12
I	4	15	17	0	0
O	3	11	13	0	0
O	3	13	16	0	0
O	3	11	15	2	6
U	2	11	16	0	0
U	2	11	17	1	2
U	2	12	13	2	4
U	2	12	17	0	0
U	2	13	14	2	4
U	2	13	15	1	2
U	2	13	17	2	4
U	2	14	16	1	2
U	2	14	17	1	2
				Layout score	46

Department 6 is then positioned adjacent to department 5, as shown in Figure 6.A3*a*. It should be noted that CORELAP adds 10 to the department number prior to it being printed, that is, department 1 will be printed as department 11, department 5 as 15, department 18 as 28, and so on. Department 7 will maximize its placing rating by next entering the layout, as shown in Figure 6.A3*b*. Department 2 has an area of 1 unit square and relationships, "I," "I," and "U" with departments 5, 6, and 7, respectively. Therefore, the highest placing rating would be obtained if department 2 could be adjacent to departments 5 and 6. Unfortunately, as noted in Figure 6.A3*b*, no such location is available. In fact, the only location that allows department 2 to be adjacent to more than one department is the location shown in Figure 6.A3*c*. The placing rating for this location is 28 (27 for department 2 being adjacent to department 6, and 1 for department 2 being adjacent to department 7). This is the highest placing rating and is the location CORELAP selects for department 2. By pursuing this same course of action, CORELAP produces the final layout shown in Figure 6.A3*d*. The last CORELAP function would be the calculation of the layout score shown in Table 6.A4.

6.A3   ALDEP

ALDEP [54], an acronym representing **automated layout design program,** has the same basic data input requirements and objectives as CORELAP. The basic proce-dural difference between CORELAP and ALDEP is that CORELAP selects the first department to enter the layout and breaks ties with the total closeness rating, where ALDEP selects the first department and breaks ties randomly. The basic philosophical difference between CORELAP and ALDEP is that CORELAP attempts to produce the one best layout, whereas ALDEP produces many layouts, rates each layout, and leaves the evaluation of the layouts to the facilities designer. ALDEP is the first model having provisions for multiple floors; it can handle up to three floors. Additionally, ALDEP allows departments to be placed in specific locations and is able to include dummy departments [54]. However, ALDEP does not accommodate flow between floors as effectively as MULTIPLE or SABLE/STAGES. Also, if there is a fixed department, ALDEP often splits departments.

As was previously mentioned, the first department ALDEP selects to enter the layout is selected randomly. The relationship chart is then scanned to determine if there is a department having an "A" relationship with the randomly selected first department. If one exists, it is selected to enter the layout. If more than one exists, one is randomly selected to enter the layout. If no departments have a relationship at least equal to the minimum acceptable closeness rating specified by the user, the second department to enter the layout will be selected randomly. Once the second department to enter the layout is selected, the selection procedure is repeated for the second department and all unselected departments. Once the third department is selected, the next department to enter the layout is determined by repeating the selection procedure. This process continues until all departments have been selected to enter the layout.

The placement routine within ALDEP begins by placing the first department in the upper left corner of the layout and extends it downward. The width of the downward extension of the department entering the layout is input by the user and is termed the **sweep width.** The department is placed in the layout using the sweep pattern illustrated in Figure 6.A4*f.* Each additional department added to the layout begins where the previous department ends and continues to follow the serpentine path. When all departments have entered the layout, ALDEP rates the layout by assigning values to the relationships among adjacent departments. If a department is adjacent to a department with which it has an "A" relationship, a value of 64 is added to the rating of the layout. An "E" relationship adds 16, "I" adds 4, and an "O" relationship adds 1 to the rating of the layout. A "U" relationship has no effect on the rating of the layout and if two adjacent departments have an "X" relationship, 1024 is subtracted from the rating of the layout. ALDEP prints the layout and the rating and then returns to randomly generate the first department to be selected for the next layout. The entire procedure is repeated. ALDEP can be used in this manner to generate up to 20 layouts and ratings per run.

ALDEP will not print layouts whose rating is less than an initially input minimal score. For this reason, users of ALDEP should be aware of the very high penalty that is built into ALDEP for the "X" relationship. Relationship charts having more than a few "X" relations often will not result in layouts being printed. Additionally, users should not accept ALDEP answers without experimenting with the input values for the specified level of importance and the sweep width. The specified level of importance basically segregates all relationships into an important or an unimportant classification. The decision as to which relationships should be considered important and which should be unimportant must be made

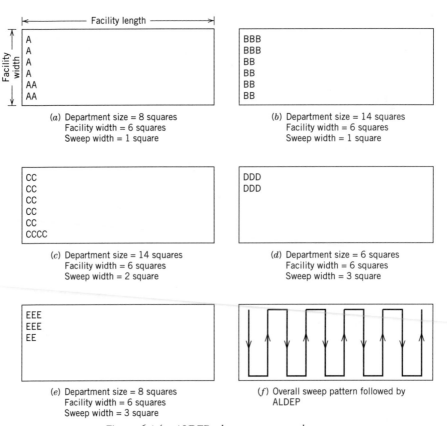

Figure 6.A4   ALDEP placement procedure.

for each problem and can only be determined by experimentation. The sweep width that provides the best layout is problem dependent and must be determined via experimentation.

Example 6.A2

In addition to the input data for Example 6.A1, ALDEP requires the specification of the sweep width and the specified level of importance. For the iteration to be described, the sweep width will be assigned a value of 2, and the minimum acceptable level of importance will be input as an "E" relationship. Assume that ALDEP randomly selects department 4 to be the first department to enter the layout. The relationship chart is scanned to determine if a department has either an "A" or "E" relationship with department 4. Department 2 has an "E" relationship with department 4 and is selected to be the second department to enter the layout. The unselected department having an "A" or "E" relationship with department 2 is department 1; therefore, department 1 will enter the layout next. Of the remaining unselected departments, none have either an "A" or an "E" relationship with department 1, so the next department to enter the layout is selected randomly. Suppose department 6 is selected; it has an "A" relationship with department 5, so department 5 is the fifth department to enter the layout. Next, department 7 is randomly selected to enter

```
4 4                              4 4 2 2 2 2
4 4                              4 4 2 2 2 2
4 4                              4 4 2 2 2 2
4 4                              4 4 2 2 2 2
4 4                              4 4 2 2 2 2
4 4 4 4                          4 4 4 4
4 4 4 4                          4 4 4 4
4 4 4 4                          4 4 4 4
4 4 4 4                          4 4 4 4
4 4 4 4                          4 4 4 4
```
(*a*) Partial layout consisting   (*b*) Partial layout after placing
    of department 4   department 2

```
4 4 2 2 2 2 1 1                  4 4 2 2 2 2 1 1 6 6 5 5 5 5 7 7 3 3
4 4 2 2 2 2 1 1                  4 4 2 2 2 2 1 1 6 6 5 5 5 5 7 7 3 3
4 4 2 2 2 2 1 1                  4 4 2 2 2 2 1 1 6 6 5 5 5 5 7 7 3 3
4 4 2 2 2 2 1 1                  4 4 2 2 2 2 1 1 6 6 5 5 5 5 7 7 3 3
4 4 2 2 2 2 1 1                  4 4 2 2 2 2 1 1 6 6 5 5 5 5 7 7 3 3
4 4 4 4 1 1 1 1                  4 4 4 4 1 1 1 1 6 6 6 6 7 7 7 7 3 3
4 4 4 4 1 1 1 1                  4 4 4 4 1 1 1 1 6 6 6 6 7 7 7 7 3 3
4 4 4 4 1 1 1 1                  4 4 4 4 1 1 1 1 6 6 6 6 7 7 7 7 3 0
4 4 4 4 1 1 1 1                  4 4 4 4 1 1 1 1 6 6 6 6 7 7 7 7 0 0
4 4 4 4 1 1 1 1                  4 4 4 4 1 1 1 1 6 6 6 6 7 7 7 7 0 0
```
(*c*) Partial layout after   (*d*) Final layout
    placing department 1

Figure 6.A5 ALDEP layout construction for Example 6.A2.

the layout. Department 7 is followed by the only remaining department, department 3. The order of departments to enter the layout for this iteration is 4-2-1-6-5-7-3.

Given the overall layout is to be 10 units wide and 18 units in length, the placement of department 4 in the layout is shown in Figure 6.A5*a*. (Each "digit" in the layout represents 400 ft^2.) The addition of department 2 is given in Figure 6.A5*b*. The inclusion of department 1 in the layout is depicted in Figure 6.A5*c*. The final layout is shown in Figure 6.A5*d*. The rating for the final layout given in Figure 6.A5*d* is 240. The calculation procedure followed by ALDEP to determine this rating is shown in Table 6.A5. Once the

Table 6.A5 *ALDEP Scoring (Rating) Procedure Applied to Example 6.A2*

Adjacent Departments	Relationship	Value	Rating
4-2 and 2-4	E	16	32
4-1 and 1-4	I	4	8
2-1 and 1-2	E	16	32
1-6 and 6-1	U	0	0
6-5 and 5-6	A	64	128
6-7 and 7-6	E	16	32
5-7 and 7-5	I	4	8
7-3 and 3-7	U	0	0
	Total		240

layout and rating are printed, ALDEP begins the next layout by once again randomly selecting the first department to enter the layout.

PROBLEMS

6.1   What are some of the important factors that should be taken into consideration when a layout is being designed?

6.2   What kind of impacts do the material handling decisions have on the effectiveness of a facility layout in a manufacturing environment?

6.3   What kind of manufacturing environments are the following types of layout designs best suited for?

    a. Fixed Product Layout   **b.** Product Layout

    c. Group Layout   **d.** Process Layout

6.4   Compare the primary layout design objectives for the following situations:

    a. Soda bottler   **b.** Printing shop

    c. Meat-processing plant   **d.** Furniture manufacturing plant

    e. Computer chip maker   **f.** Shipyard

    g. Refinery plant   **h.** College campus

6.5   What are the basic differences between construction type and improvement type layout algorithms?

6.6   Contrast and compare the plant layout procedures proposed by Apple, Reed, and Muther.

6.7   Four departments are to be located in a building of 600 ft × 1000 ft. The expected personnel traffic flows and area requirement for departments are shown in the tables below. Develop a block layout using SLP.

Dept.	A	B	C	D
A	0	250	25	240
B	125	0	400	335
C	100	0	0	225
D	125	285	175	0

Department:	Department Dimension:
A	200 ft. × 200 ft.
B	400 ft. × 400 ft.
C	600 ft. × 600 ft.
D	200 ft. × 200 ft.

6.8   XYZ Inc. has a facility with six departments (A, B, C, D, E, and F). A summary of the processing sequence for ten products and the weekly production forecasts for the products are given in the tables below.

    a. Develop the from-to chart based on the expected weekly production.

    b. Develop a block layout using SLP.

Product	Processing Sequence	Weekly Production
1	A B C D E F	960
2	A B C B E D C F	1,200
3	A B C D E F	720
4	A B C E B C F	2,400
5	A C E F	1,800
6	A B C D E F	480
7	A B D E C B F	2,400
8	A B D E C B F	3,000
9	A B C D F	960
10	A B D E F	1,200

Dept.	Dimension (ft. × ft.)
A	40 × 40
B	45 × 45
C	30 × 30
D	50 × 50
E	60 × 60
F	50 × 50

6.9 A toy manufacturing company makes ten different types of products. There are fifteen equal sized departments involved. Given the following product routings and production forecasts,

a. Construct a from-to chart for the facility.

b. Develop a block layout using SLP.

Product	Processing Sequence	Weekly Production
1	A B C D B E F C D H	500
2	M G N O N O	350
3	H L H K	150
4	C F E D H	200
5	N O N	100
6	I J H K L	150
7	G N O	200
8	A C F B E D H D	440
9	G M N	280
10	I H J	250

6.10 Shown below is the activity relationship chart along with the space reqirements for each of the six cells in a small auto parts manufacturing facility.

    a. Construct relationship and space relationship diagrams.

    b. Design the corresponding block layout using SLP.

CELL A	2,100
CELL B	2,100
CELL C	2,100
CELL D	2,800
CELL E	1,500
CELL F	1,500
CELL G	2,900

6.11 An activity relationship chart is shown below for the American Mailbox Company. Construct a relationship diagram for the manufacturing facility. Given the space requirements (in sq. ft.), construct a block layout using SLP.

RECEIVING	2,500
PUNCH PRESS	5,500
PRESS BENDING	2,500
PRESS FORMING	2,500
RIVETING	1,500
POWER SAWING	2,500
POWER DRAW	2,000
WELDING ROBOT	1,000

6.12 Suppose five departments labeled A through E are located as shown in the layout below. Given the corresponding flow-between chart, compute the efficiency rating for the layout.

	A	B	C	D	E
A	—	5	0	4	−3
B		—	6	−1	2
C			—	−6	0
D				—	3
E					—

A	B
C	
D	E

6.13 In an assembly plant, material handling between departments is performed using a unidirectional closed-loop conveyor. The figure below shows the layout for the modular

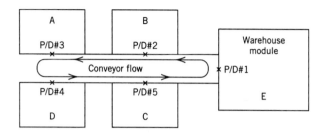

facility, which consists of three equal-sized assembly modules (A, B, and C), one administrative module (D), and one warehouse module (E). P/D points for each module are also shown in the figure. The administrative and warehouse activities are not to be moved; however, assembly areas A, B, C can be relocated. The distance between P/D points and the number of pallet loads moved between departments are given below.

Distance between P/Ds				Pallet flow per day				
From	To	Distance		From/To A	B	C	D	E
P/D1	P/D 2	60 ft.		A	0	0	5	0	30
P/D2	P/D 3	90 ft.		B	10	0	25	0	0
P/D3	P/D 4	30 ft.		C	25	5	0	0	0
P/D4	P/D 5	90 ft.		D	0	0	0	0	0
P/D5	P/D 1	60 ft.		E	5	20	5	0	0

Using the pairwise exchange method, determine new locations for assembly modules A, B, and C that minimizes the sum of the products of pallet flows and conveyor travel distances.

6.14 Four equal-sized machines are served by an automated guided vehicle (AGV) on a linear bidirectional track as shown in the figure below. Each machine block is 30 ft × 30 ft. The product routine information and required production rate are given in the table below. Determine a layout arrangement based on the pairwise exchange method. Assume that the pickup/delivery stations are located at the midpoint of the machine edge along the AGV track.

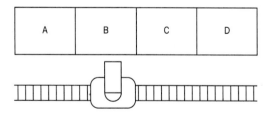

Product	Processing Sequence	Weekly Production
1	B D C A C	300 units
2	B D A C	700 units
3	D B D C A C	900 units
4	A B C A	200 units

6.15 Using the data from Problem 6.14 but assuming that the locations of the P/D points for machines A and B are 5 ft from the lower-right-hand corner of each machine and the P/D points for machines C and D are 10 ft from the lower-left-hand corner of machines C and D, develop an improved layout using the pairwise exchange method.

6.16 A mobile robot is serving two cells located at either sides of the AGV track as shown by the figure below. There are three machines placed in each cell. Given the from-to chart in the table below, find the best machine arrangements for both cells. Rearrangement is limited only to machines within each cell. Assume that the P/D point of each machine is located at the midpoint of the machine edge along the AGV track.

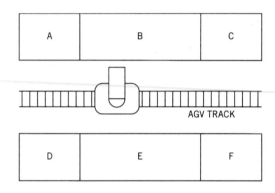

M/C	A	B	C	D	E	F
A	—	10	50	30	0	60
B	5	—	45	40	30	0
C	40	30	—	35	5	20
D	40	25	50	—	40	50
E	0	55	40	50	—	0
F	20	0	60	20	10	—

M/C	Distance	M/C	Distance
A-B	30	D-E	30
A-C	60	D-F	60
B-C	30	E-F	30

6.17 Five machines located in a manufacturing cell are arranged in an "U" configuration as shown in the layout below. The material handling system employed is a bidirectional conveyor system. Determine the best machine arrangement given the product routing information and production rates in the table.

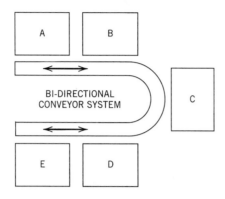

Product	Machine Sequence:	Prod. Rate
1	B-E-A-C	100
2	C-E-D	200
3	B-C-E-A-D	500
4	A-C-E-B	150
5	B-C-A	200

M/C	Distance (ft.)	M/C	Distance (ft.)
A-B	20	B-D	100
A-C	70	B-E	120
A-D	120	C-D	50
A-E	140	C-E	70
B-C	50	D-E	20

6.18 The ABC Cooling and Heating Company manufactures several different types of air conditioners. Five departments are involved in the processing required for the products. A summary of the processing sequences required for the five major products and the weekly production volumes for the products are shown in the table below along with the department area. Based on the graph-based construction method, develop a block layout.

Product:	Process Sequence:	Weekly Production
1	A B C	150
2	A B E D	200
3	A C E	50
4	A C B E	200
5	A D E	250

Department	Area (sq. ft.)
A	1,500
B	1,500
C	1,000
D	2,000
E	2,000

6.19 The activity relationship chart for Walter's machine shop is shown in the figure below. The space requirements are in square feet. Construct the relationship diagram and develop a block layout using the graph-based method.

6.20 The from–to material flow matrix for an eight-department facility is given in the table below. Construct a relationship diagram based on the material flow matrix and construct a block layout using the graph-based method.

Dept.	A	B	C	D	E	F	G	H
A	—	302	0	0	0	66	0	68
B	0	—	504	20	136	154	56	40
C	0	0	—	76	352	0	122	94
D	0	0	0	—	0	0	180	8
E	0	0	0	0	—	122	0	282
F	0	0	0	0	0	—	188	24
G	0	0	0	0	0	0	—	296
H	0	0	0	0	0	0	0	—

Dept.	Area Required (ft^2)
A	2,800
B	2,100
C	2,600
D	400
E	600
F	400
G	2,300
H	1,800

6.21 Consider the layout of five equal-sized departments. The material flow matrix is given in the figure below.
a. Develop the final adjacency graph using the graph-based procedure.
b. Develop a block layout based on the final adjacency graph obtained in part a.

	A	B	C	D	E
A	—	0	5	25	15
B	0	—	20	30	25
C	0	25	—	40	30
D	30	5	20	—	0
E	20	30	5	10	—

6.22 The material flow matrix for ten departments is given below.

	A	B	C	D	E	F	G	H	I	J
A	—	0	12	0	132	16	0	220	20	24
B	0	—	176	0	216	0	144	128	0	0
C	0	0	—	0	0	184	0	0	28	0
D	212	136	240	—	36	0	236	0	164	0
E	0	0	140	0	—	0	192	0	0	160
F	0	180	0	188	108	—	248	228	0	0
G	172	0	156	0	0	0	—	112	224	152
H	0	0	32	40	204	0	0	—	0	0
I	0	168	0	0	104	156	0	148	—	200
J	0	124	196	120	0	116	0	108	0	—

The area requirements are

Dept.	Area (sq. ft.)
A	400
B	1,000
C	2,600
D	400
E	2,400
F	1,000
G	3,600
H	1,200
I	400
J	2,400

    a. Determine a final adjacency graph using the graph-based procedure.

    b. Construct a block layout based on the adjacency graph in part a.

6.23 Suppose the efficiency rating is designated by E and the cost computed by CRAFT is designated by K. Further suppose that four departments (labeled A, B, C, and D) are given for a layout problem. Each department is assumed to be of *equal area,* and each department is represented by a unit square. The unit cost used in computing K is assumed to be equal to 1.0 for all department pairs. (That is, $c_{ij} = 1.0$ for all i, j.) The flow-between the above departments is give as follows:

	A	B	C	D
A	—	10	10	0
B		—	0	4
C			—	4
D				—

    a. Compute the values of E and K for the following layout.

A	C
B	D

    b. Compute the values of E and K for the following layout.

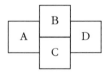

c. Compute the values of E and K for the following layout.

A	B	
	C	D

d. What can you say regarding the **consistency** of the two measures. That is, if a layout is "good" when measured by its E value, would it still be "good" when it is measured by its K value or vice versa? Justify your answer.

6.24 Consider four departments labeled A, B, C, and D. Each department is represented by a 1×1 square. The following data are given:

Initial Layout

A	B
C	D

Flow-Between Matrix

	A	B	C	D
A	—	6	0	3
B		—	5	0
C			—	0
D				—

Unit Cost Matrix

	A	B	C	D
A	—	2	0	3
B	2	—	1	0
C	0	1	—	0
D	3	0	0	—

Location of department A is **fixed.** Answer the following questions using CRAFT with **two-way** exchanges only.

a. List all the department pairs that CRAFT would consider exchanging. (Do **not** compute their associated cost.)

b. Compute the **actual** cost of exchanging departments C and D.

c. Given that department A is fixed and that each department must remain as a 1×1 square, is the layout obtained by exchanging departments C and D optimum? Why or why not? (Hint: examine the properties of the resulting layout and consider the objective function of CRAFT.)

6.25 The following layout is an illegal CRAFT layout. Nevertheless, given that the volume of flow from A to B is 4, A to C is 3, and B to C is 9, and that all move costs are 1, what is the layout cost?

C	C	C	C	C	C
C	B	B	B	B	C
C	B	A	A	B	C
C	B	B	B	B	C
C	C	C	C	C	C

6.26 When CRAFT evaluates the exchange of departments, instead of actually exchanging the departments, it only exchanges the centroids of departments.

a. What is the impact of this method of exchanging if all departments are the same size?

b. Given the following from-to chart and scaled layout (each square is 1 × 1), what does the evaluation of the exchange of departments B and C indicate should be saved over the existing layout and what is actually saved once this exchange is made?

To From	A	B	C
A	—	10	6
B	2	—	7
C			—

From-to chart

A	A	A	C	B	B	B	B
A	A	A	C	B	B	B	B
A	A	A	C	B	B	B	B

Initial layout

6.27 Explain the steps CRAFT would take with the following problem and determine the final layout. Only two-way exchanges are to be considered.

To From	A	B	C	D	E
A	—	3	2	1	
B		—	1	3	
C	1		—	4	
D				—	
E					—

From-to chart

A	A	A	B	B	B
A	A	A	C	C	C
A	A	A	C	C	C
D	D	D	E	E	E
D	D	D	E	E	E

Initial layout

6.28 A manufacturing concern has five departments (labeled A through E) located in a rectangular building as shown below:

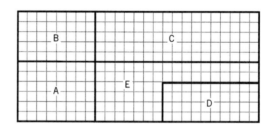

Suppose the flow data, the unit cost data, and the distance matrix are given as follows:

FLOW-BETWEEN MATRIX					
	A	B	C	D	E
A	—	0	5	0	5
B		—	6	2	0
C			—	3	0
D				—	7
E					—

UNIT COST ($/UNIT DIST.)					
	A	B	C	D	E
A	—	0	1	0	1
B		—	1	4	0
C			—	3	0
D				—	1
E					—

DISTANCE MATRIX					
	A	B	C	D	E
A	—	6	20	18	11
B		—	12	22	15
C			—	10	8
D				—	7
E					—

a. Using the CRAFT two-way exchange procedure, indicate all the department pairs CRAFT would consider exchanging in the above layout.

b. Compute the *estimated* cost of exchanging departments A and E.

6.29 Suppose the following layout is provided as the initial layout to **CRAFT.** The flow-between matrix and the distance matrix are given as follows. (All the c_{ij} values are equal to 1.0)

Flow-Between Matrix					
	A	B	C	D	E	F
A	—	0	8	0	4	0
B		—	0	5	0	2
C			—	0	1	0
D				—	6	0
E					—	4
F						—

Distance Matrix					
	A	B	C	D	E	F
A	—	30	25	55	50	80
B		—	45	25	60	50
C			—	30	25	55
D				—	45	25
E					—	30
F						—

a. Given the above data and initial layout, which department pairs will **not** be considered for exchange.

b. Compute the cost of the initial layout.

c. Compute the *estimated* layout cost assuming that departments E and F are exchanged.

d. In general, when the same data and initial layout are supplied to CRAFT and MULTIPLE, why would one expect MULTIPLE to often (but not always) outperform CRAFT? Also, what type of data and/or initial layout would allow CRAFT to consistently generate layout costs that are comparable to those obtained from MULTIPLE?

6.30 Answer the following questions for CRAFT.

a. State two principal weaknesses and two principal strengths of CRAFT.

b. CRAFT uses the estimated cost in evaluating the potential impact of exchanging two or three department locations. Suppose we modify the original computer code to obtain a new code, say, NEWCRAFT, where we *always* use the *actual cost* in evaluating the potential impact of exchanging two or three departments. Assuming that we start with the *same initial layout* and that we use the *same exchange option* (such as two-way exchanges only, or any other exchange option), the cost of the final layout obtained from NEWCRAFT will *not necessarily be less than* the cost of the final layout obtained from CRAFT. True or false? Why?

6.31 Using BLOCPLAN's procedure, convert the following from-to chart to a relationship chart.

TO FROM	A	B	C	D	E	F	G	H
A	—	8		3		6		
B	1	—			5			
C			—				4	
D		9		—			18	
E			4	1	—			
F	4			4		—		
G				2			—	20
H				7				—

6.32 Consider BLOCPLAN. Suppose the following REL-chart and layout are given for a five-department problem. (It is assumed that each grid in the layout represents a unit square.) Further suppose that the following scoring vector is being used:

A = 10, E = 5, I = 2, O = 1, U = 0, and X = −10.

	1	2	3	4	5
1	—	A	U	E	U
2		—	U	U	I
3			—	U	X
4				—	A
5					—

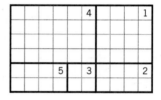

a. Compute the "efficiency rating."
b. Compute the REL-DIST score.
c. In improving a layout, BLOCPLAN can exchange only those departments that are either adjacent or equal in area. True or false? Why? (If true, then explain why there is such a limitation. If false, then explain how two departments that do not meet the above constraint are exchanged.)
d. In constructing and improving a layout, BLOCPLAN maintains rectangular department shapes. Discuss the advantages and limitations of maintaining such department shapes in facility layout.

6.33 Use the MIP model to obtain an optimal layout with the data given for Problem 6.28. (You may assume that each grid measures 20' × 20'.)

6.34 Use the MIP model to obtain an optimal layout with the data given for Problem 6.29. (You may assume that each grid measures 10' × 10'.)

6.35 Consider the layout shown in Figure 6.26*d*. Use LOGIC and the cut-tree shown in Figure 6.27 to exchange departments B and F.

6.36 Re-solve Problem 6.35 by exchanging departments D and H in the layout shown in Figure 6.26*d*.

6.37 Re-solve Problem 6.35 by exchanging departments G and H in the layout shown in Figure 6.29.

6.38 Answer the following questions for **MULTIPLE.**
a. Consider four departments (labeled A through D) with the following area requirements: A = 7 grids, B = 3 grids, C = 4 grids, and D = 2 grids. Using the spacefilling curve shown below, show the layout that would be obtained from the sequence A-B-C-D.

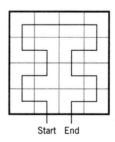

Start   End

    b. Using the data given for part a and the spacefilling curve shown above, show the layout that would be obtained by exchanging departments B and D.

    c. Discuss the advantages and limitations of using spacefilling curves in the manner they have been used in MULTIPLE. Your discussion should include the treatment of fixed departments, dummy departments, unusable floor space (i.e., obstacles), and extra (i.e., empty) floor space.

6.39 Consider the initial layout and flow/cost data given for Problem 6.28.

    a. Using the following conforming curve and MULTIPLE, improve the initial layout via two-way department exchanges.

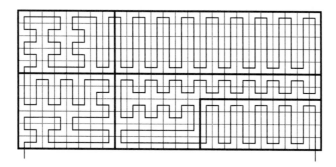

    b. In general, what is the disadvantage of using conforming curves?

6.40 Consider the initial layout and flow data given for Problem 6.29. Use MULTIPLE and the following conforming curve to obtain a two-opt layout.

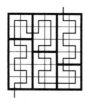

Part Three

FACILITY DESIGN FOR VARIOUS FACILITIES FUNCTIONS

7

WAREHOUSE
OPERATIONS

7.1 INTRODUCTION

At the mercy of myriad business, logistics, and government initiatives—including just-in-time (JIT) production, quick response, efficient consumer response, continuous flow distribution, enhanced customer satisfaction, operator safety, and environmental protection—warehouse operations have been and are continuously being revolutionized.

With the passing of time come more complex and complicated problems. Supply chains are shorter and, hopefully, more integrated, the world is smaller, customers are more demanding, and technology changes occur rapidly.

In addition to the complexity of these problem statements, consider the breadth of the following problems, which further impact the challenge of exceeding client expectations:

- Company A has had an explosion of stock-keeping units (SKUs), resulting in a major shortage of warehouse space. A disagreement exists as to whether this problem should be solved by the manufacturing group's producing in smaller lot sizes or by the warehousing group's adding square footage.

- Company B has a capacity problem on a new, hot-selling item. A disagreement exists as to whether this problem should be resolved by the manufacturing group's adding capacity, by the quality group's increasing yield, or by the maintenance group's increasing uptime.

- Company C has a customer satisfaction problem. A disagreement exists as to whether the problem should be resolved by the logistics group's reconfiguring the distribution network, by the information technology group's installing a new *warehouse management system* (WMS), or by the organizational excellence group's implementing a continuous improvement process.

A renewed emphasis on the supply chain and *customer satisfaction* has increased the number and variety of value-added services in the warehouse. The extra services may include kitting, special packaging, or label application. For example, a large fine-paper distributor counts and packages individual sheets of paper for overnight shipment. A large discount retailer requires vendors to provide slipsheets between each layer of cases on a pallet to facilitate internal distribution.

Increased emphasis on customer satisfaction and evolving patterns of consumer demand in the United States have increased the number of unique items in a typical warehouse or distribution center (DC)—the result is *SKU proliferation.* Each item stored is assigned a numeric identifier, or SKU, associated with its unique qualities, such as size, color, or packaging. SKU proliferation is perhaps best illustrated by the beverage industry. Not many years ago, the beverage aisle in a typical grocery store was populated with two or three flavors of 12-ounce bottles in six packs. Today, the typical beverage aisle is populated with colas (regular and diet, with caffeine, and caffeine-free), clear drinks, water, and fruit-flavored drinks in 6-, 12-, and 24-pack plastic bottles and cans, and 1-, 2-, and 3-liter bottles.

Finally, increased concern about the preservation of the environment, the conservation of natural resources, and human safety has brought more stringent government regulations into the design and management of warehousing operations.

The traditional response to increasing demands is to acquire additional resources. In the warehouse, those resources include people, equipment, and space. Unfortunately, those resources have been difficult to obtain and maintain. During the recent economic expansion, finding and maintaining qualified labor was difficult and costly. As the economy slows, this will become easier for a period of time, but the next expansion will likely be more challenging, as more demand for service and responsiveness will be placed on operations. In addition, we will have to adjust to a workforce characterized by advancing age, minority and non-English-speaking demographics, and declining technical skills. New standards for workforce safety and composition through OSHA's lifting standards and the Americans With Disabilities Act also makes it difficult to rely on an increased workforce as a way to address the increased demands on warehousing operations.

When labor is not the answer, we typically turn to mechanization and automation to address increasing demands. Unfortunately, our history of applying technology as a substitute for labor in warehousing operations has not been distinguished. In many cases, we have over-relied on technology as a substitute for labor. We must balance the appropriate levels of technology and systems to make sure the logical labor savings today do not interfere with future business requirements.

In the face of rapidly increasing demands on warehouse operations and without a reliable pool of additional resources to turn to, the planning and management of today's warehousing operations are very difficult. To cope, we must turn to simplification and process improvement as a means of managing warehouses and DCs. Toward that end, this chapter is meant to serve as a guide for warehouse operations improvement through the application of best-practice procedures and available material handling systems for warehousing operations. We begin with an introduction to the missions of the warehouse. We then turn our attention to individual functions and activities within the warehouse. Then, each function is described in detail and best-practice principles and systems for executing each function are defined. One way to reengineer warehousing operations is to justify each warehouse function and handling step relative to the mission of the supply chain. If a function is not clearly serving the supply chain mission, it should be eliminated. Likewise, one or more functions may need to be added to bring the warehouse operations more closely in line with the mission of the supply chain. For example, to improve response time within the warehouse, a cross-docking function may need to be included.

7.2 MISSIONS OF A WAREHOUSE

The warehouse plays a critical role in supporting a company's supply chain success. The mission of a warehouse is to effectively ship product in any configuration to the next step in the supply chain without damaging or altering the product's basic form. Numerous steps must be accomplished and hence there are key warehousing opportunities to address. Doing that will optimize the methods used to achieve the mission.

If the warehouse cannot process orders quickly, effectively, and accurately, then a company's supply chain optimization efforts will suffer. Information technology and physical distribution play a significant role in making warehousing operations more effective, but the best information system will be of little use if the physical systems necessary to get the products out the door are constraining, misapplied, or outdated. All warehousing opportunities, including order picking, cross-docking, productivity, space utilization, and value-added services, allow the warehouse to process and ship orders more effectively. In more detail, these opportunities are:

1. *Improving order picking operations.* Traditionally, order picking is the operation where a company spends or misspends most of its time and money to improve productivity. Successful order picking is critical to a warehouse's success, and supply chain requirements today are driving warehousing operations to develop better order picking solutions.

2. *Utilizing cross-docking.* Cross-docking can occur at the manufacturer, distributor, retailer, and transportation carrier levels. Each participant has different requirements, depending on whether they are shipping the goods to be cross-docked or preparing to receive cross-docked goods. The receiver typically requests that the cross-docked goods be sorted and prelabeled. The shipper, to meet these requirements, must perform a more detailed picking process. For example, if 100 items are ordered, the warehouse must pick the 100 items and also separate those items for the different store orders.

3. *Increasing productivity.* In the past, productivity has meant "to do it faster with fewer people." The first objective of warehousing has always been to maximize the effective use of space, equipment, and labor. This objective implies that productivity is not just labor performance but also includes space and equipment and a combination of factors that all contribute to increased productivity.

4. *Utilizing space.* The old rule of thumb has always been that when a warehouse is more than 80% full, more space is needed. This rule is based on the fact that when a warehouse reaches this capacity, it takes longer to put something away. As the time to find a storage location increases, the proper slotting of product starts to disappear. Slow-moving items are stored in fast-moving locations, so then fast-moving items must be stored in slow-moving locations. The end result

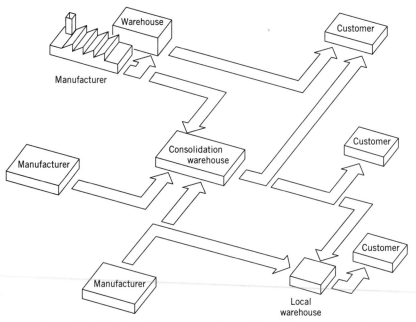

Figure 7.1 Warehousing opportunities within a logistics network.

is a decline in productivity and an increase in damage and mispicks, all due to poor space utilization.

5. *Increasing value-added services.* Warehouses are no longer just picking and shipping locations. Their role has extended to include services that facilitate more efficient operations in the receiving warehouse and therefore benefit the customer. Whether it is presorting and prelabeling goods for eventual cross-docking or the actual customization of the outbound product, customers' demands are becoming more strenuous.

Any one of the above warehousing opportunities or a combination of them can be found within most warehouses today. The old definition of a warehouse as a place to store, reconfigure, and shorten lead times has become much more complex and technology driven. (See Figure 7.1.)

7.3   FUNCTIONS IN THE WAREHOUSE

Although it is easy to think of a warehouse as being dominated by product storage, there are many activities that occur as part of the process of getting material into and out of the warehouse. The following list includes the activities found in most warehouses. These tasks, or functions, are also indicated on a flow line in Figure 7.2 to make it easier to visualize them in actual operation.

The warehouse functions—roughly in the order in which they are performed—are defined briefly as follows:

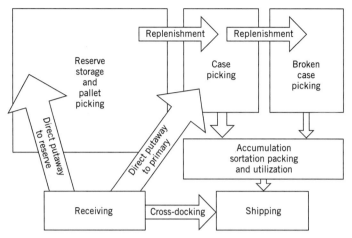

Figure 7.2 Typical warehouse functions and flows.

1. *Receiving* is the collection of activities involved in (a) the orderly receipt of all materials coming into the warehouse, (b) assuring that the quantity and quality of such materials are as ordered, and (c) disbursing materials to storage or to other organizational functions that need them.

2. *Inspection and quality control* are an extension of the receiving process and are done when suppliers are inconsistent in quality or the product being purchased is heavily regulated and must be inspected at all steps in the process. Inspections may be as simple as a visual check or as complex as lab testing.

3. *Repackaging* is performed in a warehouse when products are received in bulk from a supplier and subsequently packaged singly, in merchandisable quantities or in combinations with other parts to form kits or assortments. An entire receipt of merchandise may be processed at once, or a portion may be held in bulk form to be processed later. The latter may be done when packaging greatly increases the storagecube requirements or when a part is common to several kits or assortments. Relabeling is done when product is received without markings that are readable by systems or humans for identification purposes.

4. *Putaway* is the act of placing merchandise in storage. It includes material handling and placement.

5. *Storage* is the physical containment of merchandise while it is awaiting a demand. The form of storage will depend on the size and quantity of the items in inventory and the handling characteristics of the product or its container.

6. *Order picking* is the process of removing items from storage to meet a specific demand. It represents the basic service that the warehouse provides for the customer and is the function around which most warehouse designs are based.

7. *Postponement* may be done as an optional step after the picking process. As in the repackaging function, individual items or assortments are boxed for more convenient use. Waiting until after picking to perform these functions has the advantage of providing more flexibility in the use of on-hand inventory. Individual items are available for use in any of the packaging configurations right up to the time of need. Pricing is current at the time of sale. Prepricing at

manufacture or at receipt into the warehouse inevitably leads to some repricing activity as price lists are changed while merchandise sits in inventory. Picking tickets and price stickers are sometimes combined into a single document.

8. *Sortation* of batch picks into individual orders and *accumulation* of distributed picks into orders must be done when an order has more than one item and the accumulation is not done as the picks are made.

9. *Packing and shipping* may include the following tasks:
   - Checking orders for completeness
   - Packaging merchandise in an appropriate shipping container
   - Preparing shipping documents, including packing list, address label, and bill of lading
   - Weighing orders to determine shipping charges
   - Accumulating orders by outbound carrier
   - Loading trucks (in many instances, this is a carrier's responsibility)

10. *Cross-docking* inbound receipts from the receiving dock directly to the shipping dock.

11. *Replenishing* primary picking locations from reserve storage locations.

In this chapter, *receiving* includes those activities described above as receiving, prepackaging, and putaway; *order picking* includes those activities described above as order picking, packaging, and sortation/accumulation; and *shipping* includes those activities described as packing and shipping. The activities, best-practice operating principles, and space planning methodologies for each of the functional areas are described in subsequent sections.

7.4 RECEIVING AND SHIPPING OPERATIONS

Problems can occur in planning receiving and shipping facilities if the carriers that interface with receiving and shipping activities are not properly considered. Positioning of the carriers and their characteristics are important to receiving and shipping operations. It is useful to think of the carriers that interface with the receiving and shipping functions as a portion of the receiving and shipping facility. Hence, all carrier activities on the site are included in receiving and shipping facility planning. Receiving and shipping functions will be defined to begin and end when carriers cross the property line.

The activities required to receive goods include:

- Inbound trucker phones the warehouse to get a delivery appointment and provides information about the cargo.
- Warehouse receiving person verifies the *advance shipping notice* (ASN) and confirms it with information received by phone from inbound trucker.
- Trucker arrives and is assigned to a specific receiving door (similar dock location is selected for boxcar receipts).
- Vehicle is safely secured at the dock.
- Seal is inspected and broken in presence of carrier representative.

- Load is inspected and either accepted or refused.
- Unitized merchandise is unloaded.
- Floor-loaded or loose merchandise is unloaded.
- All unloaded material is staged for count and final inspection.
- Proper disposal is made of carrier damage.
- Load is stored in an assigned location.

The facility requirements to perform these receiving activities include:

- Sufficient area to stage and spot carriers
- Dock levelers and locks to facilitate carrier unloading
- Sufficient staging area to palletize or containerize goods
- Sufficient area to place goods prior to dispatching
- A host information system for ASN/EDI on purchase orders to allow for report generation

The activities required to ship goods include:

- Accumulate and pack the order.
- Stage and check the order.
- Reconcile shipping release and customer order.
- Spot and secure the carrier at the dock.
- Position and secure dock levelers and locks.
- Load the carrier.
- Dispatch the carrier.

The facility requirements to perform these shipping activities include:

- Sufficient area to stage orders
- An in-house host information system for shipping releases and customer orders
- Sufficient area to stage and spot carriers
- Dock levelers to facilitate carrier loading

Some desirable attributes of receiving and shipping facilities plans include:

- Directed flow paths among carriers, buffer or staging areas, and storage areas
- A continuous flow without excessive congestion or idleness
- A concentrated area of operation that minimizes material handling and increases the effectiveness of supervision
- Efficient material handling
- Safe operation
- Minimizing damage
- Good housekeeping

Requirements for people, equipment, and space in receiving and shipping depend on the effectiveness of programs to incorporate prereceiving and postshipping considerations. For example, by working with vendors and suppliers, peak loads at receiving can be reduced. Scheduling inbound shipments is one method of reducing the impact of randomness on the receiving workload.

Another reason for being concerned with prereceiving activities is the opportunity to influence the unit load configurations of inbound material. If, for example, cases are hand-stacked in the trailer by the vendor, they will probably have to be unloaded by hand. Likewise, if either slipsheets or clamp trucks were used by the vendor to load the carrier and receiving does not have similar material handling equipment, the shipment probably will have to be unloaded by hand. Finally, if materials are received in unit loads not compatible with the material handling system, then additional loading, unloading, or both might be necessary.

A third reason for trying to influence prereceiving activities is to provide a smooth interface between the vendor and receiver's information systems. Where automatic identification systems are in use in receiving, some firms supply their vendors with the appropriate labels to be placed on inbound material to facilitate the receiving activity. The faster and more accurately receiving is performed, the more both vendor and buyer benefit. Information is provided to accounting and payments are made sooner and materials are available for use sooner, rather than sitting idle.

Just as the receiver wishes to influence the vendor, the customer wishes to influence the shipper. Hence, post-shipping activities must be considered. In addition to the reasons cited for considering prereceiving activities, the following are issues in postshipping: returnable containers, returned goods, returning carriers, and shipping schedules.

If goods are shipped to customers in returnable containers, a system must be developed to keep track of the containers and to ensure they are returned. Additionally, whether or not the shipping container or support is returnable, there will be a natural attrition for which replacements must be planned.

Goods are returned because the goods failed to meet the customer's quality specifications, mistakes were made in the type and amount of material shipped, or the customer just simply decided not to accept the material. Regardless of the reason, returned goods must be handled and an appropriate system for handling the material must be designed.

If supplier-owned equipment is used to deliver materials to customers, consideration should be given to utilizing the carrier's capacity on the return trip. The "backhaul" of the carrier could be used for returning the returnable containers or for other transportation purposes.

Schedules can have a significant impact on the resource requirements for shipping. Hence, close coordination is required between the shipper and the shipping department. The carrier might be the customer's carrier, the shipper's carrier, a contract hauler, or a commercial carrier. If shipping activities are to be planned, then shipping schedules must be accurate and reliable.

Shipping systems have taken on an increasingly important role in the operation of the supply chain. Customer initiatives such as just-in-time (JIT) and Efficient Consumer Response (ECR) have resulted in expanded responsibilities for the warehouse/traffic manager. No longer is it acceptable to simply ensure that the product is shipped on time; now the warehouse/traffic manager often assumes responsibility for when the product arrives at the customer's location. Fueled primarily by government deregulation, customer requirements have increased, changes in shipping modes have occurred, use of next-day and second-day delivery services has increased, and business has become global.

In addition to the need for closely coordinating vendor and receiving activities and shipping and customer activities, it is equally important to coordinate the activities of receiving and production, production and shipping, and receiving and shipping.

The natural sequence for the flow of materials is: vendor, receiving, storage, production, warehousing, shipping, customer. However, in some cases, materials might go directly from receiving to production and from production to shipping. Hence, such possibilities should be included in the system design.

Why should receiving and shipping be coordinated? Common space, equipment, and/or personnel might be used to perform receiving and shipping. Additionally, when slave pallets are used in a manufacturing or warehousing activity, empty pallets will accumulate at shipping and must be returned to the loading point at either receiving or production.

A key decision in designing the receiving and shipping functions is whether to centralize the two functions. As depicted in Figure 7.3, the location of receiving and shipping depends on access to transportation facilities.

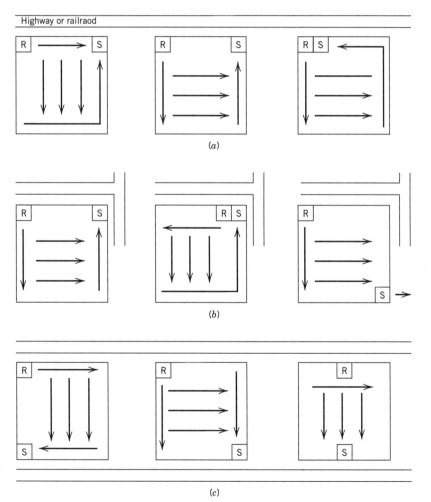

Figure 7.3 Possible arrangements of shipping and receiving areas. (*a*) Transportation facilities on one side of building. (*b*) Transportation facilities on two adjacent sides of building. (*c*) Transportation facilities on two opposite sides of building. (*With permission from Apple [5]*)

The decision to centralize receiving and shipping depends on many factors, including the nature of the activity being performed. For example, if receiving is restricted to the morning and shipping is restricted to the afternoon, then it would be appropriate to utilize the same docks, personnel, material handling equipment, and staging space for both. Alternately, if they occur simultaneously, closer supervision is required to ensure that received goods and goods to be shipped are not mixed.

The opportunity to centralize receiving and shipping should be investigated carefully. The pros and cons of centralization should be enumerated and considered before making a decision.

Receiving and Shipping Principles

Receiving Principles

The following principles serve as guidelines for streamlining receiving operations. They are intended to simplify the flow of material through the receiving process and to insure the minimum work is required. In order, they are:

1. *Don't receive.* For some materials, the best receiving is no receiving. Often, drop shipping—having the vendor ship to the customer directly—can save the time and labor associated with receiving and shipping. Large, bulky items lend themselves to drop shipping. An example is a large camp and sportswear mail-order distributor drop-shipping canoes and large tents.

2. *Prereceive.* The rationale for staging at the receiving dock, the most time and space-intensive activity in the receiving function, is often the need to hold the material for location assignment, product identification, and so on. With today's electronic capabilities, it is relatively easy to obtain a detailed manifest with every receipt of merchandise. The popular terminology is ASN (advance shipping notice). A growing number of warehouses have established a policy that no receipt will be unloaded without an ASN. This document represents the best way to be sure that the correct product has been delivered to the correct place. There is little excuse for surprises with today's capabilities for information transmission. The ASN should include seal numbers to verify that the load was not tampered with.

3. *Cross-dock "cross-dockable" material.* Since the ultimate objective of the receiving activity is to prepare material for the shipment of orders, the fastest, most productive receiving process is cross-docking—and the simplest kind of cross-docking activity is one in which an entire inbound load is sorted and then reloaded onto one or more outbound vehicles. In some cases, the sorting was done previously, so the amount of additional handling by the warehouse is minimized. Sometimes material from more than one inbound vehicle must be assembled to complete outbound loads. In other cases, cross-docking involves the blending of material on an inbound vehicle with material that is already in the warehouse.

4. *Put away directly to primary or reserve locations.* When material cannot be cross-docked, material handling steps can be minimized by bypassing receiving stages and putting material away directly to primary picking locations, essentially

replenishing those locations from receiving. When there are no severe constraints on product rotation, this may be feasible. Otherwise, material should be directly put away to reserve locations. In direct putaway systems, the staging and inspection activities are eliminated. Hence, the time, space, and labor associated with those operations are eliminated. In either case, vehicles that serve the dual purpose of truck unloading and product putaway facilitate direct putaway. For example, counterbalanced lift trucks can be equipped with scales, cubing devices, and on-line RF terminals to streamline the unloading and putaway function. The most advanced logistics operations are characterized by automated, direct putaway to storage. The material handling technologies that facilitate direct putaway include roller-bed trailers and extendable conveyors. In addition to prereceiving, prequalifying vendors helps eliminate the need for receiving staging.

5. *Stage in storage locations.* If material has to be staged, the floor space required for staging can be minimized by providing storage locations for receiving staging. Often, storage locations may be live storage locations, with locations blocked until the unit is officially received. Storage spaces over dock doors may be a good option.

6. *Complete all necessary steps for efficient load decomposition and movement at receiving.* The most time we will ever have available to prepare a product for shipping is at receiving. Once the demand for the product has been received, there is precious little time available for any preparation prior to shipment. Hence, any material processing that can be accomplished ahead of time should be accomplished. Those activities include:

   a. *Prepackage in issue increments.* At a large office supplies distributor, quarter- and half-pallet loads are built at receiving in anticipation of orders being received in those quantities. Customers are encouraged to order in those quantities by quantity discounts. A large distributor of automotive aftermarket parts conducted an extensive analysis of likely order quantities. Based on that analysis, the company is now prepackaging in those popular increments.

   b. *Apply necessary labeling and tags.*

   c. *Cube and weigh for storage and transport planning.*

7. *Sort inbound materials for efficient putaway.* Just as zone picking and location sequencing are effective strategies for improving order picking productivity, inbound materials can be sorted for putaway by warehouse zone and by location sequence. At the U.S. Defense Logistics Agency's Integrated Materiel Complex, receipts are staged in a rotary rack carousel and released and sorted by warehouse aisle and location within the aisle to streamline the putaway activity. Automated guided vehicles take the accumulated loads to the correct aisle for putaway.

8. *Combine putaways and retrievals when possible.* To further streamline the putaway and retrieval process, putaway and retrieval transactions can be combined in a dual command to reduce the amount of empty travel on industrial vehicles. This technique is especially geared for pallet storage-and-retrieval operations. Again, counterbalance lift trucks, which can unload, putaway, retrieve, and load, are a flexible means for executing dual commands.

9. *Balance the use of resources at receiving by scheduling carriers and shifting time-consuming receipts to off-peak hours.* Through computer-to-computer EDI links and Internet communication, companies have improved access to schedule information about inbound and outbound loads. This information can be used to schedule receipts and to provide ASN information.

10. *Minimize or eliminate walking by flowing inbound material past workstations.* An effective strategy for enhancing order picking productivity, especially when a variety of tasks must be performed on the retrieved material (e.g., packaging, counting, labeling), is to bring the stock to an order picking station equipped with the aids and information to perform the necessary tasks. The same strategy should be employed for receipts that, by their nature, require special handling. At the flagship distribution center of a large retailer, receipts flow out of inbound trailers on a powered roller conveyor and past a stationary scanning station equipped with an in-motion scale. At the scanner stations, inbound cases are weighed, cubed, and tagged with a bar code label, using a print-and-apply device, describing all necessary product and warehouse location information. In addition, if the material is needed for an outbound order, a conveyor diverts the inbound case down a separate line for cross-docking.

Shipping Principles

Many of the best-practice receiving principles also apply in reverse to shipping, including direct loading (the reverse of direct unloading), advanced shipping notice preparation (prereceiving), and staging in racks. In addition to those principles, best-practice principles for unitizing and securing loads, automated loading, and dock management are introduced.

1. **Select cost- and space-effective handling units.**
   a. *For loose cases.* The options for unitizing loose cases include wood (disposable, returnable, and rentable) pallets, plastic, metal, and "nestable" pallets. The advantage of plastic pallets over wood pallets includes durability, cleanliness, and color-coding. The Japanese make excellent use of colored plastic pallets and totes in creating appealing work environments in factories and warehouses. Metal pallets are primarily designed for durability and weight capacity. Nestable pallets offer good space utilization during pallet storage and return, but do not have a high weight capacity. Other options for unitizing loose cases include slipsheets and roll carts. Slipsheets improve space utilization in storage systems and trailers, but require special lift truck attachments. Roll carts facilitate the containerization of multiple items in case quantities and facilitate material handling throughout the shipping process from order picking to packaging to checking to trailer loading. The selection factors for unitizing loose cases include initial purchase cost, maintenance costs and requirements, ease of handling, environmental impact, durability, and product protection.

b. *For loose items.* The options for unitizing loose items include totes (nestable and collapsible) and cardboard containers. As was the case with unitizing loose cases, the selection factors include impact on the environment, initial purchase cost, life cycle cost, cleanliness, and product protection. An excellent review and comparison of carton and tote performance is provided in [36].

2. **Minimize product damage.**

   a. *Unitize and secure loose items in cartons or totes.* In addition to providing a unit load to facilitate material handling, a means must be provided to secure material within the unit load. For loose items in totes or cartons those include foam, "peanuts," "popcorn," bubble wrap, newsprint, and air packs. The selection factors include initial and life-cycle cost, environmental impact, product protection, and reusability. An excellent review of alternative packaging methods is provided in [36].

   b. *Unitize and secure loose cases on pallets.* Though the most popular alternative is stretch wrapping, wrapping loose cases with Velcro belts and adhesive tacking are gaining in popularity as environmentally safe means for securing loose cases on pallets.

   c. *Unitize and secure loose pallets in outbound trailers.* The most common methods are foam pads and plywood.

3. **Eliminate shipping staging, and direct-load outbound trailers.** As was the case in receiving, the most space- and labor-intensive activity in shipping is staging. To facilitate the direct loading of pallets onto outbound trailers, pallet jacks and counterbalance lift trucks can serve as picking and loading vehicles, allowing a bypassing of staging. To go one step further, automating pallet loading can be accomplished with pallet conveyor interfacing with specially designed trailer beds to allow pallets to be automatically conveyed onto outbound trailers by automated fork trucks and/or automated guided vehicles. Direct, automated loading of loose cases is facilitated with extendable conveyors.

4. **Use storage racks to minimize floor space requirements for shipping staging.** If shipping staging is required, staging in storage racks can minimize the floor space requirements. A large automotive aftermarket supplier places racks along the shipping wall and above dock doors to achieve this objective.

5. **Route on-site drivers through the site and minimize paperwork and time.** A variety of systems are now in place to improve the management of shipping and receiving docks and trailer drivers. At one brewery, drivers use a smart card to gain access throughout the DC site, to expedite on-site processing, and to ensure shipping accuracy. At another brewery, terminal stands are provided throughout the site to allow drivers on-line access to load status and dock schedules.

6. **Use small-parcel shipping.** The design of the shipping and staging area for small-parcel shipments will look decidedly different from the one dedicated to unit loads. Stretch/shrinkwrap stations may be replaced by packing lines. Conveyors often replace the use of a lift truck. The carrier's bar-coded tracking labels are customarily affixed in the warehouse before the goods are shipped.

Receiving and Shipping Space Planning

The steps required to determine the total space requirements for receiving and shipping areas are:

- Determine what is to be received and shipped.
- Determine the number and type of docks.
- Determine the space requirements for the receiving and shipping area within the facility.

1. Determine What Is to Be Received and Shipped.

The first seven columns of the receiving and shipping analysis chart in Figure 7.4 include information on the *what, how much,* and *when* of items received or shipped. For an existing receiving or shipping operation or one that is to have similar objectives to an existing operation, this information may be obtained from past receiving reports or shipping releases. For a new receiving or shipping operation, the parts lists and the market analysis information for all products must be analyzed to determine reasonable unit loads and order quantities. Once this information is obtained, the first seven columns of the receiving and shipping analysis chart may be completed.

The eighth and ninth columns of the receiving and shipping analysis chart identify the types of carriers used for receiving and shipping materials. At a minimum, the type of carrier, the overall length, width, and height, and the height to the carrier's dock should be specified.

2. Determine the Number and Type of Docks.

Waiting-line analysis may be used to determine the number of docks that will provide the required service if arrivals and/or services are Poisson-distributed and the arrival and service distributions do not vary significantly over time. If arrival and service distributions vary with the time of day, day of week, or number of trucks waiting at the dock, simulation may be used.

The last two columns on the chart concern the handling of materials on and off a carrier. If the chart is being completed for an existing receiving or shipping area, the current methods of unloading or loading materials should be evaluated and, if acceptable, recorded on the chart. If the chart is completed for a new operation, methods of handling materials should be determined by following the procedure (described in Chapter 5). The time required to unload or load a carrier may be determined for an existing operation from historical data, work sampling, or time study. The accuracy of the answers and the effort required to determine time standards using these approaches typically result in the facility planner's using predetermined time elements. Predetermined time elements are used for new receiving or shipping operations.

Predetermined time elements vary from very general (macro) standards to very detailed (micro) standards. Probably the most general standard is one in which an individual utilizing the proper equipment can unload or load 7500 lb/hr. Much more detailed standards have been compiled by the U.S. Department of Agriculture (USDA) [27]. An application of the standard developed via USDA standards is given

Company ________ Date ________ Raw Materials ________ Finished Goods ________
Prepared by ________ Sheet ________ of ________ Plant Supplies ________

1	2	3	4	5	6	7	8	9	10	11
	UNIT LOADS				Size of Shipment (unit loads)	Frequency of Shipment	TRANSPORTATION		MATERIAL HANDLING	
Description	Type	Capacity	Size	Weight			Mode	Specifications	Method	Time

Figure 7.4 Receiving and shipping analysis chart.

TIME STANDARD WORK SHEET

Company BCD Prepared by JA

Process Unload cartons, palletize, and store

Sarting Point Spot carrier Ending point Close truck door dispatch truck

Date_____Sheet 1 of 1

Setp	Descripktion	Crew size	USDA Reference Table Number	Productive Labor (Hours)
1	Spot carrier.	1	XII	.1666
2	Open truck door	1	XII	.0163
3	Remove bracing.	1	XII	.1101
4	Get bridge plate.	1	XII	.0310
5	Position bridge plate	1	XII	.0168
6	Place empty pallet at truck tailgate. Repeat 40 times.	1	XX	.2333
7	Unload cartons onto pallet. REpeat 640 times.	1	XIV A	1.5360
8	Pick up loaded pallet and move out of truck. Repeat 40 times.	1	VI	.3800
9	Transport pallet to storage area (100 ft) and return to truck. Repeat 40 times.	1	XI	.6040
10	Store paller. Repeat 40 times.	1	VI	.1760
11	Remove bridge plate.	1	XII	.0048
12	Close truck door and dispatch truck	1	XII	.0068
	Total			3.2817 hr

Figure 7.5 Example of use of U.S. Department of Agriculture (USDA) predetermined time elements to determine truck unloading standard time.

in Figure 7.5. Even more detailed are standards developed by material handling equipment manufacturers for specific types of equipment.

Once all the information for the receiving and shipping analysis chart has been obtained for all materials to be received or shipped, the receiving or shipping function is fully defined. Although considerable effort is required to obtain the information, it must be obtained if receiving and shipping facilities are to be properly planned.

Once the number of docks is determined, the dock configuration must be designed. The first consideration in designing the proper dock configuration is the flow of carriers about the facility. For rail docks, location and configuration of the railroad spur dictate the flow of railroad cars and the configuration of the rail dock. For truck docks, truck traffic patterns must be analyzed. Truck access to the property should be planned so that trucks need not back onto the property. Recessed or "Y" approaches, as shown in Figure 7.6, should be utilized if trucks turn into the facility from a narrow street. Other truck guidelines to be taken into consideration are:

1. Two-directional service roads should be at least 24 ft wide.

2. One-way service roads should be at least 12 ft wide.

3. If pedestrian travel is to be along service roads, a 4-ft-wide walk physically separated from the service road should be included.

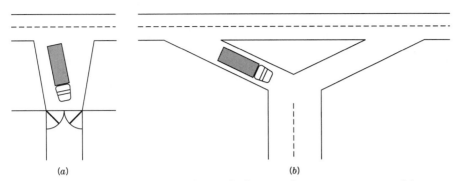

Figure 7.6 Recessed and "Y" approaches to facilities truck access to property. (*a*) Recessed truck entrance. (*b*) "Y" truck entrance.

4. Gate openings for two-directional travel should be at least 28 ft wide.
5. Gate openings for one-way travel should be at least 16 ft wide.
6. Gate openings should be 6 ft wider if pedestrians will also use the gate.
7. All right-angle intersections must have a minimum of a 50-ft radius.
8. If possible, all traffic should circulate counterclockwise because left turns are easier and safer to make than right turns (given that steering is on the left of the vehicle).
9. Truck waiting areas should be allocated adjacent to the dock apron and need to be big enough to hold the maximum expected number of trucks waiting at any given time.

After considering the guidelines given above, the overall flow of trucks about a facility may be determined. Care must be taken to ensure that adequate space exists for 90° docks. Space requirements for 90° docks are illustrated in Figure 7.7 and given in Table 7.1. If adequate apron depth does not exist for a 90° dock, a finger dock

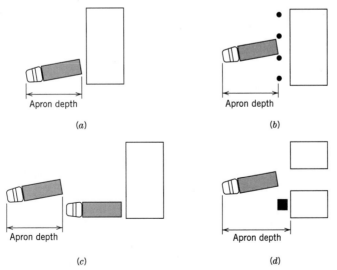

Figure 7.7 Apron depth definition of 90° docks. (*a*) Unobstructed dock. (*b*) Post-supported canopy. (*c*) Alongside other trucks. (*d*) Driveways and stalls.

Table 7.1   *Space Requirements for 90° Docks*

Truck Length (feet)	Dock width (feet)	Apron Depth (feet)
40	10	46
	12	43
	14	39
45	10	52
	12	49
	14	46
50	10	60
	12	57
	14	54
55	10	65
	12	63
	14	58
60	10	72
	12	63
	14	60

must be utilized. As can be seen in Figure 7.8, 90° docks require greater apron depth but less bay width. Drive-in docks are typically not economical; 90° docks require a greater outside turning area, and finger docks require greater inside maneuvering area. Because outside space costs considerably less to construct and maintain than inside space, 90° docks are used whenever space exists. Furthermore, when finger docks must be utilized, the largest-angle finger dock should be utilized. Space requirements for finger docks are given in Table 7.2.

Even though dock widths of 10 ft are adequate for spotting trucks, the potential for accidents, scratches, and increased maneuvering time has resulted in an accepted width of 12 ft. An exception exists for extremely busy docks, where 14-ft dock widths are recommended.

The following procedure can determine overall space requirements outside a facility for truck maneuvering:

- Determine the required number of docks.
- Determine truck flow patterns.

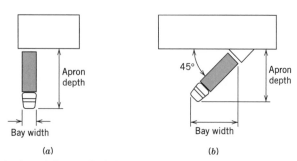

(a)   (b)

Figure 7.8   90° dock and finger dock bay width and apron depth tradeoffs. (*a*) 90° dock. (*b*) 45° finger dock.

Table 7.2 *Finger Dock Space Requirements for a 65-ft Trailer*

Dock Width (feet)	Finger Angle (degrees)	Apron Depth (feet)	Bay Width (feet)
10	10°	50	65
12	10°	49	66
14	10°	47	67
10	30°	76	61
12	30°	74	62
14	30°	70	64
10	45°	95	53
12	45°	92	54
14	45°	87	56

- Determine whether 90° docks may be used, and if not, select the largest-angle finger dock for which space is available.
- Specify a dock width.
- Determine apron depth for that dock width.
- Establish the overall outside space requirement by allocating the space determined in the previous step for the number of docks determined in the first step.

3. Determine Internal Receiving and Shipping Area Requirements.

Receiving and shipping department area requirements within a facility may include space allocations for the following:

- *Personnel convenience/offices* must be provided for receiving and shipping supervision and for clerical activities. Approximately 125 square ft of office space should be provided for each dock employee who will regularly work in the office. The supervisor's office space will often be located within the dock area, and many of the receiving and shipping, clerical, and data-processing activities will be combined with similar activities in the remainder of the warehouse.
- *A receiving hold area* is essential for accumulating received material that has been rejected during a receiving or quality-control inspection and is awaiting return to the vendor or some other form of disposition. Rejected material should never be allowed to accumulate in the receiving buffer area. To do so will surely cause unsatisfactory merchandise to be accepted into the warehouse. A separate and distinct receiving hold area must be allocated. The amount of space required for the receiving hold area depends on the type of material likely to be rejected, the specific inspection process followed, and the timeliness of the disposition of the rejected merchandise.
- *Trash disposal and recycling bins.* Dock operations, particularly receiving functions, generate a tremendous amount of waste materials, including corrugated boxes, binding materials, broken and disposable pallets, bracing, and various other packing materials. Space must be allocated within the receiving and shipping areas for the disposal and recycling of these items. Failure to do so will result in poor housekeeping, congestion, unsafe working conditions,

and a loss of productivity. Receiving and shipping functions are often planned without concern for waste material disposal and recycling and so must sacrifice space allocated for some other function to hold these materials. Dock operations generate trash and other waste materials, so be prepared.

- *Pallet and packaging material storage/palletizing equipment.* In most warehouses, loads often arrive unpalletized, on pallets with odd dimensions, or on disposable pallets, and they require palletizing or repalletizing. A store of empty pallets must be readily available to the dock area for this activity.

- *The truckers' lounge* is an area to which truck drivers are confined when not servicing their trucks. The truckers' lounge should include seating, magazines, and, ideally, refreshment facilities, telephones, and private restrooms to provide adequate facilities to meet normal workers' needs while waiting for their trucks to be serviced. General space requirements for a basic truckers' lounge are approximately 125 sq ft for the first trucker, and an additional 25 sq ft for each additional trucker expected in the lounge at the same time. Consequently, a truckers' lounge designed for an average of three truckers would require approximately 175 square ft.

- *Buffer or staging areas* are areas within receiving departments where materials removed from carriers may be placed until dispatched. If the operating procedure is to deliver all merchandise to stores, inspection, or the department requesting the merchandise immediately upon receipt, a buffer area is not required. If the operating procedure is to remove materials from the carrier and place them into a holding area prior to dispatching, space must be allocated to store the merchandise. In a similar manner, staging areas are areas within shipping departments where merchandise is placed and checked prior to being loaded into a carrier. If merchandise is loaded into carriers directly after being withdrawn from the warehouse, no staging area is required.

    The space required for buffer or staging areas may be determined by considering the number of carriers for which merchandise is to be stored in these areas and the space required to store the merchandise for each carrier. Typically, when buffer and staging areas are utilized, sufficient space should be allocated for one full carrier for each dock. When fluctuations in hourly unloading or loading rates become pronounced, space for storing two or more carriers of merchandise in buffer or storage areas should be considered. The cost of the buffer or staging area should be compared with the truck unloading and loading costs to determine the proper amount of storage space. If a simulation model is used to determine the number of docks, it may be used to determine the cost tradeoff.

    The space required to store merchandise unloaded from or to be loaded onto a carrier depends on the size of the carrier, the cube utilization in the carrier, and the cube utilization in the buffer or staging area. Merchandise can typically be stored at least at carrier height and often much higher. The inverse is true for staging areas where order checking must take place. Storage equipment, as described in Chapter 5, may be utilized in buffer or staging areas, but typically lacks the versatility required for temporary storage in these areas. Aisle spacing between buffer areas for various docks and between staging areas for various docks should follow the guidelines given in Chapter 3.

Table 7.3 *Minimum Maneuvering Allowances for Receiving and Shipping Areas*

Material Handling Equipment Utilized	Minimum Maneuvering Allowance (ft)
Tractor	14
Platform truck	12
Forklift	12
Narrow-aisle truck	10
Handlift (Jack)	8
Four-wheel hand truck	8
Two-wheel hand truck	6
Manual	5

- *Material handling equipment maneuvering* space is provided between the backside of the dockboard and the beginning of the buffer or staging areas. Maneuvering space is dependent on the type of material handling equipment, as indicated in Table 7.3.

Example 7.1

A simulation study has been completed for a new facility, and four docks have been specified. (See Figure 7.9.) Two docks are to be allocated to receiving and two to shipping. Sixty-foot tractor-trailers are the largest carriers serving the facility. The new plant will be positioned to the north of a road that is oriented east–west. All trucks will be unloaded and loaded with lift trucks. The docks are located at the northwest corner of the facility. Buffer areas are not required, but one staging area must be included for each dock. Each staging area must be capable of holding 52 pallet-loads of merchandise that are

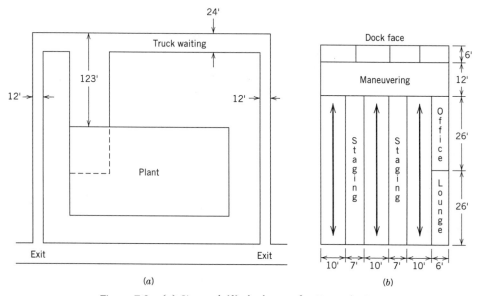

Figure 7.9 (*a*) Site and (*b*) dock area for Example 7.1.

each $48 \times 40 \times 42''$. What are the roadway requirements to the east, north, and west of the facility and what space requirements are needed within the facility?

All trucks should enter the property from the east of the facility and exit to the west. Service roads to the east and west of the facility should be 12 ft wide. The area to the north of the facility on the eastern side should be used for truck waiting and should be at least 24 ft wide. All docks should be 90°, 12-ft docks. A dock face of 48 ft with an apron depth of 63 ft is required. The apron depth of 63 ft is in addition to the 60-ft length of the trucks; therefore, a minimum of 123 ft should be allowed between the northern extreme of the building and the property line. The dockboards will extend 6 ft into the building; therefore, a maneuvering distance of 18 ft (6 ft + 12 ft) is required for the entire 48-ft dockface for a total area of 864 ft². Two staging areas are required, assuming loads are stacked two high and are stored back-to-back for 13 rows, and each staging area will require an area that is 7 ft wide and 52 ft long. Ten-foot aisles should be on each side of the staged material. A total staging area of:

$$\{3(10) + 2(7)\} \times 52 = 2{,}288 \text{ ft}^2$$

is required. A truckers' lounge of 150 ft² should be included, as well as 150 ft² for a receiving and shipping office. Personnel convenience, material handling equipment maintenance, trash disposal, pallet storage, and packaging material storage areas will be integrated with similar areas within the facility. The total space requirement within the facility is:

Maneuvering area	864 ft²
Staging area	2288 ft²
Truckers' lounge	150 ft²
Office	150 ft²
	3452 ft²
or more realistically,	3500 ft²

Dock Operations Planning

Equipment requirements for receiving and shipping areas consist of the equipment required to properly interface between the carriers and the docks. The methods of handling materials through this interface are described in Chapter 5. The dock equipment needed to perform this interface includes:

- Dock levelers as the interface between a dock at a given height and variable-height carriers
- Bumper pads as the interface between a fixed dock and a moveable carrier
- Dock shelters as the interface between a heated/air-conditioned dock and an unheated/non-air-conditioned carrier

Dock Levelers

Viewing the carrier as a portion of the receiving and shipping functions leads to an interesting question. If a temporary storage area (the carrier) is to be used in conjunction with the receiving and shipping functions and the storage area is at a different height than the dock, how should this difference in height be treated?

Five possible answers are:

1. Walking up or down a step to accommodate the difference
2. A portable "ramp" between the dock and carrier
3. A permanent adjustable "ramp" between the dock and carrier
4. Raising the carrier to the height of the dock
5. Raising the dock to the height of the carrier

The first option of walking up or down a step to accommodate the difference in carrier and dock height is acceptable for any type of load not requiring an industrial truck to enter the carrier. Stepping into a carrier to load or unload small packages is acceptable. If a large quantity of small packages is to be loaded or unloaded in this manner, telescoping belt, skatewheel, or roller conveyors may be helpful. If materials are to be loaded or unloaded via crane, the difference in height need only be negotiated by the personnel aligning the materials, and step up or down to the carrier is acceptable. If materials are to be lifted off either side of a carrier using an industrial truck, the difference in carrier and dock height is also of little consequence. Unfortunately, most materials are loaded and unloaded with an industrial truck that enters the carrier; therefore, a step up or down is usually unacceptable.

The second option of a portable "ramp" suggests the use of dockboards (also referred to as dock plates or bridge plates) or yard ramps. Dockboards are typically made of aluminum or magnesium and are no more than ramps that can be placed between a dock and a carrier so that an industrial truck may drive into and out of the carrier. In an attempt to minimize dockboard weight, they are often less than 4 ft long. Therefore, for even small height differentials, inclines can result that are unsafe for industrial truck travel. The major use of dockboards is for loading or unloading rail cars where rail car spotting locations are variable. Even for rail car applications, unless volume of movement is very low, dockboards should be replaced with a permanent device to adjust for height differentials.

Yard ramps or portable docks are ramps that are sometimes used so that carriers can be unloaded using an industrial truck at a facility without a permanent dock. Yard ramps are useful for unloading carriers when inadequate dock space exists, when carriers are to be unloaded in the yard, or for ground-level plants. Figure 7.10 illustrates yard ramps.

The third option, the permanent adjustable dockboard or dock leveler, is the most frequently used approach to compensate for the height differential between carriers and docks. Permanent levelers may be manually, mechanically, or hydraulically activated. Because permanent adjustable levelers are fastened to the dock and not moved from position to position, the dockboard may be longer and wider than portable dock levelers. The extra length results in a smaller incline between the dock and the carrier. This allows easier and safer handling of handcarts, reduced power drain on electrically powered trucks, and less of a problem with fork and undercarriage fouling on the dockboard. The greater width allows for safer and more efficient carrier loading and unloading. Permanent adjustable dock levelers also eliminate safety, pilferage, and alignment problems associated with other dock levelers. For these reasons, permanent adjustable dock levelers, as shown in Figure 7.11, should always be given serious consideration.

Figure 7.10   Yard ramps. [*Part (a) courtesy of Brooks and Perkins, Inc.; part (b) courtesy of MagLine, Inc.*]

The fourth option of raising the carrier to the height of the dock may be utilized when the variation in height of the trucks to be loaded or unloaded is very large. Truck levelers, as shown in Figure 7.12, are platforms installed under the rear wheels of a truck that lift the rear of the truck to the height of the dock. Typical truck levelers can raise a truck bed up to 3 ft. Although the flexibility and functional aspects of truck levelers are desirable, the installation cost of truck levelers limits their use.

The last approach to matching the height of the dock to the carrier is raising or lowering the level of the dock to that of the carrier. Scissors-type dock elevators,

Figure 7.11 Permanent adjustable dock levelers. (*a*) Kelly FX leveler with Star Restraint; (*b*) Serco Versa dock leveler. (*Courtesy of the Material Handling Industry of America*)

(b)

Figure 7.11   (continued)

Figure 7.12   Truck leveler. (*Courtesy of Autoquip Corporations*)

as shown in Figure 7.13, are the most common method of raising or lowering the entire dock. Dock elevators may be permanently installed or may be mobile. Dock elevators are typically used when the plant is at ground level and room does not exist for a yard ramp or when the existing dock is too low or too high and sufficient space does not exist for a dockboard.

Figure 7.13 Scissors-type lifting docks. (_a_) Roller table; (_b_) ED dock. (_Courtesy of the Material Handling Industry of America_)

Bumper Pads

When hitting a dock, a trailer carrying a load of 40,000 pounds and traveling at four miles per hour transmits 150,000 pounds of force to a building. Reinforced concrete docks will disintegrate under this type of constant impact. The addition of one inch of cushioning to the front of the dock will reduce the force transmitted to the building to 15,000 pounds for the same truck traveling at the same speed. Bumper pads are nothing more than molded or laminated rubber cushions that, when fastened to a dock, allow the force to be absorbed.

Dock Shelters

A dock shelter is a flexible shield that, when engaged by a carrier, forms a hermetic seal between the dock and the carrier. Figure 7.14 illustrates dock shelter. The advantages of dock shelters include:

Figure 7.14 Dock shelters. (*Courtesy of Serco Company and the Material Handling Industry of America*)

- *Energy saving.* Often, the heat-loss and cooling-load reductions that result from enclosing the docks more than pay for the cost of installation and maintenance.
- *Increased safety.* Dock shelters protect dock areas from rain, ice, snow, dirt, and debris that enter through open, unprotected dock doors.
- *Improved product protection.* Materials entering and leaving a facility are better protected. For foodstuffs and electronic equipment, dock shelters are often required.
- *Better security.* Eliminating openings around the carriers reduces the opportunity for pilferage.
- *Reduced maintenance.* Dock shelters prevent leaves, dirt, debris, and water from entering the facility, so less housekeeping is required.
- *Reduced spotting time.* Most dock shelters have guidance stripes to allow quicker and better truck positioning.

7.5   DOCK LOCATIONS

The designer's first goal is to get the trucks off the highway safely. The first step is determining where the facility and docks should be positioned on the building site and designing safe access roads to the dock area. Before positioning the building and docks, answer these questions:

1. Where is the material needed or produced in the plant?
2. How much space will the dock and maneuvering area require?
3. Will the operation need more dock positions soon?
4. What type of dock will best serve the operation?
5. What is the safest design for access roads?
6. Will the operation receive or ship deliveries by rail?
7. How will site topography affect traffic and dock operations?
8. What are the dimensions of a safe approach?

Before locating the plant and docks on the site, determine where the docks should be within the building. This depends on where deliveries are needed or where shipments originate in the plant. The in-plant destination of the material influences whether the facility will have one central dock or several point-of-use docks. Each would require different clearances from the boundaries of the building site.

Traditionally, buildings had one dock area. In small facilities, shipping and receiving were combined, and in large plants, shipping and receiving might have been separate but adjacent. The central dock reduced supervision costs and efficiently used material handling people and equipment. The external maneuvering and parking area for this type of dock has to accommodate the most trucks serving the operation at any moment. It also has to serve both the largest and smallest trucks.

Point-of-use docks are becoming more popular as just-in-time inventory demands efficient flow of material or components to production departments. With

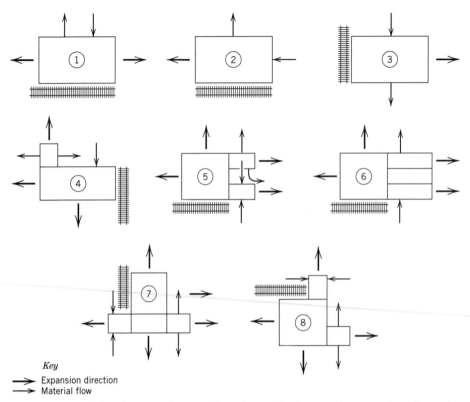

Key
→ Expansion direction
→ Material flow

Figure 7.15 Modular designs of storage/warehouse facilities with expansion alternatives shown.

this arrangement, several docks are located around the plant perimeter. Each is designed to serve a particular production line or operating area. One might receive frequent, small deliveries from light-duty vans. Another might receive deliveries from trucks that are longer, lower, and wider than previous state or federal transportation regulations ever allowed. The design decisions for each type of dock are clearly different and affect placement of the building and docks on the site.

As has been noted, docks are among the first requirements at a site and are vital for smoothly functioning operations. Additionally, plans for facility expansion must incorporate receiving and shipping. An important rule of thumb concerning expansion is to expand the warehouse without disrupting dock operations. As shown in Figure 7.15, numerous expansion alternatives exist for a distribution center. The small arrows represent material flows, and large, open arrows represent directions for expansion that will minimize disruptions to receiving and shipping activities.

7.6 STORAGE OPERATIONS

The objective of storage and warehousing functions is either to maximize resource utilization while satisfying customer requirements or to maximize customer satisfaction subject to a resource constraint. Storage and warehousing resources are space, equipment, and personnel. Customer requirements for storage and warehousing functions are to be able to obtain the desired goods quickly and in good condition. Therefore, in designing storage and warehousing systems, it is desirable to maximize:

- Space utilization
- Equipment utilization
- Labor utilization
- Material accessibility
- Material protection

Planning storage and warehousing facilities follows directly from these objectives. Planning for maximum equipment utilization requires the selection of the correct equipment. The third objective, maximizing labor, involves providing the needed offices and other services for personnel. Planning for the maximum accessibility of all materials is a layout issue. Planning for the maximum protection of items follows directly from having well-trained personnel store goods in adequate space with the proper equipment in a properly planned layout.

Storage Space Planning

A chart that can be used to facilitate the calculation of space requirements for storage and warehousing is illustrated in Figure 7.16. The first five columns of the storage analysis chart are identical to the first five columns of the shipping and receiving analysis chart in Figure 7.4. If materials are stored in the same method whether received or shipped, data may be extracted directly from the shipping and receiving analysis chart. If changes are made in the unit loads before shipping, the changes should be reflected in the storage analysis chart. If, for an existing facility, a shipping and receiving analysis chart has not been completed, a physical survey of the items stored will provide the data required in the first five columns.

The maximum and average quantities of unit loads stored (the sixth and seventh columns on the storage analysis chart) are directly related to the method of controlling inventory and inventory control objectives. These quantities should be provided as inputs to the facilities planner by the inventory control function.

The planned number of unit loads for each material to be stored may be determined by considering the receiving schedule and the method of assigning materials to storage locations. If all materials to be stored in a particular manner are to be received together, the planned quantity of unit loads stored must be equated to the maximum quantity of unit loads. If materials to be stored in a particular manner are to arrive over time, then the method of assigning materials to storage locations will determine the planned quantity of unit loads stored.

There are two major material storage philosophies: fixed, or assigned location storage; and random, or floating location storage. In fixed-location storage, each individual SKU is stored in a specific location, and no other SKU may be stored there, even though the location may be empty.

With random-location storage, any SKU may be assigned to any available storage location. A SKU stored in location *A* one month might be stored in location *B* the following month, and a different SKU stored in location *A*.

Date September 14, 1997   Raw Materials X

In-Process Goods
Company J.D.S., Inc.
Prepared by B. Hudock   Sheet 1 of 1
Finished Goods   Plant Supplies

Space _____   Storage _____

	Unit Loads				Quantity of Unit Loads Stores			Storage			
Description (1)	Type (2)	Capacity (3)	Size (4)	Weight (5)	Maximum (6)	Average (7)	Planned (8)	Method (9)	Specs (10)	Area ft^2 (11)	Ceiling Height Required ft (12)
Aluminum rails, three runner 288"	Bundles	50 pcs.	18" × 28" × 288"	1250 lbs.	14	5	12	Cantilever rack	Four-arm dual rack, 4' × 12' × 6'	192	24
Glass, 1/4" thick, 8' × 4' sheets	Racks	4 sheets	8' × 4' × 4'	400 lbs.	20	13	15	Pallet rack	4' × 4' × 22' (4 levels)	240	24
Rubber Stripping 1/8" square	Cartons	500 feet	12" × 4" × 12"	20 lbs.	20	12	8	Industrial Shelving	5 level 42" × 18" shelves	12	9
Adhesive, water sealant	Drum	350 lbs.	2' diameter × 4'	375 lbs.	7.	3	6	Floor simple	2.5' × 2.0' floor loaded	60	8
Packing boxes 2' × 4' × 4"	Pallets	500	40" × 48" × 48"	600 lbs.	10.	8	6	Bulk Stack	4' × 4' × 12' (3 levels)	80	15

Figure 7.16 Storage analysis chart.

432

The amount of space planned for a SKU is directly related to the method of assigning space. If fixed-location storage is used, a given SKU must be assigned sufficient space to store the maximum amount that will ever be on hand. For random-location storage, the quantity of items on hand at any time will be the average amount of each SKU. In other words, when the inventory level of one item is above average, another item will likely be below average. The sum of the two items will be close to the average. However, in random-location storage, if a large safety stock of SKUs is maintained, the quantity of items on hand at any time will be the safety stock plus one-half the maximum replenishment quantity of each individual SKU.

Example 7.2

At the ABC manufacturing firm, the average daily withdrawal rate for product A is 20 cartons per day, safety stock is five days, order leadtime is 10 days, and the order quantity is 45 days. What are the maximum and average quantities of unit loads to be stored?

Reorder point	= (safety stock in days)(demand/day) + (leadtime in days)(demand/day)
	= (5 days)(20 cartons per day) + (10 days) (20 cartons per day)
	= 300 cartons
Maximum quantity to be stored	= (safety stock) + (order quantity)
	= (100 cartons) + (45 days)(20 cartons per day)
	= 1000 cartons
Average quantity to be stored	= $\frac{1}{2}$ (order quantity) + (safety stock)
	= $\frac{1}{2}$ (900) + (100)
	= 550 cartons

Example 7.3

The number of openings assigned to a SKU must accommodate its maximum inventory level. Hence, the planned quantity of unit loads required for dedicated storage is equal to the sum of the openings required for each SKU. With randomized storage, however, the planned quantity of unit loads to be stored in the system is the number of openings required to store *all* SKUs. Since typically all SKUs will not be at their maximum inventory levels at the same time, randomized storage will generally require fewer openings than dedicated storage.

There are two reasons that randomized storage results in less storage space than required for dedicated storage. First, if an "out-of-stock" condition exists for a given SKU in dedicated storage, the empty slot remains "active" and won't be used for anything else, whereas it could be used in randomized storage. Second, when there are multiple slots for a given SKU, then empty slots will develop as the inventory level decreases, even if the SKU is not "out of stock."

In order to illustrate the effect of the storage method on the storage space required, suppose six products are received by a warehouse according to the schedule in Table 7.4. By summing the inventory levels of the six products, the **aggregate inventory** level is obtained.

With dedicated storage, the required space, as given in Table 7.4, equals the sum of the maximum inventory level for each product, or 140 pallet positions. With randomized

Table 7.4 *Inventory Levels for Six Products in a Warehouse; Expressed in Pallet Loads of Product*

| | PRODUCTS | | | | | | |
Period	1	2	3	4	5	6	Aggregate
1	24	12	2	12	11	12	73
2	22	9	8	8	10	9	66
3	20	6	6	4	9	6	51
4	18	3	4	24	8	3	60
5	16	36	2	20	7	24	105
6	14	33	8	16	6	21	98
7	12	30	6	12	5	18	83
8	10	27	4	8	4	15	68
9	8	24	2	4	3	12	53
10	6	21	8	24	2	9	70
11	4	18	6	20	1	6	55
12	2	15	4	16	24	3	64
13	24	12	2	12	23	24	97
14	22	9	8	8	22	21	90
15	20	6	6	4	21	13	75
16	13	3	4	24	20	15	84
17	16	36	2	20	19	12	105
18	14	33	8	16	13	9	98
19	12	30	6	12	17	6	83
20	10	27	4	8	16	3	68
21	8	24	2	4	15	24	77
22	6	21	8	24	14	21	94
23	4	18	6	20	13	18	79
24	2	15	4	16	12	15	64

Maximum of aggregate inventory level   = 105 pallet loads
Sum of individual maximum inventory levels = 140
Average inventory level   = 77.5
Minimum of aggregate inventory level   = 51

storage, the required amount of space equals the maximum aggregate inventory level, or 105 pallet positions. In this example, dedicated storage requires one-third more pallet positions than does randomized storage.

If inventory shortages seldom occur and single slots are assigned to SKUs, then there are no differences in the storage space requirements for randomized and dedicated storage. Interestingly, many carousel and miniload systems meet these conditions.

To maximize throughput when using dedicated storage, SKUs should be assigned to storage locations based on the ratio of their activity to the number of openings or slots assigned to the SKU. The SKU having the highest ranking is assigned to the preferred openings and so on, with the lowest-ranking SKU assigned to the least-preferred openings. Because "fast movers" are up front and "slow movers" are in back, throughput is maximized.

In ranking SKUs, it is important to define *activity* as the number of storages/ retrievals per unit time, not the quantity of materials moved. Also, it is important to think of "part families" as well. "Items that are ordered together should be stored together" is a maxim of activity-based storage.

Despite the greater throughput of dedicated storage, it is not used as often as it should be. One reason is that it requires more *information* to plan the system for maximum efficiency. Very careful estimates of activity levels and space requirements must be made. Also more *management* is required in order to continue to realize the benefits of dedicated storage after the system is installed. When conditions change significantly, items must be relocated to achieve the benefits of dedicated storage. Hence, randomized storage is more appropriately used under highly seasonal and dynamic conditions.

When many SKUs exist, dedicated storage based on each SKU may not be practical. Instead, SKUs can be assigned to classes based on their activity-to-space ratios. Class-based dedicated storage, with randomized storage within the class, can yield the throughput benefits of dedicated SKU storage and the space benefits of randomized SKU storage. Depending on the activity-to-space ratios, three to five classes might be defined.

Example 7.4

To illustrate the effect on space and throughput of the storage method used, suppose the storage area for the warehouse is designed as shown in Figure 7.17. A single input/output (I/O) point serves the storage area. All movement is in full-pallet quantities. The storage area is subdivided into 10-ft × 10-ft storage bays. Three classes of products (A, B, and C) will be stored. Class A items represent 80% of the input/output activity and have a dedicated storage requirement of 40 storage bays, or 20% of the total storage. Class B items generate 15% of the I/O activity and have a dedicated storage requirement of 30% of the

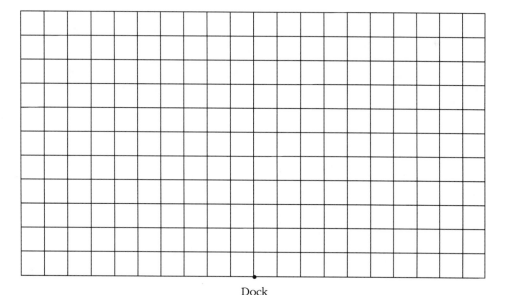

Dock

Figure 7.17 Warehouse layout example.

190	180	170	160	150	140	130	120	110	100	100	110	120	130	140	150	160	170	180	190
180	170	160	150	140	130	120	110	100	90	90	100	110	120	130	140	150	160	170	180
170	160	150	140	130	120	110	100	90	80	80	90	100	110	120	130	140	150	160	170
160	150	140	130	120	110	100	90	80	70	70	80	90	100	110	120	130	140	150	160
150	140	130	120	110	100	90	80	70	60	60	70	80	90	100	110	120	130	140	150
140	130	120	110	100	90	80	70	60	50	50	60	70	80	90	100	110	120	130	140
130	120	110	100	90	80	70	60	50	40	40	50	60	70	80	90	100	110	120	130
120	110	100	90	80	70	60	50	40	30	30	40	50	60	70	80	90	100	110	120
110	100	90	80	70	60	50	40	30	20	20	30	40	50	60	70	80	90	100	110
100	90	80	70	60	50	40	30	20	10	10	20	30	40	50	60	70	80	90	100

Dock

Figure 7.18   Average distances traveled.

total, or 60 storage bays. Class C items account for only 5% of the throughput for the system, but represent 50% of the storage requirement.

Assuming lift truck travel between the I/O point and individual storage bays can be approximated by rectilinear distances between the dock and the centroid of each bay, the distances are as shown in Figure 7.18. Based on the ratio of I/O activity to dedicated storage requirement, the product classes will be placed in the layout in the rank order A, B, and C to obtain the product layout.

The expected average distance traveled for the dedicated storage layout shown in Figure 7.19 is 53.15 ft. If randomized storage is used such that each bay is equally likely to be used for storage, then the expected average distance traveled will be 100 ft. However, with randomized storage, a total storage requirement less than 200 bays is anticipated for the reasons discussed previously. The exact storage requirement will depend on the demand and replenishment patterns for the three product classes.

Even though the storage requirement for randomized storage is not known, it is possible to compute an upper bound for storage that will yield an expected distance traveled equal to or less than that for dedicated storage. To do this, storage bays are

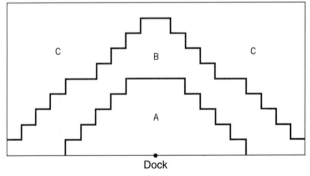

Dock

Figure 7.19   "Optimum" dedicated storage layout.

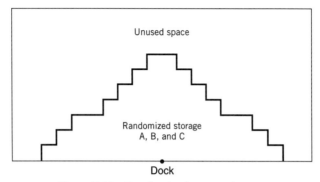

Figure 7.20   Randomized storage layout.

eliminated from consideration in reverse order of their distances from the I/O point and expected distance values are computed. The process continues until a sufficient number have been eliminated. For the example, 138 storage bays, or 69%, must be eliminated for randomized storage to yield an expected distance traveled equal to 52.90 ft. The resulting randomized storage layout is shown in Figure 7.20.

If storage space is to be rectangular in shape, and if 10-ft × 10-ft storage bays are used, then it can be shown [15] that the layout having the minimum expected distance will be as given in Figure 7.21. The resulting expected average distance traveled for Figure 7.21 is 50 ft. In addition, only 50 storage bays are required for storage, rather than the 62 required in Figure 7.20. (The comparison also serves to demonstrate the effect building design can have on expected distance traveled.)

From the bound obtained for randomized storage, it is seen that it is not likely that randomized storage will yield a reduction in space for this example sufficient to obtain throughput values comparable to those obtained from dedicated storage. However, if space costs are significantly greater than handling or throughput costs, then randomized storage might be preferred, regardless of the impact on throughput.

Often, the storage philosophy chosen for a specific SKU will not be strictly fixed-location or random-location storage. Instead, it will be a combination of the two. A grocery store is an excellent example of combination or hybrid location storage. Fixed-location storage is used in the sales area of a grocery store where the consumers shop. Pickles are assigned a fixed location, and only pickles are stored there. Pickles will not be found in any other location of the sales area of the grocery store. Excess or overstock merchandise, however, is often stored randomly in the storeroom of a grocery store. Pickles may be found in one location one week and in a different location the next. Because combination-location storage is based on a mixture of fixed-location storage and random-location storage, its planned

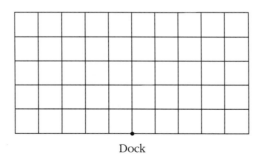

Dock

Figure 7.21   Rectangular-shaped randomized storage layout.

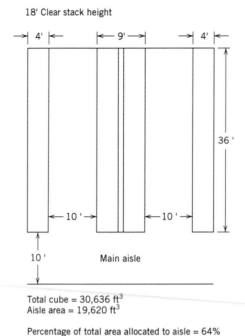

Figure 7.22   An example of the calculation of the percentage of the total warehouse area allocated to aisles.

inventory level falls between the fixed-location quantity and random-location quantity. Where it falls depends upon the percentage of inventory to be assigned to fixed locations.

Once the planned quantity of unit loads to be stored is determined, the method of storing unit loads must be specified. For each method of storage, the loss of cube utilization due to aisles and to honeycombing must be determined. For an existing facility, the loss of cube utilization due to aisles may be determined by drawing various layouts.

Example 7.5

To calculate the loss in cube utilization because of aisles, suppose a counterbalance lift truck will be used to store and retrieve materials from a pallet rack. A 10-ft service aisle is required. The pallet rack is to be located in an area that is 46 ft deep and has a clear stock height of 18 ft. The warehouse layout is shown in Figure 7.22. As shown, 64% of the total warehouse cube is allocated to aisles.

Honeycombing is wasted space that results when a partial row or stack cannot be utilized because adding other materials would result in blocked storage. Figure 7.23 demonstrates honeycombing. The percentage of storage space lost because of honeycombing may be calculated using the analytical models given in Chapter 11.

Once the method of storage and the losses in cube utilization due to aisles and honeycombing are determined, space standards for all unit loads to be stored may be calculated. A **space standard** is the volume requirement per unit load stored to include allocated space for aisles and honeycombing. By multiplying the space standard for an item times the planned quantity of unit loads stored, the space

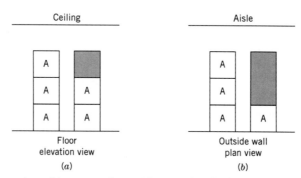

Figure 7.23 Examples of honeycombing. The crosshatched area cannot be used to store other material. (*a*) Vertical honeycombing. (*b*) Horizontal honeycombing.

requirement for an item may be determined. The sum of the space requirements for all items to be stored is the total storage space requirement. By adding to the total storage space requirement areas for receiving and shipping, offices, maintenance, and plant services, the total area requirement for storage departments or warehouses may be determined.

Storage Layout Planning

Before layout planning can begin, the specific objectives of a warehouse layout must be determined. In general, the objectives of a warehouse layout are:

- To use space efficiently
- To allow for the most efficient material handling
- To provide the most economical storage in relation to costs of equipment, use of space, damage to material, handling labor, and operational safety
- To provide maximum flexibility in order to meet changing storage and handling requirements
- To make the warehouse a model of good housekeeping

These objectives are similar to the overall objectives of storage and warehousing planning. This should not be surprising, as layout planning involves the coordination of labor, equipment, and space. To accomplish the objectives, several storage area principles must be integrated. The principles relate to the following:

- Popularity
- Similarity
- Size
- Characteristics
- Space utilization

Popularity. Vilfredo Pareto was an Italian sociologist and economist who discovered an interesting relationship between wealth and individuals. His law was stated as "85% of the wealth of the world is held by 15% of the people." Pareto's law often applies to the popularity of materials stored. Typically, 85% of the turnover will be a result of 15% of the materials stored. To maximize throughput, the most popular

15% of materials should be stored such that travel distance is minimized. In fact, materials should be stored so that travel distance is inversely related to the popularity of the material. Travel distances may be minimized by storing popular items in deep storage areas and by positioning materials to minimize the total distance traveled. As illustrated in Figure 7.24, by storing popular materials in deep storage areas, the travel distance to other materials will be less than if materials were stored in shallow areas. In addition, Figure 7.25 illustrates the popularity issue in combination with the decentralized/centralized shipping and receiving issue (see Section 7.5).

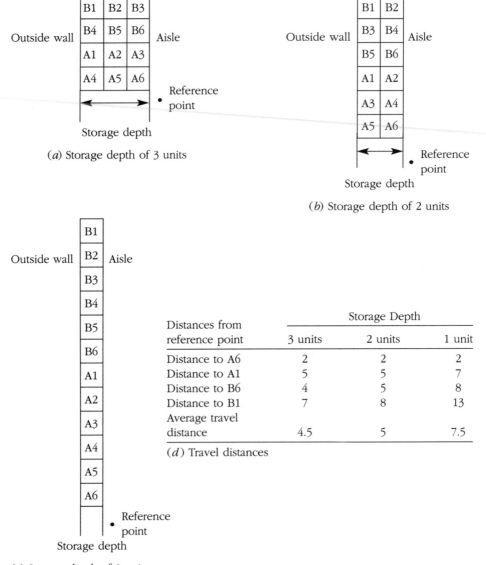

Distances from reference point	Storage Depth		
	3 units	2 units	1 unit
Distance to A6	2	2	2
Distance to A1	5	5	7
Distance to B6	4	5	8
Distance to B1	7	8	13
Average travel distance	4.5	5	7.5

(*d*) Travel distances

(*c*) Storage depth of 1 unit

Figure 7.24 The impact of storage depth on travel distances.

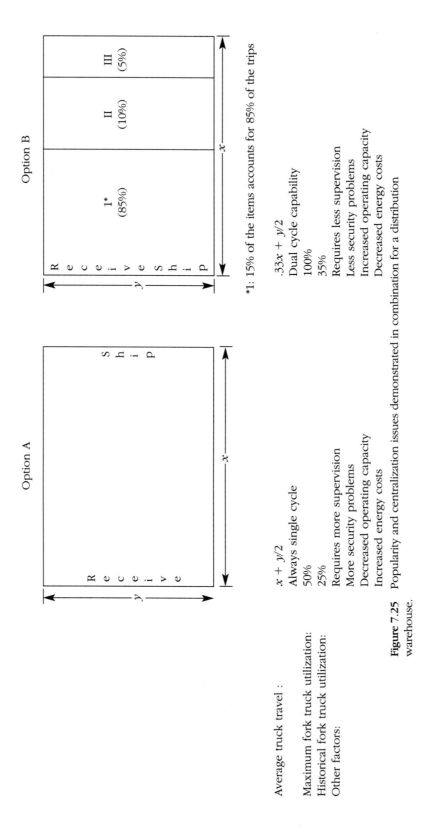

Option A

Option B

*1: 15% of the items accounts for 85% of the trips

Average truck travel :

$x + y/2$

$.33x + y/2$

Maximum fork truck utilization:

Always single cycle

Dual cycle capability

50%

100%

Historical fork truck utilization:

25%

35%

Other factors:

Requires more supervision

Requires less supervision

More security problems

Less security problems

Decreased operating capacity

Increased operating capacity

Increased energy costs

Decreased energy costs

Popularity and centralization issues demonstrated in combination for a distribution

Figure 7.25 warehouse.

441

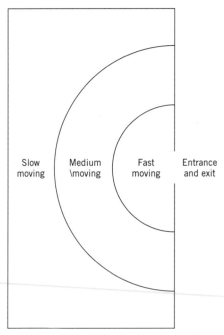

Figure 7.26 Material storage by popularity.

From the previous discussion (and subsequent discussion in Chapter 11), a number of stock location rules of thumb can be developed. Namely, if materials enter and leave the stores department or warehouse from the same point, popular materials should be positioned as close to this point as possible. Figure 7.26 illustrates how to position materials with respect to such a point. Further, if materials enter and leave a storage area from different points and are received and shipped in the same quantity, the most popular items should be positioned along the most direct route between the entrance and departure points. Also, if materials enter and leave a storage area from different points and are received and shipped in different quantities, the most popular items having the smallest receiving/shipping ratio should be positioned close to the shipping point along the most direct route between the entrance and departure points. Finally, the most popular items having the largest receiving/shipping ratio should be positioned close to the receiving point along the most direct route between the entrance and departure points. (The receiving/shipping ratio is no more than the ratio of the trips to receive and the trips to ship a material.)

Example 7.6

The products given in Table 7.5 are the most popular in the warehouse shown in Figure 7.27*a*. How should these items be aligned along the main aisle?

The receiving/shipping ratios may be calculated and are given in Table 7.6. A ratio of 1.0 indicates that shipping and receiving require the same number of trips. For products,

Table 7.5 *Receiving and Shipping information on the Most Popular Products*

Product	Quantity per Receipt	Trips to Receive	Average Customer Order Size	Trips to Ship
A	40 pallets	40	1.0 pallets	40
B	100 pallets	100	0.4 pallets	250
C	800 cartons	200	2.0 cartons	400
D	30 pallets	30	0.7 pallets	43
E	10 pallets	10	0.1 pallets	100
F	200 cartons	67	3.0 cartons	67
G	1000 cartons	250	8.0 cartons	125
H	1000 cartons	250	4.0 cartons	250

Table 7.6 *Receiving/Shipping Ratios for Example 7.6*

Product	Receiving/Shipping Ratio
A	1.0
B	0.4
C	0.5
D	0.7
E	0.1
F	1.0
G	2.0
H	1.0

A, F, and H, the travel distance will be the same no matter where along the main aisle the products are stored. A receiving/shipping ratio less than 1.0 indicates that fewer trips are required to receive the product than to ship. Therefore, products having ratios less than 1.0 should be located closer to shipping. In order of importance of being close to shipping, the products with ratios less than 1.0 are E, B, C, D. Product G has a ratio greater than 1.0, indicating its need to be close to receiving. One of the assignments of products to locations that will result in minimal travel is given in Figure 7.27*b*.

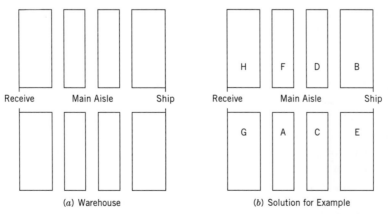

Figure 7.27 Warehouse and product assignment for Example 7.6.

Similarity. Items that are commonly received and/or shipped together should be stored together. For example, consider a retail lawn and garden supply distributor. Chances are that a customer who requires a spreader will not buy, at the same time, a chain saw. Chances are good, however, that a customer who buys the spreader might also require lime, grass seed, and fertilizer. The chain saw should be stored in the same area that the handsaws, clippers, and hand tools are stored. Sometimes, certain items are commonly received together, possibly from the same vendor; they should be stored together. They will usually require similar storage and handling methods, so their consolidation in the same area results in more efficient use of space and more efficient material handling.

An exception to the similarity philosophy arises whenever items are so similar that storing them close together might result in order picking and shipping errors. An example of items that are too similar is two-way, three-way, and four-way electrical switches; they look identical but function quite differently.

Size. The size philosophy suggests that heavy, bulky, hard-to-handle goods should be stored close to their point of use. The cost of handling these items is usually much greater than that of handling other items. That is an incentive to minimize the distance over which they are handled. In addition, if the ceiling height in the warehouse varies from one area to another, the heavy items should be stored in the areas with a low ceiling, and the lightweight, easy-to-handle items should be stored in the areas with a high ceiling. Available cubic space in the warehouse should be used in the most efficient way while meeting restrictions on floor-loading capacity. Lightweight material can be stored at greater heights within typical floor-loading capacities than heavy materials.

The size philosophy also asserts that the size of the storage location should fit the size of the material to be stored. Do not store a unit load of 10 cubic ft in a storage location capable of accommodating a unit load of 30 cubic ft. A variety of storage location sizes must be provided so that different items can be stored differently. In addition to looking at the physical size of an individual item, one must consider the total quantity of the item to be stored. Different storage methods and layouts will be used for storing two pallet loads of an item than will be used for storing 200 pallet loads of the same material.

Characteristics. Characteristics of materials to be stored often require that they be stored and handled contrary to the method indicated by their popularity, similarity, and size. Some important materials characteristics include:

1. *Perishable materials.* Perishable materials may require that a controlled environment be provided. The shelf life of materials must be considered.

2. *Oddly shaped and crushable items.* Certain items will not fit the storage areas provided, even when various sizes are available. Oddly shaped items often create significant handling and storage problems. If such items are encountered, open space should be provided for storage. If items are crushable or become crushable when the humidity is very high, unit load sizes and storage methods must be appropriately adjusted.

3. *Hazardous materials.* Materials such as paint, varnish, propane, and flammable chemicals require separate storage. Safety codes should be checked and strictly followed for all flammable or explosive materials. Acids, lyes, and other dangerous substances should be segregated to minimize exposure to employees.

4. *Security items.* Virtually all items can be pilfered. However, items with high unit value and/or small size are more often the target of pilferage. These items should be given additional protection within a storage area. With the increasing need to maintain traceability of materials, both pilferage and incorrect stock withdrawal must be prevented. Security of storage areas will be a problem if the layout is not designed to specifically secure the materials stored.

5. *Compatibility.* Some chemicals are not dangerous when stored alone, but become volatile if allowed to come into contact with other chemicals. Some materials do not require special storage, but become easily contaminated if allowed to come in contact with certain other materials. Therefore, the items to be stored in an area must be considered in light of other items to be stored in the same area. For example, butter and fish require refrigeration, but if refrigerated together, the butter quickly absorbs the fish odor.

Space Utilization. Space planning includes the determination of space requirements for the storage of materials. While considering popularity, similarity, size, and material characteristics, a layout must be developed that will maximize space utilization as well as the level of service provided. Some factors to be considered while developing the layout are:

1. *Space conservation.* This means maximizing concentration and cube utilization and minimizing honeycombing. Maximizing space concentration enhances flexibility and the capability of handling large receipts. Cube utilization will increase by storing to a height of 18″ below the trusses or sprinklers if the ceiling is 15 ft or less and 36″ below the trusses or sprinklers if the ceiling is greater than 15 ft. Storing materials at the proper height and depth for the quantity of materials typically stored minimizes honeycombing. Honeycombing also occurs from improper withdrawal of materials from storage.

2. *Space limitations.* Space utilization will be limited by the truss, sprinkler, and ceiling heights; floor loads; posts and columns; and the safe stacking heights of materials. Floor-loading is of particular importance for multistory storage facilities. The negative impact of posts and columns on space utilization should be minimized by storing materials compactly around them. Safe stacking heights of materials depend on the crushability and stability of materials stored, as well as the safe storage and retrieval of materials. In particular, materials to be manually picked should be stacked so that operators can safely pick loads without excessive reaching.

3. *Accessibility.* An overemphasis on space utilization may result in poor materials accessibility. The warehouse layout should meet specified objectives for material accessibility. Main travel aisles should be straight and should lead to doors in order to improve maneuverability and reduce travel times. Aisles should be wide enough to permit efficient operation, but they should not waste space. Aisle widths should be tailored to the type of handling equipment using the aisle and the amount of traffic expected. These concepts are illustrated in Figure 7.28. Avoid locked stock by planning for the rotation of all inventories.

4. *Orderliness.* The orderliness principle emphasizes the fact that good "warehouse keeping" begins with housekeeping in mind. Aisles should be well-marked with aisle tape or paint. Otherwise, materials will begin to infringe on the aisle space and accessibility to material will be reduced. Empty spaces

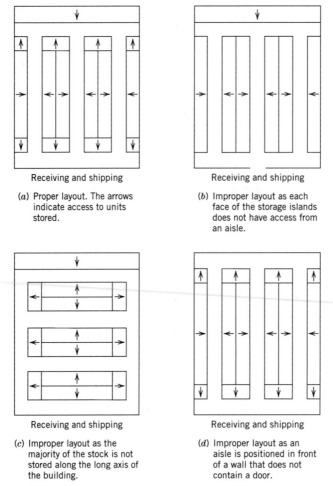

Receiving and shipping

(*a*) Proper layout. The arrows indicate access to units stored.

Receiving and shipping

(*b*) Improper layout as each face of the storage islands does not have access from an aisle.

Receiving and shipping

(*c*) Improper layout as the majority of the stock is not stored along the long axis of the building.

Receiving and shipping

(*d*) Improper layout as an aisle is positioned in front of a wall that does not contain a door.

Figure 7.28   Illustration of storage area accessibility considerations.

within a storage area must be avoided, and they must be corrected where they do occur. If a storage area is designed to accommodate five pallets and, in the process of placing material into that area, one pallet infringes on the space allocated for an adjacent pallet, a void space will result. Because of this, only four pallets can actually be stored in the area designed for five pallets. The lost pallet space will not be regained until the entire storage area is emptied.

Developing and Maintaining a Stores Department or Warehouse Layout

The easiest method of developing a layout is to develop alternatively scaled layouts and to compare these layouts with the principles of popularity, similarity, size, characteristics, and space utilization. The steps required to develop a scaled layout are:

1. Draw the overall area to scale.
2. Include all fixed obstacles, such as columns, elevators, stairs, plant services, etc.

3. Locate the receiving and shipping areas.
4. Locate various types of storage.
5. Assign materials to storage locations.
6. Locate all aisles for equipment and access.

Maintenance of the layout requires that materials be stored in an orderly manner and that stock locations be known. Disorderly storage can cause significant losses in cube utilization and accessibility. All materials should be stored in a uniform and neat manner. All materials should be accessible and should be placed properly in the assigned area. Wasted space often results when materials are not properly placed within allotted areas. Aisle marking should be maintained to indicate where loads are to be placed.

All materials within a storage area must be able to be quickly located and picked. If materials are assigned to a fixed location, stock location is easily performed. If materials are assigned to locations randomly, a stock-location system must be used to keep track of storage locations. The stock-location system should include the quantity and location of all materials within the storage area. It will serve as the basis for assigning, locating, picking, and changing locations of all materials. A stock-location system is essential if a dynamic layout is to be provided in the storage area. It is the interface between planning and managing storage facilities and warehouses.

7.7 ORDER PICKING OPERATIONS

Order picking is the most critical function in distribution operations. It is at the center of the flow of products from suppliers to customers. In fact, it is where customer expectations are actually filled.

Warehousing professionals identify order picking as the highest-priority activity in the warehouse for productivity improvements. There are several reasons for their concern. First, and foremost, order picking is the most costly activity in a typical warehouse. A 1988 study in the United Kingdom [12] revealed that 55% of all operating costs in a typical warehouse can be attributed to order picking (Figure 7.29). Second, the order picking activity has become increasingly difficult to manage. The difficulty

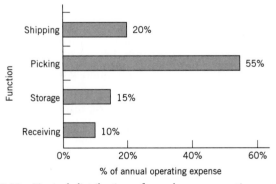

Figure 7.29 Typical distribution of warehouse operating expenses.

arises from the introduction of new operating programs such as *JIT, cycle-time reduction, and quick response,* and new marketing strategies such as *micromarketing* and *megabrand.* These programs require that smaller orders be delivered to warehouse customers more frequently and more accurately and that more SKUs be incorporated in the order picking system. As a result, throughput, storage, and accuracy requirements have increased dramatically. Also, renewed emphasis on quality improvements and customer service have forced warehouse managers to reexamine order picking from the standpoint of minimizing product damage, reducing transaction times, and further improving picking accuracy. Finally, the conventional responses to these increased requirements, to hire more people or to invest in more automated equipment, have been stymied by labor shortages and high hurdle rates due to uncertain business environments. Fortunately, there are a variety of ways to improve order picking productivity without increasing staffing or making significant investments in highly automated equipment. Fifteen ways to improve order picking productivity in light of the increased demands now placed on order picking systems are described below.

Principles of Order Picking

Regardless of size, mission, volume, inventory, customer requirements, or type of control system of a warehouse operation, there are certain principles that apply equally well to the order picking function.

1. **Apply Pareto's law.** We have made many useful modern applications of Pareto's law, which we difined earlier. In a warehouse operation, a small number of SKUs constitutes a large portion of the inventory. This may be measured in value or cube. Similarly, a small number of SKUs will represent a large portion of the throughput in a warehouse. This may be measured in cubic volume shipped or times sold. If we group together items that are popular, we can reduce travel time in the warehouse during picking. This is a very powerful law.

2. **Use a clear, easy-to-read picking document.** A picking document should provide specific instructions to the order picker, making the job easier than it otherwise would be. A common pitfall is to leverage the shipping paperwork as picking paperwork. The problem is that it includes extraneous information and is designed to aid the customer's receiving function, not the picker. Information should be presented in the order that it is required: location, stock number, description, unit of material, and quantity required. Additionally, any special labeling or packaging may be noted. The font should be easy to read. This implies that it is large enough and that the printer is well-maintained. If it is part of a multipage form, the picker should use the top copy. Double-spacing of lines and using horizontal rules are advisable. Of course, elimination of the paperwork by using radio frequency terminals presents the pickers with line-at-a-time order picking documentation.

3. **Use a prerouted, preposted picking document.** A picking document will control the order picking process. It will cause the operator to travel throughout the warehouse in a random manner if the order is not prerouted (sorted according to stock location to minimize travel time). It will also allow the order picker to travel to locations with insufficient stock if no consideration is given to other orders that have requirements for the available inventory.

4. **Maintain an effective stock-location system.** It is not possible to have an efficient order picking system without an effective stock-location system. To pick an item, you first have to find it. If you don't have a specific location, then time must be spent searching for the product. This is neither value-added nor productive. Without an address, it is impossible to take advantage of Pareto's law. Without a stock-location system, it is not possible to have a prerouted picking document.

5. **Eliminate and combine order picking tasks when possible.** The human work elements involved in order picking may include:

   - **Traveling** to, from, and between pick locations
   - **Extracting** items from storage locations
   - **Reaching** and **bending** to access pick locations
   - **Documenting** picking transactions
   - **Sorting** items into orders
   - **Packing** items
   - **Searching** for pick locations

   A typical distribution of the order picker's time among these activities is provided in Figure 7.30. Means for eliminating the work elements are outlined in Table 7.7.

   When work elements cannot be eliminated, they can often be combined to improve order picking productivity.

   a. **Traveling and Extracting Items.** Stock-to-order (STO) systems such as carousels and the miniload automated storage/retrieval system are designed to keep order pickers extracting while a mechanical device travels to, from, and between storage locations, bringing pick locations to the order picker. As a result, a man–machine balancing problem is introduced. If the initial design of stock-to-operator systems is not accurate, a significant portion of the order picker's time may be spent waiting for the storage/retrieval machine to bring pick locations forward.

   b. **Traveling and Documenting.** Since a person-aboard storage/retrieval machine is programmed to automatically transport the order picker between successive picking locations, the order picker is free to document picking transactions, sort material, or pack material while the S/R machine is moving.

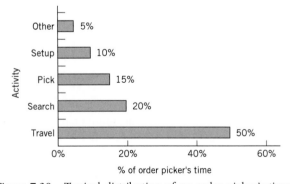

Figure 7.30 Typical distribution of an order picker's time.

Table 7.7 *Order Picking Work Elements and Means for Elimination*

Work Element	Method of Elimination	Equipment Required
Traveling	Bring pick locations to picker	Stock-to-Picker System Miniload AS/RS Horizontal Carousel Vertical Carousel
Documenting	Automate information flow	Computer-Aided Order Picking Automatic Identification Systems Light-Aided Order Picking Radio-Frequency Terminals Headsets
Reaching	Present items at waist level	Vertical Carousels Person-aboard ASRS Miniload ASRS
Sorting	Assign one picker per order and one order per tour	
Searching	Bring pick locations to picker Take picker to pick location Illuminate pick locations	Stock-to-Picker Systems Person-Aboard AS/RS Pick-to-Light Systems
Extracting	Automated dispensing	Automatic Item Pickers Robotic Order Pickers
Counting	Weigh count Prepackage in issue increments	Scales on picking vehicles

c. **Picking and Sorting.** If an order picker completes more than one order during a picking tour, picking carts equipped with dividers or totes may be designed to allow the picker to sort material into several orders at a time.

d. **Picking, Sorting, and Packing.** When the cube occupied by a completed order is small, say less than a shoebox, the order picker can sort directly into a packing or shipping container. Packing or shipping containers must be set up ahead of time and placed on picking carts equipped with dividers and/or totes.

6. **Batch orders to reduce total travel time.** By increasing the number of orders (and therefore items) picked by an order picker during a picking tour, the travel time per pick can be reduced. For example, if an order picker picks one order with two items while traveling 100 ft, the distance traveled per pick is 50 ft. If the picker picked two orders with two items on each order, the distance traveled per pick is reduced to 25 ft.

A natural group of orders to put into a batch is single-line orders. Single-line orders can be batched by small zones in the warehouse to further reduce travel time. A profile of the number of lines requested per order helps identify the opportunity for batching single-line orders. An example profile is illustrated in Figure 7.31.

Other order-batching strategies are enumerated in Figure 7.32. Note that when an order is assigned to more than one picker, the effort needed to reestablish order integrity is significantly increased. The additional cost of sortation must be evaluated with respect to the travel time saved generated by batch picking.

a. **Discrete (Order) Picking.** One person picks one order, one line (product) at a time. There is only one order-scheduling window during

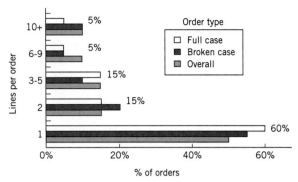

Figure 7.31 Lines per order distribution.

a shift. This means that orders are not scheduled and may be picked at any time on a particular day. This method is the most common because of its simplicity. It is also known as "order picking" in the pure sense.

Discrete order picking has several advantages. It is the simplest for the order picker in paper-based systems, as only one picking document must be managed. As a result, the risk of omitting merchandise from an order is reduced. In a service-window environment, it provides the fastest response to the customer. Accountability for accuracy is clear-cut, unless checkers are used. On the downside, it is the least productive procedure. Because the picker must complete the total order, travel time is likely to be excessive when compared with other methods.

b. **Batch Picking.** One picker picks a group of orders (batch) at the same time, one line at a time. When a product appears on more than one order, the total quantity required for all the orders combined is picked at one time, and then segregated by order. The segregation may take place while picking into totes (small items in small quantities) or transported to a designated area where the products are sorted and grouped by individual order. There is only one order-scheduling window per shift.

Picking more than one order at a time has a significant effect on picking productivity for case and broken-case picking. In fact, the more orders a picker can effectively manage at once, the greater the productivity gain. Of course, there is a point of diminishing returns. The best candidates for batch picking are orders with few (one to four) lines and small cube. Once

Procedure	Pickers Per Orders	Line Items Per Pick	Periods Per Shift
Discrete	Single	Single	Single
Zone	Multiple	Single	Single
Batch	Single	Multiple	Single
Wave	Single	Single	Multiple
Zone-Batch	Multiple	Multiple	Single
Zone-Wave	Multiple	Single	Multiple
Zone-Batch-Wave	Multiple	Multiple	Multiple

Figure 7.32 Methods of order picking.

again, the reason for productivity improvement is the reduction in travel time. Instead of traveling throughout the warehouse to pick a single order, the picker completes several orders with a single trip. It is critical, however, that measures be taken to minimize the risk of picking and sorting errors. Computer control systems and automatic sortation are very effective.

c. **Zone Picking.** The total pick area is organized into distinct sections (zones), with one person assigned to each zone. The picker assigned to each zone picks all the lines, for each order, that are located within that zone. The lines from each zone are brought to an order consolidation area where they are combined into a complete order before shipment. If the required lines are spread out across three zones, then three pickers work on that order. Each picker only works on one order at a time, and there is only one scheduling period per shift.

There are two variations of zone picking. Sequential zone picking is picking one zone at a time. The order is passed to the next zone, where it may (or in some cases, may not) have lines to be picked. This is sometimes referred to as a "pick and pass" method. *Simultaneous* zone picking is picking from all applicable zones independently and then consolidating the order in a designated location as it is completed.

Zone picking is often used because of the different skills or equipment associated with a hybrid warehouse. Pallet picking with narrow-aisle equipment, case picking from selective rack, broken-case picking from static high-rise shelving, and broken-case picking from horizontal carousels lend themselves to zones. In order to reduce travel time, a large equipment zone may be subdivided into several separate zones. A highly active storage equipment zone may be further subdivided into picking zones to reduce congestion and associated delays. In fact, it is often a good idea to size zones to balance the workload between them.

Bucket brigade is a version of zone picking where the volume, not a fixed break point, defines the zone. A picker travels back to the previous picker or to the start to get the next order rather than waiting for another order to enter the zone. This has the effect of eliminating wait time or backlogs within a zone. This takes training to assure that all items are picked during multiple handoffs.

d. **Wave Picking.** This method is similar to discrete picking in that one picker picks one order, one line at a time. The difference is that a selected group of orders is scheduled to be picked during a specific planning period. There is more than one order scheduling period during each shift. This means that orders may be scheduled to be picked at specific times of the day. Typically, this is done to coordinate the picking and shipping functions. The following three methods are combinations of the basic procedures. They are more complicated and thus require more control.

e. **Zone-Batch Picking.** Each picker is assigned a zone and will pick a part of one or more orders, depending on which lines are stocked in the assigned zones. Where orders are small in terms of lines, the picker may pick the order complete in the zone. Still, only one scheduling period is used each shift.

    f. **Zone-Wave Picking.** Each picker is assigned a zone and picks all lines for all orders stocked in the assigned zone, one order at a time, with multiple scheduling periods per shift.

    g. **Zone-Batch-Wave Picking.** Each picker is assigned a zone and picks all lines for orders stocked in the assigned zone, picking more than one order at a time, with multiple scheduling periods in each shift.

7. **Establish separate forward and reserve picking areas.** Since a minority of the items in a warehouse generates a majority of pick requests, a condensed picking area containing some of the inventory of popular items should be established. The smaller the allocation of inventory to the forward area (in terms of the number of SKUs and their inventory allocation), the smaller the forward picking area, the shorter the travel times, and the greater the picking productivity. However, the smaller the allocation, the more frequent the internal replenishment trips between forward and reserve areas and the greater the staffing requirement for internal replenishments.

    Some typical approaches for making forward reserve decisions include allocating an equal time supply of inventory for all SKUs in the forward area or allocating an equal number of units of all SKUs in the forward area. A near-optimal procedure for solving the forward-reserve problem was developed in 1992 [22]. The procedure makes use of math programming techniques to decide for each SKU whether it should receive a location in the forward area, and if it should, its proper allocation. The annual savings in picking and replenishment costs is typically between 20% and 40%.

    A simplified approach to the configuration of a forward/reserve system may be determined as follows [21]:

    a. **Determine which items should be in the forward picking area.** Because most inventories include many slow-moving items that have relatively small storage cube requirements, a forward picking area may include the entire inventory of each of the slow movers and only a representative quantity of the fast movers. Alternatively, to provide the fastest possible picking, very slow-moving items may be stored elsewhere in the warehouse in a less accessible but higher-density storage mode.

    b. **Determine the *quantities* of each item to be stored in the forward picking area.** As mentioned, slow movers may have their entire on-hand inventory located in the forward picking area. Storage allocation for other products may be determined by either (1) an arbitrary allocation of space, as much as one case or one shelf, or (2) space for a quantity sufficient to satisfy the expected weekly or monthly demand. (It is common to select large quantities from reserve storage and smaller quantities from forward storage.)

    c. **Size the total storage cube requirement for items in the forward picking area.** Space planned for each item must be adequate for the expected receipt and/or replenishment quantity, not just for the average balance on-hand.

    d. **Identify alternative storage methods that are appropriate for the total forward picking cube and that meet the required throughput.**

e. **Determine the operating methods within each storage alternative in order to project personnel requirements.** The description of the operating method must also include consideration of the storage location assignment (i.e., random, dedicated, zoned, or a combination of these), because this will have a significant impact on picking productivity. It is also necessary to evaluate the opportunity for batch picking (picking multiple orders simultaneously).

f. **Estimate the costs and savings for each alternative system described and implement the preferred system.**

8. **Assign the most popular items to the most easily accessed locations in the warehouse.** Once items have been assigned to storage modes and space has been allocated for their forward and reserve storage locations, the formal assignment of items to warehouse locations can commence. In a typical warehouse, a minority of the items generates a majority of the picking activity. This phenomenon can be used to reduce order picking travel time and reaching and bending. For example, by assigning the most popular items close to the front of the warehouse, an order picker's or S/R machine's average travel time can be significantly reduced. In automated storage/retrieval systems, average dual command travel time can be reduced by as much as 70% over random storage [18]. In miniload automated storage/retrieval systems and carousel picking systems, order picking productivity can be improved by as much as 50%, depending on the number of picks per bin retrieval and the other tasks assigned to the order picker (e.g., packaging, counting, weighing).

Popularity storage can also be used to reduce stooping and bending, consequently reducing fatigue and improving picking accuracy. Simply, the most popular items should be assigned to the picking locations at or near waist height. In an order picking operation for small vials of radio-pharmaceuticals, a stock location assignment plan was devised which concentrated over 70% of the picks in locations at or near waist level.

The most common mistake in applying this principle in the design of stock-location systems is overlooking the size of the product. The objective in applying the principle is to assign as much picking activity as possible to the locations that are easily accessible. Unfortunately, there are a limited number of picking locations that are easy to access—those near the front of the system and/or about waist height. Consequently, the amount of space occupied by an item must be incorporated into the ranking of products for stock assignment. A simple ranking of items based on the ratio of pick frequency (the number of times the item is requested) to shipped cube (the product of unit demand and unit cube) establishes a good baseline for item assignment. Items with high rankings should be assigned to the most easily accessible locations.

Two helpful profiles for stock-location assignment are a distribution illustrating items ranked by popularity and the portion of total picking activity they represent (Figure 7.33) and a distribution illustrating items ranked by popularity and the portion of orders those items can complete Figure 7.34. The first profile may reveal a small grouping of items that completes a large number of orders. Those items become candidates for assignment to a small picking zone dedicated to high-density, high-throughput order picking.

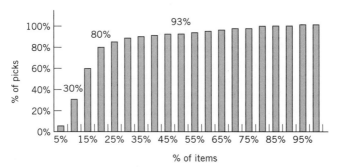

Figure 7.33   ABC analysis: items and picks.

9. **Balance picking activity across picking locations to reduce congestion.** In assigning popular items to concentrated areas in operator-to-stock systems, congestion can reduce potential productivity gains. Care must be taken to distribute picking activity over areas large enough to avoid congestion, yet not so large as to significantly increase travel times. This is often achieved in horseshoe configurations of walk-and-pick systems. A typical picking tour will require the picker to traverse the entire horseshoe, but the most popular items are assigned locations on or near the horseshoe. In stock-to-operator systems, system designers must be careful not to overload one carousel unit or one miniload aisle. Balanced systems are more productive.

10. **Assign items that are likely to be requested together to the same or nearby locations.** Much as a minority of items in a warehouse generates a majority of the picking frequencies, there are items in the warehouse that are likely to be requested together. Examples include items in repair kits, items from the same supplier, items in the same subassembly, and items of the same size. Correlations can be identified from order profiles [18] and can be capitalized on by storing related items in the same or nearby locations. Travel time is reduced because the distance between pick locations for an order is reduced. In a carousel or miniload AS/RS application, storing items that are likely to be requested together in the same location minimizes the number of location visits required to complete an order and therefore helps reduce order picker idle time and wear and tear on the system.

    At a major mail-order apparel distributor, nearly 70% of all orders can be completed from a single size (e.g., small, medium, large, extra large) regardless

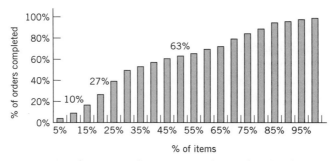

Figure 7.34   ABC analysis: items and completed orders.

of the type of item ordered (shirts, pants, or belts). At a major distributor of healthcare products, a majority of the orders can be filled from a single vendor's material. Since material is also received that way, correlated storage by vendor improves productivity in picking and putaway.

A computerized procedure for jointly considering the popularity and correlation of demand for items in developing intelligent stock assignment plans was developed in 1989 [18]. In the example, order picking productivity was improved by nearly 80%. The procedure suggests that items be clustered into families containing items that are likely to be requested together and the families assigned to warehouse locations on the basis of the pick frequency and space associated with them. A simple way to begin the process of identifying demand families is to rank pairs of items based on the number of times the pair appears together on an order. The pairs at the top of the list often reveal the rationale for demand family development.

11. **The order picker must be accountable for order accuracy.** Suppose a warehouse operation has experienced too many customer complaints about order accuracy. Of course, the complaints are about low counts and missing items. Management understands that there are probably just as many high counts and extra items that are not reported. Action must be taken. The manager decides to add labor to the process in the form of "checkers." It is a checker's job to ensure that orders are shipped as ordered. Is this value-added? Does this solve the problem? Who is accountable for order accuracy, the picker or the checker? In manufacturing, we learned that it is not possible to inspect quality into the product. Why try to inspect quality into an order? Checkers should be used only as a short-term, stopgap measure. Long-term quality requires that the order picker is held accountable for picking at the correct time and in the correct quantity and for delivering the order to the correct warehouse location.

12. **Avoid counting.** Where appropriate, measure instead of counting. Let's face it; counting can be boring, especially when we count large quantities. To eliminate counting errors, we must simplify the counting problem. Packaging can be designed to hold a reasonable quantity of product relative to the quantity ordered. If the packaging holds 1,000 units, and the typical order quantity is 100 units, then the package is too large. Packages of 25 would drastically reduce the counting requirement. Similarly, if the product is packaged in individual units, and the typical customer order quantity is 100 units, the package count is too small. Another approach to solving the counting problem is to measure instead. Electronic weigh scales can be both accurate and can enhance productivity, especially for very small items.

13. **Require pick confirmation.** It is critical for order accuracy that the order picker actively verify that the quantity picked is the quantity required or report the actual quantity picked if it differs from that requested. This will eliminate confusion in shipping and at the customer's receiving operation. Moreover, it is a part of the routing that must be followed to ensure the picker's accountability.

14. **Design picking vehicles to minimize sorting time and errors and to enhance the picker's comfort.** The order-picking vehicle is the order picker's workstation. Just as workstation design is critical to the productivity and comfort of assembly and office workers, the design of the picking vehicle is critical

to the productivity and morale of order pickers. The vehicle should be tailored to the demands of the job. If sorting is required, the vehicle should be equipped with dividers or tote pans. If picking occurs above a comfortable reaching height, the vehicle should be equipped with a ladder. If the picker takes documents on the picking tour, the vehicle should help the picker organize the paperwork. Unfortunately, the design of the picking vehicle is often of secondary concern, yet it is at this workstation that order picking really takes place.

A major wholesale drug distributor recently installed picking vehicles that are powered and guided by rails running in the ceiling between each picking aisle. The vehicle automatically takes the order picker to the correct location. An on-board CRT communicates the correct pick location, quantity, and order or container in which to place the pick quantity. The vehicle can accommodate multiple containers to allow batch picking and is equipped with on-board scales for on-line weigh-counting and pick-accuracy verification.

15. **Eliminate paperwork from the order picking activity.** Paperwork is one of the major sources of inaccuracies and productivity losses in the order picking function. Pick-to-light systems, radio frequency data communication, and voice input/output have been successfully used to eliminate paperwork from the order picking function. Each technology was described in Chapter 5.

7.8 SUMMARY

It is old-fashioned to think of warehousing as a non-value-added activity. Traditionally, though, that's how warehousing has been perceived—a cost-adding burden to the supply chain. Additionally, warehousing has not been afforded the same type of quantitative scrutiny as other functions of the supply chain. Transportation, for example, is typically the hotbed for inspection, while warehousing has been left alone without enjoying the benefits of continuous improvement strategies and operational scrutiny. Today, the Internet, e-commerce, supply chain integration, and the growth of third-party logistics have placed warehousing in management's spotlight. As a result, the emphasis on planning and managing these operations has never been greater. In covering the operating principles and space planning methodologies for receiving, storage, picking, shipping, and dock operations, this chapter demonstrates that the true value of warehousing lies in its ability to have the right product in the right place at the right time. Warehousing ultimately provides the utility of time and place that companies need to satisfy their customers.

BIBLIOGRAPHY

1. Ackerman, K. B., Gardner, R. W., and Thomas, L. P., *Understanding Today's Distribution Center*, Traffic Service Corp., Washington, D.C., 1972.
2. Allred, J. K., "Automated Storage Systems: How They Are Being Justified and Applied Today," paper presented to the Society of Manufacturing Engineers, Moline, IL, 1975.

3. Allred, J. K., "Large Automated Warehousing Systems: New Methods of Justifying, Buying and Applying Them," *Proceedings of the 1975 MHI Material Handling Seminar and MHI Material Handling Symposium,* Material Handling Institute, Pittsburgh, PA, 1975.

4. Allred, J. K., "How to Plan and Buy a Computer Controlled Storage and Handling System," *Material Handling Engineering,* vol. 29, no. 5, May 1974, pp. 69–73.

5. Apple J. M., *Material Handling Systems Design,* Ronald Press, New York, 1972.

6. Apple J. M., *Plant Layout and Material Handling,* Ronald Press, 3rd ed., New York, 1977.

7. Apple J. M. Jr., and Strahan, B. A., "Proper Planning and Control-The Keys to Effective Storage," *Industrial Engineering,* vol. 13, no. 4, April 1981, pp. 102–112.

8. Barrett, J. P., "Modern Dock Design Concepts," International Material Management Society, Monroeville, PA, 1976.

9. Bolz, H. A., and Hageman, G. E. (eds.), *Material Handling Handbook,* Ronald Press, New York, 1958.

10. Bozer, Y. A., and White, J. A., "Optimum Designs of Automated Storage/Retrieval Systems," presentation to the TIMS/ORSA Joint National Meeting, Washington, D.C., May 1980.

11. Briggs, A. J., *Warehouse Operations Planning and Management,* Wiley, New York, 1960.

12. Drury, J., *Towards More Efficient Order Picking,* IMM Monograph Number 1, The Institute of Materials Management, Cranfield, United Kingdom, 1988.

13. Elsayed, E. A., "Algorithms for Optimal Material Handling in Automatic Warehousing Systems," *International Journal of Production Research,* vol. 19, no. 5, 1981, pp. 525–535.

14. Foley, R. D., and Frazelle, E. H., "The Effect of Class-Based Storage on End-of-Aisle Order Picking," unpublished working paper, Material Handling Research Center, Georgia Institute of Technology, 1989.

15. Francis, R. L., and White, J. A., *Facility Layout and Location: An Analytical Approach,* Prentice-Hall, Englewood Cliffs, NJ, 1974.

16. Frazelle, E. H., "Automated Storage Retrieval, and Transport Systems in Japan," U.S. Departments of Commerce Technical Report, December 1990.

17. Frazelle, E. H., "Small Parts Order Picking: Equipment and Strategy," Material Handling Research Center Technical Report Number 01-88-01, Georgia Institute of Technology, Atlanta, Georgia, 30332-0205.

18. Frazelle, E. H., *Stock Location Assignment and Order Picking Productivity,* Ph.D. Dissertation, Georgia Institute of Technology, December 1989.

19. Frazelle, E. H., and Apple, J. M., Jr., "Warehouse Operations," *The Distribution Management Handbook* (J. A. Tompkins, ed.), McGraw-Hill, New York, 1993.

20. Frazelle, E. H., and Ward, R. E., "Material Handling Technologies in Japan," National Technical Information Service Report #PB93-128197, Washington, D.C., 1992.

21. Frazelle, E. H., Hackman, S. T., and Platzman, L. K., "Intelligent Stock Assignment Planning," *Proceeding of the 1989 Council of Logistics Management's Annual Conference,* St. Louis, MO, October 1989.

22. Frazelle, E. H., Hackman, S. T., Passy, U., and Platzman, L. K., *Solving the Forward-Reserve Problem,* Material Handling Research Center Technical Report, Georgia Institute of Technology, May 1992.

23. Goetschalckx, M., *Storage and Retrieval Policies for Efficient Order Picking Operations,* Ph.D. Dissertation, Georgia Institute of Technology, Atlanta, GA (1983).

24. Goetschalckx, M., and Ratliff, H. D., "Sequencing Picking Operations in a ManAboard Order Picking System," *Material Flow,* vol. 4, no. 4, 1988, pp. 255–263.

25. Graves, S. C., Hausman, W. H., and Schwarz, L. B., "Storage Retrieval Interleaving in Automatic Warehousing Systems," *Management Science,* vol. 23, no. 9, May 1977, pp. 935–945.

26. Han, M., McGinnis, L. F., Shieh, J. S., and White, J. A., "On Sequencing Retrievals in an Automated Storage/Retrieval System," *IIE Transactions,* vol. 19, no. 2, March 1987, pp. 56–66.

27. Jenkins, C. H., *Modern Warehouse Management,* McGraw-Hill, New York, 1968.

28. Kinney, H. D., "How to Size the Warehouse," Material Handling Management Course, American Institute of Industrial Engineers, Norcross, GA, June 1979.

29. Kinney, H. D., "A Total Integrated Material Management System," presentation to the 1980 Annual Conference of the International Material Management Society, Cincinnati, OH, June 1980.

30. Mullens, M. A., "Use a Computer to Determine the Size of a New Warehouse Particularly in Storage and Retrieval Areas," *Industrial Engineering,* vol. 13, no. 6, June 1981, pp. 24–32.

31. Tompkins, J. A., *Facilities Design,* North Carolina State University, Raleigh, NC, 1975.

32. Tompkins, J. A., "Problem Solving Techniques in Material Handling," *Proceedings of the 1976 Material Handling and the Industrial Engineer Seminar,* American Institute of Industrial Engineers, Norcross, GA, 1976.

33. White, J. A., "Justifying Material Handling Expenditures," *Proceedings, 1977 Spring Annual Conference,* American Institute of Industrial Engineers, pp. 103–111.

34. White, J. A., "Randomized Storage or Dedicated Storage?," *Modern Materials Handling,* vol. 35, no. 1, January 1980, p. 19.

35. White, J. A., and Kinney, H. D., "Storage and Warehousing," G. Salvendy (ed.), *Handbook of Industrial Engineering,* Wiley, New York, 1982.

36. Wilde, J. A., "Container Design Issues," *Proceedings of the 1992 Material Handling Management Course,* sponsored by the Institute of Industrial Engineers, Atlanta, GA, June 1992.

37. National Safety Council, Chicago, IL.

38. *Warehousing Education and Research Council's Annual Membership Survey,* Warehousing Education and Research Council, Oak Brook, IL, 1988.

39. *Warehouse Modernization and Layout Planning Guide,* U.S. Naval Supply Systems Command Publication 529, U.S. Naval Supply Systems Command, Richmond, VA, 1988.

40. *Warehousing Education and Research Council's Annual Membership Survey,* Warehousing Education and Research Council, Oak Book, IL, 1988.

PROBLEMS

7.1 You are employed as an industrial engineer by the Form Utility Co. and feel an analysis of the dock area would prove beneficial. Before undertaking such a project, you must convince your boss, Mr. Save A. Second, that the receiving and shipping areas are worth the study. Prepare a written report justifying your proposed dock study and outlining the types of savings you feel may result.

7.2 If trucks arrive in a Poisson fashion at a rate of 20 vehicles per 8-hour day, and truck-unloading time is exponentially distributed with a mean of 40 minutes, how many docks should be planned if the goal is an average total truck turnaround time of less than 50 minutes?

7.3 A new musical textbook production plant is to be located along a street running north and south. The plant is to be 800 ft long and 500 ft wide. The property is 900 ft long and 700 ft deep. Four receiving docks are to be located in one rear corner of the plant and four shipping docks on the other rear corner. The trucks servicing the plant are all 55 ft. Sufficient truck buffer area must be planned for three trucks for both receiving and shipping. Draw a plot plan for the site.

7.4   A hospital is to have two receiving docks for receipt of foodstuffs, paper forms, and supplies. Forty-foot carriers will serve the hospital. A narrow-aisle lift truck will be used to unload the carriers. All foodstuffs, paper forms, and supplies are stored immediately upon receipt. What will be the impact on space requirements if 45° finger docks are used instead of 90° docks?

7.5   A warehousing consultant has recommended that your firm eliminate portable dockboards and install permanent adjustable dockboards to achieve considerable space savings. What is your reaction to this recommendation?

7.6   Perform a survey of the receiving areas on campus and determine how to handle the problem of differences between dock and carrier heights. Also determine if any potential for dock shelters exists.

7.7   The legal limits on trucks continue to increase. The length, width, and height have all increased in the past 10 years. What impact does this have on the planning of receiving and shipping areas?

7.8   What fraction of the total cost of operating a warehouse is typically associated with order picking?

7.9   What are four order-batching procedures?

7.10   What are five tasks typically performed by an order picker?

7.11   What are three principles of intelligent stock-assignment planning?

7.12   List five key data elements to consider in designing an order picking system.

7.13   Batch picking, as opposed to strict order picking, reduces the _____ between picks, but increases _____ requirements.

7.14   By reducing the amount of stock in the forward picking area, forward picking costs (a) increase, (b) decrease and the cost to replenish the forward picking area (c) increases, (d) decreases.

7.15   What are the three key data elements in determining if an item should be located in the forward picking area?

7.16   What are the two key data elements in determining the location assignment for an item when using popularity-based storage?

7.17   Popularity-based storage generally causes _____ to decrease and _____ to increase.

7.18   List four concepts that make the order picking function difficult to manage.

7.19   What are the two principal errors made in order picking?

7.20   In a correlated stock assignment policy, items frequently requested together are stored in _____.

7.21   Visit a warehouse and categorize the type of storage used (dedicated vs. randomized vs. a combination of the two). Identify the technologies used to transport material to and from storage, to place material in and retrieve material from storage, and to store the material. Assess the age and condition of the technologies identified in the warehouse. Identify the bar-code symbols used, as well as the types of bar-code readers used. How are the bar-code labels printed and applied?

Describe the approach used to perform receiving and shipping. Identify the various pallet dimensions and designs found in the warehouse. Assess the quality of housekeeping in the warehouse. Determine the fraction of floor space devoted to the various functions performed in the warehouse. Compute the fraction of floor space devoted to aisles. If racked storage is used, determine the fraction of the rack openings that are empty. Also, estimate the fraction of cube space within a storage opening that is utilized (i.e., how much is material, including pallets). Assess the use of cube space within the warehouse.

7.22 Consider three items with the following profiles:

Item	Annual Demand	Annual Requests	Unit Cube
A	100 units	25 requests	1 ft^3
B	50 units	5 requests	0.5 ft^3
C	200 units	16 requests	2 ft^3

a. What is the total demand in cube for items A, B, and C?

b. If 20 ft^3 of space is allocated for item C, how many times will the item be replenished in 1 year?

c. If each item is to receive a location in the forward area and an equal-time-supply allocation policy is used, how much space will be allocated for each item assuring a one-week time supply (assume 50 weeks)?

d. If the items are to be assigned locations in the forward area on the basis of popularity storage, and they are allocated space as in part c, rank the items so as to maximize the number of picks in the most accessible storage space.

7.23 Under what circumstances or conditions would you prefer to store goods in a warehouse by:

a. Placing the product directly on the floor

b. Palletizing and stacking pallet loads (block storage)

c. Palletizing and storing in conventional pallet rack

d. Palletizing and storing in pallet flow rack

e. Palletizing and storing in drive-in rack

f. Palletizing and storing in drive-through rack

g. Placing cases directly in flow rack

h. Storing in cantilever rack

i. Palletizing and stacking using portable stacking rack

j. Placing the product directly in a bin

7.24 Given the assignment to design a new warehouse for storing finished goods involving 1200 different stock-keeping units (SKUs), how would you determine the storage space required? Distinguish between dedicated storage, randomized storage, and class-based dedicated storage as they impact space, handling time, and the stock location system.

7.25 Given a warehouse similar to that given in Figure 6.20*a*, where would you position the following items?

Product	Monthly Throughput	Quantity per Receipt	Trips to Receive	Average Customer Order Size	Trips to Ship
A	High	300 pallets	300	2.0 pallets	150
B	Low	200 cartons	50	4.0 pallets	50
C	Low	10 pallets	10	0.2 pallets	50
D	High	400 pallets	400	0.5 pallets	800
E	High	6000 cartons	1000	10.0 cartons	600
F	Low	40 cartons	40	2.0 pallets	20
G	High	200 pallets	200	1.0 pallets	200
H	High	9000 cartons	2250	5.0 cartons	1800
I	Low	50 pallets	50	1.0 pallet	50
J	High	500 pallets	500	0.7 pallets	715
K	Low	80 pallets	80	2.0 pallets	40
L	High	400 pallets	400	1.0 pallets	400
M	High	7000 cartons	1167	3.0 cartons	2334
N	Low	700 cartons	140	7.0 cartons	100

7.26 All items to be stored in a warehouse are to be block-stacked. All loads are on 48- × 48- × 6-inch pallets. Each load is 4 ft tall. There are 30 different products to be stored, and it is planned to store 300 loads of each product. The loads may be stacked four high in a warehouse that has a 22-ft clear ceiling height. The receiving and shipping docks are to be located along one side of the warehouse. A counterbalanced lift truck is to operate in the warehouse, so 13-ft aisles are needed. Develop a layout for the warehouse.

7.27 What is the relationship between cube utilization and product accessibility?

7.28 Explain the advantages and disadvantages of each of the layouts in Figure 7.35.

7.29 Materials are stored on 42- × 48-in. pallets in selective pallet racks, *double deep*. The pallet has a height of 5″. The load stored on the pallet is 46 in. high. Two pallet loads are

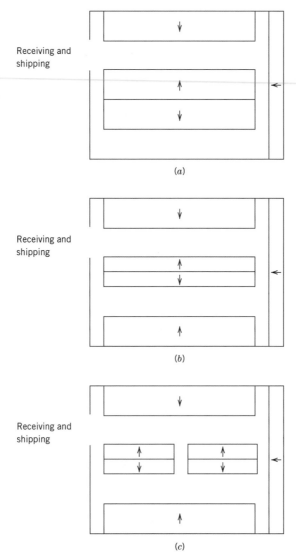

Figure 7.35 Warehouse layouts for Problem 7.28.

stored side by side in each rack opening. Likewise, two pallet loads are stored back to back in each rack opening. A 5-in. clearance is used (side to side) between loads in an opening and between a load and a vertical rack member. In an opening, the load in the first-depth position just touches the load in the second-depth position. Each vertical rack member is 4″ wide. Pallets are placed on 4-in. load beams, including the bottom load. There is a clearance of 4″ between the top of each load and the load beam above it. An 8.5-ft aisle (load to load) and a flue spacing (back to back between loads in the second-depth position) of 12″ is used. The rack has five tiers of storage. Loads do not overhang the pallet. Compute the storage space utilization assuming the loads in the second-depth position are full and the loads in the first-depth position are half full.

7.30 Materials are stored on 42- × 36-in. pallets in selective pallet racks, *one deep*. The pallet has a height of 6″. The load stored on the pallet is 48″ high. Three pallets are stored (side by side) in each rack opening. A 4-in. clearance is used between loads in an opening and between a load and a vertical rack member (upright truss). Each vertical rack member is 3″ wide. Pallets are placed on 4-in. load beams except the bottom load, which is stored on the floor. There is a clearance of 3″ between the top of each load and the load beam above it. An 8.5-ft aisle (load to load) and a flue spacing of 15″ between loads are used. The rack has four tiers of storage. Loads do not overhang the pallet. Compute the storage space utilization.

7.31 Materials are stored on 42- × 36-in. pallets in selective pallet racks, *one deep*. The pallet has a height of 6″. The load stored on the pallet is 40″ high. Two pallets are stored (side by side) in each rack opening. A 4-in. clearance is used between loads in an opening and between a load and a vertical rack member (upright truss). Each vertical rack member is 3″ wide. Pallets are placed on 4-in. load beams except the bottom load, which is stored on the floor. There is a clearance of 4″ between the top of each load and the load beam above it. A 10-ft aisle (load to load) and a back-to-back spacing of 12″ between loads are used. The rack has six tiers of storage. Loads do not overhang the pallet. Compute the storage space utilization.

8

MANUFACTURING SYSTEMS

8.1 INTRODUCTION

Facilities design for manufacturing systems is extremely important because of the economic dependence of the firm on manufacturing performance. Since manufacturing, as a whole, is a *value-adding* function, the efficiency of the manufacturing activities will make a major contribution to the firm's short- and long-run economic profitability.

Greater emphasis on improved quality, decreased inventories, and increased productivity encouraged the design of manufacturing facilities that are integrated, flexible, and responsive. The effectiveness of the facility layout and material handling in these facilities will be influenced by a number of factors, including changes in

- Product mix and design
- Processing and materials technology
- Handling, storage, and control technology
- Production volumes, schedules, and routings
- Management philosophies

Today's environment requires that companies quickly respond to varying customer requirements. Manufacturing strategies, such as just-in-time (JIT) production, lean manufacturing, and other strategies have emerged as viable and effective methods for achieving efficiency. Their successes have been rooted in the total commitment of companies to solve their problems on a daily basis.

But first, let us look at some historical perspective. Not too long ago, automation was viewed as the panacea of many U.S. manufacturers. One goal then was to build and operate a totally lights-out factory, that is, to build the *automatic factory*.

464

The concept of an automatic factory has captured the imagination of both managers and engineers throughout the world. The automatic factory can be distinguished from the *automated* factory as follows: the automatic factory is essentially a paperless factory. In the automated factory, automation and mechanization are dominant; however, people perform a limited number of direct tasks and a greater number of indirect tasks. In the automated factory, factory personnel resolve unusual situations; in the automatic factory, every effort is made to ensure that unusual situations do not arise.

Not all facilities planning efforts involving new production plants will result in automatic factory designs. To the contrary, it is likely that few will justify such levels of automation, at least in the short run. However, in developing strategic facilities plans, it is important to plan for upgrading facilities to accommodate new technology, changing attitudes concerning automation, and increased understanding concerning factory automation. Firms producing high-technology products may choose to subcontract component manufacturing and labor-intensive activities. They may focus only on high-technology design and processing activities that represent significant value-added contributions.

Many external factors affect the facility planning process. Among the factors that appear to have a significant impact are

- Volume of production
- Variety of production
- Value of each product

Each of these factors may lead to different types of manufacturing facilities (e.g., job shop, production line, cellular manufacturing, etc.). In complex facilities, all types of manufacturing systems will most likely exist.

Facilities development often involves an incremental approach to making the transition from a conventional or mechanized factory to the automated factory. While this strategy may be the only viable alternative due to limited capital investment dollars, the ultimate goal should be to bridge the pockets of automation and move toward an integrated factory system. Integration does not necessarily mean automation. In some situations, "information integration" may be the only choice with the physical system not integrated. This situation is often observed in older facilities.

From a system viewpoint, an incremental approach is not necessarily bad, so long as the steps are considered to be interim steps in a phased implementation of an automation project. However, to obtain an integrated factory system, the subsystems must be linked together. An obvious approach is to build physical bridges that join together automated subsystems. Likewise, information bridges can be provided through a factorywide control system.

The modern factory must have a brain and circulation system; a set of plantwide control system software must tie all the automated hardware subsystems together with an integrated, automated material handling system. The software requirements include the following:

- Integrate material handling information flows with shop floor control information.
- Assign and schedule material handling resources.
- Provide real-time control of material move, store, and retrieve actions.

The ability to know when and where the resources are needed is necessary to achieve proper control. Detailed information on every product and major components found in the bill of materials, as well as tools, fixtures, and other production resources, must be captured and maintained in real-time. The implementation of an automated data capture and communication system is a critical element to achieving a high-performance manufacturing system.

The critical types of information needed to integrate material handling plantwide include:

- Identification and quantification of items flowing through the system
- Location of each item
- Current time relative to some master production schedule

Such information is necessary in order to synchronize the multitude of transactions that take place in a manufacturing system.

Alternative routing paths and buffer storage are two methods that may protect the factory from catastrophic interrupts and delays. However, to make use of alternative paths a realtime dispatching decision must be made. The material handling system control must have the flexibility to determine which of several alternate paths should be taken and the physical resources needed to execute the decision.

In designing an integrated material handling system, the following objectives should be considered:

- Create an environment that results in the production of high-quality products.
- Provide planned and orderly flows of material, equipment, people, and information.
- Design a layout and material handling system that can be easily adapted to changes in product mix and production volumes.
- Reduce work-in-process and provide controlled flow and storage of materials.
- Reduce material handling at and between workstations.
- Utilize the capabilities that people have from the shoulders up, not the shoulders down.
- Deliver parts to workstations in the right quantities and physically positioned to allow automatic transfer and automatic parts feeding to machines.
- Deliver tooling to machines in the right position to allow automatic unloading and automatic tool change.
- Utilize space most effectively, considering overhead space and impediments to cross traffic.

But more importantly, products must be designed for both manufacturability and ease of handling. Specifically, the shapes and sizes of materials, parts, tooling, subassemblies, and assemblies must be carefully considered to ensure that automatic transfers, loading, and unloading can be performed.

The past decades have been characterized as the era that strived to develop the automatic factory. A noble goal, perhaps, but looking back it seems to have been too ambitious. Many of today's factories do not strive to be fully automated. The current thinking is more toward finding the right blend of machine and human performance characteristics. The primary goal is to increase customer satisfaction at a reasonable price. Achieving the economic goal requires that the company reduce work-in-process, increase performance of individual machines, reduce capital cost at the same level of production capacity, and achieve higher productivity from factory personnel.

We now look at several types of manufacturing systems. They are representative of those commonly found in industry.

8.2   FIXED AUTOMATION SYSTEMS

Transfer Line

In a transfer line, materials flow from one workstation to the next in a sequential manner. Because of the serial dependency of the transfer line, the production rate for the line is governed by the slowest operation. A transfer line is one example of hard automation.

Transfer lines are often used for high-volume production and are highly automated. In highly automated lines, the processing rates of individual machines are matched so that there is usually no need for buffer storage between machines. Or, if there is buffer storage, it would be for reasons of expected machine breakdowns.

The transfer line offers production rates unmatched by other types of manufacturing systems. But its disadvantages are:

- Very high equipment cost
- Inflexible in the number of products manufactured
- Inflexible layout
- Large deviation in production rates in case of equipment failure in the line

Many features of a transfer line can be observed on manual, paced assembly lines. Inventory banks or buffers can be placed between workstations to compensate for variations in production rates at individual workstations. Production rate variations occur due to machine failures, the inherent variation in operator performance, uncertainty in the arrival of components needed at individual workstations, and other causes.

The design of transfer lines includes both the specification of the individual processing stages and the linkage of the stages. The performance of the system is dependent on the layout of the facility, the scheduling of production, reliability of the individual stages in terms of processing variability and machine failures, and the loading of the line. Facility planning for the transfer line is relatively straightforward. The processing equipment is arranged according to the processing sequence. Buffer sizes between workstations must be determined and accommodated by several types of material handling devices such as vertical storage systems or spiral-type conveyors. Or, the spacing between machines can be increased to accommodate buffer stock.

As a result of the predictable flow sequences, straight-line flow and its derivatives are frequently used. Figure 8.1 shows several variations of the straight-line flow structure.

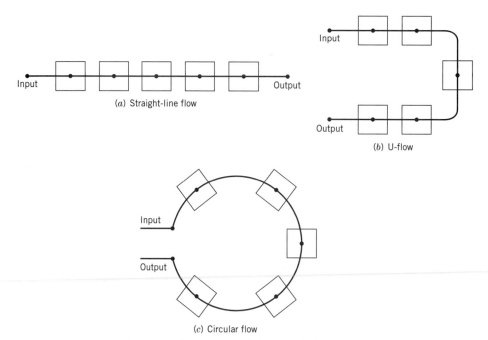

Figure 8.1 Several variations of the straight-line flow pattern.

Dial Indexing Machine

One particular implementation of the circular flow is the dial indexing machine (Groover [21]), where the workstations and the input/output stations are arranged in a circular pattern. The worktable where the parts are mounted is indexed clockwise or counterclockwise at predetermined times based on the required processing rate. This configuration corresponds to the circular pattern in Figure 8.1*c*.

Although they are not referred to as transfer lines, a number of production systems have characteristics that are similar to those of transfer lines. A few examples include

- Automated assembly systems for automotive parts
- Chemical process production systems
- Beverage bottling and canning processes
- Heat treatment and surface treatment processes
- Steel fabrication processes

The common characteristic among the above examples is that the automation is fixed and the processing is synchronous. Parts go from one machine to another without manual intervention. Materials are transferred from one machine to the next in sequence by an automated material handling system.

As with transfer lines, the development of the layout and material handling system involves the placement of workstations in processing sequence, specification of spaces between workstations, as well as the determination of the type of handling method to use.

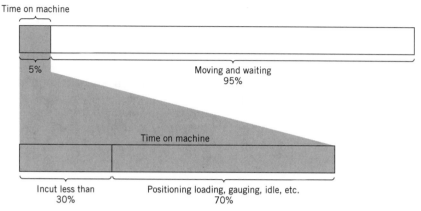

Time on machine

5%

Moving and waiting
95%

Time on machine

Incut less than
30%

Positioning loading, gauging, idle, etc.
70%

Figure 8.2 Pictorial representation of work-in-process after being released to the shop floor (From Merchant [35]).

8.3 FLEXIBLE MANUFACTURING SYSTEMS

In a study of batch-type metal-cutting production, Merchant [35] reported that the average workpiece spends only about 5% of its time on machine tools (see Figure 8.2).

Furthermore, of the time the material is loaded on a machine, it is being processed less than 30% of the time. It is argued that gaining control of the "95% of the time parts are not machined" involves linking the machines by an automated material handling system, and computerizing the entire system and operation. The term *flexible manufacturing system* (FMS) has thus emerged to describe the system.

The word "flexible" is associated with such a system since it is able to manufacture a large number of different part types. The components of a flexible manufacturing system are the processing equipment, the material handling equipment, and the computer control equipment. The computer control equipment is used to track parts and manage the overall flexible manufacturing system.

Groover [20] lists the following manufacturing situations for which a flexible manufacturing system might be appropriate:

- Production of families of workparts
- Random launching of workparts onto the system
- Reduced manufacturing lead time
- Increased machine utilization
- Reduced direct and indirect labor
- Better management control

In designing the material handling system for a flexible manufacturing system the following design requirements are recommended by Groover [20]:

- Random, independent movement of palletized workparts between workstations
- Temporary storage or banking of workparts
- Convenient access for loading and unloading
- Compatible with computer control

- Provision for future expansion
- Adherence to all applicable industrial codes
- Access to machine tools
- Operation in shop environment

Flexible manufacturing systems are used to cope with change. Specifically, the following changes are accommodated:

- Processing technology
- Processing sequence
- Production volumes
- Product sizes
- Product mixes

Flexible manufacturing systems are designed for small-batch (low-volume) and high-variety conditions. Whereas hard automation manufacturing systems are frequently justified on the basis of economies of scale, flexible automation is justified on the basis of scope. The concept of flexible manufacturing system is one that has very broad scope.

Flexibility can be accomplished through

- Standardized handling and storage components
- Independent production units (manufacturing, assembly, inspection, etc.)
- Flexible material delivery system
- Centralized work-in-process storage
- High degree of control

Because of the variety of alternatives for storing, handling, and controlling material, the specification of the material handling system and design of the layout can be quite varied. Figures 8.3*a* and *b* show two alternative configurations for the same flexible manufacturing system. Figure 8.3*a* illustrates an external centralized work-in-process storage while Figure 8.3*b* shows an internal centralized work-in-process storage. Work-in-process is needed because a part has to go through several machines before it eventually leaves the system after the last machining operation is performed. A more detailed discussion on reduction of work-in-process is provided in Section 8.5.

Yet another FMS configuration based on cellular manufacturing principles is illustrated in Figure 8.3*c*. In this configuration, the handling distances are reduced significantly as the machines are placed within the work envelope of the transfer device, a robot handler in this illustration. This configuration has some built-in scalability as only one cell can be used in case there is a reduction in the demand for the products that are machined in this system.

We conclude from these illustrations that there are multiple alternative configurations to the FMS cell design problem. The determination of which configuration is best requires a detailed comparison using analytical tools such as simulation analysis.

What makes FMS flexible? According to Groover [21], for a manufacturing system to be categorized as flexible, it must have the following capabilities:

1. Process different part styles in a nonbatch mode.
2. Accept changes in production schedule.

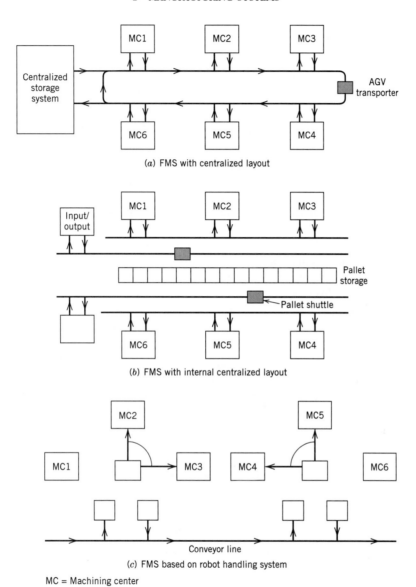

(a) FMS with centralized layout

(b) FMS with internal centralized layout

(c) FMS based on robot handling system

MC = Machining center

Figure 8.3 Alternative layouts for a flexible manufacturing system.

3. Respond gracefully to equipment malfunction and breakdowns in the system.
4. Accommodate the introduction of new part designs.

From these conditions, we can say that an automated system is not always flexible (e.g., transfer lines), and vice versa, that a flexible system need not be automated (manual assembly lines).

In the next section, we consider a special case of flexible manufacturing system, the single-stage multimachine system.

8.4 SINGLE-STAGE MULTIMACHINE SYSTEMS

Yet another alternative in automation in machining systems is what has been termed *single-stage multimachine systems* (SSMS). The discussion below is based on Koo [26] and Koo and Tanchoco [27]. SSMS is described in terms of the resources involved.

1. *Manufacturing configuration and machines.* The machining centers in SSMS are very versatile machines. They are all identical and versatile so that all operations on any part can be performed on any one machine. This being the case, there will be only one setup per part. There is a tool magazine on each machine, but with limited capacity. The tool handling system supplies the tools that are not resident on the machine. Each machine has a separate input and output spur with limited capacity. Once a part is loaded on a machine, it does not leave the machine until all required operations are performed. Thus, the part is moved to and from the machine only twice. Under this type of manufacturing system, the tool delivery system becomes the critical resource. Figure 8.4 illustrates two alternative tool delivery systems in a single-stage multimachine system.

2. *Parts.* Parts arrive according to the production requirements schedule. As mentioned above, parts visit only one machine since all operations can be performed

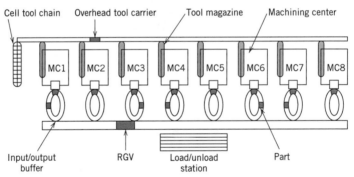

(*a*) A system with a centralized tool storage (Sharma, et al. [50])

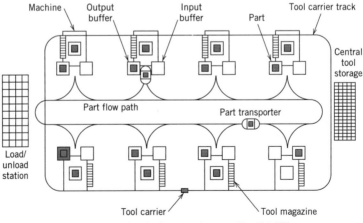

(*b*) A system with distributed tool storage (Koo [26,27])

Figure 8.4 Altenative tool handling system in single-stage multimachine systems.

in a single machine. A part transporter system handles the delivery of parts from the input station to the machine and from the machine to the output station. A part cannot be processed unless the required tool is available.

3. *Tools.* The required tools are specified by the process plan. Some tools are resident to the machine while others are stored in a centralized tool storage system. These tools are generally expensive and storing a complete tool set resident in a machine will be very expensive. Dynamic sharing of tools may be the economic choice. Still, there is a decision that must be made on what percentage of the tools should be resident or shared.

4. *Part transporter.* The part transportation activity is minimal since a part visits a machine only once. Deadheading still occurs and there must be an efficient dispatching algorithm for the vehicles. Vehicles are dispatched based on part arrivals and job completions when vehicles are idle, or when vehicles become idle with one or more parts waiting to be transported.

5. *Tool carriers.* A tool carrier handles the transport of tools to and from the centralized tool storage and the machines requiring the tool. Under dynamic tool sharing, these tools are dynamically dispatched to the machines needing them. One can also reallocate the tools when opportunities arise. Figure 8.5 illustrates a combined schedule of machines and tools. The determination of the combined schedule of machines and tools is a critical element in the operation of SSMS.

The same layout can be used for both FMS and SSMS; the difference is in the composition of the machine tools. In SSMS, the machine tools are identical, versatile machines. Transfer lines, flexible manufacturing systems, and single-stage multimachine systems address only one aspect of manufacturing; the machining operations. Furthermore, they are all automated systems, typically operating without manual intervention. What about manufacturing systems with people and machines working together? Is it possible to attain the efficiency of transfer lines with the flexibility of the flexible manufacturing system? We will refer to this question again, but first we will look at the work-in-process and discuss how control of work-in-process may be accomplished. The topic of Section 8.5 is reduction of work-in-process.

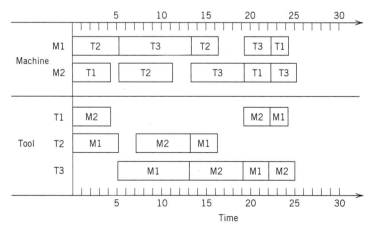

Figure 8.5 A combined schedule of machines and tools. (*From Koo [26, 27]*)

8.5   REDUCTION OF WORK-IN-PROCESS

In this section we focus on issues related to handling and storing operations in a manufacturing environment. In-process handling includes the movement of material, tooling, and supplies to and from production units, as well as the handling that occurs at a workstation or machine center. The term *in-process storage* can include the storage of material, tooling, and supplies needed to support production. However, as shown in Figure 8.6, the term normally applies to the storage of material in a semifinished state of production.

Several rules of thumb can be used in designing in-process handling and storage systems. Among them are the following:

- Handling less is best.
- Grab, hold, and don't turn loose.
- Eliminate, combine, and simplify.
- Moving and storing material incur costs.
- Preposition material.

Handling less is best suggests that handling should be eliminated if possible. It also suggests that the number of times materials are picked up and put down and the distances materials are moved should be reduced.

Grab, hold, and don't turn loose emphasizes the importance of maintaining physical control of material. Too often parts are processed and dumped into tote boxes or wire baskets. Subsequently, someone must handle each part individually to orient and position it for the next operation.

Eliminate, combine, and simplify suggests the principles of work simplification and methods improvement appropriate in designing in-process handling and storage systems. Handling and storage can frequently be completely eliminated by making changes in processing sequence or production scheduling. Certainly, it is possible to combine handling tasks through the use of standardized containers.

Moving and storing material incur costs serves as a reminder that inventory levels should be kept as small as possible. Reducing inventories is one of the goals of JIT production and lean manufacturing. The underlying principle is to move material only when it is needed and store it only if you have to. Moving materials incurs personnel and equipment time and costs, and it increases the likelihood of product damage. Finally, moving materials requires a corridor of space for movement and there are costs associated with building and maintaining aisle space. Basically, the longer material stays in the plant the more costly the product will be since no value is added to the product while material is moved and/or stored.

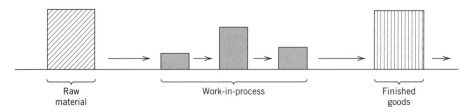

Figure 8.6   Inventories in the manufacturing cycle. (*From Apple and Strahan [4]*)

Prepositioning material has two aspects to be considered. First, parts should be prepositioned to facilitate automatic load/unload, insertion, inspection, and so on. Second, when material is delivered to a workstation and/or machine center, it should be placed in a prespecified location with a designated orientation. Too often, direct labor personnel need to spend non-value-adding time to preposition the material for machine loading.

The application of the rules of thumb discussed above can be used to improve operations so that non-value-adding time is reduced and consequently the manufacturing cycle time is shortened. Reduction in wasted time increases available machine time. Reducing cycle time means that products are completed early and therefore the company can receive revenue dollars earlier. Thus, the turnover of capital is increased, resulting in lower capital costs and higher profits to the company.

Control of work-in-process has been the focus of many manufacturing firms. Demand-driven manufacturing systems (produce only what the customer needs at the time it is needed) have gained popularity spawned by the success of just-in-time production at Toyota Motor Company in Japan.

8.6 JUST-IN-TIME MANUFACTURING

The *just-in-time* (JIT) production system was developed more than three decades ago by Ohno Taiichi at the Toyota Motor Company in Japan. In a broad sense, JIT applies to all forms of manufacturing such as jobshop, process, and repetitive manufacturing.

According to Ohno [65],

> No matter how difficult it may be, take it up as a challenge to reduce man-hours as a means of improving efficiency. . . . At Toyota, in order to proceed with our man-hour reduction activities, we divide waste into the following seven categories:
>
> 1. Waste arising from overproduction
> 2. Waste arising from time on hand (waiting)
> 3. Waste arising from transporting
> 4. Waste arising from processing itself
> 5. Waste arising from unnecessary stock on hand
> 6. Waste arising from unnecessary motion
> 7. Waste arising from producing defective goods

Waste arising from overproduction is a result of a production philosophy based on achieving economies of scale; that is, as the production lot increases, the unit production cost will decrease. The classic economic order quantity lot-sizing model is an example. To eliminate waste due to overproduction, Ohno says "production lines must be reorganized, rules must be established to prevent overproduction, and restraints against overproduction must become a built-in feature of any equipment within the workplace."

Waste from time on hand (waiting) arises when a worker tends only one machine. Assigning more machines per worker will reduce this waste. A better layout may also contribute to maximizing the number of machines that one worker can tend. And if the machines are not identical, the worker must be trained to operate the different machines involved. This has resulted in workers with multifunctional skills.

Waste from transporting comes from moving items over long distances, from work-in-process storage, and from arranging and/or rearranging parts in containers and/or pallets. The principle of simplification can be beneficial in reducing transportation waste. Waste from transporting can be eliminated if the machines are placed closed to each other. By doing so, a worker can handle the transportation of parts from one machine to the next without unitizing. The transport task can be accomplished within the machine cycle time. It is also considered a waste for parts to go through several logistic steps from the warehouse to the factory and eventually to the hands of the worker.

Waste from processing occurs when an operator is unnecessarily tied up on a machine due to poor workplace design. For example, a pneumatic clamping device to hold work parts allows an operator to do more productive work.

Reducing the labor content of the cost of making products is key to increasing the productivity of the plant. This statement does not mean that we need to automate the process; it means improving the process so that unnecessary labor time is reduced.

We conclude this discussion by stating that the goal of JIT production is consistent with the definition of material handling given in this book. Standard and Davis [59] state that "Just-in-time means having the right part at the right place in the right amount at the right time." We defined material handling as "providing the right amount of the right material, in the right condition, at the right place, at the right time, in the right position, in the right sequence, and for the right costs, by using the right method."

We now look at specific ways by which the JIT philosophy is implemented. We categorize these methods into five elements: visibility, simplicity, flexibility, standardization, and organization.

1. *Visibility* can be obtained by the following technique: electronic boards for quick feedback, pull system with *kanbans*,[1] problem boards, colored standard containers, decentralized storage system, marked dedicated areas for inventory, tools, etc.

2. *Simplicity* can be achieved with a pull system with kanbans, simple setup changes, certified processes, small lot sizes, leveled production, simple machines, simple material handling, multifunctional workers, teamwork, etc.

3. *Flexibility* can be achieved with short setup times, short production times, small lot sizes, kanban carts, flexible material handling equipment, multifunctional employees, mixed-model sequencing lines, etc.

4. *Standardization* of tools, equipment, pallets, methods, containers, boxes, materials, and processes is pursued.

5. *Organization* is required for setups, for cleanliness, for work areas, for the kanban system, for the storage areas, for tools, for teamwork activities, etc.

[1]A *kanban* is a card or any signal used to request or authorize production of parts; it contains information on the part, the processes used, identification of the storage area for the part, and the number of parts to produce.

At the core of the JIT philosophy is the elimination of waste. To understand waste, it is important to recognize it first. Waste can be defined as any resource that adds cost but does not add value to the product. The following are the most common sources of waste in a manufacturing facility:

- Equipment
- Inventories
- Space
- Time
- Labor
- Handling
- Transportation
- Paperwork

Equipment can be underutilized for many reasons. Some reasons are poor scheduling methods, inadequate maintenance programs, absence of feedback mechanisms to report maintenance problems, inadequate spare parts inventory, operators not trained in the use of equipment and in basic preventive maintenance, inefficient product design, long setup times and inefficient setup procedures, and large lot sizes.

Inventories for raw material, components, parts, work-in-progress, and finished products are usually wasted for the following reasons: large purchasing and production lot sizes, poor facility layout, inefficient material handling systems, inefficient packaging systems, inadequate training for the workers, poor organization, too many suppliers, centralized warehouses, inefficient product design, lack of standardization, awkward storage and retrieval systems, and poor inventory control systems.

Space is another key resource where waste can be identified by excess inventory, inappropriate facility layout, inadequate building design, unnecessary material handling, inefficient storage systems, and inefficient product design.

Time can be wasted due to excessive waiting or delays due to inefficient scheduling, machine downtimes, long repair times, poor-quality products, unreliable suppliers, poor facility layout, inefficient material handling systems, deficient inventory control systems, large lot sizes, and long setup times.

Labor is wasted when the employee is assigned to produce unnecessary products, to move and store unnecessary inventory, to rework bad products, to wait during machine repairs, and to receive unnecessary training. Labor is also wasted when the employee makes mistakes due to lack of training, individual responsibility, or mistake-proofing mechanisms.

From the JIT requirements point of view, all the subsystems in the manufacturing system must be improved.

JIT Impact on Facilities Design

There are many concepts and techniques related to the JIT production system that impact building design, facility layout, and the material handling system, such as:

- Reduction of inventories
- Deliveries to points of use

- Quality at the source
- Better communication, line balancing, and multifunctional workers

Reduction of Inventories

One of the main objectives of the JIT production system is the reduction of inventories. Inventories can be reduced if products are produced, purchased, and delivered in small lots; if the production schedule is leveled appropriately; if quality control procedures are improved; if production, material handling, and transportation equipment are maintained adequately; if products are pulled when needed and in the quantities needed; and so on. Therefore, if inventories can be reduced:

1. *Space requirements are reduced,* justifying arranging the machines closer to each other and the construction of smaller buildings. Reducing the distances between machines reduces the handling requirements.

2. *Smaller loads are moved and stored,* justifying the use of material handling and storage equipment alternatives for smaller loads.

3. *Storage requirements are reduced,* justifying the use of smaller and simpler storage systems and the reduction of material handling equipment to support the storage facilities.

Consequently, the building could be smaller, a better plant layout could be used, fewer handling and storage requirements might be needed, and the material handling and storage equipment alternatives could move and store smaller loads.

Deliveries to Points of Use

If products are purchased and produced in smaller lots, they should be delivered to the points of use to avoid stockouts at the consuming processes. Products can be delivered to the points of use if the building has multiple receiving docks and if a decentralized storage policy is used. If products can be delivered to the points of use, the following scenarios could happen:

1. If parts are coming from many suppliers, several receiving docks surrounding the plant could be required. Receiving docks need extra space for parking of trucks plus equipment for unloading, doors, and so on.Additionally, depending on the number of loads arriving at each receiving dock, in some cases some storage equipment could be needed at each decentralized storage area. Therefore, if parts are delivered to the points of use, the building could require multiple receiving docks, storage and handling equipment could be needed at each receiving dock, and the parts could be moved shorter distances and fewer times (promoting a shorter production cycle and improving the inventory turnover, reducing the possibility of having bad quality because of excessive handling, and detecting bad-quality parts more quickly). The storage and handling equipment alternatives to support multiple receiving docks are usually simple and inexpensive (i.e., pallet racks, pallet jack, pallet truck, walkie stacker). Side-loading trailer-trucks could also be used to avoid building expensive receiving docks.

2. A decentralized storage policy could be required to support the multiple receiving docks, and computerized or manual inventory control could be used. If inventory control is manual (using cards or kanbans), a centralized kanban control system could be used to collect the kanbans and request more deliveries from the suppliers, or a decentralized kanban control system could be used with returnable containers and/or withdrawal kanbans being returned to the supplier after every delivery.

3. If plant layout rearrangements have been performed to support the JIT concepts, internal deliveries to the points of use could be carried out using material handling equipment alternatives for smaller loads and short distances. If the JIT delivery concept is applied without performing layout rearrangements, faster material handling equipment alternatives could be justified, depending on the "pulling rate" of the consuming processes.

4. If a truck is serving different receiving docks, it could be loaded based on the unloading sequence. Side-loading trucks could also be used. If the traffic around the plant is problematic, a receiving terminal could be used to sequence the deliveries.

Quality at the Source

Every supplying process must regard the next consuming process as the ultimate customer and each consuming process must always be able to rely on receiving only good parts from its suppliers. Consequently, the transportation, material handling, and storage processes must deliver to the next process parts with the same level of quality that they received from the preceding process. To achieve the quality-at-the-source concept, the following could be required:

1. Proper packaging, stacking, and wrapping procedures for parts and boxes on pallets or containers

2. Efficient transportation, handling, and storage of parts

3. A production system that allows the worker to perform his or her operation without time pressure and that is supported by teamwork

Therefore, if quality at the source is implemented, better packaging, stacking, and wrapping procedures could be required; transportation, storage, and handling must be carefully done; and nonsynchronous assembly systems, sometimes with U-configurations, could be used. The assembly systems could be supported by the team approach concept, and quick feedback procedures supported by electronic boards could be used to achieve line balancing.

Better Communication, Line Balancing, and Multifunctional Workers

In many JIT manufacturing systems, U-shaped production lines are being used to promote better communication among workers, to use the multipleabilities of workers that allow them to perform different operations, and to easily balance to production line using visual aids and the team approach. Most of these applications have been very successful and have incorporated "problem boards" to write down the problems that occur during the shift, to analyze them at the end of the shift, and to generate solution methods to eliminate the future occurrence of these problems.

Layout arrangements with U-shaped patterns minimize material handling requirements (in most cases, they eliminate them). They also support the pulling procedure with kanbans. In some companies in which this layout arrangement is in operation, inventory buffers between processes of only one part or one container are used.

U-Shaped Flow Lines

A U-shaped flow line is a variation of the straight-line flow structure (Figure 8.7). It follows the concept of group technology, where parts of similar processing characteristics are grouped together for processing in a common area, or *manufacturing cell*. The shape is unique in that it allows more efficient use of operators tending the machines, and it promotes better communication among workers.

The transition from traditional processing to processing in a U-line has brought significant benefits to many companies. In a study of 114 U.S. and Japanese U-lines, Miltenburg [36] reports that "the average U-line has 10.2 machines and 3.4 operators. About one-quarter of all U-lines are manned by one operator and so run in chase

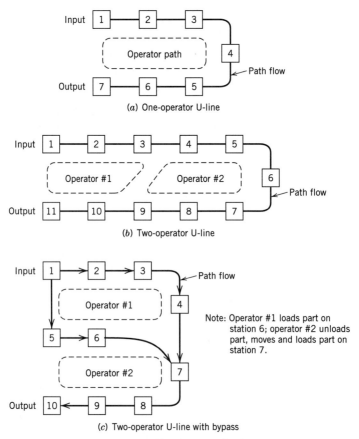

(a) One-operator U-line

(b) Two-operator U-line

(c) Two-operator U-line with bypass

Note: Operator #1 loads part on station 6; operator #2 unloads part, moves and loads part on station 7.

Figure 8.7 Illustrations of U-lines.

mode. The reported benefits are impressive. Productivity improved by an average of 76%. WIP dropped by 86%. Lead time shrank by 75%. Defective rates dropped by 83%."

The U-line comes in many different forms depending on the number of products manufactured, the number of machines in the line, and the number of operators tending the machines. For a one-operator U-line, the machines are laid out in a U-pattern and the operator works within the U-line. The machines are close together and the space between is limited by the size of the machine. Since proximity is important in reducing the material handling time and the operator walking time between adjacent machines, there are benefits to using small and simple machines. There is also the goal of minimizing the inventory between machines to a maximum of one unit. The operator in this type of U-line is said to be in "chase mode," that is, the operator is chasing the parts as they go from one machine to the next. Figure 8.7*a* illustrates a one-person U-line manufacturing or assembly cell. A two-operator U-line is illustrated in Figure 8.7*b*. A U-line with bypass is shown in Figure 8.7*c*. Other variations include the tandem-line, with separate U-lines on the same side of the aisle, and one where the second U-line, is across the aisle from the first line. See Figure 8.8 for an illustration. The configuration shown in Figure 8.8*a* illustrates independent lines while Figure 8.8*b* illustrates dependent lines. The dependency is based on using one of the operators assigned to tend machines on both lines.

More complex U-lines can be derived by using the simple U-line as a template. Miltenburg [36] shows a simple U-line with a bypass, the embedded U-lines, a figure 8

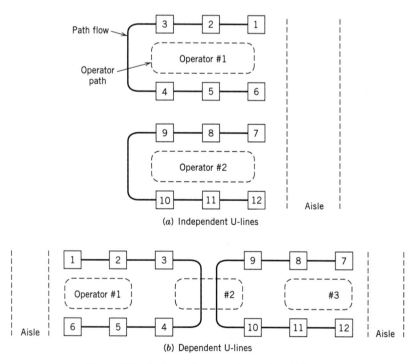

Figure 8.8 Independent vs. dependent U-lines.

pattern consisting of three lines with two operators, and the multi-U-line facility. The last category is designed to minimize the number of operators tending the machines.

One can undoubtedly come up with other types of U-line layouts based on the required number of machines and operators, and the required output rate. The objective here is to find an efficient pattern that will minimize the time to move parts between machines and the time it takes the operator to complete a cycle.

U-lines are scalable since they are able to adjust the line output to conform to an increase or decrease in production requirements based on the demand rate (i.e., the line is flexible to production volume changes). Scaling is done by increasing or decreasing the number of operators tending the line. Increasing output on short notice requires that there be excess processing capacity in the line. Rebalancing of the line is done quite often to adjust the line to varying demand quantities. Rebalancing also involves adding or removing machines on the line or changing the standard times based on the new layout configuration. The operator walking time will change depending on the new layout of the line. Rebalancing involves determining the number of operators required and assigning the machines that each operator tends.

Flexibility is achieved by making machines movable so that the machines can be re-laid out to minimize part transportation time and operator walking time. Putting machines on standard platforms that are easily lifted and moved, or using casters, are two ways of making the machines movable. Quick setup is necessary so that new parts can be immediately processed. Flexibility also requires that operators perform multiple tasks and are knowledgeable and versatile enough to tend all machines in the cell. It also requires that the tasks to be performed are clearly defined and the operator is trained to perform the tasks properly.

Scalability and flexibility make U-lines one of the most desirable manufacturing cell configurations.

Remarks

There is another interpretation of what has been the main thrust of production systems such as transfer lines, flexible manufacturing systems, and JIT production systems. From a manufacturing point of view, productive time means that a product is being transformed from one state to another (e.g., metal blocks converted to machined parts, component parts assembled to form end-products, etc.). The rest of the time is moving and waiting time. A more contemporary interpretation set within the context of lean manufacturing is the concept of value-adding and non-value-adding activities. Machining and assembly operations are value-adding activities while transportation and storage are non-value-adding activities. Reduction in non-value-adding times will lead to shorter cycle time and consequently work-in-process inventories will be lower. Reduction in non-value-adding activities leads to lean manufacturing.

The lean manufacturing concept can be summarized as follows:

- Eliminate or minimize non-value-adding activities.
- Produce only what is demanded.
- Minimize the use of time and space resources.
- Manufacture in the shortest cycle time possible.

These concepts are consistent with the just-in-time manufacturing philosophy.

Other versions of the JIT manufacturing system include:

- Stockless production
- Material as needed
- Continuous-flow manufacturing
- Zero-inventory production systems

We can expect to see the JIT manufacturing system continue to be refined by the requirements of individual companies.

8.7 FACILITIES PLANNING TRENDS

Factories are now being designed to provide the smoothest flow of materials, achieve flexibility (to adapt to proliferating product lines and volatile market demands), improve quality, increase productivity and space utilization, and simultaneously reduce facilities and operating costs. The following trends can be observed:

1. Buildings with more than one receiving dock and smaller in size (less space requirements for staging, storage, warehousing, offices, and manufacturing primarily because of JIT purchasing, small lot production, deliveries to the points of use, decentralized storage areas, better production and inventory control, visual management, manufacturing cells, EDI, focused factories, open-layout offices, decentralized functions, and simplified organizational structures). Multiple receiving docks require efficient loading/unloading of side-loading trucks when used.

2. Smaller centralized storage areas and more decentralized storage areas (supermarkets) for smaller and lighter loads.

3. Decentralized material handling equipment alternatives at receiving docks.

4. Material handling equipment alternatives for smaller and lighter loads.

5. Visible, accessible, returnable, durable, collapsible, stackable, and readily transferable containers.

6. Nonsynchronous production lines with mixed-model sequencing for "star" products.

7. Group technology layout arrangements for medium-volume products with similar characteristics to support cellular manufacturing concept.

8. U-shaped assembly lines and fabrication cells.

9. Better handling and transportation part protection.

10. Fewer material handling requirements at internal processes; more manual handling within manufacturing cells.

11. New fabrics and construction materials for dock seals and shelters.

12. Standard containers, trays, or pallets.

13. Side-loading trucks for easier access and faster loading and unloading operations.

14. The traditional function of storing changes to one of staging.

15. Use of bar codes, laser scanners, EDI, and machine vision to monitor and control the flow of units.

16. Focused factories with product and/or cellular manufacturing and decentralized support efforts.
17. Lift trucks equipped with radio data terminals and mounted scales to permit onboard weighing.
18. Flow through terminals and/or public warehouses to receive, sort, and route materials.
19. Facilities design process as a coordinated effort among many people.

8.8 SUMMARY

The objective of this chapter is to give a brief overview of manufacturing systems that may impact the facilities design function. Depending on the variety, volume, and value of products to be processed, different levels of automation, types of layout, and material handling systems will be appropriate. The design of transfer lines, flexible manufacturing systems, and single-step multimachine systems require a variety of automated material handling devices. Design of flow systems for automated production systems brings us back to the fundamental issues addressed in the layout section of this book. The continued trend toward just-in-time manufacturing puts the material handling and layout functions to the front line. Are they necessary to begin with? Layout methods based on job shop assumptions may have to be forgone in favor of cellular-based layouts. The layout of U-lines is one that will continue to be of interest. And perhaps there will be other flow structures that can augment the U-line configuration.

BIBLIOGRAPHY

1. Askins, G. G., and Standridge, C. R., *Modeling and Analysis of Manufacturing Systems,* Wiley, New York, 1993.
2. Albano, R. E., Friedman, D. J., and Hendryx, S. A., "Manufacturing Execution: Equipment," *AT&T Technical Journal,* July 1990, pp. 53–63.
3. Apple, J. M. Jr., and McGinnis, L. F., "Innovations in Facilities and Material Handling Systems: An Introduction," *Industrial Engineering,* vol. 19, no. 3, March 1987, pp. 33–38.
4. Apple, J. M. Jr., and B. B. Strahan, "In-Process Handling and Storage," presentation at the *32nd Annual Material Handling Short Course,* Georgia Institute of Technology, Atlanta, GA, March 1982.
5. Auguston, K. A., "For Just $142,000 We Doubled Picking Productivity," *Modern Materials Handling,* vol. 47, no. 2, February 1992, pp. 54–55.
6. Auguston, K. A., "Automated Handling Is Our Key to Quality," *Modern Materials Handling,* vol. 46, no. 6, May 1991, pp. 44–46.
7. Black, J. T., *The Design of the Factory with a Future,* McGraw Hill, New York, 1991.
8. Burgam, Patrick M., "JIT: On the Move and Out of the Aisles," *Manufacturing Engineering,* June 1984, pp. 65–71.
9. Chang, T., Wysk, R. A., and Wang, H., *Computer Aided Manufacturing,* Prentice Hall, Englewood Cliffs, NJ, 1991.

10. Costanza, J. R., *The Quantum Leap in Speed to Market,* JIT Institute of Technology, Inc., 1990.

11. Dauch, R. E., *Passion for Manufacturing,* Society of Manufacturing Engineers, Dearbon, MI, 1993.

12. Deming, W. E., *Quality, Productivity, and Competitive Position,* MIT, Cambridge, MA, 1982.

13. Deming, W. E., *The New Economics for Industry, Government, Education,* Massachusetts Institute of Technology, Center for Advanced Engineering Study, Cambridge, MA, 1993.

14. Forger, G., "Better Data Cuts Handling Time at McKesson," *Modern Materials Handling,* vol. 47, no. 9, August 1992, pp. 38–40.

15. Goddard, W. E., "Kanban—One Trick in a Magic Act," *Modern Materials Handling,* vol. 47, no. 5, April 1992, p. 31.

16. Goddard, W. E., *Just-In-Time, Surviving by Creating Tradition,* Oliver Wight Limited Publications, Essex Junction, VT, 1986.

17. Gould, L., "Manufacturing Cells Trim Throughout Time 80%," *Modern Materials Handling,* vol. 41, no. 10, September 1986, pp. 107–109.

18. Gould, L., "How Handling and Storage Support a FMS," *Modern Materials Handling,* vol. 42, no. 12, October 1987, pp. 69–71.

19. Greif, M., *The Visual Factory,* Productivity Press, Inc. Cambridge, MA, 1991.

20. Groover, M. P., *Automation, Production Systems, and Computer-Aided Manufacturing,* Prentice-Hall, Englewood Cliffs, NJ, 1980.

21. Groover, M. P., *Fundamentals of Modern Manufacturing,* Prentice-Hall, Englewood Cliffs, NJ, 1996.

22. Gunn, T. G., *Manufacturing for Competitive Advantage,* Harper Publishing, 1987.

23. Hirano, H., *JIT Factory Revolution,* Productivity Press, Cambridge, MA, 1988.

24. Kenfield, J. E., "A Nine-Step Approach to JIT Implementation," *1st International Electronics Assembly Conference,* Santa Clara, CA, October 7–9, 1985.

25. Kinney, H. D., Jr., and McGinnis, L. F., "Manufacturing Cells Solve Material Handling Problems," *Industrial Engineering,* vol. 19, no. 8, August 1987, pp. 54–60.

26. Koo, P. H., *Flow Planning and Control of Single-Stage Multimachine Systems,* Ph.D. dissertation, Purdue University, West Lafayette, IN 1996.

27. Koo, P. H., and J. M. A. Tanchoco, "Real-Time Operation and Tool Selection in Single-Stage Multimachine Systems," *Int. J. of Production Research,* vol. 37, no. 5, 1999.

28. Kusiak, A., *Intelligent Manufacturing System,* Prentice Hall, Englewood Cliffs, NJ, 1990.

29. Kulwiec, R., "Toyota's New Parts Center Targets Top Service," *Modern Materials Handling,* vol. 47, no. 13, November 1992, pp. 73–75.

30. Lorincz, J. A., "Machine Cells Meet JIT Need," *Tooling & Production,* May 1992, pp. 50–54.

31. Lubben, R. T., *Just-In-Time Manufacturing,* McGraw-Hill, New York, 1988, p. 57.

32. Majima, I., *The Shift to JIT,* Productivity Press, Cambridge, MA, 1982.

33. Maleki, R. A., *FMS: The Technology and Management,* Prentice Hall, Englewood Cliffs, NJ, 1991.

34. McGinnis, L. F., Toro-Ramos, Z., and Trevino, J., *A Review of the Toyota Production System,* Material Handling Research Center, RS-85-01, Georgia Institute of Technology, Atlanta, September 1985.

35. Merchant, M. E., "The Factory of the Future," paper presented at the *American Association of Engineering Societies Annual Meeting,* Port St. Lucie, FL, May 1982.

36. Miltenburg, J., "U-Shaped Production Lines: A Review of Theory and Practice," *Int. J. Production Economics,* vol. 70, 2001, pp. 201–214.

37. Monden, Y., *Toyota Production System,* Industrial Engineering and Management Press, Institute of Industrial Engineers, Norcross, GA, 1983.

38. Nakajima, S., *Introduction to 1PM,* Productivity Press, Cambridge, MA, 1988.

39. Nyman, L. R., *Making Manufacturing Cells Work,* SME, 1992.

40. Ohno T., and Setsuo, M., *Just-in-Time for Today and Tomorrow,* Productivity Press, Cambridge, MA, 1988.

41. Parmelee, R. A., "JIT Implementation by the Numbers," *American Production and Inventory Control Society, Conference Proceedings,* 1985, pp. 457–461.

42. Peeler, D., *QFD: Customer-Driven Engineering,* Hewlett-Packard Quality Publishing, Spring 1990.

43. Robinson, A., *Continuous Improvement in Operations,* Productivity Press, Cambridge, MA, 1991.

44. Russell, R. S., and Taylor, B. W., *Operations Management,* 3rd ed., Prentice-Hall, Englewood Cliffs, NJ, 2000.

45. Sandras, B., and Johnson, T., "Job Shop JIT," *1985 Conference Proceedings,* American Production and Inventory Control Society, 1988, pp. 448–452.

46. Schonberger, R. J., *Japanese Manufacturing Techniques,* Free Press, New York, 1982.

47. Schonberger, R. J., *World Class Manufacturing,* Free Press, New York, 1986.

48. Sepehri, M., "Case in Point: Buick City Genuine JIT Delivery," *P&IM Review with APICS News,* 1988, pp. 34–36.

49. Sepehri, M., *Just-in-Time, Not Just in Japan,* Library of American Production, 1986.

50. Sharma, R., Barash, M., and J. Talavage, *Simulation-based Evaluation of Alternative Tool Delivery Systems in a Flexible Manufacturing Cell,* Technical Report, TR-ERC 92-7, School of Industrial Engineering, Purdue University, West Lafayette, IN, 1992.

51. Sheridan, J. H., "Toward the CIM Solution," *Industry Week,* vol. 238, no. 20, October 1989, pp. 35–80.

52. Shingo, S., *A Revolution in Manufacturing: The SMED System,* Productivity Press, Cambridge, MA, 1985.

53. Shingo, S., *Zero Quality Control,* Productivity Press, Cambridge, MA, 1986.

54. Shingo, S., *Non-Stock Production,* Productivity Press, Cambridge, MA, 1988.

55. Shingo, S., *Study of Toyota Production System,* Japan Management Association, Tokyo, Japan, 1981.

56. Shingo, S., *The Sayings of Shigeo Shingo,* Productivity Press, Cambridge, MA, 1987.

57. Shinohara, I., *New Production System,* Productivity Press, Cambridge, MA, 1988.

58. Shingo, S., *Modern Approaches to Manufacturing Improvement,* Productivity Press, Cambridge, MA, 1990.

59. Standard, C., and Davis, D., *Running Today's Factory,* SME, 1999.

60. Trevino, J., *Next Generation JIT Systems,* Material Handling Focus, Atlanta, GA, September 1989.

61. Trevino, J., *The Impact of Just-in-Time on Material Handling, Facilities Layout, and Building Design,* Material Handling Focus, Atlanta, GA, September 1987.

62. Trevino, J., Irizarry-Resto, M., and Hwang, J., *Just-in-Time Impact on Industrial Logistics and Plant Design,* working paper, Department of Industrial Engineering, North Carolina State University, Raleigh, 1993.

63. Wantuck, K. A., *Just-in-Time for America,* The Forum, 1989.

64. Womack, J., Jones, D., and Roos, D., *The Machine that Changed the World,* Macmillan, New York, 1990.

65. *Kanban (Just-in-Time at Toyota),* edited by Japan Management Association, translated by D. J. Lu, Productivity Press, 1985.

66. "Containers: Now They're More than Storage Boxes," *Modern Materials Handling,* vol. 41, no. 1, January 1986, pp. 87–90.

67. "How We Gained Flexible Flow to 64 Assembly Stations," *Modern Materials Handling,* vol. 38, no. 13, November 1983, pp. 52–55.

68. "Our Flexible Systems Help Us Make a Better Product," *Modern Materials Handling,* vol. 39, no. 8, July 1984, pp. 46–49.

PROBLEMS

8.1   Is the concept of an "automatic factory" still valid in today's manufacturing environment?

8.2   What sectors of the U.S. economy would be good targets for the automatic factory? Provide an explanation.

8.3   What are the advantages and disadvantages of operating an automatic warehouse?

8.4   Can a cross-docking facility be totally automated? Prepare a list of what would be required to fully automate a cross-docking facility.

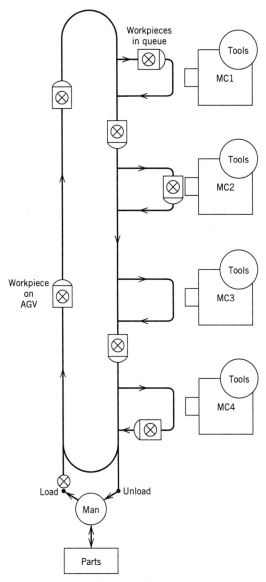

Figure 8.9   A flexible manufacturing system layout.

8.5 Develop a set of criteria for comparing transfer lines and flexible manufacturing systems.

8.6 Develop a set of criteria for comparing flexible manufacturing systems and single-step multimachine systems.

8.7 Describe the tool management problem in:

    a. Flexible manufacturing systems

    b. Single-stage multimachine systems

8.8 Consider the flexible manufacturing system depicted in Figure 8.9. What alternatives exist for moving parts as shown? Under what conditions should each be considered?

8.9 Prepare a review of a nontechnical paper on JIT.

8.10 Prepare a review of a technical paper on JIT.

8.11 Describe the differences, if any, between JIT and lean manufacturing.

8.12 Is JIT applicable to all types of manufacturing systems?

8.13 Prepare a review of a nontechnical paper on U-lines.

8.14 Prepare a review of a technical paper on U-lines.

8.15 Describe the U-line balancing problem and discuss the implementation issues associated with balancing U-lines.

9

FACILITIES SYSTEMS

9.1 INTRODUCTION

The objective of this chapter is not to prepare the facilities planner to become an architect, mechanical engineer, structural engineer, or builder. Rather, it is to provide the facilities planner with a unified picture of the building technology and the interrelationship of the facility systems so that the interior handling system and the layout will not be made without recognizing the reality of the facility constraints. The intent is to provide an understanding of the various system elements within the facility, not to prepare the facilities planner to design a structure or its heating, ventilation and air conditioning, or other systems. The intent is to provide an overview of how the systems' elements impact the overall process of facilities planning.

Planning facility systems is typically not the responsibility of the facilities planner. However, specifying *what* systems are required *where* and integrating those systems into the overall facility are the responsibility of the facilities planner. The facilities planner must be aware that:

1. The cost of constructing, operating, and maintaining a facility is significantly impacted by the facility systems.
2. A critical factor affecting facility flexibility is the facility's systems.
3. A facility plan is not complete until all systems are specified.
4. Facility systems have an important impact on employee performance, morale, and safety.
5. Facility systems have an important impact on the fire protection, maintenance, and security of a facility.

The facility systems considered in this chapter are:

1. Structural systems
2. Atmospheric systems
3. Enclosure systems
4. Lighting and electrical systems
5. Life safety systems
6. Sanitation systems
7. Building automation systems
8. Computerized maintenance management systems

9.2 STRUCTURAL SYSTEM PERFORMANCE

The most common structural types for industrial facilities are the steel skeleton frame or reinforced concrete skeleton frame. Many factors will impact the choice of the structural type or the choice of materials, for example, fire protection (steel is notorious for losing its strength when heated above 1000°F), environment (steel becomes brittle below −20°F and fails easily), and the overall planning grid or module. For facilities planners, the structure and planning needs should mesh, as the column configuration will impact the layout and the layout affects the structural design. The reverse is also true. The structural design and the grid configuration chosen (e.g., 20 ft × 30 ft or 25 ft × 25 ft) will impact the layout, which will impact the planner's options with respect to interior flow. The facilities planner must insist that the structural grid spacing optimize the function of the facility. For example, in a warehouse, the column spacing should be dictated by the rack dimensions and the access aisles between racks. That is, the clear spacing between columns must be compatible with the storage system. In a parking garage, the grid element should be a multiple of the car bay width and length and the circulation needs. The facilities planner must recognize that the structural dimensions of the building grid can be secondary to the optimal layout of the facility without affecting its structural integrity. Given today's mill capability, specific structural members other than 20-ft and 40-ft lengths can be obtained without a significant penalty cost. Therefore, if a 24-ft × 36-ft bay will optimize the layout, this grid configuration should be specified instead of the more traditional 20-ft × 40-ft grid. If alternatives for the structural framing system are considered, then the depth of these members usually drives the need for a taller building. But the added skin area is usually offset by a much lighter structure. Preengineered metal buildings can also provide a very economical structure and achieve greater open spans than can be economically achieved with a conventional steel or open-web system.

Clear height is important in the planning process because lower clear height may demand a larger footprint. Higher clear heights can impact racking systems and be restricted by local zoning and ordinance requirements.

A common error in designing warehousing facilities is using building cost criteria to define column spacing and determine the grid configuration. Steel mills generally produce structural members in 20- and 40-ft lengths. Architects and structural engineers habitually use these lengths in designing a storage system. However, the structural dimensions of the steel beams and girders or other structural members should be secondary to the storage system for which the facility is intended. For example, if as shown in Figure 9.1 a 48- by 40-in. pallet is used and a normal 4-in. clearance is allowed between pallets and between uprights and pallets, a minimum of 92 in. clear distance is required between the faces of the uprights in a pallet rack. If the pallet rack is a conventional four-high unit capable of carrying 2000-lb unit loads, the uprights will normally be 4 in. Thus, the overall width of the pallet rack is 100 in., or 96 in. centerline-to-centerline of the columns. Racking should drive the column spacing, but you must not allow the spacing to get excessive and impact the economy of the structure.

Carrying this analysis a step further, a typical 26-ft-clear storage building will require 12-in. heavy wall pipe columns or 12- by 12-in. flange beam or box columns. If pallet racks will be used and three sections of pallet racks are placed between each pair of columns, the clear space between the faces of the columns must be 24 ft-4 in. or (96 in. × 3) + 4 in. A four-unit bay would require 32 ft-4 in. or (96 in. × 4) + 4 in. Placement of six double-pallet-width units requires a clear

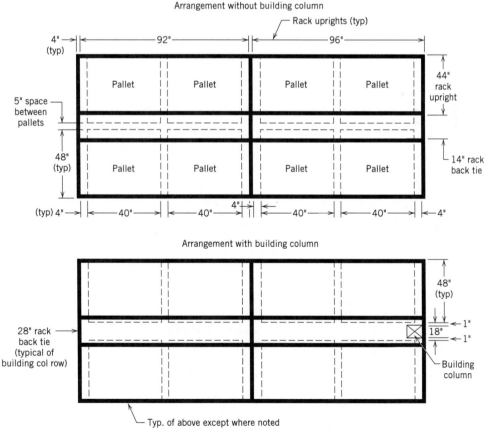

Figure 9.1 Plan view rack detail. (*Courtesy of Tompkins et al. [10]*)

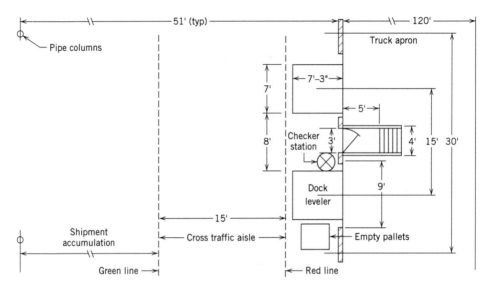

Figure 9.2   Basic dock layout. (*Courtesy of Tompkins et al. [10]*)

space of 49 ft-8 in. or [(96 in. × 3) + 4 in.] × 2 + 12 in. for the column. Thus, in a typical warehouse design, a 33- by 51-ft center-to-center bay spacing with 12-in. columns would probably accommodate rack storage in either direction and provide optimum flexibility of layout with elimination of column loss.

In such a design pattern, arranging the 33-ft bay parallel with the truck docks provides further advantages. The placement of the truck docks on 15-ft centers, two docks in each bay, and the 51-ft span allows a column-free truck service area behind the docks (Figure 9.2). If, as is normally recommended, truck docks and rail docks are at right angles on one corner of the building, putting rail docks along the 51-ft span would allow optimum flexibility for placement of cars varying in length from 40 ft to 60 ft or even 90 ft. The 33-ft bay would provide approximately 20 ft from the face of an inside rail dock to the column line, or 33 ft from the outer wall with clear spacing if building face doors were used.

Needless to say, different pallet sizes affect the dimensions of the racks, and rack dimensions affect column spacing. Thus, before designing a building, it is essential to determine how the facility is to be used. Since the 48- × 40-in. pallet is the Grocery Manufacturers Association's standard and the most commonly used size in consumer and many industrial distribution systems, the dimensional analysis above will probably provide a suitable basis for most warehouse designs where conventional rack systems are used. This dimensional pattern eliminates the need to modify the building in accordance with vehicle dimensions, since the racks will fit between the columns in either configuration and aisle spacing is at the option of the layout designer. In the event that the storage machine or forklift truck requires a different aisle width, a modification in column dimensions may be appropriate.

In this analysis, the size and character of the columns must also be considered. Heavy-wall round or square tubular columns should be used in warehouses. These types of columns eliminate rodent and vermin nesting places, ensure easy maintenance and cleanliness, provide the ability to place downspouts within the columns, and minimize the effect of denting on column strength. Such columns are a little more expensive than H beams, but provide a much better facility in the long run. They also eliminate the corners that can pull packages from pallets when fork-lift truck operators move too close to the column. In an all-concrete structure, a square column is normally preferable. Because of their rectangular shape, tube columns provide a finished edge all around. They also can provide lateral bracing in the structure, thereby eliminating the need for "X bracing."

Wall selections are normally architectural in nature. They are usually driven by local building restrictions or by a desired look for "curb appeal." However, the insulation value must be considered because it can significantly impact the heating and cooling requirements for the building.

In the final analysis, however, the building's structural integrity is the objective of the structural engineer and the architect. And the building must exhibit stability under the following various load conditions:

- Gravity: continuous, dead loads—roofs, floors
- Gravity: intermittent, live loads—snow, equipment, people
- Wind: hours, days
- Seismic: seconds, minutes

In short, the objective of the structural system is to ensure that the building will be stable. The degree of stability required and designed into the building is dictated by the load conditions. The facilities planner must be cognizant of the environment where the facility will be located and the types of load conditions that might render the building unstable.

9.3 ENCLOSURE SYSTEMS

The enclosure system provides a barrier against the effects of extreme cold or heat, lateral forces (wind), water, and undesirable entries (humans, insects). It is also used as a controlling mechanism. Doors and windows act as filters to provide not only a barrier that controls access to desirables, but also ventilation, light transmission (privacy), and sound. The facility's enclosure is dictated by different and conflicting influences.

The enclosure elements are floor, walls, and roof. All three elements are intended to provide the facility with a specified comfort level, which is often impacted by the thermal performance of the building enclosure. The tendency, however, is to ignore/underestimate the enclosure's thermal role and rely on environmental systems to make it right. In fact, the following generalization holds true:

Building Enclosure + Services Input = Comfort

Thermal performance is usually required to rectify the heat transmission imbalance between inside and outside areas of the enclosure. Because this transmission is contingent on the temperature differential, climate is the major determinant. One of the major problems in thermal performance, therefore, is how to make effective use of solar gain. Solar transmission, if effectively controlled, can significantly reduce the building's dependence on artificial atmospheric systems. Figure 9.3 shows how solar gain can be controlled. In Figure 9.3*a*, 90% of the incident solar rays are transmitted due to the poor absorption and reflection qualities of single-glazed element. Note in Figure 9.3*b* that a double-pane configuration with a heat-absorbing element reduces the amount of transmission from 90% to 45%. The solar gain can be reduced additionally from 90% to 25% if a reflective glass member is substituted for a heat-absorbing member in the double-pane configuration shown in Figure 9.3*c*.

Typically for manufacturing and warehousing facilities, keeping undesirables out is key. As such, the material selected (metal, plastic, or masonry cladding) should be more impervious to penetration. Thermal performance coupled with water exclusion forms the backbone of most enclosure systems. There are two areas of consideration: aboveground and below the ground. Aboveground, facilities planners must be aware of the problems that a poorly designed roof can cause. The primary performance need of the roof is water exclusion. However, the importance of thermal comfort necessitates the need for adequate insulation. Vapor

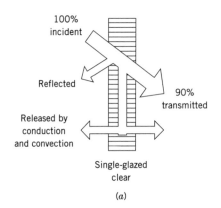

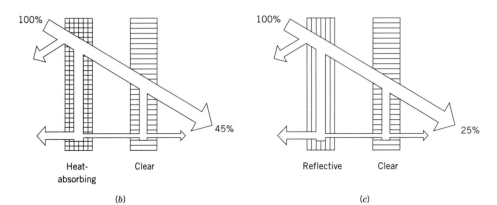

Figure 9.3 Controlling solar transmission. (*Courtesy of Reid [6]*)

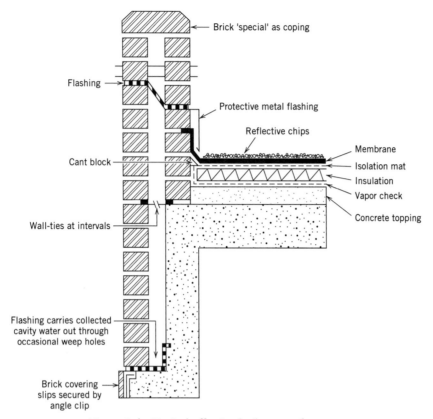

Brick 'special' as coping

Flashing

Protective metal flashing

Reflective chips

Membrane

Cant block

Isolation mat

Insulation

Vapor check

Concrete topping

Wall-ties at intervals

Flashing carries collected cavity water out through occasional weep holes

Brick covering slips secured by angle clip

Figure 9.4 Typical effective built-up-roof system.

migration, good insulation, and water barrier and removal are essential to an effective roofing system.

The built-up roof, shown in Figure 9.4, details an effective flat roof design. Note that a membrane layer to prevent water penetration, an insulation layer to assist with thermal comfort, and a vapor check to stop vapor migration are integral elements of the system. Other roofing alternatives include metal roof systems and membrane roofs that are ballasted, mechanically attached, or fully adhered.

The ground and below-ground condition for industrial facilities is typically a concrete slab sitting directly on the ground. The primary concerns with ground conditions are water penetration and vapor migration.

As swimming pools must be sealed to keep water in, similarly, the enclosure elements making contact with the ground must be sealed to keep water out. There are two principal ways to accomplish this: integral waterproofing and applied membrane. Integral waterproofing consists of an additive to the concrete to make it watertight. This method, however, is not very effective in controlling vapor migration. Typically, it can take 12 in. or more of concrete to prevent vapor penetration inward from the surrounding wet soil.

Membranes have the potential to be 100% effective in handling vapor and water penetration, but poor design and shoddy installation can make them ineffective.

When membranes are applied to the internal surface, problems may occur with adherence, as the hydrostatic pressure increases. A more effective method is to apply the membrane to the external surface. This position will allow the membrane to work in conjunction with the hydrostatic pressure of the wet soil surface, and the membrane will be pressed more firmly against the structure as the hydrostatic pressure increases. However, regardless of the design method, poor installation resulting in cracked or punctured membranes will make them ineffective.

9.4 ATMOSPHERIC SYSTEMS

Atmospheric systems provide for the health and comfort of occupants and the conditions for building equipment and machinery. The criteria for equipment and machinery are dictated by manufacturer's specification, but the criteria for human comfort were vague and intuitive until 1900. However, as new technologies were developed for controlling building atmospheres, the criteria became more precisely defined.

The definitive study of human comfort criteria was done in the early 1960s at Kansas State University using large groups of college students. The data gathered was used to develop equations for the comfort parameters using the following variables:

- Temperature
- Humidity
- Clothing
- Metabolic activity

Atmospheric systems respond to these criteria by heating and cooling building air and controlling humidity. Air speed is usually held very low, so it is not readily perceived. Clothing and metabolic activity are assumed to be constant.

Codes require only minimum indoor temperatures in winter. Market standards for different building types provide more precise responses involving cooling and humidification.

One example of such a standard is found in healthcare facilities, where humidity is held close to 50%. This humidity level is the optimum for inhibiting the growth of viruses and bacteria. Oxygen in air is essential for building occupants to breathe, but oxygen depletion would occur only at very high population densities. This is usually not a consideration in building air quality.

A more significant problem is the introduction of inorganic chemical pollutants into building atmospheres from sources in the building. One important pollutant particularly found in single-story buildings is radioactive radon gas, which occurs naturally in the earth and ground water and seeps into buildings. Such pollutants are best handled by removing the sources from the building or blocking the seepage.

The major air-quality problem in most buildings is the buildup of odor and airborne particulate matter. By definition, the rate of dilution is expressed as $ft^3/min-ft^2$ of floor area. One can see that the way to eliminate unwanted substances

Table 9.1   *Typical Air Changes Rates*

Use	Supply Air Ft^3/Min-Ft^2	Exhaust Air Ft^3/Min-Ft^2	Air Changes/Hr.
Residence			
Toilet room	—	1.5	10
Kitchen	—	1.5	10
Other rooms	Natural ventilation—5% of floor area		—
Office/commercial	0.6	0.3	4
School	1.5	0.75	9
Public assembly	3.0	1.5	—
Hospital	2.0	1.0	13
Labs	1.2	1.2	8
Animal laboratory	—	3.0	20

is to dilute them both by mixing stale air that is in the building and by adding fresh outdoor air. This air exchange (dilution rate) is typically expressed in air changes per hour and is stipulated by local code requirements as it pertains to the building use. Table 9.1 shows typical air change rates for several building use types.

Because of the importance of eliminating unwanted substances, the equipment to bring in fresh air and remove stale air is a primary consideration in planning a facility. This space requirement is governed by the size of the equipment room, the air speed capabilities of main and branch ducts, and the louver size needed to accommodate the required exhaust and intake needs as follows:

Equipment Rooms.   Central equipment rooms can be estimated roughly from Table 9.2. Note that as the air change rate increases, the handling equipment area also increases.

Duct sizes.   Duct sizes can be estimated from air quantities and air speeds in ducts.

Main duct—1800 ft/min
Branch ducts—900 to 1110 ft/min

Louver Sizes.   These areas are also estimated from air quantities and air speeds.

Exhaust louvers: Recommended air speed of 2000 ft/min
Intake louvers: Recommended air speed of 1000 ft/min

The following example illustrates how atmospheric space is allocated.

Table 9.2   *Central Equipment Room Area by Use*

Use	Area of Equipment Rooms
Residential	2%
Office/industrial	5–7%
Public assembly	10–15%
Hospital	25%
Laboratory	25–50%
Animal laboratory	50%

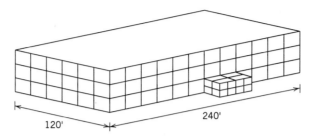

Figure 9.5   Light assembly plant dimensions for Example 9.1.

Example 9.1

Considering the light assembly operation shown in Figure 9.5, what space requirements should be considered for atmosphere systems? First, allocate the mechanical equipment room space. Then:

1. Calculate total gross area: 240 ft × 120 ft -28,800 ft². From Table 9.2, using office/industrial use type, assume 5% of gross area for mechanical equipment space.
2. Assume the system will be roof-mounted.

Second, determine the air-handling requirements:

1. Minimum air supply rate: 0.6 ft³/min-ft²
    Maximum air supply rate: 1.0 ft³/min-ft²
    Exhaust 0.3 ft³/min-ft²
2. Air supply: 28,800 ft² × 1.0 ft³/min-ft² = 28,800 ft³/min (CFM)
    Air exhaust: 28,800 ft² × 0.3 ft³/min-ft² = 8640 ft³/min (CFM)
3. Determine louver cross-section using respective CFM requirements and allowable air speeds:

    Supply: $\dfrac{28,800 \text{ ft}^3/\text{min}}{1000 \text{ ft/min}} = 28.8 \text{ ft}^2$

    Exhaust: $\dfrac{8640 \text{ ft}^3/\text{min}}{2000 \text{ ft/min}} = 4.3 \text{ ft}^2$

    Note supply and exhaust louvers must be 15 ft apart to avoid "short circuit" of exhaust air back into building.
4. Main duct calculation and layout.
    Main duct sizing is determined as follows:
    Two main ducts, each covering half of the building's area, multiplied by the air supply rate:

    0.5 × 28,800 ft² × 1 ft³/min-ft² = 14,400 ft³/min CFM. See Figure 9.6.

    Divide this airflow rate by the main duct air speed to obtain the cross-sectional area:

    $\dfrac{14,400 \text{ CFM}}{1800 \text{ ft/min}} = 8 \text{ ft}^2$

    Main duct can be 2 ft × 4 ft. See Figure 9.7

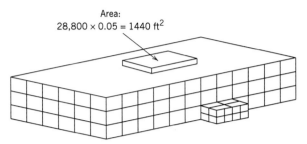

Area:
28,800 × 0.05 = 1440 ft²

Figure 9.6 Assembly plant with roof-mounted air handling unit for Example 9.1.

5. Branch duct calculation and layout. For each main duct, assume 10 branch ducts. Divide flow rate for each branch duct by the number of branches to obtain the branch flow rate:

$$\frac{14,400 \text{ ft}^3/\text{min}}{10} = 1,440 \text{ ft}^3/\text{min}$$

Divide the branch flow rate by the allowable branch duct air speed.

$$\frac{1,440 \text{ ft}^3/\text{min}}{1000 \text{ ft/min}} = 1.4 \text{ ft}^2$$

For the example problem, two main ducts 2 ft-0 in. by 4 ft-0 in. with ten 1 ft-4 in. by 1 ft-0 in. branch ducts would provide the required air dilution to meet the air handling needs (Figure 9.8).

Although the air dilution method is prevalent and widely accepted, it has some drawbacks. The major drawback to this method is that when you exhaust the air, you are removing heat and/or air conditioning. Therefore, there is a substantial loss of energy.

To reduce the volume of makeup air required, many operations use mechanical filtration or electrostatic precipitation. Mechanical filtration removes contaminants by passing air through filtering material, sometimes called media or impingement filters, to remove particulate matter. Electrostatic precipitation removes particulates by charging the air particles and collecting them on oppositely charged plates. The

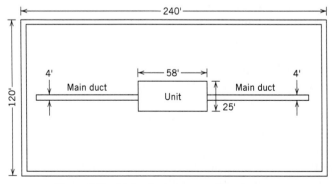

Figure 9.7 Main duct cross-sectional area.

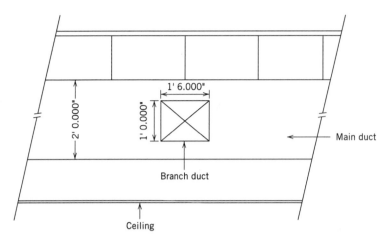

Figure 9.8   Cross section of main and branch ducts.

air passes through a prefilter and then into the ionizer section of the unit. Here the particles receive a positive charge. These positively charged particles are then attracted to and collected on high-voltage, negatively charged collecting plates. The cleaner air is then returned to the area.

Facilities planners are often called upon to design carefully controlled environments for electronics, pharmaceuticals, or food-processing operations. These controlled environments are known as *clean rooms*. There are four official clean-room classifications: class 100,000; class 10,000; class 1000; and class 100. These rooms are kept under positive air pressure to prevent dust infiltration. The classifications refer to the number of particles (0.5 micron and larger) in a cubic foot of air. One widely recognized method used to achieve these air-quality levels is with the use of HEPA (high-efficiency particulate air) filters. These filters are used in a wide variety of industrial, commercial, and institutional installations where the highest possible degree of air cleaning is required. They are also used when it is necessary to remove radioactive or toxic dust, pathogenic organisms, or mold spores. They are available in efficiencies of 95.00%, 99.97%, and 99.999% guaranteed minimum in removing 0.3-micron particles or larger. In actual operation, the HEPA filter is 100% efficient in removing all known airborne organisms since bacteria, pollens, yeast, and fungi range up in size from 0.5 micron.

The purpose of heating, ventilation, and air conditioning (HVAC) is to control the temperature, humidity, and cleanliness of the environment within a facility. HVAC is important for employee comfort, process control, and/or product control. HVAC is an extremely complex field and is undergoing continuous change. A facilities planner must realize the complex and dynamic nature of HVAC and not make the mistake of attempting to design HVAC for a facility.

The input required by an HVAC expert to design an HVAC system includes the facility layout and the construction specifications. The facilities planner should be able to provide the HVAC expert with this input as well as understand how the HVAC expert translates this input into HVAC requirements. The heating requirement for a facility is referred to as *heat loss* and is calculated using Equation 9.1:

$$Q_H = Q_F + Q_R + Q_G + Q_D + Q_W + Q_I \tag{9.1}$$

where:

Q_H = total heat loss
Q_F = heat loss through the floor
Q_R = heat loss through the roof
Q_G = heat loss through glass windows
Q_D = heat loss through doors
Q_W = heat loss through walls
Q_I = heat loss due to infiltration

Each of the individual heat losses is calculated using Equation 9.2:

$$Q = AU(t_i - t_o) \tag{9.2}$$

where:

Q = heat loss for facility component
A = area of facility component
U = coefficient of transmission of facility component
t_i = temperature inside the facility
t_o = temperature outside the facility

The air conditioning requirement for a facility is referred to as the *cooling load* and is obtained from Equation 9.3:

$$Q_C = Q_F + Q_R + Q_G + Q_D + Q_W + Q_V + Q_S + Q_L + Q_P \tag{9.3}$$

where:

Q_C = total cooling load
Q_F = cooling load due to the floor
Q_R = cooling load due to the roof
Q_G = cooling load due to glass windows
Q_D = cooling load due to doors
Q_W = cooling load due to walls
Q_V = cooling load due to ventilation
Q_S = cooling load due to solar radiation
Q_L = cooling load due to lighting
Q_P = cooling load due to personnel

The cooling loads Q_F, Q_R, Q_G, Q_D, Q_W, and Q_V are determined using Equation 9.4:

$$Q = AU(t_o - t_i) \tag{9.4}$$

where:

Q = cooling load for facility component
A, U, t_o and t_i are as defined for Equation 9.2

The cooling load from solar radiation Q_S is computed using Equation 9.5:

$$Q = (A)(H)(S) \tag{9.5}$$

Table 9.3 *Heat Gains for Males and Females Performing Light and Heavy Work*

Activity	Heat Gains (Btu/hr)
Male—light work	800
Male—heavy work	1500
Female—light work	700
Female—heavy work	1300

where:

A = area of glass surface

H = heat absorption of the building surface

S = shade factor for various types of shading

The cooling load from lighting Q_L may be calculated by using the following conversion factors:

$$Q_L \text{ (incandescent)} = \left(\frac{\text{Incandescent}}{\text{wattage}}\right)\left(3.4 \frac{\text{Btu}}{\text{(hr)(watt)}}\right) \qquad (9.6)$$

$$Q_L \text{ (fluorescent)} = \left(\frac{\text{Fluorescent}}{\text{wattage}}\right)\left(4.25 \frac{\text{Btu}}{\text{(hr)(watt)}}\right) \qquad (9.7)$$

The cooling load from personnel Q_P may be calculated by determining the number of males and females performing light and heavy work and multiplying that times the heat gain factors given in Table 9.3.

Example 9.2

An 18-ft-tall facility having the dimensions 250 ft × 100 ft has eight 4-ft × 8-ft double-pane glass windows and two 3-ft × 8-ft glass doors on both the front and back sides. The facility is made of 8-in. solid cinder block and 1-in. metal, insulated roof with an insulated ceiling and an uninsulated slab floor. The facility is located in Chicago, Illinois. What is the heat loss?

The following data needed to calculate the heat loss may be obtained from a variety of sources ([2] and [3]).

$U_F = 0.81$ Btu/(hr)(ft^2)(°F) $t_i = 70°F$

$U_R = 0.20$ Btu/(hr)(ft^2)(°F) $t_o = 0°F$

$U_G = 0.63$ Btu/(hr)(ft^2)(°F)

$U_D = 1.13$ Btu/(hr)(ft^2)(°F)

$U_W = 0.39$ Btu/(hr)(ft^2)(°F)

$U_I = 0.20$ Btu/(hr)(ft^2)(°F)

Then, using Equation 9.2:

$Q_F = (25,000 \text{ ft}^2)[0.81 \text{ Btu/(hr)(ft}^2)(°F)](70°F - 0°F)$

$Q_F = 14.175 \times 10^5$ Btu/hr

$Q_R = (25,000 \text{ ft}^2)[0.20 \text{ Btu/(hr)(ft}^2)(°F)](70°F - 0°F)$

$Q_R = 3.5 \times 10^5$ Btu/hr

$Q_G = (16)(32 \text{ ft}^2)[0.63 \text{ Btu/(hr)(ft}^2)(°F)](70°F - 0°F)$

$Q_G = 0.226 \; 10^5$ Btu/hr

$Q_D = (4)(24 \text{ ft}^2)(1.13 \text{ Btu/(hr)(ft}^2)(°F)](70°F - 0°F)$

$Q_D = 0.076 \times 10^5 \text{ Btu/hr}$

$Q_W = [(2)(4500 \text{ ft}^2) + (2)(1800 \text{ ft}^2) - (16)(32\text{ft}^2) - (4)(24 \text{ ft}^2)][(0.39 \text{ Btu/(hr)(ft}^2)(°F)]$
$(70°F - 0°F)$

$Q_W = 3.274 \times 10^5 \text{ Btu/hr}$

$Q_I = [(25,000 \text{ ft}^2) + (12,600 \text{ ft}^2)](0.20 \text{ Btu/(hr)(ft}^2)(°F)](70°F - 0°F)$

$Q_I = 0.526 \times 10^5 \text{ Btu/hr}$

Then, using Equation 9.1, the total heat loss may be calculated as follows:

$Q_H = 14.175 \times 10^5 \text{ Btu/hr} + 3.5 \times 10^5 \text{ Btu/hr} + 0.226 \times 10^5 \text{ Btu/hr} + 0.076 \times$
$10^5 \text{ Btu/hr} + 3.274 \times 10^5 \text{ Btu/hr} + 0.526 \times 10^5 \text{ Btu/hr}$

$Q_H = 21.777 \times 10^5 \text{ Btu/hr}$

Example 9.3

For the facility described in Example 9.2, given the following additional information, what is the cooling load? The front of the facility faces west and all windows and doors have awnings. The lighting within the facility consists of 575 fluorescent luminaries, each containing two 60-watt lamps. Within the facility, 75 men performing light work and 50 women performing light work will be employed. The outside design temperature is 97°F, and the inside design temperature is 78°F [3].

Using Equation 9.4 results in the following cooling loads for facility components:

$Q_F = (25,000 \text{ ft}^2)[0.81 \text{ Btu/(hr)(ft}^2)(°F)](97°F - 78°F)$

$Q_F = 3.85 \times 10^5 \text{ Btu/hr}$

$Q_R = (25,000 \text{ ft}^2)[0.20 \text{ Btu/(hr)(ft}^2)(°F)](97°F - 78°F)$

$Q_R = 0.95 \times 10^5 \text{ Btu/hr}$

$Q_G = (16)(32 \text{ ft}^2)[0.63 \text{ Btu/(hr)(ft}^2)(°F)](97°F - 78°F)$

$Q_G = 0.06 \times 10^5 \text{ Btu/hr}$

$Q_D = (4)(24 \text{ ft}^2)(1.13 \text{ Btu/(hr)(ft}^2)(°F)](97°F - 78°F)$

$Q_D = 0.02 \times 10^5 \text{ Btu/hr}$

$Q_W = [(2)(4500 \text{ ft}^2) + (2)(1800 \text{ ft}^2) - (16)(32 \text{ ft}^2) - (4)(24 \text{ ft}^2)]$
$[0.39 \text{ Btu/(hr)(ft}^2)(°F)] (97°F - 78°F)$

$Q_W = 0.88 \times 10^5 \text{ Btu/hr}$

$Q_V = [(25,000 \text{ ft}^2) + (12,600 \text{ ft}^2)](0.20 \text{ Btu/(hr)(ft}^2)(°F)](97°F - 78°F)$

$Q_V = 1.43 \times 10^5 \text{ Btu/hr}$

The heat absorption factor for the west side of a facility in Chicago, Illinois, is 100 Btu/(hr)(ft^2) and for the east side 75 Btu/(hr)(ft^2) [3]. The shade factor awnings are 0.3 [3]. Using Equation 9.5, the cooling load from solar radiation may be calculated as:

$Q_S = [(8)(32 \text{ ft}^2) + (2)(24 \text{ ft}^2)][100 \text{ Btu/(hr)(ft}^2)(°F)(0.3)] + [(8)(32 \text{ ft}^2) + (2)(24 \text{ ft}^2)]$
$[75 \text{ Btu/(hr)(ft}^2)(°F)(0.3)]$

$Q_S = 0.16 \times 10^5 \text{ Btu/hr}$

The cooling load from lighting may be calculated via Equation 9.7, as follows:

$$Q_L \text{ (fluorescent)} = (575)(2)(60 \text{ watts}) (4.25) \frac{\text{Btu}}{(\text{hr})(\text{watt})}$$

$Q_I = 2.93 \times 10^5 \text{ Btu/hr}$

With the aid of Table 9.2, the cooling load from personnel may be calculated as:

$$Q_P = 75(800 \text{ Btu/hr}) + 50(700 \text{ Btu/hr})$$

$$Q_P = 0.95 \times 10^5 \text{ Btu/hr}$$

Then, using Equation 9.3, the total cooling load may be calculated as follows:

$$Q_C = 3.85 \times 10^5 + 0.95 \times 10^5 + 0.06 \times 10^5 + 0.02 \times 10^5 + 0.88 \times 10^5 + 0.14 \times 10^5 + 0.16 \times 10^5 + 2.93 \times 10^5 + 0.95 \times 10^5$$

$$Q_C = 9.94 \times 10^5 \text{ Btu/hr}$$

Or, utilizing the conversion factor 12,000 Btu/hr = 1 ton, a total cooling load of 83 tons is required to air condition the total facility.

9.5 ELECTRICAL AND LIGHTING SYSTEMS

The facilities planner's responsibility is to specify what level of service is required and where it is required. The power company needs the electrical load requirement for a facility well before construction begins. This usually requires a facilities planner to have a good preliminary estimate, since the detailed design of the facility is often not finalized when this information is required. In general, every electrical system should have sufficient capacity to serve the loads for which it is designed, plus spare capacity to meet anticipated growth.

Allowance for load growth is probably the most neglected consideration in the planning and design of the electrical system. The facilities planner must ensure that major load requirements, both present and future, are included in the design data so that the design engineer can adequately size mains, switchgear, transformers, feeders, panel boards, and circuits to handle the growth of the load. The process begins with analyzing the building type and its traditional electrical loads. The facilities planner should then determine all unusual and special electrical requirements for the building type under consideration, as well as current trends and practices. An idea of the average load requirement for several building types is shown in Table 9.4.

Once this activity is complete, a list of load categories can be generated including:

- Lighting
- Miscellaneous power, convenience outlets, and small motors
- HVAC
- Plumbing and sanitary equipment
- Vertical transportation equipment
- Kitchen equipment
- Special equipment

The point of service, service voltage, and metering location should then be estimated with the local utility company. In addition, the facilities planner should determine with the client the type and rating of all equipment to be used in the facility. The space required for electrical equipment and electrical closets is then

Table 9.4 *Average Load Requirements*

Building Type	Load Requirements
Plants	20 watts per s.f.
Office buildings	15 watts per s.f.
Hospitals	3000 watts per bed
Schools	3–7 watts per s.f.
Shopping centers	3–10 watts per s.f.

determined. Finally, the facilities planner should provide detailed input for the lighting requirements for the facility.

Facilities lighting should not be planned by the same person planning the electrical services because illumination should be custom designed for the areas within a facility. A facilities planner fully understands the tasks to be performed within a facility and is in the best position to custom design the lighting system.

Lighting systems may be designed using the following eight-step procedure:

Step 1: *Determine the Level of Illumination*

The minimum levels of illumination for specific tasks are given in Table 9.5. These levels may be achieved using general lighting or a combination of general lighting and supplementary lighting. When supplementary lighting is used, general lighting should contribute at least one-tenth of the total illumination at no less than 20 foot-candles [1]. Supplementary lighting, particularly when used in conjunction with modular office furniture, results in significant energy savings.

Step 2: *Determine the Room Cavity Ratio (RCR)*

The RCR is an index of the shape of a room to be lighted. The higher and narrower a room, the larger the RCR and the more illumination is needed to achieve the required level of lighting. The RCR may be calculated using Equation 9.8:

$$\text{RCR} = \frac{(5)\left(\begin{array}{c}\text{Height from the working} \\ \text{surface to the luminaris}\end{array}\right)\left(\begin{array}{cc}\text{Room} & \text{Room} \\ \text{length} + \text{width}\end{array}\right)}{\left(\begin{array}{c}\text{Room} \\ \text{length}\end{array}\right)\left(\begin{array}{c}\text{Room} \\ \text{width}\end{array}\right)} \tag{9.8}$$

Step 3: *Determine the Ceiling Cavity Ratio (CCR)*

If luminaries are ceiling mounted or recessed into the ceiling, the reflective property of the ceiling will not be impacted by the luminaries' mounting height, so the CCR need not be considered. The greater the distance from the luminaries to the ceiling, the greater will be the CCR and the reduction in ceiling reflectance. The CCR may be calculated using Equation 9.9.

$$\text{CCR} = \frac{\left(\begin{array}{c}\text{Height from} \\ \text{luminaries to ceiling}\end{array}\right)(\text{RCR})}{\begin{array}{c}\text{Height from the working} \\ \text{surface to the luminaries}\end{array}} \tag{9.9}$$

Table 9.5   *Minimum Illumination Levels for Specific Visual Tasks*

Task	Minimum Illumination Level (footcandles)
Assembly	
Rough easy-seeing	30
Rough difficult-seeing	50
Medium	100
Fine	500
Extra Fine	1000
Inspection	
Ordinary	50
Difficult	100
Highly difficult	200
Very difficult	500
Most difficult	1000
Machine Shop	
Rough bench and machine work	50
Medium bench and machine work, ordinary automatic machines, rough grinding, medium buffing and polishing	100
Fine bench and machine work, fine automatic machine, medium grinding, medium buffing and polishing	500
Extra fine bench and machine work, fine grinding	1000
Material Handling	
Wrapping, packing, labeling	50
Picking stock, classifying	30
Loading	20
Offices	
Reading high or well-printed material; tasks not involving critical or prolonged seeing, such as conferring, interviewing, and inactive files	30
Reading or transcribing handwriting in ink or medium pencil on good-quality paper; intermittent filing	70
Regular office work; reading good reproductions, reading or transcribing handwriting in hard pencil or on poor paper; active filing; index references; mail sorting	100
Accounting, auditing, tabulation, bookkeeping, business machine oepration; reading poor reproductions; rough layout drafting	150
Cartography, designing, detailed drafting	200
Corridors, elevators, escalators, stairways	20
Paint Shops	
Dipping, spraying, rubbing, firing, and ordinary hand painting	50
Fine hand painting and finishing	100
Extra fine hand painting and finishing	300
Storage Rooms and Warehouses	2
Inactive	5
Active	
Rough bulky	10
Medium	20
Fine	50
Welding	
General illumination	50
Precision manual arc welding	1000

Table 9.6 *Approximate Reflectance for Wall and Ceiling Finishes*

Materials	Approximate Reflectance (%)
White paint, light-colored paint, mirrored glass, and porcelain enamel	80
Aluminum paint, stainless steel, and polished aluminum	65
Medium-colored paint	50
Brick, cement, and concrete	35
Dark-colored paint, asphalt	10
Black paint	5

Source: From [14].

Step 4: *Determine the Wall Reflections (WR) and the Effective Ceiling Reflectance (ECR).*
The WR and a base ceiling reflectance (BCR) value can be obtained from Table 9.6.
If the luminaries are to be ceiling mounted or recessed into the ceiling, the ECR is
equal to the BCR. If luminaries are to be suspended, the ECR is determined from
the CCR, WR, and BCR using Table 9.7.

Step 5: *Determine the Coefficient of Utilization (CU).*
The CU is the ratio of lumens reaching the work plane to those emitted by the lamp.
It is a function of the luminaries used, the RCR, the WR, and the ECR. The CU for
standard luminaries is given in Table 9.8.

Step 6: *Determine the Light Loss Factor (LLF).*
The two most significant light-loss factors are lamp lumen depreciation and luminary dirt depreciation. Lamp lumen depreciation is the gradual reduction in lumen
output over the life of the lamp. Typically, lamp lumen depreciation factors are
expressed as the ratio of the lumen output of the lamp at 70% of rated life to the
initial value. The lumen outputs for various lamps given in Table 9.9 include the
lamp lumen depreciation; therefore, no additional consideration is needed.
Luminary dirt depreciation is the light loss associated with the conditions under
which the luminary operates. The luminary dirt depreciation varies with conditions
and the length of time between cleanings.

Step 7: *Calculate the Number of Lamps and Luminaries.*
Utilize Equation 9.10 to calculate the number of lamps and Equation 9.11 to calculate the number of luminaries.

$$
\begin{array}{c} \text{Number} \\ \text{of} \\ \text{Lamps} \end{array} = \frac{\left(\begin{array}{c}\text{Required level}\\\text{of illumination}\end{array}\right)\left(\begin{array}{c}\text{Area to}\\\text{be lit}\end{array}\right)}{(\text{CU})(\text{LIF})\left(\begin{array}{c}\text{Lamp output at}\\70\%\text{ of rated life}\end{array}\right)}
\tag{9.10}
$$

$$
\begin{array}{c} \text{Number} \\ \text{of} \\ \text{luminaries} \end{array} = \frac{\left(\begin{array}{c}\text{number}\\\text{of}\\\text{lamps}\end{array}\right)}{\left(\begin{array}{c}\text{lamps}\\\text{per}\\\text{luminary}\end{array}\right)}
\tag{9.11}
$$

Table 9.7 *The Percent (%) Effective Ceiling Reflectance (ECR) for Various Combinations of Ceiling Cavity Ratios (CCR), Wall Reflectances (WR), and Base Ceiling Reflectances (BCR)*

	BCR																																				
	80						65						50						35						10						5					
	WR						WR						WR						WR						WR						WR					
CCR	80	65	50	35	10	5	80	65	50	35	10	5	80	65	50	35	10	5	80	65	50	35	10	5	80	65	50	35	10	5	80	65	50	35	10	5
.5	76	74	72	69	67	65	64	60	58	56	54	52	49	47	46	44	42	41	36	34	32	31	29	28	12	12	11	11	9	8	8	8	7	6	5	5
1.0	74	71	67	63	57	56	60	55	53	49	45	43	48	45	43	39	36	35	35	33	31	29	26	25	14	13	12	11	9	8	10	7	8	7	5	5
1.5	72	67	62	55	49	47	58	52	49	44	38	36	47	44	40	35	31	28	35	33	28	26	23	20	16	15	12	11	8	7	14	11	9	7	5	4
2.0	69	63	56	49	41	39	55	49	44	38	32	30	46	42	37	31	26	25	35	32	26	23	18	17	18	17	13	10	8	6	15	12	10	7	4	4
2.5	67	60	51	43	35	33	54	45	40	33	26	25	46	40	35	28	22	21	35	31	23	20	16	13	20	19	13	10	7	5	17	14	10	8	4	4
3.0	65	57	47	38	30	28	53	42	38	29	22	21	45	39	32	25	19	18	35	31	21	18	14	12	21	20	13	10	7	4	19	15	11	8	3	3
3.5	63	54	43	34	26	25	52	39	33	26	18	17	44	38	30	23	17	16	35	31	20	16	12	10	22	21	13	10	7	4	20	16	11	8	3	3
4.0	61	52	40	31	22	21	50	37	31	23	15	14	44	38	28	21	15	13	34	30	18	14	10	8	23	22	14	10	7	3	20	17	12	8	3	3
5.0	58	46	35	26	18	15	48	33	26	18	9	8	42	35	25	18	12	10	34	29	17	12	9	7	25	23	14	10	6	3	23	18	12	8	3	2
8.0	50	36	25	17	11	6	41	24	18	11	5	3	40	30	19	13	7	5	34	28	11	5	5	4	27	24	13	10	5	2	26	19	12	6	3	1

Source: From [14].

Table 9.8 *Coefficients of Utilization for Standard Luminaries*

Luminaire	Spacing Not to Exceed	RCR	ECR 80% — WR 80%	WR 50%	WR 30%	WR 10%	ECR 50% — WR 80%	WR 50%	WR 30%	WR 10%	ECR 10% — WR 80%	WR 50%	WR 30%	WR 10%	ECR 0% — WR 0%
Filament reflector lamps	1.5 × mounting height	1	1.11	1.09	1.07	1.03	1.04	1.02	1.00	.98	.96	.95	.94	.93	.91
		2	1.04	1.00	.95	.92	.99	.95	.92	.88	.92	.90	.87	.85	.83
		3	.95	.92	.88	.82	.92	.88	.84	.80	.85	.83	.80	.77	.75
		4	.90	.85	.79	.73	.86	.81	.76	.71	.79	.77	.73	.70	.68
		5	.82	.77	.71	.65	.80	.75	.69	.64	.75	.71	.67	.63	.61
		10	.58	.50	.43	.38	.54	.49	.43	.38	.51	.47	.42	.38	.36
High-intensity discharge lamps (mercury, metal halide or sodium)	1.3 × mounting height	1	.89	.87	.84	.82	.81	.80	.78	.77	.74	.73	.72	.71	.70
		2	.82	.79	.75	.72	.77	.74	.71	.68	.70	.68	.66	.64	.63
		3	.76	.72	.67	.63	.72	.68	.64	.61	.65	.63	.60	.58	.56
		4	.70	.66	.61	.57	.67	.63	.58	.55	.57	.55	.54	.53	.51
		5	.64	.60	.55	.51	.62	.58	.53	.49	.56	.54	.52	.49	.46
		10	.45	.40	.34	.30	.42	.38	.34	.30	.40	.36	.41	.20	.28
Fluorescent lamps in uncovered fixtures	1.3 × mounting height	1	.88	.85	.82	.79	.79	.65	.72	.71	.65	.64	.63	.62	.59
		2	.78	.75	.70	.65	.71	.67	.63	.59	.60	.57	.55	.52	.50
		3	.69	.66	.60	.55	.63	.59	.54	.50	.54	.51	.48	.45	.42
		4	.61	.59	.52	.46	.56	.52	.47	.43	.48	.45	.41	.38	.36
		5	.53	.51	.44	.39	.51	.46	.40	.36	.43	.40	.36	.33	.30
		10	.35	.30	.23	.19	.32	.27	.21	.18	.26	.23	.19	.16	.14
Flourescent lamps in prismatic lens fixture	1.2 × mounting height	1	.65	.63	.61	.59	.60	.59	.58	.56	.56	.55	.54	.53	.52
		2	.60	.57	.54	.51	.56	.54	.51	.49	.51	.50	.49	.47	.46
		3	.54	.51	.48	.44	.51	.49	.46	.43	.47	.46	.44	.42	.41
		4	.49	.46	.42	.39	.48	.44	.41	.38	.44	.42	.39	.37	.36
		5	.45	.42	.37	.34	.44	.40	.36	.34	.40	.38	.35	.33	.32
		10	.31	.26	.21	.18	.29	.25	.21	.18	.27	.24	.20	.18	.17

Source: From [14].

Table 9.9   *Lamp Output at 70% of Rated Life*

Lamp Type	Watts	Lamp Output at 70% of Rated Life (lumens)
Filament	100	1,600
	150	2,600
	300	5,000
	500	10,000
	750	15,000
	1000	21,000
High-intensity discharge	400	15,000
	700	28,000
	1000	38,000
Fluorescent	40	2,500
	60	3,300
	60	3,300
	85	5,400
	110	7,500

Source: From [14].

Step 8: *Determine the Location of the Luminaries.*

The number of luminaries calculated via Equation 9.11 will result in the correct *quantity* of light. In addition to the quantity of light, the *quality* of light must be considered. The most important factors affecting the quality of light are glare and diffusion. Glare is defined as any brightness that causes discomfort, interference with vision, or eye fatigue. The brighter the luminary, the greater the potential for glare.

Several design considerations result from observations concerning glare: overly bright luminaries should not be used, light sources should be mounted above the normal line of vision, ceilings and walls should be painted with light colors to reduce contrast, and background lighting should be provided if supplementary lighting is used.

Diffusion of light indicates that illumination results from light coming from many directions. The greater the number of luminaries, the more diffuse the light. Fluorescent luminaries provide more diffuse light than incandescent luminaries. Also, light will not diffuse properly if luminaries are spaced too far apart.

Because shadows produce eye fatigue, many luminaries should be used to illuminate an area. The selection of fluorescent versus diffusion and the available diffusion panels and hoods should be considered during lighting design. To obtain properly diffused light, use the maximum permissible ratio of luminary spacing to mounting height above the working surface as given in Table 9.8 in the column labeled "Spacing Not to Exceed."

Example 9.4

A machine shop 100 ft × 40 ft with a 13-ft ceiling will be illuminated for automatic machining and rough grinding. Uniform lighting is required through the machine shop. If the luminaries are to be ceiling mounted, all ceilings and walls are to be painted white, and all luminaries are to be cleaned every 24 months, what lighting should be specified?

Step 1: *Determine the level of illumination.* From Table 9.5, a minimum illumination level of 100 footcandles is required.

Step 2: *Determine the room cavity ratio.* Assume all working surfaces are 3 ft from the floor. Using Equation 9.8 to calculate the RCR results in the following:

$$RCR = \frac{(5)(13 - 3)(100 + 40)}{(100)(40)} = 1.75$$

Step 3: *Determine the ceiling cavity ratio (CCR).* The CCR need not be considered as the luminaries are ceiling mounted.

Step 4: *Determine the wall reflections (WR) and the effective ceiling reflectance (ECR).* According to Table 9.6, the WR and BCR are 80%. Because the luminaries are to be ceiling mounted, the ECR is also 80%.

Step 5: *Determine the coefficient of utilization (CU).* If fluorescent lamps in uncovered fixtures are to be utilized, the CR may be interpolated in Table 9.8 as being between 0.88 and 0.78. A CU of 0.80 will be utilized.

Step 6: *Determine the light-loss factor (LLF).* For fluorescent lamps in uncovered fixtures in a "medium-dirty environment" that are cleaned every 24 months, a lamp luminary dirt depreciation factor of 0.85 may be obtained from Table 9.10. Therefore, the LLF is 0.85.

Step 7: *Calculate the number of lamps and luminaries.* According to Equation 9.10, the number of lamps required if 60-watt fluorescent lamps are used is:

$$\text{Number of lamps} = \frac{(100)(100)(40)}{(0.8)(0.85)(3300)} = 179$$

If two lamps are placed in each luminary, according to Equation 9.11, then 90 luminaries are required.

Step 8: *Determine the location of the luminaries.* Table 9.8 indicates that for a mounting height 10 ft above the work surface, the luminaries should be spaced no more than 13 ft apart. Each fluorescent fixture is 4 ft long. By placing nine rows of 10 luminaries across the room with the first and last rows 6 ft from the wall and the other rows 11 ft apart, the illumination level within the room will be evenly distributed and adequate to perform the tasks within a machine shop. The lighting layout is shown in Figure 9.9.

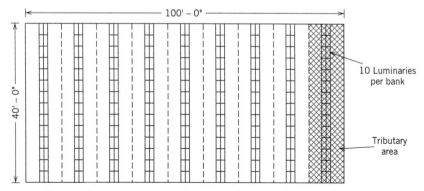

Figure 9.9 Luminaries configuration for 10 ft above work surface.

Table 9.10 *Lamp Luminary Dirt Depreciation Factors*

DIRT—CONDITION[a]

Luminarie	Clean—Offices, Light Assembly, or Inspection Months Between Cleaning					Medium—Mill Offices Paper Processing, or Light Machining Months Between Cleaning					Dirty—Heat Treating, High-Speed Printing, or Medium Machining Months Between Cleaning					Very Dirty—Foundary or Heavy Machining Months Between Cleaning				
	6	12	24	36	48	6	12	24	36	48	6	12	24	36	48	6	12	24	36	48
Filament reflector lamps	0.95	0.93	0.89	0.86	0.83	0.94	0.89	0.85	0.81	0.78	0.87	0.84	0.79	0.74	0.70	0.83	0.74	0.60	0.56	0.52
High-intensity discharge lamps	0.94	0.90	0.84	0.80	0.75	0.92	0.88	0.80	0.74	0.69	0.90	0.83	0.76	0.68	0.64	0.86	0.79	0.69	0.63	0.57
Fluorescent lamps in uncovered fixtures	0.97	0.94	0.89	0.87	0.85	0.93	0.90	0.85	0.83	0.79	0.93	0.87	0.80	0.73	0.70	0.88	0.83	0.75	0.70	0.64
Fluorescent lamps in prismatic lens fixtures	0.92	0.88	0.83	0.80	0.78	0.88	0.84	0.77	0.73	0.71	0.82	0.78	0.71	0.67	0.62	0.78	0.72	0.64	0.60	0.57

[a]Information under this heading from [14].

Example 9.5

Do the lighting requirements change for the machine shop described in Example 9.4 if the luminaries are still to be hung at 13 ft but suspended from a 41-ft ceiling?

Steps 1 and 2. The same as Example 9.4.

Step 3. According to Equation 9.9, the CCR may be calculated as:

$$CCR = \frac{(41-13)}{(13-3)} \times 1.75 = 4.9$$

Step 4. As in Example 9.4, the WR and BCR are 80%. According to Table 9.7, the ECR is 58%.

Step 5. Given that fluorescent lamps in uncovered fixtures are still to be utilized, the CU may be interpolated from Table 9.8 as 7.5%.

Step 6. Unchanged from Example 9.4.

Step 7. According to Equation 9.10, the number of lamps required if 60-watt fluorescent lamps are utilized is:

$$\text{Number of lamps} = \frac{(100)(100)(40)}{(0.75)(0.85)(3300)} = 191$$

If two lamps are placed in each luminary according to Equation 9.11, then 96 luminaries are required.

Step 8. Table 9.8 indicates that for a mounting height 10 ft above the work surface, the luminaries should be spaced apart no more than 13 ft. Each fluorescent fixture is four ft long. By placing four rows of 24 luminaries along the long axis of the machine shop so that each row begins and ends two ft from the wall and the first and last rows are five ft from the wall and all other rows are spaced 10 ft apart, the illumination level within the room will be evenly distributed and adequate to perform the required tasks. This lighting layout is shown in Figure 9.10.

The lighting information given to the person planning the electrical services should include a lighting layout and a description of the type of luminaries and lamps to be used. With this information and the location of the equipment required for electrical services, as described on the department service and area requirement sheets, the total electrical service requirement can be planned.

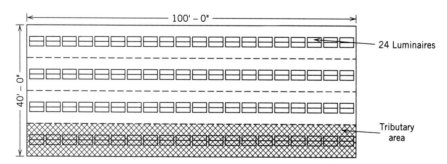

Figure 9.10 Luminaries configuration for 13 ft above work surface.

9.6   LIFE SAFETY SYSTEMS

Life safety systems are designed to control emergency situations that will disrupt normal operations. These emergencies are created primarily by:

1.   Fire
2.   Seismic events
3.   Power failure

Fire is the most pervasive of the three and accounts for the majority of costs associated with disaster. Fire resistance is therefore critical in the design of any facility. Fire protection for buildings is governed by the Uniform Building Code (UBC). UBC is a model building code that outlines the protection features that must be included in the building design. The features that are typically covered are fire ratings of walls, floors and roofs as well as egress, sprinklers, and standpipe requirements.

The first objective of the facilities planner is to determine the building's function and construction type as defined by the occupancy classification. This occupancy classification consists of over 20 UBC specific classifications. In general, however, the following seven facility types typically account for all occupancy classifications:

Group	Type
A	Assemblies, theaters
E	Educational facilities
I	Institutional occupancies
H	Hazardous occupancies (fuel, paint, chemicals)
B	Business, office, factory, commercial facilities
R	Residential
M	Miscellaneous

The type of structure also governs the degree of fire resistance. These construction types range from Type I (nearly fireproof) to Type D (conventional wood-frame construction). Fire resistance, therefore, refers to the ability of a structure to act as a barrier that will not allow the fire to spread from its point of origin. Even with these design efforts, there is no such thing as a fire-immune building. Given this fact, facilities planners must be cognizant of the need to provide for safety routes as an integral part of the layout. The Uniform Building Code requires that means of egress be provided in every facility from every part of every floor to a public street or alley. The minimum requirement, therefore, is at least one exit from every facility and, in cases when the rated number of occupants exceeds a certain number, two or more exits. In general, most buildings require two or more exits. This is done so that other exits are available if one is blocked by fire. In addition, most local building codes will require that maximum allowable times to exit the building not be violated. This time could be the time to reach a protected area, that is, outside the building, or get to another compartment if the facility has fully fireproof, segregated compartments. Also, no point can be more than 150 ft from an

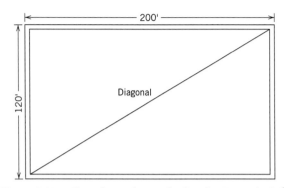

Figure 9.11 Plan of warehouse facility for Example 9.6.

exit (200 ft in a building with sprinklers), and access must be provided for handicapped persons. This may entail ramps or an exit passageway to a protected compartment. It is important to note that the use of elevators is to be avoided. In
general, elevator shafts collect smoke from the floor that is on fire, so their use is
strongly discouraged. In any event, facilities planners must recognize that when a
fire occurs, the order of priority is:

1. Lives of occupants in the facility
2. The building
3. The goods and equipment in the building

As an example, consider the determination of the exits for a warehouse shown in
Figure 9.11.

Example 9.6

Using BOCA (Building Officials and Code Administrators) code from Table 9.11 (maximum
floor area allowances per occupant) for industrial areas, 100 square ft per occupant is the
recommended area:

$$\text{Maximum population} = \frac{120 \text{ ft} \times 200 \text{ ft}}{100 \text{ square ft/person}}$$
$$= 240 \text{ people}$$

From Table 9.12 (minimum number of exits for occupant load), given the population, the
minimum number of exits required is two.

From Table 9.13, the remoteness requirements can be addressed. That is, how far
apart should these doors be to safely move occupants out of the facility?

For the building shown in Figure 9.11,

$$\text{The distance between exits} = \frac{\text{the diagonal dimension}}{4} = \frac{d}{4}$$
$$d = ((200)^2 + (120)^2)^{1/2} = 233$$
$$\text{Minimum distance between exits} = \frac{233}{4} = 58$$

Note: If the building does not have sprinklers, exits should be d/2.

Table 9.11 (BOCA Table 806) Maximum Floor Area Allowances per Occupant

Use	Floor Area per Occupant (ft^2)
Assembly with fixed seats	See Section 806.1.6
Assembly without fixed seats	
Concentrated (chairs only—not fixed)	7 net
Standing space	3 net
Unconcentrated (tables and chairs)	15 net
Bowling alleys allow five persons for each alley,	7 net
including 15 ft of runaway, and for additional areas	100 gross
Business areas	
Court rooms—other than fixed seating areas	40 net
Educational	
Classroom area	20 net
Shops and other vocational room areas	50 net
Industrial areas	100 gross
Institutional areas	
Inpatient treatment areas	240 gross
Outpatient areas	100 gross
Sleeping areas	120 gross
Library	
Reading rooms	50 net
Stack areas	100 gross
Mercantile, basement and grade floor areas	30 gross
Areas on other floors	60 gross
Storage, stock, shipping areas	300 gross
Parking garages	200 gross
Residential	200 gross
Storage areas, mechanical equipment room	300 gross

Note 1: ft = 304.8 mm; 1 ft^2 = 0.093 m.
Source: BOCA [12].

Table 9.12 (BOAC Table 809.2) Minimum Number of Exits for Occupant Load

Occupant Load	Minimum Number of Exits
500 or less	2
501–1000	3
Over 1000	4

Source: BOCA [12].

Table 9.13 BOCA Remoteness Requirements

Where two exit access doors are required, they shall be placed a distance apart equal to not less than one-half the length of the maximum overall diagonal dimension of the building or area to be served. Where exit enclosures are provided as a portion of the required means of egress and are interconnected by a corridor conforming to the requirements for corridor construction, the exit separation distance shall be measured along the line of travel within the corridor. In all other cases, the separation distance shall be measured in a straight line between exits or exit access doors.

Source: BOCA [12].

Table 9.14 *Length of Exit Access Travel (ft)*

Use Group	Without Fire Suppression System	With Fire Suppression System
A, B, E, F-1, I-1, M, R, S-1	200	250
F-2, S-2	300	400
H	—	75
I-2, I-3	150	200

Source: BOCA [12].

Using Table 9.12, we can now determine the allowable minimum width for each door.

$$\text{Capacity per exit} = \frac{240 \text{ people}}{2 \text{ exits}} = 120 \text{ people/exit}$$

$$\text{Minimum width} = 120 \times 0.2 \text{ in} = 24''$$

Since the minimum width is below the 2 ft-8 in. recommended for barrier-free access, each door should be increased to 2 ft-8 in. in width (see Figure 9.12). Finally, check the recommended solution to determine if the allowable length of travel exiting the building is satisfied. From Table 9.14, for use group (I) with fire suppression, travel distance to exits is okay.

The planner must balance the number of exits with the maximum-allowable travel distance required by code. From Table 9.14 the maximum-allowable travel distance for industrial use groups with fire a suppression system is 250 feet. This suggests that both doors could be on the same elevation of the warehouse.

Industrial engineers are often concerned with fire protection systems for rack storage facilities. It is necessary to adequately protect rack storage facilities with the correct fire suppression system. The National Fire Protection Association (NFPA) addresses these types of storage systems in NFPA Standard 230, "Standard for the Fire Protection of Storage." NFPA Standard 230 also references NFPA Standard 13, "Standard for the Installation of Sprinkler Systems," for protection of rack storage facilities. It is important to note there are four classes of fire. A, B, C, and D. These classifications are determined by what is burning and the extinguishment method. Class A fires involve organic material such as wood, paper, and cloth and can be

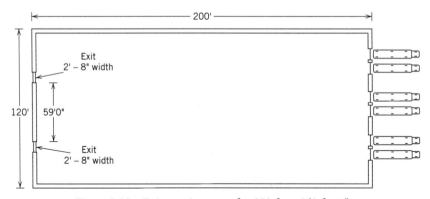

Figure 9.12 Exit requirements for 200 ft × 240 ft w/h.

easily extinguished with water. Class B fires involve flammable liquids and gases such as gasoline, paint thinner, propane, and acetylene. NFPA Standard 30 discusses the protection of flammable liquid storage facilities. An aqueous film-forming foam is the recommended extinguishing agent for Class B fires. Class C fires involve energized electrical equipment such as motors, computers, and panel boxes. Water must not be used on these fires due to its conductive properties. However, if the equipment can be de-energized then the fire can be fought with water. Class D fires involve exotic metals such as magnesium, sodium, and titanium. Burning metal fires are rare and must be dealt with by professionals. A chemical called Met-L-X is the recommended extinguishing agent for burning metal fires.

NFPA 231C's hazard categories are established based on the design of automatic sprinkler systems. These automatic sprinkler systems are a widely used and very effective means of extinguishing or controlling fires in their early stages. This effectiveness is due principally to the fact that being automatic, they can provide protection during periods when the facility is unoccupied. Automatic sprinklers,

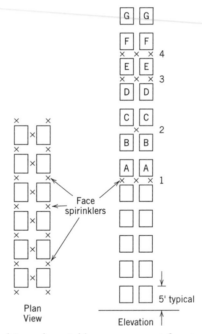

Figure 9.13 Typical in-rack sprinkler arrangement for storage over 25 ft high.

Notes:

1. Sprinklers labeled 1 are required when loads labeled A or B represent top of storage.
2. Sprinklers labeled 1 and 2 are required when loads labeled C or D represent top of storage.
3. Sprinklers labeled 1 and 3 are required when loads labeled E represent top of storage.
4. Sprinklers labeled 1 and 4 are required when loads labeled F or G represent top of storage.
5. For storage higher than represented by loads labeled G, the cycle defined by notes 2, 3, and 4 is represented.
6. The symbol X indicates face and in-rack sprinklers.

Source: NFPA [15].

therefore, because of their effectiveness, are mandatory in certain occupancy classifications. They are not mandatory in rack storage systems, but their usefulness is quite evident. Within a rack storage system, sprinklers are generally located along the longitudinal flue spaces (the space between rows of storage perpendicular to the direction of loading), or transverse flue spaces (the space between rows of storage parallel to the direction of loading). If these sprinklers are located along the longitudinal flue passages, they are generally known as in-rack sprinklers. However, if they are located in transverse flue spaces along the aisle or in the rack within 18 in. of the aisle face, they are known as face sprinklers. This particular sprinkler location is designed to oppose vertical development of a fire on the external face of the rack.

The design requirements of a rack storage sprinkler system are quite involved. The designer must consider a variety of design elements such as height of storage, sprinkler spacing, sprinkler pipe size, water shield location, discharge pressure, water demand requirements, and so on in order to properly determine the in-rack sprinkler arrangement. One such arrangement is shown in Figure 9.13. The objective here is not to make facilities planners fire protection experts but to provide them with a sense of the complex nature of providing an adequate fire protection system.

9.7 SANITATION SYSTEMS

Sanitation systems consist of a hot and cold water supply, a distribution network to supply potable and cleansing water, as well as a collection network for refuse. Refuse handling goes beyond collecting and disposing of wastewater and "soil solids" (known as sewage). It also includes refuse chutes, incinerators, and shredders. The plumbing system, which typically handles the supply of hot and cold water, and the sewage system form the bulk of the sanitation system. A discharge system typically consists of a series of lateral branches tied to a vertical stack. The system is designed to handle both soil (solid waste material) and wastewater (discharge from basins, baths, and washing machines). The primary problem with drainage networks is the problem of preventing odor passing back into the facility. The water seal trap shown in Figure 9.14b is the most common way of alleviating this problem. This method, however, is not foolproof, and even with design precautions, such as making the trap deeper, it is wise to use venting. Venting is widely used to ensure that odors are effectively removed from the facility and vented to the atmosphere. Figure 9.14c shows how the trap can be made more effective by introducing an air vent to break the siphon. A typical trap with air vent arrangement is shown in Figure 9.15.

Plumbing systems within a facility may be required for personnel, processing, and fire protection. Personnel plumbing systems include drinking fountains, toilets, urinals, sinks, and showers. The facilities planner should supply the person planning the overall plumbing systems with the location of all personnel plumbing systems and the number of personnel who will use these services.

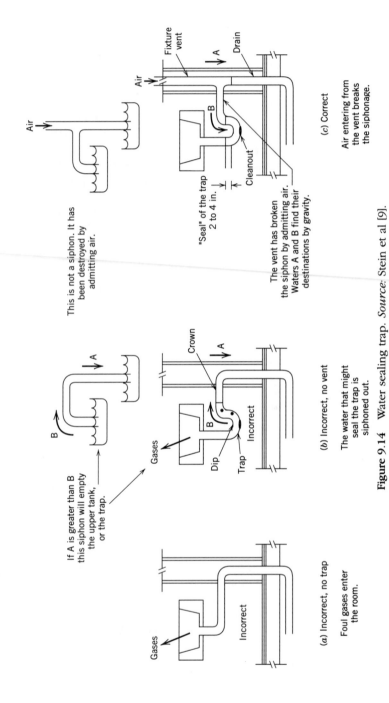

This is not a siphon. It has been destroyed by admitting air.

If A is greater than B this siphon will empty the upper tank, or the trap.

"Seal" of the trap 2 to 4 in.

The vent has broken the siphon by admitting air. Waters A and B find their destinations by gravity.

(a) Incorrect, no trap

Foul gases enter the room.

(b) Incorrect, no vent

The water that might seal the trap is siphoned out.

(c) Correct

Air entering from the vent breaks the siphonage.

Figure 9.14 Water sealing trap. *Source:* Stein et al [9].

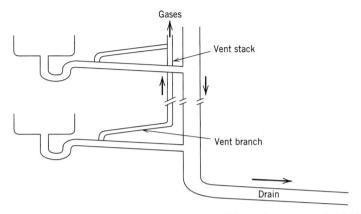

Figure 9.15 Vent used to improve water seal integrity. *Source:* Reid [6].

Example 9.4

For the office building with a floor area as shown in Figure 9.16, the facilities planner is required to determine the plumbing requirements for each floor.

Area of floor = 120 ft × 240 ft
  = 28,800 square ft

Using BOCA code Table 9.11

Minimum recommended space is 100 square ft/person. For business use type, the maximum-allowable floor population is:

$$\frac{28,800 \text{ square ft}}{100 \text{ square ft/person}} = 288 \text{ persons}$$

Assume 144 men, 144 women

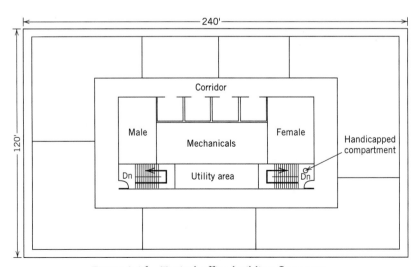

Figure 9.16 Typical office building floor area.

Table 9.15   *Plumbing Fixture Requirements for the Example Problem*

Plumbing Fixture	Women	Men	
	Option 1	Option 1	Option 2*
Water closet	6	6	4
Lavatories	7	7	7
Urinals	—	—	2

*Note up to 1/3 water closets can be replaced by urinals.

Using Table 4.2, the minimum number of fixtures needed was obtained. The adjusted requirement for urinals is as shown in Table 9.15.

For industrial or plant operations, the facilities planner must provide process water and drain requirements to the person planning the plumbing services. In addition, the facilities planner must also provide the composition of all process liquids. The treatment or filtering of these liquids must be carefully planned.

Plumbing services for fire protection require not only that the proper quantity of water is available, but also that the proper water pressure is provided. The most common approach to fire protection is automatic sprinkler systems. The facilities planner's input into the planning of fire protection plumbing services is a facility layout and a description of the activities to be performed in various areas within the facility.

9.8 BUILDING AUTOMATION SYSTEMS

In many cases, the facilities planner will also play a key role in the transition from construction to the actual operations of the facility to perform its design function as a warehouse, manufacturing plant, service center, or research facility. Building automation systems link most of the previously discussed systems together via systems integration into a central control point of process control devices, temperature and humidity measurements, alarm and security systems, lighting controls, and HVAC monitoring and control. Building automation systems in hospitals, research facilities, and refrigerated warehouses all impact the employee as well as product quality within many operations. "Smart facilities" can be monitored on-site from a central location or monitored remotely via modem connectors.

One of the things that a facilities planner must consider is how to link separate facility systems controls and achieve an effective computer-integrated building operation.

Often there are a number of possible systems: the building management system for HVAC, humidity control, and HVAC power monitoring; energy management systems for lighting; security systems that control access to the facility; or a computerized maintenance management system to control and manage the facility maintenance process. In addition, there can be condition-monitoring systems that create alarms for taking action for repairs, for preventive maintenance (PM) services, or simply for emergency shutdown of specific equipment. Process controls, which can have significant impact on throughput and quality, add one more standalone system also needing effective monitoring.

The next step beyond modem connectivity is integration software that allows multiple systems to appear as a single system accessible over the Internet from a single location. Processes for monitoring and controlling the building automation system, energy management, the facility maintenance management system, and other standalone systems software can now be integrated and viewed over the Internet.

Via the Internet, facility operations can start and stop equipment, turn lights on and off, adjust air flows based on outside air temperatures, reset alarms, operate an irrigation system, and schedule occupancy times. Multiple standalone systems can now be integrated into a total facility solution by third parties that provide true computer-integrated building operations. Today's information technology can provide a one-stop Web site that contains documentation about facility operations, such as access to current drawings, system operating and repair procedures, preventive maintenance schedules, maintenance records, safety records, MSDS records, room utilization, real-time graphs and trending, and the ability to make and track appointments and to make work requests on-line.

During the planning process for construction of all the facilities systems discussed, the facilities planner must consider how best to have the facility fully ready for transition to operations and commissioning. Facility operators no longer have the time to learn five or more different monitoring and control systems, yet they want to maintain the flexibility to chose the best control system for each application involved. The first need is the ability to easily understand and operate the facility. A second need is the capability to monitor it and to operate it from anywhere, on any computer, without expensive proprietary software. Today's successful facilities planner will understand facility systems and understand the typical challenges during transition to facilities operations. A successful facilities planner will consider how best to integrate standalone systems into an effective operating facility and to use today's best practices for building automation and information technology.

9.9 FACILITIES MAINTENANCE MANAGEMENT SYSTEMS

Successful transition to facility operation comes after good facilities planning, effective systems design, and construction. The design-build phase is extremely short compared with the long-term operation of the facility. Therefore the successful facilities planner understands and considers one more essential system—the facilities maintenance management system (FMMS). Often termed CMMS (computerized maintenance management systems) or EAM (enterprise asset management), the FMMS supports long-term physical asset management and day-to-day maintenance. This will include all building-related assets—the facilities systems that were discussed previously. It will encompass maintenance of internal equipment, whether manufacturing, warehousing, or service related. It includes all mobile equipment associated with the facility, and it can be used to monitor and control contracted maintenance services as well. Simply stated, FMMS is the business management system for effective facilities management and maintenance.

Briefly, the FMMS as a business management system may include modules with system functionality for:

- Asset data and equipment history

- Parts and material inventory management
- Work management and control
- Budget control
- Preventive maintenance procedures
- Materials requisitioning and purchasing

- Planning, scheduling, work dispatching
- Facility specifications and documentation
- Occupancy management
- Project management
- Regulatory compliance data
- Life-cycle cost data

The facilities planner may or may not be directly involved with the selection and implementation of FMMS, but should understand basic functionalities and requirements for implementing an effective FMMS. To facilitate a future FMMS and transition to operations, the facilities planner should support these key areas:

- Ensure asset/systems documentation is complete.

- Ensure that spare parts are in place and material inventory is established.

- Validate that regulatory issues and all life-safety issues have been resolved.

- Strive to standardize system components.

- Ensure that a facility PM program is in place on day of commissioning.

- Training on all facility systems has been established and occurs before commissioning.

9.10   SUMMARY

Planning facility systems is typically not the responsibility of the facilities planner. Nevertheless, the facilities planner will be required to interact with the architects and engineers responsible for planning and designing a facility's systems and should therefore be familiar with the approaches used to plan these systems. Facilities planners may not be at the facility during actual operation. However, they must envision the transition to facility operations and take steps during the planning process to consider what maintenance will be required during operations. This chapter presents an introduction to the approaches used to plan structural, atmospheric, enclosure, lighting and electrical, life-safety, and sanitation systems. It has shown how these systems, once designed and operating, can be monitored, controlled, and maintained with today's information technology.

BIBLIOGRAPHY

1.   Jennings, B. J. *Environmental Engineering: Analysis and Practice,* International Textbook, Scranton, PA, 1970.

2. Kaufman, J. E., *IES Lighting Handbook,* Illuminating Engineering Society of North America, New York, 1981.

3. McKinnon, G. P., *Fire Protection Handbook,* 15th ed., NFPA, Quincy, MA, 1981.

4. Olivieri, J. B., *How to Design Heating-Cooling Comfort Systems,* Business News Publishing, Birmingham, MI, 1971.

5. Ramsey, C. G., and Sleeper, H. R., *Architectural Graphics Standards,* John Wiley & Sons, New York, 1988.

6. Reid, E., *Understanding Buildings,* MIT Press, Cambridge, MA, 1989.

7. Schiller, M., *Architectural License Seminars,* ALS Publishing, Los Angleles, CA, 1988.

8. Stanford, H.W., *Analysis & Design of Heating, Ventilating and Air Conditioning Systems,* Prentice Hall, Englewood Cliffs, NJ, 1988.

9. Stein, B., Reynolds, J.S., and McGuinness, W.J., *Mechanical And Electrical Equipment For Buildings,* 7th ed., John Wiley, New York, 1986.

10. Tompkins, J.A., and Smith, J.D., *The Warehouse Management Handbook,* McGraw–Hill, New York, 1988.

11. *ASHRAE Handbook of Fundamentals,* American Society of Heating, Refrigerating and Air Conditioning Engineers, New York, 1977.

12. *BOCA, National Building Code,* 11th ed., Building Officials & Code Administrators International, Inc., Country Club Hills, IL, 1990.

13. *Fire Resistance Directory,* Underwriters Laboratories, Northbrook, IL, 1993.

14. *Life Safety Code Handbook,* NFPA, Boston, MA, 1990.

15. *Lighting Handbook,* Westinghouse Lamp Division, Westinghouse Electric Corp., Bloomfield, NJ, 1974.

16. National Fire Protection Association—Section 231C, NFPA Publications, Quincy MA, 1992.

PROBLEMS

9.1a A general office that is 250 ft × 150 ft will be used for regular office work. The ceiling height is 10 ft and all working surfaces will be 3 ft from the floor. If the luminaries will be ceiling mounted, all ceilings and walls painted white, and all luminaries cleaned every six months, what lighting should be specified? Use fluorescent lamps in prismatic lens fixtures.

9.1b What would be the changes in light specifications for the facility described in Problem 9.1a if the office were to be used for accounting and bookkeeping work?

9.1c What would be the changes in lighting specifications for the facility described in Problem 9.1a if the ceiling height were 15 ft?

9.1d What would be the changes in lighting specifications for the facility described in Problem 9.1a if the ceiling was 15 ft and the luminaries were to be mounted at a height of 9 ft?

9.1e What would be the changes in lighting specifications for the facility described in Problem 9.1a if all ceilings and walls were painted a medium color?

9.1f What would be the changes in lighting specifications for the facility described in Problem 9.1a if the fixtures were cleaned only once every 36 months?

9.2 What would be the heat loss of a facility having the following characteristics if the inside design temperature is 72°F and the outside design temperature is 12°F?

- 400 ft × 300 ft
- 20 ft tall
- No windows

- Six glass doors measuring 3 ft × 8 ft
- 8 ft solid cinder block construction
- 1 in. metal insulated roof with an insulated ceiling
- Uninsulated slab floor

9.3   What would be the heat loss of the facility described in Problem 9.2 if 20 4-ft × 8-ft double-pane glass windows were installed?

9.4   What would be the heat loss of the facility described in Problem 9.2 if the roof were replaced with a 2-in. metal roof having an insulated ceiling [$U_R = 0.13$ Btu/(hr)(ft^2)(°F)]?

9.5   What would be the heat loss of the facility described in Problem 9.2 if the slab floor were insulated [$U_F = 0.55$ Btu/(hr)(ft^2)(°F)]?

9.6   Given the following additional information, what is the cooling load for the facility described in Problem 9.2?

- Three doors on 400-ft east side
- Three doors on 400-ft west side
- Three doors on 400-ft east side = 55 Btu Btu/(hr)(ft^2)
- Three doors on 400-ft west side = 110 Btu/(hr)(ft^2)
- Lighting consists of 700 60-watt fluorescent lamps
- Fifty men perform heavy work within the facility and 60 women perform light work
- Outside design temperature = 98°F
- Inside design temperature = 78°F

9.7   If ten 4-ft × 8-ft double-pane glass windows were installed on the east and west sides of the facility described in Problem 9.6, what would be the cooling load of the facility?

9.8   The shade factor for awnings is 0.3 and for blinds is 0.7. What change in cooling load would occur if awnings were placed on the windows described in Problem 9.7? What if blinds were placed in these windows?

9.9   On what basis are building occupancy groups determined?

9.10 For the building shown in Figure 9.6, if the building occupancy were considered for educational/classroom use, determine the following:
  a. Maximum population
  b. Maximum number of exits
  c. Minimum distance between exits (assume building has no sprinkler system)
  d. Provide a sketch showing possible location of the exits

9.11 What is the major benefit of using a sprinkler system?

9.12 Rank the following design objectives in descending order of priority:
- Safety of firefighters
- Safety of regular building occupants
- Salvage of building
- The goods and equipment in the building

9.13 What is the main purpose of face sprinklers in an in-rack sprinkler system?

9.14 What is the purpose of a trap in a sewage system?

9.15 In the design of a plumbing system, what are the two principal reasons why a vent stack is critical?

9.16 As a facilities planner, is the following statement true or false: "The grid structure should defer to the function of the facility"? Give the reasons for your answer.

9.17 How can the thermal performance of a facility be improved? What is the primary purpose of an enclosure system for a manufacturing facility?

9.18 Using Figure 9.6, if the floor area of the building is 120 ft × 180 ft, provide recommendations for the main duct, louver supply, and exhaust sizes.

Assume:

Air supply rate	$1.0 \ ft^3/min\text{-}ft^2$
Exhaust rate	$0.3 \ ft^3/min\text{-}ft^2$
Air speed	1800 ft/min
Louver air speed	
• Exhaust	2000 ft/min
• Intake	1000 ft/min

## Part Four

DEVELOPING ALTERNATIVES: QUANTITATIVE APPROACHES

10

QUANTITATIVE FACILITIES PLANNING MODELS[1]

10.1 INTRODUCTION

Part 4 of the text consists of a single chapter, which provides a number of quantitative models that can be used to facilitate the development of alternative facilities plans. We make no claim that all relevant models are contained in the chapter. (Recall, computerized layout and graph-based layout models were presented in Chapter 6.) Instead, we claim that we have found the models presented in this chapter to be useful in both teaching and practicing facilities planning.

In this chapter, we present, first, a number of prescriptive models for solving some relatively simple facilities location and facilities layout problems. Next, we present a number of descriptive and prescriptive models that can be used to design a warehouse or storage system.

Our presentation of warehouse design models begins with a consideration of the classical storage method of block stacking; specifically, prescriptive models are presented for minimizing the average amount of floor space required to store a product. Based on the results obtained for block stacking, similar models are provided for several varieties of racked storage.

The focus on storage systems continues with a consideration of automated storage and retrieval systems. In addition to presenting models for performing rough-cut estimations of the acquisition cost for an AS/RS, a number of models of the operating performance of an AS/RS are presented.

---

[1]This chapter is intended for advanced undergraduate students and first-year graduate students who are familiar with the concepts of classical optimization, branch and bound, linear programming, and probability theory.

Our consideration of storage systems design concludes with a presentation of prescriptive models for use in designing order picking systems. Both in-the-aisle and end-of-aisle order picking are considered.

Following the storage systems models, we present a number of descriptive models that can be used to design conveyor systems. First, we model a recirculating trolley conveyor; next, we model the horsepower requirements for unit load and package conveyors, as well as belt conveyors for bulk material.

Since the design of conveyor systems is influenced by random variation, following the presentation of deterministic models of recirculating trolley conveyors and horsepower requirements, we present a number of waiting-line models. The subject of many books and papers, our coverage of waiting lines or queues is limited to relatively simple models that have obvious applications to facilities planning. Although several "expected value" models are presented in the chapter, the section on waiting-line models is the first place where we give explicit consideration to the impact of random phenomena on facilities planning.

Following our presentation of waiting-line models, we consider the use of simulation models to address a broader set of probabilistic phenomena than can be addressed using waiting-line models. As with waiting lines, entire books have been devoted to simulation modeling. Hence, our treatment of the subject is very abbreviated. Beyond the rudiments of simulation, we provide an overview of simulation software that is specifically designed for material handling systems.

As noted above, we have not attempted to provide a comprehensive treatment of quantitative models for use in facilities planning. In addition to the computerized layout and graph-based layout models cited, we note our omission of assembly line balancing models, models of flexible manufacturing systems and transfer machines, and simulation models of a variety of elements of the facilities planning regime. Further, many more facilities layout and location models are available—the same is true for storage systems, conveyors, and waiting lines.

In summary, there is no shortage of quantitative models that have been applied to gain some insight or facilitate a design decision in facilities planning. However, there is a shortage of space available within this text on the subject. For that reason, we encourage you to maintain a currency in the facilities planning literature, with the references for this chapter being a good starting point.

10.2   FACILITY LOCATION MODELS

In this section, a number of analytical models of facility location problems are presented. The subject of location analysis is sufficiently vast to warrant one or more texts devoted entirely to it. Our objective in this section is to illustrate its applications to facilities planning problems. Hence, the models and solution procedures we present are selected for their illustrative value.

Location analysis can be applied to many problems including locating an airport, a school, a machine tool, a warehouse, a sewage treatment plant, a production facility, a post office, a hospital, a library, and so on. Because of the breath of application of location analysis, there exists a strong interdisciplinary interest in the subject.

There are various objectives used in facility location decisions, for example, minimizing the sum of the weighted distances between the new facility and the other existing facilities (denoted as the *minisum location problem*), and minimizing the maximum distance between the new facility and any existing facility (denoted as the *minimax location problem*).

The distance measures involved in a facility location problem are an important element in formulating an analytical model. Distance measures can be categorized as:

1. Rectilinear where distances are measured along paths that are orthogonal (or perpendicular) to each other. This measure is also known as Manhattan distance due to the fact that many streets in the city are perpendicular or parallel to each other. An example would be a material transporter moving along rectilinear aisles in a factory.

2. Euclidean (or straight line) where distances are measured along the straight line path between two points. A straight conveyor segment linking two workstations illustrates Euclidean distance.

3. Flow path distance where distances are measured along the actual path traversed between two points. For instance, in an automated guided vehicle system a vehicle on a transport mission must follow the guide path network. Thus, the flow path distance may be longer compared to rectilinear or Euclidean distance.

We will discuss rectilinear models below and defer the coverage of flow path distance models to Section 10.8. Discussion of Euclidean models can be found in Francis et al. [23].

Rectilinear Facility Location Problem

We use the following notation:

$X = (x, y)$ — location of the new facility

$P = (a_i, b_i)$ — location of existing facility i, $i = 1, 2, \ldots m$

w_i — "weight" associated with travel between the new facility and existing facility i

$d(X, P_i)$ — distance between the new facility and existing facility i

The annual cost of travel between the new facility and existing facility i is assumed to be proportional to the distance between the points X and P_i, with w_i denoting the constant of proportionality.

The objective is to

$$\text{Minimize } f(X) = \sum_{i=1}^{m} w_i \, d(X, P_i) \tag{10.1}$$

In a rectilinear model, the distances are measured by the sum of the absolute difference in their coordinates, that is,

$$d(X, P_i) = |x - a_i| + |y - b_i| \tag{10.2}$$

Single-Facility Minisum Location Problem

The minisum location problem is formulated as follows:

$$\text{Minimize } f(X) = \sum_{i=1}^{m} w_i |x - a_i| + \sum_{i=1}^{m} w_i |y - b_i| \tag{10.3}$$

Since Equation 10.3 is written in such a way that terms involving x are separate from terms involving y, the optimum values of x and y can be obtained independently.

In order to find the optimum value of x, two mathematical properties of such a solution are employed. Namely, the x-coordinate of the new facility will be the same as the x-coordinate of some existing facility; and the optimum x-coordinate will be such that no more than half the total weight is to the left of x and no more than half the total weight is to the right of x. The latter condition is referred to as the *median condition*. Both properties also apply in determining the optimum value of y. A justification for these conditions is given in Askin and Standrige [2] and Francis and White [24].

Example 10.1

In order to illustrate how one determines the optimum solution to the minisum problem, consider an example involving the minisum location of a new machine tool in a maintenance department. Suppose there are five existing machines that have a material handling relationship with the new machine. The existing machines are located at the points $P_1 = (1, 1)$, $P_2 = (5, 2)$, $P_3 = (2, 8)$, $P_4 = (4, 4)$, and $P_5 = (8, 6)$. The cost per unit distance traveled is the same between the new machine and each existing machine. The number of trips per day between the new machine and the existing machines are 5, 6, 2, 4, and 8, respectively.

Ordering the x-coordinates of the new facilities gives the facilities sequenced 1, 2, 4, 5, and 8, with the corresponding sequence of weights being 5, 2, 4, 6, and 8. The total

Table 10.1   x-Coordinate Solution for Example 10.1

Machine i	Coordinate a_i	Weight w_i	$\sum_{j=1}^{i} w_j$
1	1	5	5
3	2	2	7
4	4	4	$11 < 25/2$
2	5	6	$17 > 25/2$
5	8	8	25
	$x^* = a_2 = 5$		

Table 10.2   y-Coordinate Solution for Example 10.1

Machine i	Coordinate b_i	Weight w_i	$\sum_{j=1}^{i} w_j$
1	1	5	5
2	2	6	$11 < 25/2$
4	4	4	$15 > 25/2$
5	6	8	23
3	8	2	25
	$y^* = b_4 = 4$		

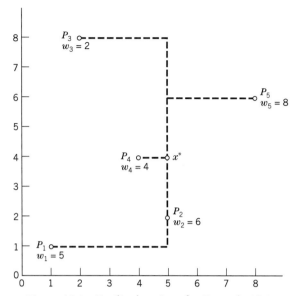

Figure 10.1   Facility locations for Example 10.1.

weight is 25. As shown in Table 10.1, the partial sum first equals or exceeds one-half the total for $i = 2$; hence $x^* = a_2 = 5$. In a similar fashion, as depicted in Table 10.2, the optimum y coordinate is found to be $y^* = b_4 = 4$. Thus, $X^* = (5, 4)$, as depicted in Figure 10.1. Rectilinear paths between the new facility and each existing facility are indicated by dashed lines in Figure 10.1.

The total weighted distance resulting from the location $x = (5, 4)$ is:

$$f(5, 4) = 5(|5 - 1| + |4 - 1|) + 6(|5 - 5| + |4 - 2|) + 2(|5 - 2| + |4 - 8|)$$
$$+ 4(|5 - 4| + |4 - 4|) + 8(|5 - 8| + |4 - 6|)$$
$$= 35 + 12 + 14 + 4 + 40 = 105$$

If one were unable to locate the new machine at the point $(5, 4)$ because the location coincided with, say, a heat treatment furnace, then alternate sites could be evaluated by computing the value of $f(X)$ for each site and selecting the site with the smallest value of $f(X)$.

It should be noted that in the case where the partial sum is equal to one-half of the sum of all the weights, then the solution includes all points between the coordinate where the equality occurred and the next coordinate value.

Example 10.2

In the previous example suppose it is not possible to place the new machine at a point other than the following candidate sites: $Q_1 = (5, 6)$, $Q_2 = (4, 2)$ and $Q_3 = (8, 4)$. Which would be preferred? Computing the value of $f(X)$ for $X = Q_k$, $k = 1, 2, 3$ yields:

$$f(5, 6) = 45 + 24 + 10 + 12 + 24 = 115$$
$$f(4, 2) = 20 + 6 + 16 + 8 + 64 \quad = 114$$
$$f(8, 4) = 50 + 30 + 20 + 16 + 16 = 132$$

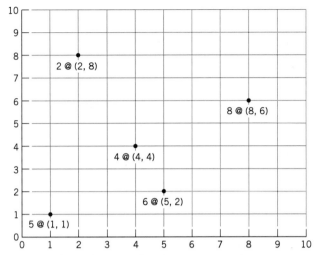

Figure 10.2   Coordinate locations for the existing machines in Example 10.1.

Hence, the best site would be Q_2; however, Q_1 has very nearly the same value of $f(X)$. Qualitative considerations, as well as quantitative considerations not reflected in $f(X)$, might indicate Q_1 is preferred to Q_2.

In general, one can construct iso-cost contour lines as an aid in determining an appropriate location for the new facility. Software tools such as Matlab [43], Mathematica, and other similar tools may be used for this purpose. If such tools are not available, then contour lines may be constructed manually using the following procedure for the rectilinear metric:

1.   Plot the locations of the existing facilities and designate the weights associated with each, as shown in Figure 10.2.

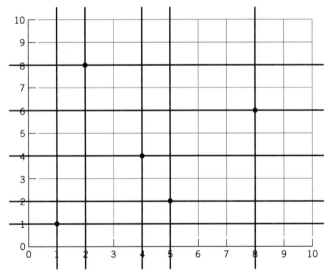

Figure 10.3   Horizontal and vertical lines intersecting the coordinates of existing facilities in Example 10.1.

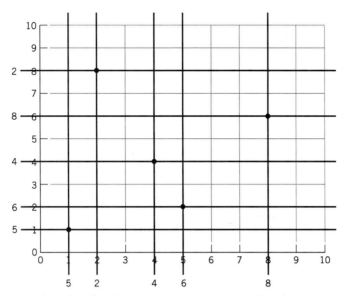

Figure 10.4 Sums of weights along lines intersecting existing machine locations in Example 10.1.

2. Draw vertical and horizontal lines through the coordinate points for the existing facilities, as shown in Figure 10.3.

3. Sum the weights for all existing facilities having the same *x*-coordinate and enter the total at the bottom of the vertical line passing through that coordinate; perform similar calculations for the *y*-coordinates, as shown in Figure 10.4.

4. Consider only the weights for the *x*-coordinates. If you are located to the left of the leftmost existing facility, how much force will be pulling you to the right? It is the

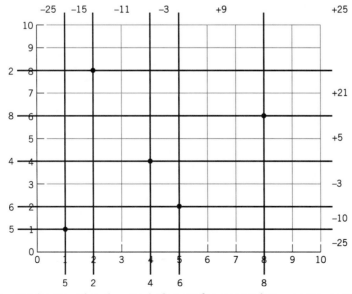

Figure 10.5 Net horizontal and vertical "forces" for regions between intersecting lines for Example 10.1.

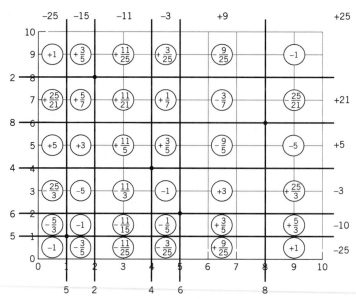

Figure 10.6   Resultant "force" for each grid region defined by the intersecting horizontal and vertical lines for Example 10.1.

sum of the weights for all facilities. Now, suppose you move along the x axis until you pass the coordinate location for the nearest existing facility. What is the net "pull" you will feel? It is the sum of the pulls to the right, less the sum of the pulls to the left. For each region between x-coordinates, determine the net pull. Designate pulls to the right as positive, and designate pulls to the left as negative. Perform the same calculations for movement along the y-axis. For the example, the net pulls felt along the x-axis are shown at the top of Figure 10.5; the net pulls felt along the y-axis are shown along the right side of Figure 10.5.

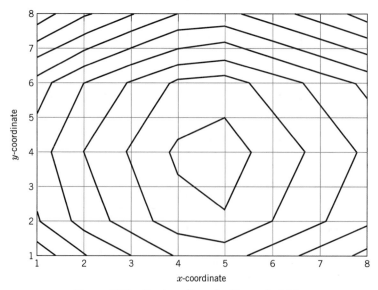

Figure 10.7   Contour lines for Example 10.1.

5. Iso-cost contour lines designate movement that does not change the value of the objective function. For each grid region enclosed by the horizontal and vertical lines passing through the coordinate locations of existing facilities, the slope of the contour line equals the negative of the ratio of the net horizontal pull and the net vertical pull. The slopes for the example are shown in Figure 10.6.

6. An iso-cost contour line can be constructed from any coordinate point by drawing a line through that point with the calculated slope. When the grid region boundary is met, the slope of the contour line changes to that of the grid region entered. Continuing to draw the contour line and changing the slope as different grid regions are entered will result in a closure at the beginning point for the contour line. Sample iso-cost contour lines are shown in Figure 10.7 for the rectilinear function.

$$f(X) = 5(|x - 1| + |y - 1|) + 6(|x - 5| + |y - 2|) + 2(|x - 2| + |y - 8|)$$
$$+ 4(|x - 4| + |y - 4|) + 8(|x - 8| + |y - 6|)$$

Single-Facility Minimax Location Problem

In the case of rectilinear distances, the minimax location problem is given by

$$\text{Minimize } f(X) = \max [(|x - a_i| + |y - b_i|), i = 1, 2, \ldots m] \quad (10.4)$$

In order to obtain the minimax solution, let

$$c_1 = \text{minimum } (a_i + b_i) \quad (10.5)$$
$$c_2 = \text{maximum } (a_i + b_i) \quad (10.6)$$
$$c_3 = \text{minimum } (-a_i + b_i) \quad (10.7)$$
$$c_4 = \text{maximum } (-a_i + b_i) \quad (10.8)$$

As shown in [23], optimum solutions to the minimax location problem can be shown to be all points on the line segment connecting the point

$$(x_1^*, y_1^*) = 0.5(c_1 - c_3, c_1 + c_3 + c_5) \quad (10.9)$$

and the point

$$(x_2^*, y_2^*) = 0.5(c_2 - c_4, c_2 + c_4 - c_5) \quad (10.10)$$

where $c_5 = \max (c_2 - c_1, c_4 - c_3)$. The maximum distance will be equal to $c_5/2$.

Table 10.3 *Data for Example 10.3*

i	a_i	b_i	$a_i + b_i$	$-a_i + b_i$
1	0	0	0	0
2	4	6	10	2
3	8	2	10	-6
4	10	4	14	-6
5	4	8	12	4
6	2	4	6	2
7	6	4	10	-2
8	8	8	16	0
$c_1 = 0$	$c_2 = 16$	$c_3 = -6$	$c_4 = 4$	$c_5 = 16$

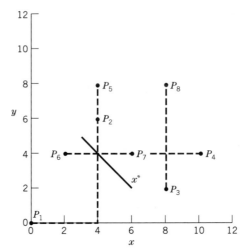

Figure 10.8   Locations for existing facilities and the new facility for Example 10.3.

Example 10.3

In order to illustrate the use of the minimax solution procedure, consider the problem of locating a maintenance department in a production area. It is desirable to locate the maintenance facility as close to each machine as possible, in order to minimize machine downtime.

Eight machines are to be maintained by crews from the central maintenance facility. The coordinate locations of the machines are (0, 0), (4, 6), (8, 2), (10, 4), (4, 8), (2, 4), (6, 4), and (8, 8). From Table 10.3 it is seen that $c_1 = 0$, $c_2 = 16$, $c_3 = -6$, $c_4 = 4$, and $c_5 = 16$. Hence, the optimum solutions lie on the line segment connecting the point.

$$(x_1^*, y_1^*) = \tfrac{1}{2}(6, 10) = (3, 5)$$

and the point

$$(x_2^*, y_2^*) = \tfrac{1}{2}(12, 4) = (6, 2)$$

as shown in Figure 10.8. The point (3, 5) is 8 distance units away from P_1, P_3, P_4, and P_8; the point (6, 2) is 8 distance units away from P_1, P_5, and P_8; the remaining points on the line segment are 8 distance units away from P_1 and P_8.

n-Facility Location Problem

The general formulation of the *n*-facility location problem is given by

$$\text{Minimum } f(X) = \sum_{j=1}^{n} \sum_{k=1}^{n} v_{jk} d(X_j, X_k) + \sum_{j=1}^{n} \sum_{i=1}^{m} w_{ji} d(X_j, P_i) \qquad (10.11)$$

where $X = (x_j, y_j)$ location of new facility j, $j = 1, 2, \ldots, n$

$\qquad P = (a_i, b_i)$ location of existing facility i, $i = 1, 2, \ldots m$

$\qquad v_{jk}$ $\qquad\qquad$ cost per unit distance travel between j and k

$\qquad w_{ji}$ $\qquad\qquad$ cost per unit distance travel between j and i

$\qquad d(X_j, X_k)$ $\qquad$ distance between j and k

$\qquad d(X_j, P_i)$ $\qquad$ distance between j and i

i.   with rectilinear distance

$$d(X_j, X_k) = |x_j - x_k| + |y_j - y_k|, \text{ and} \tag{10.12}$$

$$d(X_j, P_i) = |x_j - a_i| + |y_j - b_i|. \tag{10.13}$$

ii.   with Euclidean distance

$$d(X_j, X_k) = [(x_j - x_k)^2 + (y_j - y_k)^2]^{1/2}, \text{ and} \tag{10.14}$$

$$d(X_j, P_i) = [(x_j - a_i)^2 + (y_j - b_i)^2]^{1/2}. \tag{10.15}$$

With a transformation of terms with absolute values, the n-facility rectilinear distance location problem can be solved using linear programming methods. The Euclidean distance problem is more difficult to solve since the function to be minimized is continuous but nondifferentiable over the entire range of both the x- and y-coordinates.

We refer you to [23] for detailed discussions on solution methods to the n-facility location problem.

Location-Allocation Models

An important class of facility location problems is that involving not only the determination of where new facilities are to be located but also which customers (existing facilities) will be served by each new facility. Such a problem is referred to as a **location-allocation problem.** In its most general form, the location-allocation problem also involves a determination of the optimum number of new facilities.

A mathematical formulation of the location-allocation problem is given as follows:

$$\text{Minimize } \Psi = \sum_{j=1}^{n} \sum_{i=1}^{m} z_{ji} \, w_{ji} \, d(X_j, P_i) + g(n) \tag{10.16}$$

subject to

$$\sum_{j=1}^{n} z_{ji} = 1 \qquad i = 1, \ldots, m \tag{10.17}$$

where

Ψ = total cost per unit time

n = number of new facilities ($n = 1, 2, \ldots, m$)

$X_j = (x_j, y_j)$, the coordinate location of new facility j

w_{ji} = cost per unit time per unit distance if new facility j interacts with existing facility i

$z_{ji} = 1$, if new facility j interacts with existing facility i; 0, otherwise

$g(n)$ = cost per unit time of providing n new facilities

The decision variables in the location-allocation problem are n, the number of new facilities, Z, the allocation matrix, and X_j, $j = 1, \ldots, n$, the locations of the new facilities. Constraints of Equation 10.17 ensure that each existing facility interacts with only one new facility.

Because of the difficulty in solving the general problem, the location-allocation problem is solved by using an enumeration procedure. Namely, for a given value of n, the resulting problem is solved and the value of the objective function, Ψ_n, recorded. Then, the optimum value of n is obtained by searching for the minimum value of Ψ_n.

For the purposes of this chapter, we will simply enumerate all of the allocation combinations for each value of n, determine the optimum location for each new facility for each allocation combination, and specify the minimum cost solution. The number of possible allocations, given m customers and n new facilities is given by

$$S(n, m) = \sum_{k=0}^{n-1} \frac{(-1)^k (n-k)^m}{k! \, (n-k)!} \qquad (10.18)$$

Example 10.4

To illustrate an approach to solving location-allocation problems consider four machines which are located along a common aisle as follows: $P_1 = (0, 0)$, $P_2 = (3, 0)$, $P_3 = (6, 0)$, and $P_4 = (12, 0)$. For simplicity, we assume that $w_1 = w_2 = w_3 = 1$ and $w_4 = 2$. The cost of setting up n new facilities is $g(n) = 5n$.

For $n = 1$, the problem reduces to a single-facility location problem with $x = 6$. The total cost is computed as $5(1) + [1(6) + 1(3) + 1(0) + 2(6)] = 26$.

For $n = 2$, the following allocations can be made

New Facility 1	New Facility 2
P_1	P_2, P_3, P_4
P_1, P_2	P_3, P_4
P_1, P_2, P_3	P_4

For $n = 3$,

New Facility 1	New Facility 2	New Facility 3
P_1	P_2	P_3, P_4
P_1	P_2, P_3	P_4
P_1, P_2	P_3	P_4

For $n = 4$,

New Facility 1	New Facility 2	New Facility 3	New Facility 4
P_1	P_2	P_3	P_4

The number of assignments given above is limited since the y-coordinates of all existing facilities are equal to zero. Without this constraint, the total number of possible assignments will be that given by Equation 10.18.

We note that for each assignment, a single-facility location problem is solved. Based on these allocations, the minimum total cost location-allocation can be found.

Plant Location Model

The plant location problem and/or warehouse location problem has received considerable attention in the research literature. A mathematical formulation of one version of the plant location problem is given as follows:

$$\text{Minimize } z = \sum_{i=1}^{m} \sum_{j=1}^{n} c_{ij} y_{ij} + \sum_{j=1}^{n} f_j x_j \tag{10.19}$$

subject to

$$\sum_{i=1}^{m} y_{ij} \leq m \, x_j \quad j = 1, \ldots, n \tag{10.20}$$

$$\sum_{j=1}^{m} y_{ij} = 1 \quad i = 1, \ldots, m \tag{10.21}$$

$y_{ij} \geq 0$ for all i, j

$x_j = (0, 1)$ for all j

where

m = number of customers

n = number of plant sites

y_{ij} = proportion of customer i demand supplied by a plant at site j

$x_j = 1$, if a plant is located at site j; otherwise, 0

c_{ij} = cost of supplying all demand of customer i from a plant located at site j

f_j = fixed cost of locating a plant at site j

The objective function gives the cost of locating plants at the sites corresponding to the positive-valued x_j. The first set of n constraints indicates that the total proportion of customer demand supplied by a plant at site j either equals zero when x_j equals zero or cannot exceed the number of customers when x_j equals one. The second set of m constraints ensures that all demand must be met for customer i by some combination of plants. Non-negativity restrictions on y_{ij} and zero-one restrictions on x_j complete the mixed-integer programming formulation.

A branch and bound solution procedure can be used to solve exactly the plant location problem formulated above, as well as more general formations. Because branch-and-bound methods are beyond the scope of this text, rather than solve the plant location problem exactly we will use a quick-and-dirty method. In particular, we employ an "add" or "construction" heuristic in which one continues to add new facilities until no additional new facilities are justified. Alternatively, we could use a "drop" or "elimination" heuristic in which one drops new facilities until no additional facilities can be dropped.

Example 10.5

To motivate the discussion of the plant location problem, suppose there exist five potential sites for a new warehouse A, . . . , E and the firm's major customers are located in five cities 1, . . . , 5. The annual costs of meeting the customer's demands in a city from each site are given in Table 10.4. The annual fixed cost of providing a new warehouse is also given for each site.

If only one warehouse is to be built, which site would you prefer? Selecting site A yields a total annual cost of $19,900, including the cost of the warehouse. The total annual

Table 10.4 *Data for Example 10.5*

Customer Location	Warehouse Sites				
	A	B	C	D	E
1	100	500	1,800	1,300	1,700
2	1,500	200	2,600	1,400	1,800
3	2,500	1,200	1,700	300	1,900
4	2,800	1,800	700	800	800
5	10,000	12,000	800	8,000	900
Fixed costs	3,000	2,000	2,000	3,000	4,000

costs for the remaining sites are: $17,700, $9,600, $14,800, and $11,100. Hence, site C would be selected. Notice that if a warehouse is placed at site C, then cities 4 and 5 will be served by the warehouse at site C regardless of any subsequent decisions to locate warehouses at other sites.

Suppose we decide to place a warehouse at site A, in conjunction with the warehouse at site C. What will be the reduction in total annual cost, if any? Cities 1 and 2 would be served from site A at a cost savings of $1,700 + $1,100 = $2,800, which is less than the fixed cost of $3,000 for providing a warehouse at A. Placing a warehouse at site B yields cost savings of $1,300 + $2,400 + $500 = $4,200 for cities 1, 2, and 3; since the cost savings is $2,200 greater than the fixed cost of $2,000, site B is a feasible candidate for an additional warehouse. Placing a warehouse at site D yields annual savings of $3,100 to be offset by the fixed cost of $3,000. Site E yields annual savings of $100 for city 1 at a fixed cost of $4,000. A second warehouse will be placed at site B as the greatest net annual savings occurs for site B. The new total annual cost is ($9,600 − $4,200) + $2,000 = $7,400.

Currently, we have assigned warehouses to sites B and C. If a third one is added, the net annual savings resulting from the placement of the third warehouse at site j, denoted NAS(j), will be

$$\text{NAS(A)} = 400 - 3,000 = -2,600$$

$$\text{NAS(D)} = 900 - 3,000 = -2,100$$

$$\text{NAS(E)} = 0 - 4,000 = -4,000$$

Hence, no additional warehouses are justified and warehouses will be placed only at sites B and C for a total annual cost of $7,400.

The approach of continuing to add warehouses to sites having the greatest positive net annual savings is not guaranteed to yield an optimum solution. In particular, it may occur that a subsequent addition of warehouses causes some previous decision to add a warehouse to be nonoptimum. One approach that can be used to partially overcome such a possibility is to drop the warehouse site that produces the greatest annual savings. A check for feasible "drops" would be made after each check for feasible "adds."

10.3 SPECIAL FACILITY LAYOUT MODELS

In this section, we describe two models for special types of facility layout problems. You will recall that in our discussion on facility location problems, we were mostly concerned with finding the optimum values of x and y in continuous space. Here, the solutions will be restricted to discrete locations. In addition, area-consuming facilities will be accounted for and their relative placements considered in attempting to find the best layout. Examples of area-consuming facilities include the work envelope of a machine tool and the area required in the storage of materials in a warehouse.

Quadratic Assignment Problem

Location-allocation problems involve a determination of the number and location of new facilities. In some situations the number of new facilities is known a priori and the decision problem involves the location of the new facilities. In this section we consider the problem of locating multiple new facilities in discrete space. Such problems are referred to as assignment problems because the problem reduces to assigning new facilities to sites.

If there exists no interaction between new facilities such that one is concerned only with locating new facilities relative to existing facilities, the problem is a **linear assignment problem.** When interaction exists between new facilities, the problem is a **quadratic assignment problem** (QAP). Quadratic assignment problems are treated in this section. A mathematical formulation of the problem is as follows:

$$\text{Minimize } z = \sum_{j=1}^{n} \sum_{k=1}^{n} \sum_{h=1}^{n} \sum_{\ell=1}^{n} c_{jkh\ell} x_{jk} x_{h\ell} \tag{10.22}$$

subject to

$$\sum_{j=1}^{n} x_{jk} = 1 \qquad k = 1, \ldots, n \tag{10.23}$$

$$\sum_{k=1}^{n} x_{jk} = 1 \qquad j = 1, \ldots, n$$

$$x_{jk} = (0, 1) \text{ for all } j \text{ and } k \tag{10.24}$$

where $c_{jkh\ell}$ is the cost of assigning new facility j to site k when new facility h is assigned to site ℓ.

Heuristic procedures are often used to solve quadratic assignment problems. Generally, such heuristic solution procedures can be categorized as **construction** and **improvement** procedures. Construction procedures develop a solution "from scratch"; namely, facilities are located one at a time until all are located. Improvement procedures begin with all facilities located and seek ways to improve on the solution by interchanging or switching locations for facilities. We show a simple improvement procedure in this chapter.

The improvement procedure we present is based on the CRAFT procedure described in Chapter 6. It is referred to as a steepest-descent, pairwise (two-way) exchange method. The procedure begins with an *initial solution* where each facility is assigned to one of the sites. (Of course, no two facilities can be assigned to the same site.) Next, all pairwise exchanges are considered and the exchange that yields the greatest reduction in total cost is performed. The process continues until no pairwise exchanges can be found that will yield a reduction in total cost. The resulting solution, which is not necessarily globally optimal, is known as a 2-opt solution since no pairwise exchanges can further reduce the total cost.

The improvement procedure we show here can be applied to the general QAP formulation (with the objective function given by Equation 10.22). However, in most facilities planning problems, the objective function is expressed as a product of flows and distances. More specifically, the term $c_{jkh\ell}$ equals the product of f_{jh} (the flow from facility j to h) and $d_{k\ell}$ (the distance from site k to ℓ). Hence, assuming $c_{jkh\ell} = f_{jh}d_{k\ell}$, we present the procedure through the following example.

Example 10.6

Suppose four machines are to be placed in a job shop. The from-to flow matrix, F, for the machines (labeled A through D), and the distance matrix, D, for the four sites (numbered 1 through 4) are given as follows:

Flow Matrix (F)

	A	B	C	D
A	—	5	2	0
B	0	—	2	3
C	3	4	—	0
D	0	0	5	—

Distance Matrix (D)

	1	2	3	4
1	—	5	10	4
2	4	—	6	7
3	8	5	—	5
4	6	6	5	—

Note that the distance matrix is asymmetric. Therefore, instead of converting the from-to flow values to flow-between values, we will leave the flow matrix F in its current form (i.e., the direction of flow is relevant). Suppose the initial solution is given by (A:1, B:2, C:3, D:4); that is, facility A is assigned to site 1, facility B is assigned to site 2, and so on. The first nonzero flow value in F is a flow of 5 units from facility A to B, which is multiplied with 5 distance units (i.e., the distance from site 1 to 2). The next nonzero flow value in F is a flow of 2 units from facility A to C, which is multiplied by 10 distance units (i.e., the distance from site 1 to 3), and so on. The resulting total cost is given by $(5 \times 5) + (2 \times 10) + (2 \times 6) + (3 \times 7) + (3 \times 8) + (4 \times 5) + (5 \times 5) = 147$ units, which is also shown under the column labeled "Initial Solution" in Table 10.5

Pairwise machine exchange leads to six possible pairs. As shown in Table 10.5, these pairs are (AB), (AC), (AD), (BC), (BD), and (CD). We stress that once we exchange, say, machines A and B, and compute the new total cost, we put machines A and B back in their current locations before we compute the cost of exchanging machines A and C. That is, the exchanges are not cumulative exchanges; rather, each exchange is investigated one

Table 10.5   *Pairwise Exchange Results for the Initial Solution to Example 10.6*

Flows	Facility Pairs	Initial Solution	AB	AC	AD	BC	BD	CD
								Pairwise Exchanges (Distances)
5	AB	5	4	5	6	10	4	5
2	AC	10	6	8	5	5	10	4
2	BC	6	10	4	6	5	5	7
3	BD	7	4	7	4	5	6	6
3	CA	8	5	10	5	4	8	6
4	CB	5	8	5	5	6	5	6
5	DC	5	5	6	10	6	6	5
	Total cost	147	136	150	149	151	142	132

Table 10.6 *Pairwise Exchange Results for the First Improved Solution to Example 10.6*

Flows	Facility Pairs	Initial Solution	Distances					
			Pairwise Exchanges					
			AB	AC	AD	BC	BD	CD
5	AB	5	4	6	5	4	10	5
2	AC	4	7	6	5	5	4	10
2	BC	7	4	4	7	6	5	6
3	BD	6	10	6	4	5	5	7
3	CA	6	6	4	5	4	6	8
4	CB	6	6	5	6	7	5	5
5	DC	5	5	8	4	5	7	5
	Total cost	132	139	140	120	122	156	147

at a time relative to the current (or initial) solution. Exchanging machines A and B we obtain (A:2, B:1, C:3, D:4), which yields the distance values shown in Table 10.5 under the column labeled "AB." The resulting total cost is equal to 136 units, which is better than the initial solution but we need to explore the remaining pairs. With the next pair, that is (AC), we obtain (A:3, B:2, C:1, D:4), which yields a total cost of 150 units. The total cost values for the remaining four pairs are shown in Table 10.5.

Following the direction of steepest-descent (i.e., the pair that yields the largest reduction in total cost), we exchange machines C and D to obtain a new solution given by (A:1, B:2, C:4, D:3) for a total cost of 132 units. Taking this solution as the current solution, we evaluate all possible pairwise exchanges again. The results are shown in Table 10.6, where the new solution we obtain is given by (A:3, B:2, C:4, D:1) with a total cost of 120 units. Taking (A:3, B:2, C:4, D:1) as the current solution and considering all possible pairwise exchanges again, we find that no other pairwise exchange yields a total cost less than 120 units. Therefore, we stop the procedure with a 2-opt solution of 120 units.

The improvement procedure described above, terminates at the first "local optimal" (in this case, 2-opt) solution it encounters. Furthermore, it does not allow even temporary increases in the total cost in anticipation of finding a better solution later on. As a result, the quality of the final solution depends very much on the initial solution the procedure is started with. Therefore, we recommend that the above procedure be executed with alternative initial solutions, though there is no guarantee that each initial solution will terminate at a different final solution. In those cases where the analyst feels a better solution is needed, and the incremental effort is warranted, instead of the above steepest-descent procedure we recommend using the simulated annealing procedure described in Chapter 6. The simulated annealing procedure described there is also based on pairwise exchanges but temporary increases in total cost are allowed with a certain probability. We refer the reader to Chapter 6 for further details.

Whether a steepest-descent or simulated annealing-based procedure is used, we need to recognize that both procedures are heuristic procedures, and unless a very large number of runs are made, we are often left wondering how good the final solution is and how far off its total cost may be from the globally optimal solution. Of course, this is a difficult concern to address since the QAP is a difficult problem to solve and we often do not know what the total cost of the globally optimal solution might be. However, in some instances, computing a *lower bound* (LB) on the total cost may give us an indication on the quality of the final solution we obtain with the above heuristic procedures. Lower bounds play an important role in many combinatorial optimization problems, but an in-depth treatment of the topic is beyond our scope. Rather, using the data for Example 10.6, we will demonstrate a fairly simple lower bound developed for the QAP. The lower bound we show is reasonably "strong" for small instances of the problem. (More sophisticated bounds have been developed for larger instances of the QAP; see, for example, [1] and [29], among others.)

In reference to Tables 10.5 and 10.6, note that the total cost is computed by multiplying a set (or vector) of flow values by a set (or vector) of distance values. Of course, the appropriate distance value that each flow value is multiplied with depends on the location of the facilities, but we are still multiplying two vectors. In mathematics, it is well-known that the multiplication (i.e., inner product) of two vectors is minimized if the two vectors are sorted in opposite directions. Suppose in Example 10.6 we sort all the flow values in *nondecreasing* order and all the distance values in *nonincreasing* order. Taking the inner product of the two vectors we obtain

$$LB_1 = f \cdot d = (0, 0, 0, 0, 0, 2, 2, 3, 3, 4, 5, 5) \cdot (10, 8, 7, 6, 6, 6, 5, 5, 5, 5, 4, 4)$$
$$= 112 \text{ units} \tag{10.25}$$

Hence, no matter where we assign each facility, we cannot find a solution with a total cost *less than* 112 units. Of course, there may be no feasible assignment of facilities to sites that produces a total cost of exactly 112 units. This should be evident from the manner in which we simply sorted the two vectors without considering the flow values associated with particular facilities and the distance values associated with particular sites. However, *achievability* (i.e., finding a solution with a total cost equal to 112 units) is not our main purpose in constructing a lower bound. In fact, if we could find a solution that achieves a total cost of 112 units, we would know it is the globally optimal solution. However, finding the globally optimal solution is itself a very difficult task as we stated earlier. Hence, we generally use a lower bound not to test achievability but rather to test the quality of a heuristic solution. For the above example, we find that our 2-opt solution (120 units) is approximately 7% above the lower bound (112 units). Since the globally optimal solution cannot be less than 112 units, we conclude that our 2-opt solution is *at most* approximately 7% larger than the globally optimal solution. Of course, our 2-opt solution may be globally optimal but we do not know that. (We remark that lower bounds are also extensively used within branch-and-bound and other implicit enumeration procedures; such use of lower bounds is beyond our scope.)

The "stronger" a lower bound is, the more useful it becomes for optimization or evaluation purposes. For example, in Example 10.6, all the distance values and the flow values are greater than or equal to zero. Therefore, any solution must have a total cost greater than or equal to zero, which is a very "weak" lower bound because all it says is that the globally optimal solution lies somewhere between 0 and 120 units. This is not useful information compared to saying the "globally optimal solution lies somewhere between 112 and 120 units." This point should also make it clear that there are many possible lower bounds for a given problem (i.e., a lower bound is not unique). In fact, generally speaking, the more information and problem structure we capture, the stronger the lower bound will be. However, we also have to keep the computational burden in mind; often capturing more information or structure in the lower bound means spending more time to compute the bound.

How can we strengthen the lower bound we computed in Example 10.6? As we stated earlier, there are numerous sophisticated bounding schemes developed for the QAP. We will use a fairly straightforward and easy-to-compute lower bound developed by Bazaraa and Elshafei [6]. Consider the cost of assigning facility A to site 1. The primary difficulty with the QAP of course is that we cannot compute the exact cost of this assignment unless we know exactly at which sites facilities B, C, and D are located. However, we can compute a lower bound on the cost of assigning facility A to site 1 even if we do not know the location of the other facilities.

If facility A is located at site 1, the flows *out of* facility A (i.e., flows of 0, 2, and 5 units) would have to be multiplied with distance values *out of* site 1 (i.e., distances of 4, 5, and 10 units). Since we do not know where the other facilities are, we do not know which distance value each flow value must be multiplied with. However, no matter which distance value each flow value is multiplied with, a lower bound on the value of this multiplication can again be obtained simply by sorting the two vectors in question in opposite directions. That is, the inner product of (0, 2, 5) with (10, 5, 4), which yields 30 units, is a lower bound. Repeating the same procedure for the flows *into* facility A and the distances *into* site 1, we obtain (0, 0, 3) · (8, 6, 4) = 12 units. Hence, no matter where facilities B, C, and D are located, the cost of assigning facility A to site 1 is greater than or equal to 30 + 12 = 42 units.

Repeating the above procedure for the remaining sites and remaining facilities we obtain the assignment matrix shown in Table 10.7*a*. Note that the resulting matrix is the well-known (linear) assignment problem, which is straightforward to solve optimally. Using the Hungarian method, we obtain the optimal solution shown in Table 10.7*b*. (The Hungarian method is shown in many operations research texts. The reader may refer to p. 195 in Taha [56] for a quick-and-practical exposure to the Hungarian method; for a more theoretical look at the Hungarian method, the reader may refer to p. 168 in Murty [45].) Note that the solution shown in Table 10.7*b* is optimal for the linear assignment problem; it is not necessarily optimal for the original quadratic assignment problem because each entry in Table 10.7*a* is only a lower bound on the actual cost instead of the actual cost itself.

The total cost of the optimal solution to the linear assignment problem is equal to 42 + 67 + 81 + 40 = 230 units. Our heuristic solution has a total cost of 120 units. Of course, it is impossible for a lower bound to be greater than the total cost of any heuristic solution. The reason we obtained 230 units is that each "interaction" (i.e., flow × distance) in Table 10.7*a* is counted twice when we add up the assignment costs. That is, when we compute the lower bound on the cost of assigning, say, facility A to site 1, we account for the flow it has with facilities B, C, and D. However, when we

Table 10.7 *Assignment-based Lower Bound for Example 10.6*

	1	2	3	4		1	2	3	4
A	42	47	50	49	A	㊷	47	50	49
B	66	69	74	67	B	66	69	74	㊅₇
C	79	81	92	82	C	79	㊆₁	92	82
D	32	35	40	37	D	32	35	㊵	37

(*a*) Lower-bound on assignment costs   (*b*) Optimal solution to linear assignment problem

later compute the lower bound on the cost of assigning, say, facility B to site 2, we include its interaction with facility A. Hence, each (flow × distance) value is counted twice. To compensate, we simply divide 230 by 2 to obtain a new (stronger) lower bound of $LB_2 = 115$ units. (Recall that $LB_1 = 112$ units.) Hence, we can now state that the globally optimal solution lies somewhere between 115 and 120 units, and that our heuristic solution is at most 4.3% above the global optimal solution.

The QAP has many applications in facilities planning (including facility location) and other areas of industrial engineering. It also has applications in electrical engineering (for example, circuit board design) and economics as well. As a result, the QAP has received considerable attention from researchers and it continues to be a very active research area. The interested reader may refer to Çela [16] for a survey on the QAP.

Warehouse Layout Models

In this section, a quantitative warehouse layout model is considered. Specifically, the determination of the location of products for storage in a warehouse is considered. To motivate the discussion, it is necessary to recall the distinction given in Chapter 9 between **dedicated** or **fixed slot storage** and **randomized** or **floating slot storage.** Recall, with dedicated storage a particular set of storage slots or locations is assigned to a specific product; hence, a number of slots equal to the maximum inventory level for the product must be provided. With a pure randomized storage system each unit of a particular product is equally likely to be retrieved when a retrieval operation is performed; likewise, each empty storage slot is equally likely to be selected for storage when a storage operation is performed.

In this section, an approach is presented for determining the optimum dedicated storage layout; rectilinear travel is assumed. The warehouse layout problem considered involves the assignment of products to storage locations in the warehouse. The following notation is used:

q = number of storage locations

n = number of products

m = number of input/output (I/O) points (docks)

S_j = number of storage locations required for product j

T_j = number of trips in/out of storage for product j, that is, throughput of product j

p_i = percentage of travel in/out of storage to/from I/O point i

d_{ik} = distance (or time) required to travel from I/O point i to storage location k

x_{jk} = 1 if product j is assigned to storage location k; otherwise, 0

$f(\mathbf{x})$ = average distance (or time) traveled

The warehouse layout problem can be formulated as follows:

$$\text{Minimize} \sum_{j=1}^{n} \sum_{k=1}^{q} \frac{T_j}{S_j} \sum_{i=1}^{m} p_i d_{ik} x_{jk} \tag{10.26}$$

subject to

$$\sum_{j=1}^{n} x_{jk} = 1 \qquad k = 1, \ldots, q \tag{10.27}$$

$$\sum_{k=1}^{q} x_{jk} = S_j \qquad j = 1, \ldots, n \tag{10.28}$$

$$x_{jk} = (0, 1) \text{ for all } j \text{ and } k$$

It is assumed that each item is equally likely to travel between I/O point or dock i and any storage location assigned to item j. Hence, the quantity $(1/S_j)$ is the probability that a particular storage location assigned to product j will be selected for travel to/from a dock. For convenience, let

$$f_k = \sum_{i=1}^{m} p_i d_{ik} \tag{10.29}$$

In words, f_k is the expected distance traveled between storage location k and the docks.

In order to minimize the total expected distance traveled the following approach is taken.

1.   Number the products according to their T_j/S_j value, such that

$$\frac{T_1}{S_1} \geq \frac{T_2}{S_1} \geq \cdots \geq \frac{T_n}{S_n}$$

2.   Compute the f_k values for all storage locations.
3.   Assign product 1 to the S_1 storage locations having the lowest f_k values; assign product 2 to the S_2 storage locations having the next lowest f_k values; and so on.

Example 10.7

As an illustration of the solution procedure for designing a warehouse layout, consider the warehouse given in Figure 10.9*a*. Storage bays are of size 20 × 20 ft. Docks P_1 and P_2 are for truck delivery; docks P_3 and P_4 are for rail delivery. Dedicated storage is used. Sixty percent of all item movement in and out of storage is from/to either P_1 or P_2, with each dock equally likely to be used. Forty percent of all item movement in and out of storage is equally divided between docks P_3 and P_4. Three products, A, B, and C, are to be stored in the warehouse with only one-type product stored in a given storage bay. Product A requires 3600 ft² of storage space and enters and leaves storage at a rate of 750 loads per month; product B requires 6400 ft² of storage space and enters and leaves storage at a rate of 900 loads per month; product C requires 4000 ft² of storage space and enters and leaves storage at a rate of 800 loads per month. Rectilinear travel is used and is measured between the centroids of storage bays.

The f_k values are shown in each storage bay in Figure 10.9*b*. As an illustration of the computation of an f_k value, suppose $k = 29$. Measuring the rectilinear distance from the centroid of storage bay 29 and each of the four docks gives $d_{1,29} = 120$, $d_{2,29} = 100$, $d_{3,29} = 100$, and $d_{4,29} = 80$. Hence,

$$f_{29} = 0.3(120) + 0.3(100) + 0.2(100) + 0.2(80)$$
$$= 102$$

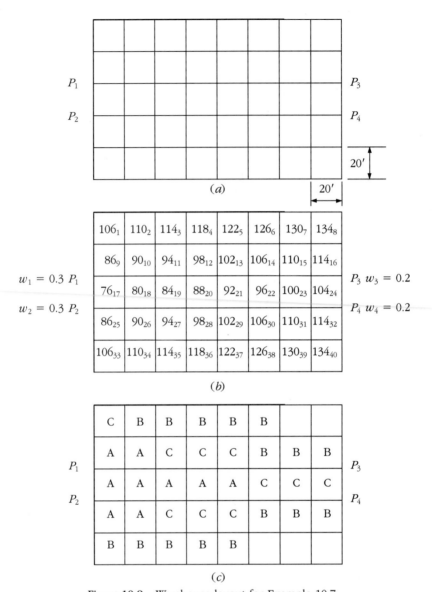

Figure 10.9   Warehouse layout for Example 10.7.

The number of storage bays required for each product equals $S_A = 3600/400 = 9$, $S_B = 16$, and $S_C = 10$. The T_j values are $T_A = 750$, $T_B = 900$, and $T_C = 800$. Therefore, the T_j/S_j values are $T_A/S_A = 83.33$, $T_B/S_B = 56.25$, and $T_C/S_C = 80$. Hence the products will be numbered 1(A), 2(C), and 3(B).

Product 1(A) requires 9 storage bays; hence, the storage bays assigned to product A include [17, 18, 19, 9, 25, 20, 10, 26, 21]. Product 2(C) requires 10 storage bays; hence, [11, 27, 22, 12, 28, 23, 13, 29, 24, 1] are assigned to product C. Product 3(B) requires 16 storage bays; hence, product B is assigned to storage locations [14, 30, 33, 2, 15, 31, 34, 3, 16, 32, 35, 4, 36, 5, 37, 6]. Storage bays 7, 8, 38, 39, and 40 are available for equipment storage, rest rooms, offices, and so on.

The layout design that minimizes expected distance traveled per unit time is given in Figure 10.9c. It is important to point out that the layout design obtained is not necessarily a final layout. Considerations other than expected distance traveled have not been incorporated; the use of the dedicated storage approach must be examined in a particular application. However, the layout does serve as a basis for evaluating other designs.

10.4 MACHINE LAYOUT MODELS

Thus far, the layout models we have covered in previous sections are aggregate in nature; that is, we considered rectangular or square blocks and distances are measured from department centroids or from fixed locations of pick-up/delivery stations. However, there are other design issues that affect machine layout problems, which are not addressed in the previous models. In particular, we mention two below.

a. The interface points for incoming and outgoing parts for individual machines are usually at fixed locations relative to the entire work envelope of the machine.

b. Minimum spaces between machines must be provided to accommodate access to machines for maintenance and service, and allow enough space for material handling devices and in-process storage areas.

Based on these two factors, we can see that the relative position of two adjacent machines would give two different values for, say, total distance parts are moved between the two machines.

A variety of formulations exist for the machine layout problem, including circular, linear single-row, linear double-row, and cluster machine layouts. We consider modifications of the linear single-row formulation of the machine layout problem and the algorithm given by Heragu [31] and Heragu and Kusiak [32] for solving it.

Let

x_i^P = location of pick-up point of machine i, $i = 1, 2, \ldots n$

x_i^D = location of delivery point of machine i

w_i = width of machine i oriented parallel to the aisle

a_{ij} = minimum clearance between machines i and j

c_{ij} = cost per trip per unit distance from machine i to j

f_{ij}^L = total loaded material handling trips from machine i to machine j

f_{ij}^E = total empty (deadhead) material handling trips from machine i to machine j

The single-row model formulation is as follows:

$$\text{Minimize TC} = \sum_{i=1}^{n} \sum_{j=1}^{n} c_{ij}(f_{ij}^L \, d_{ij}^L + f_{ij}^E \, d_{ij}^E) \tag{10.30}$$

The objective function gives the total cost of material handling where

$d_{ij}^L = |x_i^P - x_j^D|,$ loaded travel distance

$d_{ij}^E = |x_i^D - x_j^P|,$ empty (deadhead) travel distance

For simplicity, we will assume that each machine P/D point is located at the midpoint along the edge of the machine work area parallel to the aisle. Thus, $d_{ij}^L = d_{ij}^E$. Also, let $f_{ij} = f_{ij}^L + f_{ij}^E$.

The following constraints ensure that the positions of machine work areas satisfy the required clearance between machines

$$|x_i^P - x_j^D| \geq \frac{1}{2}(w_i + w_j) + a_{ij}, \quad i = 1, 2, \ldots, n - 1; \tag{10.31}$$

$j = i + 1, \ldots, n$ for loaded trips, and

$$|x_i^D - x_j^P| \geq \frac{1}{2}(w_i + w_j) + a_{ij}, \quad i = 1, 2, \ldots, n - 1; \tag{10.32}$$

$j = i + 1, \ldots, n$ for deahead trips.

Direct solution to the above model requires a transformation of the absolute value terms in both the objective function and the constraints. Instead, we describe a construction-type procedure below.

Step 1. Determine the first two machines to enter the layout by computing max $\{c_{ij}f_{ij}\}$. The rationale behind the selection of the machine pair ij with the maximum value of $c_{ij}f_{ij}$ is that this value results in the maximum cost increase when the distance between them is set farther apart. The solution is denoted as $\{i^*, j^*\}$.

Step 2. Place i^* and j^* adjacent to each other. Since the P/D points are located at the midpoint of the machine edge along its width, there would be no difference in total cost between the placement order $i^* \to j^*$ and the order $j^* \to i^*$.

Step 3. The next step is to select the next machine, denoted by k^*, to place in the layout and to determine where to locate this machine relative to those that are already in the layout. The placement order is either $k^* \to i^* \to j^*$ or $i^* \to j^* \to k^*$. The selection of k^* is based on evaluating the relative placement cost, RPC, of setting machine k to the left or to the right side of the set of machines already in the layout. The evaluation function is

$$\text{RPC} = \min_{k \in U}\left\{ \sum_{i \in A} c_{ki}f_{ki}d_{ki}, \sum_{j \in A} c_{jk}f_{jk}d_{jk} \right\} \tag{10.33}$$

where A is the set of all machines already assigned specific locations and U is the set of unassigned machines. The first summation gives the cost of placing machine k to the left and the second summation gives the cost of placing machine k to the right of all assigned machines.

Step 4. Continue Step 3 until all machines are assigned.

Example 10.8

Consider the problem of locating four machines along an aisle. There are a total of $4! = 24$ ways of arranging these four machines. Let $c_{ij} = 1$ and $a_{ij} = 1$. The machine dimensions are

Machine	1	2	3	4
Dimensions	2×2	3×3	4×4	5×5

The loaded material handling trips between machines are

		1	2	3	4
Machine	1	—	10	5	6
	2	8	—	3	8
	3	7	9	—	4
	4	5	11	13	—

The total flow-between matrix is

		1	2	3	4
Machine	1	—	18	12	11
	2	18	—	12	19
	3	12	12	—	17
	4	11	19	17	—

Steps 1 and 2. Since the maximum $\{f_{ij}\}$ value is 19, we select machines 2 and 4 to enter the layout. At this point, the placement order is not important. We arbitrarily select the order $2 \rightarrow 4$. Thus, $A = \{2, 4\}$ and $U = \{1, 3\}$.

Step 3. For this step, we will evaluate the following placement orders: $1 \rightarrow 2 \rightarrow 4$, $2 \rightarrow 4 \rightarrow 1$, $3 \rightarrow 2 \rightarrow 4$, and $2 \rightarrow 4 \rightarrow 3$.

For placement order $1 \rightarrow 2 \rightarrow 4$,

$$f_{12}d_{12} + f_{14}d_{14} = 18(3.5) + 11(8.5) = 156.5$$

For placement order $2 \rightarrow 4 \rightarrow 1$,

$$f_{21}d_{21} + f_{41}d_{41} = 18(9.5) + 11(4.5) = 220.5$$

For placement order $3 \rightarrow 2 \rightarrow 4$,

$$f_{32}d_{32} + f_{34}d_{34} = 12(4.5) + 17(9.5) = 215.5$$

For placement order $2 \rightarrow 4 \rightarrow 3$,

$$f_{23}d_{23} + f_{43}d_{43} = 12(10.5) + 17(5.5) = 219.5$$

Since the minimum relative placement cost is 156.5, we select the placement order $1 \rightarrow 2 \rightarrow 4$.

Step 4. Since there is only one machine unassigned, $k = 3$, the remaining placement orders to be evaluated are $3 \rightarrow 1 \rightarrow 2 \rightarrow 4$ and $1 \rightarrow 2 \rightarrow 4 \rightarrow 3$.

For placement order $3 \rightarrow 1 \rightarrow 2 \rightarrow 4$,

$$f_{31}d_{31} + f_{32}d_{32} + f_{34}f_{34} = 12(4) + 12(7.5) + 17(12.5) = 350.5$$

For placement order $1 \rightarrow 2 \rightarrow 4 \rightarrow 3$,

$$f_{13}d_{13} + f_{23}d_{23} + f_{43}d_{43} = 17(14) + 12(10.5) + 12(5.5) = 430$$

Thus, the final placement order is $3 \rightarrow 1 \rightarrow 2 \rightarrow 4$ with a total cost of 602.

10.5 CONVENTIONAL STORAGE MODELS

In this and the following section, we consider models of various methods of storing and retrieving product. Automated storage and retrieval systems are considered in the following section. Here, we focus on determining "optimum" configurations of storage systems in which unit loads are stored and retrieved using conventional methods. In particular, we consider storage and retrieval of unit loads by using manually operated lift trucks. Four alternative storage methods are analyzed: block stacking, deep lane storage, single-deep storage rack, and double-deep storage rack.

Block Stacking

As noted in Chapter 7, block stacking involves the storage of unit loads in stacks within storage rows. It is frequently used when large quantities of a few products are to be stored and the product is stackable to some reasonable height without load crushing. Frequently, unit loads are block stacked three high in rows that are 10 or more loads deep. The practice of block stacking is prevalent for food, beverages, appliances, and paper products, among others.

An important design question is how deep should the storage rows be. Block stacking is typically used to achieve a high space utilization at a low investment cost. Hence, it is often the case that storage rows are used with depths of 15, 20, 30, or more.

During the storage and retrieval cycle of a product lot, vacancies can occur in a storage row. To achieve first-in, first-out (FIFO) lot rotation, these vacant storage positions cannot be used for storage of other products or lots until all loads have been withdrawn from the row. The space losses resulting from unusable storage positions are referred to as "honeycomb loss"; block stacking suffers from both vertical and horizontal honeycomb loss. Figure 10.10 depicts the space losses resulting from honeycombing.

The design of the block stacking storage system is characterized by: the depth of the storage row (x), the number of storage rows required for a given product lot (y), and the height of the stack (z), where the decision variables, x, y, and z must be integer valued. If the height of the stack is fixed, then the key decision variable is the depth of the storage row.

For a single product, factors that may influence the optimum row depth include lot size, load dimensions, aisle widths, row clearances, allowable stacking heights, storage/retrieval times, and storage/retrieval distribution. For multiple products, other decision variables must be considered. For example, the optimum number of unique row depths, row depths, the assignment of products to depths, and aggregate space requirements must be determined.

The following notation is used throughout this section:

S = average amount of floor space required during the life of a storage lot

S_{BS} = average amount of floor space required, with block stacking and no safety stock

S_{BSSS} = average amount of floor space required, with block stacking and safety stock

S_{BS}^c = continuous approximation to the average amount of floor space required, with block stacking and no safety stock

S_{BSSS}^c = continuous approximation to the average amount of floor space required, with block stacking and safety stock

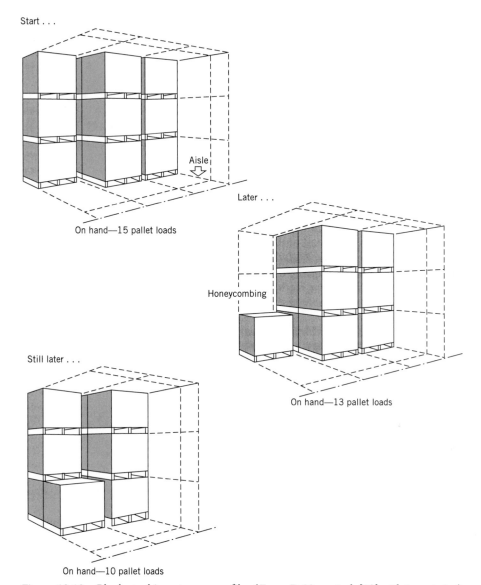

Figure 10.10 Block stacking storage profile. (*From DeMars et al. [18] with permission*)

S_{DL} = average amount of floor space required, with deep lane storage and no safety stock

S_{DLSS} = average amount of floor space required, with deep lane storage and safety stock

S_{DL}^{c} = continuous approximation to the average amount of floor space required, with deep lane storage and no safety stock

S_{DLSS}^{c} = continuous approximation to the average amount of floor space required, with deep lane storage and safety stock

S_{DD} = average amount of floor space required, with double-deep storage and no safety stock

S_{DDSS} = average amount of floor space required, with double-deep storage and safety stock

S_{SD} = average amount of floor space required, with single-deep storage and no safety stock

S_{SDSS} = average amount of floor space required, with single-deep storage and safety stock

Q = size of the storage lot, in unit loads

s = safety stock, in unit loads

W = width of a unit load

L = length or depth of a unit load

c = side-to-side clearance between unit loads and between a unit load and a vertical rack member

r = width of a vertical rack member

A = width of the storage aisle

f = depth of the back-to-back flue space between rack sections

x = depth of a storage row or storage lane, in unit loads

x_{BS} = optimum depth of a storage row, in unit loads, with block stacking

x_{BS}^{c} = continuous approximation of the optimum depth of a storage row, in unit loads, with block stacking

x_{BSSS} = optimum depth of a storage row, in unit loads, with block stacking and safety stock

x_{BSSS}^{c} = continuous approximation of the optimum depth of a storage row, in unit loads, with block stacking and safety stock

x_{DL} = optimum depth of a storage lane, in unit loads, with deep lane storage

x_{DL}^{c} = continuous approximation of the optimum depth of a storage lane, in unit loads, with deep lane storage

x_{DLSS} = optimum depth of a storage lane, in unit loads, with deep lane storage and safety stock

x_{DLSS}^{c} = continuous approximation of the optimum depth of a storage lane, in unit loads, with deep lane storage and safety stock

z = storage height, in unit loads or levels of storage

y = number of storage rows required to accommodate Q unit loads with block stacking

= smallest integer, greater than or equal to Q/xz

v = number of storage lanes required to accommodate Q unit loads with deep lane storage

= smallest integer, greater than or equal to Q/x

η = average number of storage rows required over the life of a storage lot, with block stacking

ξ = average number of storage lanes required over the life of a storage lot, with deep lane storage

To motivate the development of a model of the average amount of floor space required when using block stacking, notice that the average amount of floor space required is equal to the footprint of a storage row (including half the aisle and side-to-side clearances) times the average number of storage rows required during the life of a storage lot of a product. Thus,

$$S_{BS} = \eta(W + c)(xL + 0.5A) \tag{10.34}$$

Example 10.9

To illustrate the calculation of η, suppose the inventory level over the life of a particular storage lot is given by

Period	Inventory Level	Period	Inventory Level	Period	Inventory Level
1	15	6	10	11	5
2	14	7	9	12	4
3	13	8	8	13	3
4	12	9	7	14	2
5	11	10	6	15	1

If $x = 2$ and $z = 3$, what will be the distribution of the number of storage rows required during the life of the storage lot?

# Storage Rows	Periods	# Periods
3	1, 2, 3	3
2	4, 5, 6, 7, 8, 9	6
1	10, 11, 12, 13, 14, 15	6

Hence, the average number of storage rows will be

$$\eta = 3(3/15) + 2(6/15) + 1(6/15) = 27/15 = 1.8.$$

When the inventory level in period k, I_k, is equal to $Q + 1 - k$ (i.e., uniform withdrawal, one unit load per period), the value of η can be obtained using the following relation:

$$\eta = \{y[Q - (y - 1)xz] + (y - 1)xz + (y - 2)xz + \cdots + 2xz + xz\}/Q \tag{10.35}$$

Since the sum of the integers 1 through $n - 1$ equals $n(n - 1)/2$, then

$$\eta = [0.5y(y - 1)xz + y(Q - xyz + xz)]/Q$$

which reduces to

$$\eta = y[2Q - xyz + xz]/2Q \tag{10.36}$$

Note, $y - 1$ storage rows will be full at the time the withdrawal cycle begins. The single partial row will be depleted first; the number of unit loads in the partial row equals $Q - (y - 1)xyz$.

Substituting (10.36) in (10.34) gives

$$S_{BS} = y(W + c)(xL + 0.5A)[2Q - xyz + xz]/2Q \tag{10.37}$$

Since S_{BS} is not a convex function of x, to obtain its minimum it is necessary to enumerate over x. Notice that the optimum value of x will not depend on W or c.

Continuous Approximation

For large values of Q, a continuous approximation to S_{BS} can be used by letting $Q = xyz$ in (10.37). Replacing y with Q/xz and xyz with Q gives

$$S_{BS}^c = (W + c)(xL + 0.5A)(Q + xz)/2xz \tag{10.38}$$

Taking the derivative of S_{BS}^c with respect to x, setting the result equal to zero, and solving for x gives a continuous approximation to the optimum value of x,

$$x_{BS}^c = [AQ/2Lz]^{1/2} \tag{10.39}$$

Example 10.10

Suppose $Q = 200$, $A = 12'$, $L = 4'$, and $z = 4$. Substituting in (10.39) and solving for x_{BS}^c yields a value of 8.66. Therefore, an approximation of the optimum value of x would be either 8 or 9.

For the example, for what range of values of Q will the continuous approximation of x be equal to 25? Substituting x equal to 24.5 and x equal to 25.5 in (10.39) and solving for Q yields values 1600 and 1734 for Q, which suggests the optimum value of x is not especially sensitive to changes in the value of Q.

Safety Stock

How might the block stacking model be modified to include safety stock? It is important to recognize what conditions cause safety stock to occur. In particular, safety stock for an individual product is created by having a replacement lot arrive before depleting the inventory of the product in question. Since our optimization is based on the inventory profile for an individual lot, rather than the inventory profile for the product, we note that a safety stock implies that, for the lot that just arrived, no withdrawals occur for some period of time.

Returning to our earlier example, suppose the inventory profile for an item is as follows:

Period	Inventory Level	Period	Inventory Level	Period	Inventory Level
1	15	8	14	15	7
2	15	9	13	16	6
3	15	10	12	17	5
4	15	11	11	18	4
5	15	12	10	19	3
6	15	13	9	20	2
7	15	14	8	21	1

If uniform withdrawals normally occur, then the delay in initiating depletion is due to a safety stock condition. For the example, $s = 6$ and the distribution of storage rows is

# Storage Rows	# Periods
3	9
2	6
1	6

The average number of storage rows will be

$$\eta = 3(9/21) + 2(6/21) + 1(6/21) = 45/21 = 2.14.$$

To incorporate safety stock in the equation for η, (2) is modified as follows

$$\eta = \{y[Q + s - (y - 1)xz] + (y - 1)xz + \cdots + 2xz + xz\}/(Q + s) \quad (10.40)$$

which reduces to

$$\eta = y[2(Q + s) - xyz + xz]/2(Q + s) \quad (10.41)$$

Thus, the average amount of space required over the life of a lot with safety stock is given by

$$S_{BSSS} = y(W + c)(xL + 0.5A)[2(Q + s) - xyz + xz]/2(Q + s) \quad (10.42)$$

Notice, the denominator is twice the length of the cycle time, not twice the lot size.

Example 10.11

Suppose $L = 48''$, $W = 50''$, $A = 156''$, $c = 10''$, $z = 3$, $Q = 25$, and $s = 10$. What will be the optimum value of x? Enumerating over x, the values shown below are obtained for S_{BSSS}. Thus, $x_{BSSS} = 3$, with $x = 5$ being the next best row depth.

x	y	S_{BSSS}
1	9	310.50 ft^2
2	5	238.21
3	3	206.14*
4	3	221.79
5	2	208.21
6	2	226.57
7	2	241.50
8	2	253.00
9	1	212.50

Continuous Approximation for Safety Stock

A continuous approximation of the safety stock condition is obtained by substituting Q for xyz and substituting Q/xz for y in (10.42). The resulting expressions for average amount of space and the optimum row depth are

$$S_{BSSS}^c = Q(W + c)(xL + 0.5A)(Q + 2s + xz]/2(Q + s)xz \quad (10.43)$$

and

$$x_{BSSS}^c = [A(Q + 2s)/2Lz]^{\frac{1}{2}} \quad (10.44)$$

Deep Lane Storage

Deep lane storage is illustrated in Figure 10.11. Notice that it is very similar to block stacking, except every unit load is individually supported. Hence, there is no vertical honeycomb loss with deep lane storage. The depth of the footprint of a deep lane

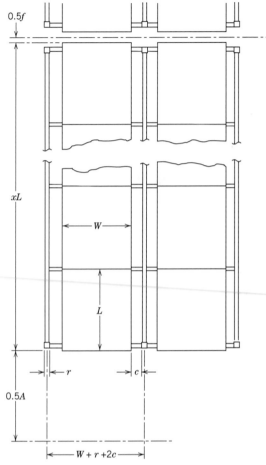

Figure 10.11 Deep lane storage. (*From DeMars et al. [18] with permission*)

is equal to half the sum of the storage aisle and the flue space at the back of the deep lane; the width of the footprint is the sum of the width of the unit load, two clearances, and the width of the vertical rack member. With deep lane storage, each deep lane is independent from every other lane, both horizontally and vertically. The square footage allocation for a lane involves a proration of the footprint over the number of levels of storage. Hence, the average amount of floor space required for a deep lane of storage is given by

$$(W + 2c + r)[xL + 0.5(A + f)]/z. \tag{10.45}$$

The average number of deep lanes of depth x required over the life of a lot is given by

$$\xi = v[2(Q + s) - xv + x]/2(Q + s) \tag{10.46}$$

Therefore,

$$S_{\text{DLSS}} = v(W + 2c + r)[xL + 0.5(A + f)][2(Q + s) \\ - xv + x]/2(Q + s)z \tag{10.47}$$

A continuous approximation for deep lane storage with safety stock yields

$$S^c_{\text{DLSS}} = Q(W + 2c + r)[xL + 0.5(A + f)](Q + 2s + x)/2(Q + s)z \quad (10.48)$$

and

$$x^c_{\text{DLSS}} = [(A + f)(Q + 2s)/2L]^{\frac{1}{2}} \quad (10.49)$$

Example 10.12

Suppose $Q = 147$, $s = 29$, $L = 50''$, $W = 42''$, $c = 10''$, $f = 6''$, $A = 144''$, and $z = 3$. Using deep lane storage, what is the optimum lane depth? Using a continuous approximation yields a value of 17.535. Therefore, it appears that 17 or 18 unit loads is the optimum lane depth. However, when (10.42) is minimized by enumerating over x, the following results are obtained:

x	v	S_{DLSS} (ft^2)
14	11	12,550.69
15	10	12,432.29
16	10	12,638.89
17	9	12,487.50
18	9	12,675.00
19	8	12,470.83
20	8	12,661.11
21	7	12,359.38*
22	7	12,565.97
23	7	12,743.40
24	7	12,891.67
25	6	12,532.29

Thus, if a lane depth of 17 is used, the resulting value of S will be only 1.04% greater than would be obtained by using the optimum lane depth of 21. (If a lane depth of 18 is used, the resulting floor space will be 2.55% greater than the minimum value.)

Pallet Rack

The single-deep and double-deep pallet racks can be considered special cases of deep lane storage with $x = 2$ and $x = 1$, respectively. The difference between standard pallet rack and deep lane rack is that two loads (not necessarily the same product or from the same lot) can be stored side-by-side using either single-deep or double-deep pallet rack. Hence, the width of a storage footprint for pallet rack will be the width of the load (W) plus one full clearance between the load and the vertical rack member (c) plus half the side-to-side clearance between loads on a common load beam ($0.5c$) plus half the width of a vertical rack member ($0.5r$), or $W + 1.5c + 0.5r$, as depicted in Figure 10.12.

Double-Deep Pallet Storage Rack

Given the results for deep lane storage, it is relatively easy to compute the value of S for double-deep storage rack installations. Double-deep storage is a special case

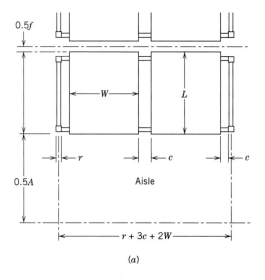

(a)

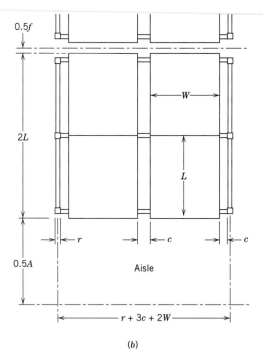

(b)

Figure 10.12   Pallet rack storage. (*a*) Single-deep rack. (*b*) Double-deep rack. (*From DeMars et al. [18] with permission*)

of deep lane storage with $x = 2$. With double-deep storage, two loads can be stored side-by-side in a pallet opening on a common load beam. Therefore, the width of a double-deep storage footprint will be $W + 1.5c + 0.5r$. The depth of a double-deep footprint will be $2L + 0.5(A + f)$. Again, the area of the footprint is prorated over the number of levels of storage. Hence, the average amount of floor space required for double-deep storage, with safety stock, is given by

$$S_{DDSS} = v(W + 1.5c + 0.5r)[2L + 0.5(A + f)][2(Q + s) \\ - 2v + 2]/2(Q + s)z \qquad (10.50)$$

without safety stock,

$$S_{DD} = v(W + 1.5c + 0.5r)[2L + 0.5(A + f)](Q - v + 1)/Qz \qquad (10.51)$$

Since the storage depth is known, v equals $Q/2$ if Q is even and $(Q + 1)/2$ if Q is odd.

Hence, if Q is even, then

$$S_{DDSS} = Q(W + 1.5c + 0.5r)[2L + 0.5(A + f)](Q \\ + 2s + 2)/4(Q + s)z \qquad (10.52)$$

$$S_{DD} = (W + 1.5c + 0.5r)[2L + 0.5(A + f)](Q + 2)/4z \qquad (10.53)$$

and, if Q is odd, then

$$S_{DDSS} = (Q + 1)(W + 1.5c + 0.5r)[2L + 0.5(A + f)](Q + \\ 2s + 1)/4(Q + s)z \qquad (10.54)$$

$$S_{DD} = (W + 1.5c + 0.5r)[2L + 0.5(A + f)](Q + 1)^2/4Qz \qquad (10.55)$$

Single-Deep Pallet Storage Rack

To determine the average floor space requirements for single-deep storage rack installations, the depth of the storage footprint will equal $L + 0.5(A + f)$. The width of the footprint is the same as for double-deep storage rack. Letting $x = 1$ and $v = Q$ and modifying appropriately the deep lane storage results, the average amount of floor space required for single-deep storage rack, with safety stock, is given by

$$S_{SDSS} = Q(W + 1.5c + 0.5r)[L + 0.5(A + f)](Q + 2s + 1)/2(Q + s)z \qquad (10.56)$$

In the absence of safety stock, (10.56) reduces to

$$S_{SD} = (W + 1.5c + 0.5r)[L + 0.5(A + f)](Q + 1)/2z \qquad (10.57)$$

Example 10.13

Let $Q = 15$

$L = 50$ in.

$W = 42$ in.

$A = 144$ in.

$r = 3$ in.

$c = 4$ in.

$f = 12$ in.

$z = 3$

Using double-deep storage rack with no safety stock,

$$S_{DD} = [42 + 1.5(4) + 0.5(3)][2(50) + 0.5(144 + 12)](15 + 1)^2/4(15)(3)$$
$$= 12,531.2 \text{ in}^2 \text{ or } 87.02 \text{ ft}^2$$

With single-deep storage rack and no safety stock,

$$S_{SD} = [42 + 1.5(4) + 0.5(3)][50 + 0.5(144 + 12)](15 + 1)/2(3)$$
$$= 16,896 \text{ in}^2 \text{ or } 117.33 \text{ ft}^2$$

10.6   AUTOMATED STORAGE AND RETRIEVAL SYSTEMS

Automated storage and retrieval systems (AS/RS) have had a dramatic impact on manufacturing and warehousing. Through the use of computer control, handling and storage systems have been integrated into manufacturing and distribution processes. While the primary emphasis originally was in storing and retrieving finished goods, more recently the focus has been on work-in-process (WIP), raw materials, and supplies.

The AS/RS has also affected the design of the warehouse. Rack-supported buildings, for example, are frequently constructed to house the AS/RS. A typical rack-supported building is shown in Figure 10.13. Such structures are usually more economical to construct than conventional buildings; additionally, they frequently receive favorable income tax treatment because of their special-purpose construction.

As noted in Chapter 5, the AS/RS consists of storage racks, storage/retrieval (S/R) machines, and input/output (I/O) or pickup/deposit (P/D) stations. A typical AS/RS layout is given in Figure 10.14. Typically, each S/R machine operates in a single aisle and services storage racks on each side of the aisle.

Various types of AS/R systems were presented in Chapter 5. In this chapter, we will first focus on the unit load AS/RS, which is designed to store and retrieve palletized loads. After we present a (deterministic) cost model to design and evaluate unit load AS/R systems, we will present cycle time results, which are based on (nondeterministic) cycles (or trips) performed by the S/R machine. Subsequently, we will consider order picking systems, which include the person-on-board AS/RS and the miniload AS/RS.

In designing an AS/RS, a number of decisions must be made. Depending on the particular situation, some of the decisions will be made by the system supplier and others will be made by the user of the system. Information must be collected on a number of aspects and values must be established for many design parameters. Among some of the most important considerations are the following [65]:

1.   Load size(s) and opening size(s).
2.   Number and location of P/D stations.
3.   Building construction; rack supported or conventional.
4.   Land availability, condition, cost, and zoning restrictions.
5.   Number, height, and length of storage aisles.

Figure 10.13 Rack-supported automated storage/retrieval systems (AS/RS). (*Part a courtesy of Clark Equipment Co.; part b courtesy of Hartman Material Handling Systems*)

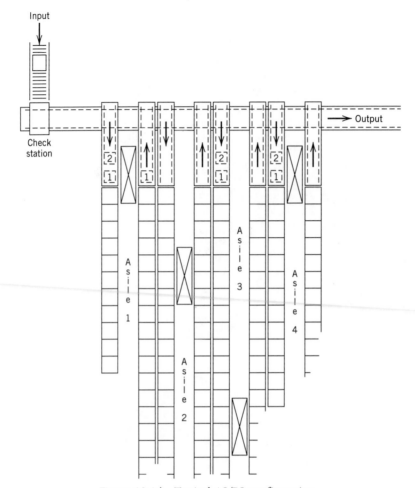

Figure 10.14 Typical AS/RS configuration.

6. Percentage of the operations (storages and retrievals) to be performed on a dual-command basis.
7. Applicability of transfer cars.
8. Randomized or dedicated storage or some combination thereof.
9. Dwell point for the S/R machine when it is idle, in aisle, or at P/D station.
10. Level of automation.
11. Level of computer control.
12. Requirement for physical inventory.
13. Replenishment requirements.
14. Requirement for maintenance.
15. Use of slave pallets versus vendor pallets.
16. Mode of input/output.
17. Plan for evolution and change.
18. Throughput requirement, peak and average.

19. Level of specifications to be developed for hardware and software.

20. Provision for interrupt and priority storage/retrieval.

21. Storage depth (single, double, or deep lane).

22. Provision for mixed loads on a pallet.

23. Use of automatic identification systems.

24. Use of simulations to support design decisions.

25. Amount of queue space required for inbound and outbound loads.

26. Impact of randomness versus scheduled operation on the requirements for the system.

27. Energy and utilities.

28. Impact of inflation and income taxes.

29. Sprinkler requirements.

30. Plan for training, startup, debugging, and postauditing.

Very early in the AS/RS design process a fixed scope of work should be developed. A schedule of management reviews, a list of deliverables, and a definition of the level of effort should be developed for the project.

The AS/RS Product Section of the Material Handling Institute, Inc. developed a listing of benefit-cost elements to be considered in AS/RS economic justification; included are the following elements [65].

1. Initial investment considerations
    a. Land value
    b. Building structure
    c. Utilities
    d. Machinery and equipment
    e. Investment tax credit
    f. Income tax
    g. Interest expense

2. Annual operating cost considerations
    a. Direct operating personnel
    b. Secondary personnel
    c. Floor space savings and cube utilization
    d. Utilities
    e. Depreciation
    f. Insurance and taxes
    g. Product damage
    h. Pilferage
    i. Inventory carrying costs
    j. Job enrichment
    k. Maintenance
    l. Improved service levels
    m. Improved materials control

For some of these factors, reasonably accurate estimates can be obtained using existing accounting data and work measurement data. Land, building, equipment, and labor costs are relatively easy to estimate. On the other hand, such items as heat, light, power, insurance, pilferage, product damage, employee satisfaction, reduced inventory levels and production delays, poor management decisions, and customer service are not easily traced and measured. However, it is readily agreed that all of these factors, as well as others, are important and do contribute to the firm's overall economic performance.

Sizing and Cost Estimation[2]

In order to develop a basis for estimating AS/RS investment costs, Zollinger [64] compiled detailed cost information for more than 60 AS/RS installations. He found that a reasonably accurate estimate of the acquisition cost of an AS/RS could be obtained by summing the costs of the storage rack, the S/R machines, and the building used to house the AS/RS.

Further, he found that the cost of the storage rack, including installation and S/R machine support rail, was a function of the number of unit loads stored in the rack, the cubic size of the unit load, the weight of the unit load, and the height of the rack. The cost of an S/R machine was found to be a function of the height of the AS/RS, the weight of the unit load, and the type and location of the S/R machine controls. The cost of the building was found to be a function of its height and the cost per square foot to construct a 25-ft-high building.

To facilitate the presentation, the following notation and estimates are used:

x = depth of the unit load (i.e., stringer dimension) in inches

y = width of the unit load in inches

z = height of the unit load in inches

v = volume of unit load in cubic feet = $xyz/1728$

w = weight of unit load in pounds

W = width of an aisle of AS/RS storage in inches

L = length of an aisle of AS/RS storage in inches

H = height of an aisle of AS/RS storage in inches

n = number of tiers or levels of storage

m = number of columns of storage per aisle side

a = number of storage aisles

BH = building height

BW = building width

BL = building length

λ = allowance, measured in feet

α = parameter for computing rack cost, expressed in dollars

---

[2]The cost-estimating procedure presented is based on one developed by Zollinger [64]. To simplify the presentation, it is assumed that all unit loads stored in the AS/RS are the same size and weight. It should be noted that the cost estimates are "quick-and-dirty" values and are presented for expository purposes only. Quotations from AS/RS suppliers should not be evaluated on the basis of the "rough cut" cost estimates provided in this section.

β = height parameter for computing S/R machine cost, expressed in dollars

γ = weight parameter for computing S/R machine cost, expressed in dollars

ϕ = control parameter for computing S/R machine cost, expressed in dollars

δ = cost per square foot to construct a 25-ft-tall building

CF = conversion factor for converting cost per sq ft to construct a 25-ft-tall building to the cost per sq ft to construct a building of height, BH

The dimensions of a storage aisle can be estimated as follows:

$$W = \begin{cases} 3(x + 6'') & \text{(with in-rack sprinklers)} \\ 3(x + 4'') & \text{(without in-rack sprinklers)} \end{cases} \tag{10.58}$$

$$L = m(y + 8'') \tag{10.59}$$

$$H = n(z + 10'') \tag{10.60}$$

The width of a storage aisle is determined by measuring from the center-line of the flue space to the right of the aisle to the center-line of the flue space to the left of the aisle. Hence, the width of a storage aisle includes the width of the storage spaces on both sides of the aisle and the width of the aisle. The width of the aisle is assumed to equal the depth of the load, plus 4 in. for clearance between the load and the rack. The depth of a storage space is equal to the depth of the load, plus 4 in. The 4-in. addition to the load depth accounts for half the flue space behind the load. In the case of in-rack sprinklers, an additional 6 in. is added to the flue space dimension.

In calculating the length of the storage aisle, it is assumed that a 4 in. upright truss is used to support the loads. Further, a clearance of 2 in. is assumed between an upright truss and the load. Hence, with a 2-in. clearance on each side of the load and a 4-in. upright truss, the width of a storage opening equals the width of the load plus 8 in.

In calculating the height of an aisle of storage, it is assumed that each load rests on cantilever arms that are 4 in. thick. Likewise, it is assumed that a clearance of 6 in. is required between the top of a load and the cantilever arm above it to allow the S/R's shuttle to retrieve the load.

In converting the dimensions of a storage aisle into building dimensions, it should be noted that the lowest storage level in an aisle cannot be at floor level. (The shuttle for the S/R machine cannot retrieve the load unless it is approximately 28 in. above floor level.) Likewise, approximately 20 in. of space must be provided above the top load for installation of the power conductor bar for the S/R machine and clearance for the S/R machine carriage in positioning the shuttle to retrieve the top-most load. Hence, the building height will be approximately 4 ft greater than the height of an aisle of storage.

In estimating the building width, one can multiply the width of a storage aisle by the number of storage aisles. However, since the outermost racks should not be installed "snug" with the wall, an addition of 2 ft should be provided. (If food is being stored, regulations might require a larger clearance in order to prevent rodent infestation. Sufficient space might be required to allow removal of trash between the wall and the rack.)

In determining the building length, space must be provided for pickup and deposit (P/D) stations, as well as S/R extension or runout at the end of the rack for maintenance access. The amount of the allowance (λ) is dependent on the width of the unit load.

Based on the above, the dimensions of the building can be estimated as follows:

$$BH = H + 48 \text{ in.} \tag{10.61}$$

$$BW = aW + 24 \text{ in.} \tag{10.62}$$

$$BL = L + \lambda \tag{10.63}$$

where λ, for values of y between 24 in. and 54 in., can be estimated by

$$\lambda = \begin{cases} 12.5 + 0.45y & \text{(without transfer cars)} \\ 29.5 + 0.45y & \text{(with transfer cars)} \end{cases} \tag{10.64}$$

In Equation 10.64, y is measured in inches while λ is measured in feet.

Example 10.14

Suppose an AS/RS is to be designed for 40 in. $\times$ 48 in. (depth $\times$ width) unit loads that are 48 in. tall. There are to be eight aisles; each aisle is 12 loads high and 80 loads long. There are to be sprinklers and no transfer cars. What will be the minimum dimensions for the building?

The building height will be 12(48 in. + 10 in.)/12 + 48 in./12 or 62 ft. The width of the building will be 8(3)(40 in. + 6 in.)/12 + 2 ft, or 94 ft. The building length, in ft, will be 80(48 in. + 8 in.)/12 + 12.5 ft + 0.45(48), or 407.43 ft. Thus, a building at least 62 ft tall, 94 ft wide, and 407.43 ft long will be required to accommodate the AS/RS. Depending on other functions to be located in the facility and the amount of staging or unit load accumulation space required "in front" of the AS/RS, additional space might be required.

Rack Cost Calculation

To estimate the acquisition cost of the storage rack for an AS/RS, we multiply the cost per rack opening (CRO) by the number of rack openings. The cost per rack opening can be estimated as follows:

$$CRO = \alpha[0.92484 + 0.025v + 0.0004424w - (w^2/82{,}500{,}000) + 0.23328n - 0.00476n^2] \tag{10.65}$$

Example 10.15

Suppose a storage rack is to be provided to accommodate 10,000 unit loads, weighing 2500 pounds, and having dimensions of 42 in. $\times$ 48 in. $\times$ 46.5 in. Further, suppose the rack will support 10 unit loads vertically, that is, 10 tiers of storage will be accommodated by the rack. If the storage rack cost parameter, α, equals $30, what will be the cost of the storage rack?

From the problem statement, $x = 42$ in., $y = 48$ in., $z = 46.5$ in., $w = 2500$ lb; $n = 10$ tiers; and $\alpha = \$30$. A calculation establishes that $v = 54.25$ ft^3. Substituting the parameters in (10.65) yields a value of $155.0442 per rack opening. Hence, the storage rack will cost approximately $1,550,442, installed.

S/R Machine Cost Calculation

The cost of an S/R machine is obtained by summing the cost contribution due to the height of the system, the weight of the unit load, and the type of machine control used. Specifically, CSR, the cost per S/R machine, is given by

$$\text{CSR} = A + B + C \tag{10.66}$$

where

    A = S/R machine cost based on the height of the aisle
    B = S/R machine cost based on the weight of the unit load
    C = S/R machine cost based on the type of machine control used

H	A	w	B
<35'	β	<1000#	γ
[35',50')	2β	[1000#,3500#)	2γ
[50',75')	3β	[3500#,6500#)	3γ
[75',110')	4β	≥6500#	4γ
≥110'	5β		

Control Logic	C
manual	ϕ
on-board	2ϕ
end-of-aisle	3ϕ
central console	4ϕ

Example 10.16

Suppose six S/R machines are to be used to lift 2500 pound loads to a height of 55 ft and a central console is to be used to control the S/R machines. What will be the cost of the S/R machines? Based on current market prices, the values of all three cost parameters are estimated to equal $25,000.

    From the problem statement, H = 55 ft; w = 2500 lb, central console, and $\beta = \gamma = \phi = \$25,000$. Hence,

$$\text{CSR} = 3\beta + 2\gamma + 4\phi = 9(\$25,000) = \$225,000 \text{ per S/R machine}$$

Building Cost Calculation

To estimate the cost of the building, we first compute the size of the footprint of the building in square feet. Then, we multiply the footprint of the building by the product of CF (the conversion factor) and δ (the cost per square foot to construct a 25-ft-tall building). Hence, the building cost (BC) can be estimated to be

$$\text{BC} = (\text{BW})(\text{BL})(\text{CF})\delta \tag{10.67}$$

BH	CF
25'	1.00
40'	1.25
55'	1.50
70'	1.90
85'	2.50

Example 10.17

Suppose a 60-ft-tall building is to be used to house an AS/RS and it costs $30/ft^2 to construct a 25-in.-tall building. Further, suppose the footprint of the building to house the AS/RS is 150 ft $\times$ 440 ft. What will be the cost of the building?

Interpolating to obtain the value of CF yields a value of 1.633. Hence,

$$BC = 150(440)(1.633)(30)$$
$$= \$3,234,000$$

Example 10.18

Suppose storage space of 1,540,000 ft^3 must be provided. A 55-ft-tall building will yield a footprint of 28,000 ft^2 at a cost of 1.5(28,000)δ, or 42,000δ. If a 25-ft-tall building is built, a footprint of 61,600 ft^2 is required and the cost will be 61,600δ. Likewise, a 70-ft-tall building will cost approximately 41,800δ and an 85-ft-tall building will cost approximately 45,294δ. Thus, a 70-ft-tall building would minimize the cost of the building required to store a constant volume of material.

S/R Machine Cycle Times

To determine the time required to perform an operation (either a storage or a retrieval), expected S/R machine cycle time formulas will be used. Recall, the S/R machine is capable of traveling simultaneously in the aisle both vertically and horizontally. Hence, the time required to travel from the P/D station to a storage or retrieval location is the maximum of the horizontal and vertical travel times, which is also known as Chebyshev travel.

Because of the importance of the cycle time in estimating the throughput capacity of the system, considerable emphasis is placed on the time for the S/R machine to perform a **single command** or a **dual command** cycle. A single command cycle consists of either a storage or a retrieval, but not both; whereas a dual command cycle involves both a storage and a retrieval.

In determining the cycle times, it is assumed that a single command storage cycle begins with the S/R at the P/D station; it picks up a load, travels to the storage location, deposits the load, and returns empty to the P/D station. A single command retrieval cycle is also assumed to begin with the S/R at the P/D station; it travels empty to the retrieval location, picks up the load, travels to the P/D station, and deposits the load. One pickup and one deposit occur during a single command cycle.

A dual command is assumed to begin with the S/R at the P/D station; it picks up a load, travels to the storage location, deposits the load, travels empty to the retrieval location, picks up the load, travels to the P/D station, and deposits the load. A total of two pickups and two deposits are performed during a dual command cycle.

In developing models of expected single command cycle times and expected dual command cycle times, the following notation is used. Let

$E(SC)$ = the expected travel time for a single command cycle

$E(TB)$ = the expected travel time from the storage location to the retrieval location during a dual command cycle

$E(DC)$ = the expected travel time for a dual command cycle

L = the rack length (in feet)

H = the rack height (in feet)

b_v = the horizontal velocity of the S/R machine (in ft/min)

v_v = the vertical velocity of the S/R machine (in ft/min)

t_h = time required to travel horizontally from the P/D station to the furthest location in the aisle, ($t_h = L/b_v$ mins)

t_v = time required to travel vertically from the P/D station to the furthest location in the aisle, ($t_v = H/v_v$ mins)

In order to determine the above expected travel times in a compact form, we first "normalize" the rack by dividing its shorter (in time) side by its longer (in time) side [12]. That is, we let

$$T = \max(t_h, t_v), \tag{10.68}$$

$$Q = \min(t_h/T, t_v/T), \tag{10.69}$$

where T (i.e., the "scaling factor") designates the longer (in time) side of the rack, and Q (i.e., the "shape factor") designates the ratio of the shorter (in time) side to the longer (in time) side of the rack. Note that the normalized rack is Q units long in one direction and one unit long in the other direction, where $0 < Q \le 1$. If $Q = 1$, the rack is said to be "square-in-time" (SIT).

Assuming randomized storage, an end-of-aisle P/D station located at the lower left-hand corner of the rack, constant horizontal and vertical S/R machine velocities (i.e., no acceleration or deceleration is taken into account), and a continuous approximation of the storage rack (which is sufficiently accurate), the expected travel times for the S/R machine are given in [12] as:

$$E(SC) = T\left[1 + \frac{Q^2}{3}\right] \tag{10.70}$$

$$E(TB) = \frac{T}{30}[10 + 5Q^2 - Q^3] \tag{10.71}$$

$$E(DC) = \frac{T}{30}[40 + 15Q^2 - Q^3] \tag{10.72}$$

Note that $E(DC)$ in Equation 10.72 is obtained by observing that $E(DC) = E(SC) + E(TB)$. Furthermore, in [12] it is shown that $E(SC)$ and $E(DC)$ are minimized at $Q = 1$; i.e., given a rack with a fixed area (but variable dimensions), the expected single and dual command travel times are minimized when the rack is SIT.

The expected total cycle time for the S/R machine is obtained by adding the load handling time to the expected travel time. That is, letting

T_{SC} = the expected single command cycle time

T_{DC} = the expected dual command cycle time

$T_{P/D}$ = time required to either pickup or deposit the load

we have

$$T_{SC} = E(SC) + 2T_{P/D} \tag{10.73}$$

$$T_{DC} = E(DC) + 4T_{P/D} \tag{10.74}$$

Example 10.19

For the rack system given in Example 10.14 determine the cycle times for single and dual command operations. Assume the S/R requires 0.35 min to perform a P/D operation, travels horizontally at an average speed of 350 fpm, and travels vertically at an average speed of 60 fpm.

The dimensions (inches) of the storage rack are 56 m (length) by 58 n (height), where m is the number of horizontal addresses and n is the number of vertical addresses. For the example, $m = 80$ and $n = 12$. Therefore, the rack dimensions are

$$L = 56(80)/12 = 373.33 \text{ ft}$$
$$H = 58(12)/12 = 58 \text{ ft}$$

(*Note:* It is assumed the S/R machine travels a maximum of 373.33 ft horizontally and 58 ft vertically. Actually, the shuttle mechanism must be lifted only to the load support position of the twelfth storage level, rather than to the top of the storage rack. The horizontal travel distance is probably underestimated from 2 to 10 ft, depending on the exact location of the P/D station.)

The maximum horizontal and vertical travel times are determined as follows:

$$t_b = L/b_v = 373.33/350 \text{ fpm} = 1.067 \text{ min}$$
$$t_v = H/v_v = 58/60 \text{ fpm} = 0.967 \text{ min}$$

Thus,

$$T = \max(1.067, 0.967) = 1.067 \text{ min}$$

and

$$Q = \frac{0.967}{1.067} = 0.906$$

Therefore, the single command and dual command cycle times are

$$T_{SC} = T\left(1 + \frac{Q^2}{3}\right) + 2T_{P/D}$$
$$= 1.067\left[1 + \frac{(0.906)^2}{3}\right] + 2(0.35)$$
$$= 2.06 \text{ min per single command cycle}$$

$$T_{DC} = \frac{T}{30}(40 + 15Q^2 - Q^3) + 4T_{P/D}$$
$$= \frac{1.067}{30}[40 + 15(0.906)^2 - (0.906)^3] + 4(0.35)$$
$$= 3.23 \text{ min per dual command cycle}$$

Thus, the average cycle time per operation is 2.06 min with a single command cycle and 1.615 min with a dual command cycle.

Example 10.20

For the situation considered in Examples 10.14 and 10.19, suppose 40% of the storages and 40% of the retrievals are performed as single command operations; the remaining are performed as dual command operations. If 120 storages per hour and 120 retrievals per hour are to be handled by the AS/RS, what will be the percent utilization of the S/R machines?

Assuming no transfer cars and assuming the 8 S/R machines are loaded uniformly, then each S/R must perform 15 storages per hour and 15 retrievals per hour. Since 6 storages per hour and 6 retrievals per hour are performed using a single command operation (i.e., 40%), there will be 9 dual command trips per hour. The workload on the S/R machine is obtained as follows:

$$\text{Workload per S/R} = 2(6)(2.06) + 9(3.23)$$
$$= 53.79 \text{ min/hr}$$

Thus, each S/R will be utilized 53.79/60, or 89.65%. The average cycle time per operation is 53.79/30, or 1.793 min per operation, which implies that the maximum throughput capacity of the system is equal to (60/1.793)8, or 267.71 operations/hr.

Since the times between requests for storages and/or retrievals are often random variables and since the locations of the storage and/or retrieval addresses to be visited on a single or dual command cycle are random variables, waiting line analysis can be used to aid in the AS/RS design. However, due to the mixture of single and dual command operations and the general probability distribution for travel times, a simulation approach may be necessary.

Transfer cars are utilized when the activity level in an aisle is not sufficient to justify an S/R machine dedicated to each aisle. In order to determine the cycle time for an AS/RS having transfer cars, the time required to perform the transfer must be included. Additionally, S/R waiting lines can occur when multiple S/R machines are served by a single transfer car. Because of the network of queues involved in the design of an AS/RS with transfer cars, simulation analysis is generally used to aid in the determination of the number of S/R machines, the number of transfer cars, and associated operating disciplines. However, since the transfer car cost might be as large as half the S/R machine cost, most AS/RS systems are installed without transfer cars; hence, there is generally an S/R machine for each aisle.

With the increasing implementation of lean manufacturing techniques, and other similar measures aimed at reducing cost/waste in production facilities as well as in the supply chain, there is no doubt that building large AS/R systems to hold substantial quantities of inventories (in raw material, work-in-process, or finished goods form) is no longer a desirable approach. Rather, the reduction of inventories in manufacturing facilities has led to relatively small AS/R systems being used to offer essentially a controlled and limited form of "staging" between processes. Such AS/R systems typically have only 1 or 2 aisles and are now being referred to as "automated buffers."

Since the rack can hold only a small number of loads, and since the inventory is under computer control (i.e., one can keep track of fluctuations in inventory and how long the parts have been in storage), we believe "automated buffers" will continue to serve as an effective way to manage work-in-process inventories and provide high "visibility" for the parts. We also note that "automated buffers" (or "stockers" as they are known) have also been used successfully in many semiconductor fabrication facilities, where they are used to stage containers (or cassettes) of wafers ahead of each "bay" and act as a buffer point between bays. As long as the AS/RS is not used as a "dumping ground" for excess or obsolete inventories, and the levels of inventories reported by the computer are monitored by humans, AS/R systems can serve a critical role in manufacturing as well as distribution facilities where accuracy and timely response are becoming critical performance parameters.

10.7   ORDER PICKING SYSTEMS

With (unit load) storage/retrieval systems, each storage or retrieval operation involves one unit load; that is, the parts are stored as one unit load and retrieved as one unit load. In some storage systems, however, the parts may be stored as a unit load but retrieved in less-than-unit-load quantities. Such systems are generally known as *order picking* (OP) systems, where each order typically calls for less-than-unit-load quantities. For example, in a mail-order music warehouse, many copies of the same title are stored together as one "unit load," while the customers almost never order two copies of the same title (but they may order one copy of several titles). Hence, parts of the same type (in our example, CDs or tapes of a particular title) are stored as one unit load but they are retrieved in less-than-unit-load quantities. Naturally, the system is replenished periodically either by restocking the empty containers or by removing the empty containers and replacing them with full containers.

There are two principal approaches to order picking. The first one is based on the picker traveling to each container to be visited in order to pick the parts on one or more orders. Since the containers are typically stored along aisles, the general term used for such systems is "in-the-aisle OP." When we prepare a shopping list and go to the grocery store, for example, we are, in essence, performing in-the-aisle OP as we walk up and down the aisles and fill our cart with one or two pieces of each grocery item we need. In fact, this type of in-the-aisle OP is generally known as "walk-and-pick" systems, which we discuss in the next section.

The second principal approach (which we will discuss later) is known as end-of-aisle OP. With such systems, the containers are brought to the end-of-the aisle, where the picker removes the appropriate number of items. Each container is then stored back in the system until it is needed again. An end-of-aisle OP system may at first resemble a unit load storage/retrieval system in that we store and retrieve unit loads. However, with an end-of-aisle OP system, a container is typically retrieved several times (until it is empty). In a unit load storage/retrieval system, once a container is retrieved, it leaves the system.

In-the-Aisle Order Picking

Various configurations and material handling equipment are available for performing in-the-aisle OP. The grocery shopping example we gave above (i.e., walk-and-pick systems) applies to many warehouses where the pickers fill their carts with one or more orders as they walk along the aisles. In some in-the-aisle OP systems, however, each picker is dedicated to a particular aisle and, to reduce fatigue and improve travel times, the picker travels to each container via an automated or semi-automated device. The person-on-board AS/RS (described in Chapter 5), for example, is one such case. With a person-on-board AS/RS, the S/R machine automatically stops in front of the appropriate container and waits for the picker to perform the pick. The parts that are picked are usually taken to a P/D station (which is typically located at the end of the aisle).

To avoid unnecessary trips between the containers in the rack and the P/D station, the picker performs several picks (i.e., he or she visits several containers) between successive stops at the P/D station. (Although the picker may actually pick several items from a particular container, here we define it as one "pick.") In other words, the picker starts at the P/D station, visits the appropriate containers in the rack, returns to the P/D station and discharges the parts before starting the next trip. The throughput capacity of such systems, that is, the number of picks performed per hour, depends on the expected time it takes to complete one trip. The time required per trip, in turn, depends on the travel speed of the S/R machine, the rack size, and rack shape (i.e., the scaling factor and the shape factor as defined by Equations 10.68 and 10.69, respectively), the time it takes to complete one pick (once the S/R machine has stopped), the time it takes to discharge the parts at the P/D station, and the sequence in which the containers are visited. Note that unnecessary S/R travel may occur if the containers to be visited on one trip, that is, the "pick points" in the rack, are not carefully sequenced. (The reader familiar with operations research would readily recognize the above sequencing problem as the traveling salesman problem [42]).

Except for the total S/R machine travel time for the trip, it is relatively straightforward to estimate the above parameters (i.e., the time spent at each pick point and the time spent at the P/D station between two successive trips). In [8] it is empirically shown that the expected time required to travel from the P/D station, make n stops (or picks) in the aisle, and return to the P/D station can be estimated by the following expression:

$$E(P, X_1, X_2, \ldots, X_n, D) \approx$$

$$T\left[\frac{2n}{(n+1)} + 0.114703n\sqrt{Q} - 0.074257 - 0.041603n + 0.459298Q^2\right]$$

$$(10.75)$$

for $3 \leq n \leq 16$. The above expression is based on the assumption that the P/D station is located at the lower left-hand corner of the rack and that the pick points, which are randomly and uniformly distributed over the rack, are sequenced in an optimum fashion (i.e., the total S/R machine travel time is minimized). Of course, we remind the reader the S/R machine travels according to the Chebyshev metric and that the travel time is not affected by whether a pick point is on the left-hand or right-hand side of the aisle. Also, no acceleration or deceleration effects have been taken into account.

Example 10.21

Consider a storage rack that is 360 ft long and 60 ft tall. Suppose that the S/R machine travels at a constant speed of 400 fpm and 80 fpm in the horizontal and vertical direction, respectively. Further suppose that 10 stops are made during each order picking trip, i.e., $n = 10$. From Equation 10.68, the scaling factor, T, is equal to max(360/400, 60/80) = 0.90 min. The shape factor, Q, is equal to 0.8333 (by Equation 10.69). Hence, from Equation

10.75, the expected travel time for the above trip with 10 picks is estimated to be equal to 2.4245 min. That is, the travel time required to start at the P/D station, pick parts from 10 containers (according to the optimum sequence), and return to the P/D station equals approximately 2.4 min.

Example 10.22

Suppose in Example 10.21 it takes the picker, on the average, 15 sec to pick one or more items from each container. Further suppose that the picker spends 1.2 min at the P/D station between successive trips (to discharge the parts he or she picked and to pre-pare for the next trip). The throughput capacity of the system, expressed in number of picks completed per hour, is computed as follows. The total approximate time required to complete one trip is equal to $2.4 + 10(15/60) + 1.2 = 6.1$ min. Since 10 picks are performed in about 6.1 min, the maximum throughput capacity is approximately equal to 98 picks/hour. In other words, the picker will visit approximately 98 containers/hour on the average. Given the expected number of items picked from each container, the above figure can also be expressed as the number of items picked per hour.

Sequencing the pick points in an optimum fashion may impose a heavy com-putational burden. In fact, although the number of points is quite small (generally no more than 20 to 24 points per trip), obtaining the optimum sequence with the Chebyshev metric can require a large amount of computer time (see [9] for details). Hence, assuming Chebyshev travel, a number of "fast" heuristic procedures (see [9] and [27], among others) have been empirically shown to generate optimum or near-optimum solutions for $3 \leq n \leq 24$, the one most often encountered in practice is a simple heuristic which is known as the "band heuristic."

With the band heuristic, the rack is divided into k equal-sized horizontal bands, where k is an even number. Starting with the first band, the S/R machine

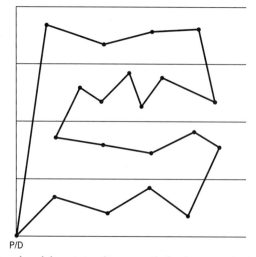

P/D

Figure 10.15   The four-band heuristic. (Due to Chebyshev travel, the solid lines do not necessarily reflect the travel path of the S/R machine.)

travels in a serpentine fashion, picking sequentially along the *x*-axis until all the picks are performed. An example with four bands is shown in Figure 10.15. Note that the S/R machine completes all the picks in the current band before proceeding to the next band. Consequently, depending on the number of bands (i.e., the band width), some "zigzagging" may occur in each band. Since the band heuristic is a heuristic approach, the expected S/R machine travel time per trip will be greater than that obtained using an exact (i.e., optimum seeking) solution procedure. However, the band heuristic performs reasonably well and it is simple to implement.

Example 10.23

Suppose the 360 ft × 60 ft rack used in Example 10.21 is divided into two bands and the coordinate locations of the 10 picks to be made are as given in Table 10.8. Using the band heuristic the items would be visited in the following sequence: 5, 7, 3, 1, 4, 2, 8, 10, 6, and 9. A variation of the band heuristic divides the rack into *k* equal-sized bands, but determines the band width by dividing *k* into the largest *y*-coordinate among the pick points. In this example, the division of the two bands would be at *y* = 27.5' and the same sequence would be used to pick the 10 items. Note that, with the above variation, the bands are problem specific; that is, they generally vary from one trip to the next.

Using the band heuristic, in [8] it is shown that the expected time required to perform *n* picks (that are randomly and uniformly distributed over the rack) can be approximated as follows:

$$E(k, n, T, Q) \cong T\left[2kA + kB(n - 1) + \frac{k - 1}{k}Q\right] \tag{10.76}$$

where *k* designates the number of bands, and

$$A = \frac{C}{2} + \frac{1 - (1 - C)^{n+2}}{C(n + 1)(n + 2)} \tag{10.77}$$

$$B = \frac{C}{3} + \frac{2nC + 6c - 2 + 2(1 - C)^{n+3}}{C^2(n + 1)(n + 2)(n + 3)} \tag{10.78}$$

$$C = \frac{Q}{k^2} \tag{10.79}$$

The above approximation works reasonably well except for small *n* values.

Table 10.8   *Pick Locations for Example 10.23*

Pick #	Coordinates	Pick #	Coordinates
1	(200, 20)	6	(50, 50)
2	(350, 30)	7	(100, 10)
3	(150, 15)	8	(250, 40)
4	(300, 25)	9	(25, 55)
5	(75, 5)	10	(225, 35)

Example 10.24

Consider the same data given for Example 10.21; that is, $T = 0.90$ min and $Q = 0.8333$, with $n = 10$ picks per trip. Using Equations 10.76 through 10.79, and assuming that the rack is divided into two bands (i.e., $k = 2$), we obtain $C = 0.2083$, $B = 0.1166$, and $A = 0.1383$, which yields $E(k, n, T, Q) \approx 2.7617$ min. Compared with the optimum expected travel time computed in Example 10.21 (i.e., 2.4245 min), the two-band heuristic results in an increase of approximately 14% in the expected travel time per trip. However, the above increase does not imply that the throughput capacity of the system using the band heuristic will be proportionally less. This is primarily due to the time required per pick and the time required at the P/D station between two successive trips, both of which are generally independent of the sequence in which the picks are made. Using the same data given for Example 10.22 (i.e., 15 sec per pick and 1.2 min at the P/D station between successive trips), for the two-band heuristic we obtain a maximum throughput capacity of about 93 picks/hour, which is 5% less than that obtained with the optimum pick sequence.

Example 10.25

Continuing with the previous example, suppose the rack is divided into four bands and the number of picks performed per trip is increased to 50. (Note that the maximum number of picks performed per trip is generally dictated by the capacity of the S/R machine relative to the weight and/or volume of the parts to be picked.) From Equations 10.76 through 10.79 we obtain $C = 0.052$, $B = 0.0269$, and $A = 0.0328$, which yields $E(k, n, T, Q) \approx 5.54$ min. Interestingly, in comparison with Example 10.24, increasing the number of picks by 400% increases the expected travel time by about 100%. This is primarily due to the fact that after more than five picks per trip, the total travel time increases very slowly with the number of picks. We remind the reader, however, that the above increase may not be as (relatively) small when the S/R machine acceleration and deceleration are taken into account.

For a given n value, since it is possible to approximate the expected travel time as a function of k, the number of bands that will minimize the expected travel time can be determined. For a SIT rack, Table 10.9 shows the "optimum" number of bands as a function of the number of picks performed per trip [8]. Alternately, given the number of bands, Equation 10.76 can be used to determine the "optimum" configuration of an aisle (i.e., the rack length and height) to minimize the expected distance traveled.

Table 10.9   *"Optimum" Number of Bands for the Band Heuristic*

No. of Picks	No. of Bands
[1, 24]	2
[25, 72]	4
[73, 145]	6
[146, 242]	8
[243, 363]	10

Table 10.10 *Alternative Configurations for Example 10.26*

Alternative	Height	Length	T	Q	E(k, n, T, Q)
1	15.00	600.00	2.0000	0.1250	4.3951
2	20.00	450.00	1.5000	0.2222	3.732
3	25.00	360.00	1.2000	0.3472	3.5636
4	30.00	300.00	1.0000	0.5000	3.6586
5	40.00	225.00	0.7500	0.8889	4.2433
6	24.00	375.00	1.2500	0.3200	3.5712
7	27.00	333.30	1.1111	0.4050	3.5783
8	25.50	352.94	1.1765	0.3613	3.5638
9	24.75	363.64	1.2121	0.3403	3.5645
10	25.25	356.44	1.1881	0.3542	3.5634*

Example 10.26

Rather than continue with the previous data set, assume that a storage area of 18,000 ft^2 is to be provided within a storage aisle. Since storage racks are provided on both sides of the aisle, 9000 ft^2 of rack space will be provided on each side. Recall that the P/D station is located at the lower left-hand corner of the rack, randomized storage is used, and the S/R machine follows the Chebyshev metric.

Suppose the S/R machine travels at 300 fpm and 60 fpm in the horizontal and vertical directions, respectively. Further suppose that 30 picks will be performed on each pick trip and the rack is divided into two bands, that is, $n = 30$ and $k = 2$. As shown in Table 10.10 one alternative is to provide a 15 ft tall storage area that is 600 ft long. (The length and height of the storage area are based on the travel region for the S/R machine, not necessarily the storage rack dimensions.) The expected travel for the 15 ft × 600 ft rack is equal to 4.3951 min. Given other configurations shown in Table 10.10, in this particular case the optimum rack height is approximately 25 ft 3 in., plus or minus 3 in., and the minimum expected travel time is 3.5634 min. Note that the "optimum" shape factor for the rack is equal to 0.3542, which is not even close to being SIT. We believe this is due to the band heuristic (with $k = 2$) which tends to yield less "zigzagging" if the band width is small.

According to [9], the most desirable rack shape depends on the heuristic used for sequencing the pick points. With certain heuristics, the expected trip length *decreases* as the rack becomes SIT. Some of these heuristics generate sequences that require less travel time than those obtained from the band heuristic; the reader may refer to [9] for details. Also, the results in [20] suggest that, for sufficiently large n values, the exact shape of the region has little or no impact on the expected length of the optimum picking sequence. Furthermore, we remind the reader that the above "optimum" rack shape is determined only with respect to the expected travel time for the S/R machine. When the rack cost is included, the "optimum" shape factor may be different from the one we computed in Table 10.10.

In deriving the above results, we assumed randomized storage and Chebyshev travel. Many deviations from these assumptions occur in practice. For example, walking along an aisle and picking parts from bins or flow racks (i.e., "walk-and-pick" systems) is typically represented as a single dimensional picking problem, with negligible vertical travel. Likewise, instead of a person-on-board S/R machine, if the picker uses, say, an order picker truck or lift truck, our Chebyshev travel assumption would not apply. (Although some order picker/lift trucks are capable of simultaneous travel, for safety reasons simultaneous travel is allowed only if the forks are not fully raised.)

Also, assigning the "fast movers" to the "closest" storage positions (i.e., turnover based storage) violates the assumption of uniformly and randomly distributed pick points.

In designing in-the-aisle OP systems, the primary design decisions are: the number, length, and height of storage aisles to provide; and the number of pickers. As a first-cut in designing such systems, it is useful to consider the situation in which *the storage aisles are SIT and each picker is dedicated to one aisle.* Under such an assumption, a single design decision is required: the number of storage aisles. The resulting model is

Minimize: Number of storage aisles

Subject to: Throughput constraint
                      Storage space constraint

The following notation is used to specify the design algorithm [8]:

S = total storage space required, expressed in square footage of storage rack,

R = total throughput required, expressed in picks per hour,

n = average number of stops (i.e., picks) per trip, not including the P/D station ($3 \le n \le 24$),

p = average pick time per stop, expressed in minutes,

K = time spent at the P/D station between successive pick trips, expressed in minutes,

b_v = horizontal velocity of picking vehicle or the S/R machine, expressed in ft/min,

v_v = vertical velocity of picking vehicle or the S/R machine, expressed in ft/min,

$t = p + (K/n)$, expressed in min/pick,

$q = [S/(2b_v v_v)]^{0.5}$, and

$u = 0.073331 + (0.385321/n) + 2/(n + 1)$; assuming that the pick points are sequenced in an optimum fashion, and $b = 1$.

In the following algorithm, $[m]$ denotes the smallest integer greater than or equal to m (i.e., the "ceiling function").

ALGORITHM 1.
STEP 0.   Set $M = [Rt/60]$.
STEP 1.   If $M(60M - Rt)^2 \ge (Rqu)^2$, then go to step 3; otherwise, go to step 2.
STEP 2.   Let $M \leftarrow M + 1$, go to step 1.
STEP 3.   STOP, M is the minimum number of aisles (or pickers).

Example 10.27

Suppose $S = 72{,}000$ ft^2, $R = 400$ picks/hr, $n = 5$ picks/trip, $p = 0.25$ min/pick, $b_v = 450$ fpm, $v_v = 90$ fpm, and $K = 0.75$ min/trip. Thus, $t = 0.40$, $q = 0.9428$, and $u = 0.4837$. The right-hand side of the inequality in Step 1 has the value 33,274.48. Enumerating over M, the left-hand side of the inequality has the following values:

M	$M(60M - 160)^2$
3	$1200 < 33{,}274.48$
4	$25{,}600 < 33{,}274.48$
5	$98{,}000 > 33{,}274.48$

Hence, the minimum number of storage aisles and pickers required is 5; each aisle (or rack) will be 37.95 ft tall and 189.74 ft long. Note that the above solution is based on the rate at which the picks must be made (i.e., R); there is no explicit allowance made for replenishing the containers.

Walk-and-Pick Systems

As we remarked earlier with the grocery shopping example, walk-and-pick systems are based on one or more pickers walking "up and down the aisles" with a pick cart. Each pick trip starts and terminates at the P/D point, where the picker empties his or her cart and picks up the pick list for the next trip. Since walk-and-pick systems often require minimal capital investment and are fairly flexible (i.e., pickers can be added to or removed from the system within certain limits), they are quite popular in industry. However, in a typical walk-and-pick system, the picker spends a significant amount of time walking rather than picking. Systems with as much as 40% to 60% walking are not uncommon. Of course, walking not only contributes to picker fatigue but it is also wasted time as far as order picking is concerned.

In order to minimize the amount of walking, users of walk-and-pick systems often take the following measures: (1) the picks are sequenced so that the walk time is minimized, and (2) the picker picks more than one order on each trip. Readers familiar with the operations research literature will recognize the first problem as the well-known *Traveling Salesman Problem* (TSP)[42]. However, the picker in our case walks through aisles that have a certain structure and, as we shall shortly see, this leads to a special type of TSP. Going back to our grocery shopping example, many shoppers (even those not familiar with the TSP!) implicitly use certain strategies to minimize their walking time. For example, many shoppers start from one end of the store and shop toward the other end instead of entering the aisles in random order.

The second measure, picking more than one order at a time, is known in OP as "batch picking" or "multiple order picking." The goal is to reduce the distance the picker has to cover from one pick to the next pick within a given trip. If we were to go to a grocery store and track two shoppers, say, shopper A (with three items on her list) and shopper B (with 18 items on his list), we can easily verify that the total walking distance *per item picked* for shopper A is, in general, significantly longer than that of shopper B (assuming that both shoppers follow a pick sequence that minimizes or nearly so their total walk time). Hence, there is an incentive to increase the number of picks performed per trip. However, the picker uses a picking cart with finite capacity (in both weight and volume), and all the items picked on a particular trip must be sorted by order number when the picker returns to the P/D point. To eliminate the sorting step, most picking carts are designed with separate compartments. All the items that belong to the same order are placed by the picker in the same compartment. To minimize human errors (i.e., picking the wrong item from the shelf/rack, or placing the item in the wrong compartment), a fairly simple but clever idea is to use "light-aided" (or "light-directed") picking (see Chapter 5). A "smart" order picking cart developed in Japan relies extensively on light-aided picking and has a computer terminal on-board showing the picker his or her current location and picking route through the warehouse with a "bird's-eye view." The reader may refer to Chapter 5 for order picking carts.

There are various issues related to the design and operation of walk-and-pick systems. These issues range from the configuration or shape of the warehouse to the assignment of particular orders to each trip so that the total trip time is minimized. Since we cannot afford a comprehensive treatment of the subject, we will focus on the expected throughput performance of walk-and-pick systems as we did for other OP systems we covered in this section. The expected time to pick a particular order depends on the shape and configuration of the warehouse, the travel speed of the picker, and the policy used for sequencing the picks. Of course, the pick time itself (i.e., reaching into the shelf/rack to pick the item and placing it into the appropriate compartment) also plays a role but estimating the pick time is straightforward. Therefore, we will focus on estimating the expected walking time.

In order to do so, we will first assume that the warehouse has a simple structure with no cross-aisles. Also known as a "ladder structure," the structure we assume is shown in Figure 10.16, where all the aisles are of equal length, and cross-aisles are provided only in the front of the warehouse (where the P/D point is located) and in the back. We also assume that the aisle width is such that the picker can reach either side of the aisle with equal ease and in negligible time. Ratliff and Rosenthal [54] developed an efficient algorithm to determine the *optimum* pick sequence under the above assumptions. Although it runs fast, it is not straightforward to determine the expected travel time per trip under their algorithm. We will instead assume a heuristic policy, known as the "traversal policy," is used to sequence the picks. Under this policy, the picker enters each aisle that contains one or more picks and fully traverses the aisle to exit from the opposite end. (Note that, if we were using the optimal sequence, in some cases it may make more sense for the picker to enter an aisle, make the pick(s), turn, and exit from the same end.) Two examples of the traversal policy are shown in Figures 10.16*a* and *b* for a trip with an even and an odd number of aisles, respectively. Note that, in the latter case, the picker has to turn and exit the aisle from the same end he or she entered in order to return to the P/D point.

Let M denote the number of aisles in the warehouse. Let y denote the length of each aisle (in distance units), and x denote the width of the warehouse (in distance units); see Figure 10.16*b*. As before, n denotes the number of picks (or stops) per trip, not including the P/D point. Assuming the n picks are equally likely to be located anywhere in the warehouse, the expected distance traveled per trip under the traversal policy, $E(D)$, can be approximated by the following expressions (see Hall [30]):

$$E(D) = E(D_1) + E(D_2) \tag{10.80}$$

$$E(D_1) \approx 2x\frac{(n-1)}{(n+1)} \tag{10.81}$$

$$E(D_2) = yM\left[1 - \left(\frac{M-1}{M}\right)^n\right] + 0.5y \tag{10.82}$$

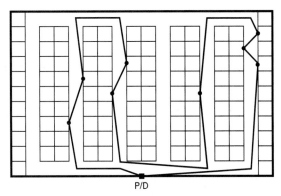

(*a*) Even number of aisles visited

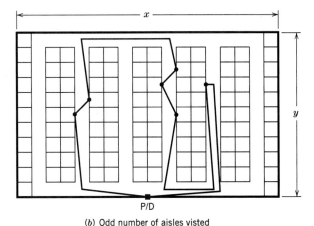

(*b*) Odd number of aisles visted

Figure 10.16   The traversal policy for walk-and-pick systems.

where $E(D_1)$ represents expected cross-aisle travel distance and $E(D_2)$ represents expected in-the-aisle travel distance. Note that the expression given for $E(D_1)$ is an approximation; it works reasonably well provided that $n \geq 5$.

Example 10.28

Suppose we use the traversal policy in a warehouse with 18 aisles ($M = 18$). Each aisle is 80 ft long ($y = 80$), and the warehouse is 180 ft wide ($x = 180$). Assuming the picker makes 12 picks (or stops) per trip (i.e., $n = 12$), the expected distance per trip, $E(D)$, is equal to approximately $304.62 + 754.82 \approx 1,060$ ft. Given the picker travel speed within and across aisles as well as the time spent per pick and the time spent at the P/D point between successive trips, we can now easily compute the expected throughput capacity of the system (expressed in number of picks completed per hour) using the same approach shown in Example 10.22. We can also investigate the impact of warehouse shape on the expected throughput capacity of the system. The reader is encouraged to compute the new values of $E(D_1)$ and $E(D_2)$ if each aisle is reduced to, say, 40 ft and M is increased to 36 aisles ($x = 360$).

End-of-Aisle Order Picking

As we described earlier, end-of-aisle OP is based on bringing the containers to the end of the aisle where the picker performs the pick and the container is returned to the storage rack. Although there are several end-of-aisle OP systems available, our analysis will be limited to the miniload AS/RS, which is used primarily for small- to medium-sized parts. We stress that the results we show here would apply to other systems as long as the device bringing the containers to the picker follows the Chebyshev metric and the storage area is rectangular.

The same assumptions used for in-the-aisle OP apply to end-of-aisle OP. Specifically, it is assumed that the containers are randomly located in the storage rack (or the S/R machine is equally likely to visit any point in the rack in order to retrieve a container), a single "pick" is made from each container (although one such pick may involve several items), the picker is dedicated to the aisle, a continuous approximation of the storage rack is sufficiently accurate, acceleration and deceleration losses are negligible, and where multiple aisles are required they are identical in activity and rack shape.

A three-aisle miniload AS/RS is shown in Figure 10.17. Note that in front of each aisle there are two *pick positions;* for the first aisle they are designated as pick position A and pick position B. The operation of, say, the first aisle may be described as follows (recall that one picker is assigned to each aisle): Suppose there are two containers—one in each pick position—and the S/R machine is idle at the P/D station. Further suppose that the picker is picking from the container in position A. After completing a pick, the picker presses a "pick complete" button, which signals

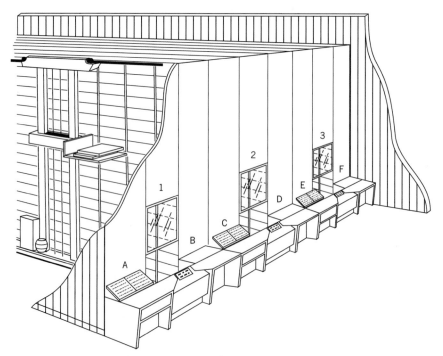

Figure 10.17   A three-aisle miniload AS/R system. (*Reprinted from [66]*)

to the S/R machine that the picker is done with the container in position A. At that point, the picker switches over to pick position B and starts picking from the container there while the S/R machine picks up the container in pick position A and stores it in the rack. The S/R machine then travels directly to another location in the rack and retrieves the next container for pick position A.

If the picker completes the pick at position B before the S/R machine delivers the next container for position A, then the picker becomes idle until the S/R machine arrives at the P/D station. On the other hand, if the S/R machine delivers the new container for position A before the picker completes the pick at position B, then the S/R machine becomes idle at the P/D station until the picker hits the "pick complete" button. In either case (i.e., whether the S/R machine waits on the picker or vice versa), the next (system) cycle begins with the picker picking from the new container in position A and the S/R machine performing another (dual command) trip to retrieve a new container for position B. Hence, the "system cycle" time is defined to be the maximum of the S/R machine cycle time and the pick time. The S/R machine cycle time is defined by a dual command cycle, which includes two pickups and two deposits.

In [13] it is noted that the above system can be modeled as a closed, two-server cyclic queuing system as shown in Figure 10.18. In such a system, each "customer" circulates in the loop indefinitely, served alternately by the picker and the S/R machine (i.e., as soon as a customer completes service at one of the servers, it joins the queue for the other server). Since there are two pick positions, the (constant) number of customers in the loop is set equal to two. The "mean service time" for the picker is set equal to the mean pick time, while the "mean service time" for the S/R machine is set equal to the expected dual command cycle time. (In the model, the same two customers circulate in the loop; however, in the actual system, a different container is retrieved by the S/R machine each time it stores one. This difference has no impact on the performance of the system from a queuing standpoint as long as the S/R machine service times are independent of the particular containers stored or retrieved on each trip.) A new "system cycle" starts whenever the two servers start serving a new customer.

The performance of the miniload AS/RS depends on the utilization of the picker. Given the expected value of the "system cycle" time, it is straightforward to obtain the expected value of the picker utilization. The determination of the expected value of the "system cycle" time, however, is complicated by either or both the S/R machine cycle time and the pick time being random variables. Specifically, the expected value of the system cycle time is the expected value of the maximum of two (independent) random variables. Even if the pick time is exponential, it is not a simple calculation because of the complex form of the probability distribution for dual command cycles (see [22]). (Note that in Equation 10.72 we presented the mean travel time for a dual command cycle but we did not show its variance.)

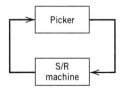

Figure 10.18   Two-server cyclic queuing system for the miniload AS/RS.

By approximating the S/R machine travel time with a uniform distribution, in [13] a design algorithm is presented for deterministic or exponential pick times. This algorithm works with a rack shape factor $(0 < Q \le 1)$ specified by the user a priori. The limits of the above uniform distribution are designated by $\hat{k}_1$ and $\hat{k}_2$ for a "normalized" rack. To specify the algorithm, we introduce additional notation as follows; let:

$E(\widehat{DC})$ = expected value of a dual command cycle in a "normalized" $(Q \times 1)$ rack

$S(\widehat{DC})$ = standard deviation of dual command cycles in a "normalized" $(Q \times 1)$ rack

$E(PU)$ = expected value of picker utilization

$E(SRU)$ = expected value of S/R machine utilization

$E(CT)$ = expected value of system cycle time

C = total container handling time per dual command cycle $(4T_{P/D})$

ALGORITHM 2.

STEP 0. Set $E(\widehat{DC}) = (4/3) + (Q^2/2) - (Q^3/30)$ and $S(\widehat{DC}) = (0.3588 - 0.1321Q)E(\widehat{DC})$. Compute the values of $\hat{k}_1$ and $\hat{k}_2$ as follows:

$$\hat{k}_1 = E(\widehat{DC}) - 1.7321S(\widehat{DC})$$

$$\hat{k}_2 = E(\widehat{DC}) + 1.7321S(\widehat{DC})$$

Lastly set $M = \lceil Rp/60 \rceil$ and $q = [S/(2Qb_v v_v)]^{0.5}$.

STEP 1. Set $r = q/\sqrt{M}$. Given the value for r, set $k_1 = r\hat{k}_1$, $k_2 = r\hat{k}_2$, and $E(DC) = rE(\widehat{DC})$. Let $t_1 = k_1 + C$ and $t_2 = k_2 + C$.

STEP 2. If the pick time is deterministic, go to step 2a; if the pick time is exponentially distributed, go to step 2b.

STEP 2a. Compute $E(CT)$ as follows:

$$E(CT) = \begin{cases} E(DC) + C & \text{for } 0 < p \le t_1 \\ \dfrac{p^2 - 2pt_1 + t_2^2}{2(t_2 - t_1)} & \text{for } t_1 < p \le t_2 \\ p & \text{for } t_2 < p \end{cases}$$

STEP 2b. Compute $E(CT)$ as follows:

$$E(CT) = E(DC) + C + \frac{p^2}{(t_2 - t_1)} [\exp(-t_1/p) - \exp(-t_2/p)]$$

STEP 3. Compute

$$E(PU) = p/E(CT)$$

$$E(SRU) = [E(DC) + C]/E(CT)$$

STEP 4. If $E(PU)(60/p)M \ge R$, go to step 6; otherwise go to step 5.

STEP 5. Let $M \leftarrow M + 1$ and go to step 1.

STEP 6. STOP, M is the "minimum" number of aisles.

Assuming that the vertical side of the rack is shorter in time, the M value obtained from Algorithm 2 can be used to determine the resulting rack dimensions as follows:

rack length, $L = [Sb_v/(2QMv_v)]^{0.5}$,

rack height, $H = QLv_v/b_v$.

Note that, unlike in-the-aisle OP systems, the number of containers per order does not play a role in the above design algorithm. This is primarily due to the fact that with in-the-aisle OP systems the number of containers per order affects the picking sequence and the resulting S/R machine travel time. With end-of-aisle OP systems, however, as long as the S/R machine operates on a dual command basis (and the containers for an order are retrieved in no particular sequence), the number of containers per order does not affect the S/R machine cycle time. Of course, we are assuming that the containers for one order are not intermixed with the containers for another order. That is, the S/R machine will not retrieve the containers for the next order until all the containers for the current order have been retrieved. To avoid unnecessary single command trips, the last container of the current order is inter-leaved with the first container of the next order. In other words, when the S/R machine picks up the last container from, say, position A, and stores it in the rack, it retrieves the first container for the next order. The same holds true for position B. This allows the S/R machine to operate on a dual command basis 100% of the time without intermixing the containers that belong to different orders.

Also note in Algorithm 2 that the expected S/R machine utilization does not play a direct role in determining the throughput capacity of the system. As indicated by the left-hand side of the inequality shown in step 4, the throughput capacity of the system is dictated by the expected picker utilization, $E(PU)$. Last, we remind the reader that no allowance has been made for "replenishment cycles." If replenishment is performed on a second shift, it should not affect the results obtained from Algorithm 2. However, if replenishment cycles are interleaved with picking cycles (i.e., "new" containers are stored in the system while existing ones are being depleted), some allowance must be made for S/R machine time lost due to replenishment cycles. Alternately, the picker may have to replenish a near-empty container when the S/R machine retrieves it. In such cases, an allowance must be made for the picker to replenish such containers.

Example 10.29

Suppose $S = 72{,}000$ ft^2, $R = 300$ picks/hr, $p = 1.20$ min/pick, $b_v = 450$ fpm, $v_v = 90$ fpm, and $C = 0.60$ min/cycle. Assuming that the user desires a SIT rack, we set $Q = 1$. We further assume that the pick time is constant. Applying Algorithm 2, we obtain the following results:

0. $E(\widehat{DC}) = 1.80$ and $S(\widehat{DC}) = 0.4081$. Therefore, $\hat{k}_1 = 1.0932$ and $\hat{k}_2 = 2.5068$. Also, $M = 6$ and $q = 0.9428$.

1. $r = 0.3849$; therefore, $k_1 = 0.4208$, $k_2 = 0.9649$, and $E(DC) = 0.6928$. Hence, $t_1 = 1.0208$ and $t_2 = 1.5649$.

2a. Since $1.0208 < p < 1.5649$, $E(CT) = 1.3223$.

3. $E(PU) = 0.9075$ and $E(SRU) = 0.9777$.

4. $E(PU)(60/p)M = 272.25 < 300$.

5. $M = 7$.

1. $r = 0.3563$; therefore, $k_1 = 0.3895$, $k_2 = 0.8932$, and $E(DC) = 0.6413$. Hence, $t_1 = 0.9895$ and $t_2 = 1.4932$.

2a. Since $0.9895 < p < 1.4932$, $E(CT) = 1.2853$.

3. $E(PU) = 0.9336$ and $E(SRU) = 0.9658$.

4. $E(PU)(60/p)M = 326.76 > 300$.

6. STOP; the "minimum" number of storage aisles and pickers required is 7. Each aisle will be 32.07 ft tall and 160.36 ft long.

Note that the throughput *capacity* of the system exceeds the throughput *requirement*. Consequently, the expected system utilization will be approximately 92% ($300/326.76 = 0.9181$). This implies that the "effective" picker and S/R machine utilizations will be approximately equal to 86% (0.9336×0.9181) and 89% (0.9658×0.9181), respectively. Further note that when $E(PU)$ increases, $E(SRU)$ decreases. In general, $E(PU)$ and $E(SRU)$ have a negative correlation; that is, as one of them decreases, the other one increases (unless, of course, it is already at 100%). This is primarily due to the closed, two-server cyclic queuing system we showed earlier in Figure 10.18. Last, the system through-put capacity (326.76 picks/hr) is not rounded off to an integer value because it reflects an hourly average.

In Algorithm 2, it is assumed that the shape factor for the rack is specified by the user. It is generally understood that the user will run Algorithm 2 for various Q values and generate several alternative configurations to evaluate them in terms of cost, throughput capacity, and space requirements.

In concluding this section on OP systems, we wish to stress that with the introduction of Web-based ordering (generally referred to as "e-tailing"), there is significant and renewed interest in the design and operation of OP systems. Many companies offer their goods through the Web with no traditional retail outlets. Even those with traditional retail outlets now offer the same goods (or an expanded set of goods) on-line through their Web sites. Customers simply log-in and order the goods on-line. The increased use of the Web to order goods has essentially shifted the OP burden from the customer to the company. Despite some of the recent failures in the dot-com sector, including on-line groceries, we expect the e-tailing trend to continue.

10.8   FIXED-PATH MATERIAL HANDLING MODELS

Our interest in fixed-path material handling models stems from the fact that fixed-path material handling systems are one of the most widely used material handling systems. They include material handling equipment such as powered belt and roller conveyors, towline and trolley-type conveyors, automated guided vehicles, and computer-dispatched lift trucks. The models presented in this section provide some basic procedures for the design of material handling systems based on the equipment types mentioned above.

We begin our discussion in this section with models for specific types of conveyors—towline and trolley conveyors. Next, we present basic procedures for calculating horsepower requirements for powered belt and roller conveyors. Finally, we present flow path design models covering both conventional systems and tandem flow systems.

Towline or Trolley Conveyors

In this section a deterministic model of a towline or trolley conveyor is considered. The towline conveyor and trolley conveyor have received considerable attention from researchers [61]. The analysis of such conveyors is referred to as **conveyor analysis.** Conveyor analysis is a term that is used to refer to the analysis of closed-loop, irreversible, or recirculating conveyors with discretely spaced carriers.

The interest in analyzing recirculating conveyors can be traced to two engineers: Kwo, with General Electric, and Mayer, with Western Electric. Kwo [39, 40] developed a deterministic model of materials flow on a conveyor having one loading station and one unloading station. He proposed a number of rules to ensure the input and output rates of materials were compatible with the design of the conveyor. Mayer [44] employed a probabilistic model in analyzing a conveyor with multiple loading stations.

Conveyor analysis was developed because of a number of operating problems that occurred with overhead trolley conveyors. Namely,

1. At a loading station, no empty carriers were available when a loading operation was to be performed.

2. At an unloading station, no loaded carriers were available when an unloading operation was to be performed.

Because of the interference problems analytical and simulation studies were undertaken in order to gain insight into the relationships among the design parameters and operating variables.

A conveyor system consists of the combination of the conveyor equipment, loading and unloading stations, and the operating discipline employed. Hence, system design parameters include not only equipment parameters, but also such considerations as waiting space allowances and the number, spacing, and sequencing of loading and unloading stations.

Conveyors provide many of the links among workstations in modern production systems. They are used not only to transport or deliver materials, but also to provide storage capacity for materials. Some conveyor systems are basically handling or transport systems, some are storage or accumulation systems, and some perform both handling and storage functions.

Kwo [39, 40] developed three principles to be used in designing closed-loop, irreversible conveyors:

1. *Uniformity principle:* Materials should be uniformly distributed over the conveyor.

2. *Capacity principle:* The carrying capacity of the conveyor must be greater than or equal to the system throughput requirements.

3. *Speed principle:* The speed of the conveyor (in terms of the number of carriers per unit time) must be within the permissible range, defined by loading and unloading station requirements and the technological capability of the conveyor.

He also developed some numerical relations based on the uniformity principle that provide **sufficient** conditions for steady-state operations. Since the work of Kwo, more general results have been developed by Muth [46–48]. Our presentations will be based on the multistation analysis of Muth [48].

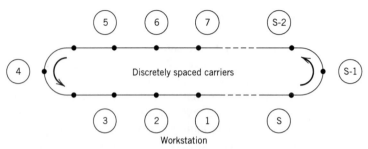

Figure 10.19 Conveyor layout considered by Muth [48].

A schematic representation of the recirculating conveyor considered by Muth is given in Figure 10.19. There are s stations located around the conveyor, *numbered in reverse sequence to the rotation of the conveyor.* Each station can perform loading and/or unloading. There are k carriers equally spaced around the conveyor. The passage of a carrier by a workstation establishes the increment of time used to define material loading and unloading sequences.

For convenience, station 1 is used as a reference point in defining time; consequently, carrier n becomes carrier $n + k$ immediately after passing station 1. The sequence of points in time at which a carrier passes station 1 is denoted (t_n), where t_n is the time at which carrier n passes station 1. The amount of material loaded on carrier n as it passes station i is given by $f_i(n)$ for $i = 1, 2, \ldots, s$.[3] The amount of material carried by carrier n immediately after passing station i is denoted $H_i(n)$.

In order for the conveyor to operate for long periods of time (achieve steady-state operations), the total amount of material loaded on the conveyor must equal the total amount of material unloaded. Since it is assumed that the conveyor will be operated over an infinite period of time, the sequences $\{f_i(n)\}$ are assumed to be periodic, with period p. Thus,

$$\{f_i(n)\} = [f_i(n + p)] \tag{10.83}$$

Additionally,

$$\sum_{n=1}^{p} \sum_{i=1}^{s} f_i(n) = 0 \tag{10.84}$$

because of the requirement that all material loaded on the conveyor must be unloaded and vice versa.

In analyzing the conveyor it is convenient to employ the following relation:

$$F_1(n) = \sum_{i=1}^{s} f_i(n) \tag{10.85}$$

Muth obtains the following results:

1. k/p cannot be an integer for steady-state operations.
2. Letting $r = k \bmod p$, r/p must be a proper fraction for general sequences $\{F_1(n)\}$ to be accommodated.[4]

---

[3]Negative values of $f_i(n)$ are permissible and denote unloading.
[4]The operation of $r = k \bmod p$ means r is given by the remainder following the division of k by p, for example, 12 mod 7 = 5 and 26 mod 8 = 2.

3.   It is desirable for p to be a prime number, as conveyor compatibility results for all admissible values of k.

The materials balance equation for carrier n can be given as

$$H_1(n) = H_1(n - r) + F_1(n) \qquad (10.86)$$

In order to determine the values of $H_i(n)$ the following approach is taken.

1.   Let $H_1^*(n)$ be a particular solution to Equation 10.86 and use the recursion

$$H_1^*(n) = H_1^*(n - r) + F_1(n) \qquad (10.87)$$

by letting $H_1^*(1) = 0$.

2.   Given $H_i^*(n)$, the value of $H_{i+1}^*(n)$ is obtained from the relation

$$H_{i+1}^*(n) = H_i^*(n) - f_i(n) \qquad (10.88)$$

3.   Given $\{H_i^*(n)\}$ for $i = 1, \ldots , s$, let

$$c = \min_{i,n} H_i^*(n) \qquad (10.89)$$

4.   The desired solution is given by

$$H_i(n) = H_i^*(n) - c \qquad (10.90)$$

5.   The required capacity per carrier is

$$B = \max_{i,n} H_i(n) \qquad (10.91)$$

Example 10.30

In order to illustrate the deterministic solution procedure, consider a situation involving a single loading station and a single unloading station, as depicted in Figure 10.20. The conveyor has nine carriers, equally spaced. The loading and unloading sequences have periods of 7 time units and are given to be

$$\{f_1(n)\} = \{1, 1, 2, 2, 2, 1, 1\}$$

and

$$\{f_2(n)\} = \{0, 0, 0, 0, 0, -5, -5\}$$

Thus,

$$\{F_1(n)\} = \{1, 1, 2, 2, 2, -4, -4\}$$

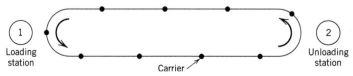

Figure 10.20   Conveyor layout for Example 10.30.

and

$$r = k \bmod p$$
$$= 9 \bmod 7$$
$$= 2$$

Notice $k/p = 9/7$ is not an integer, $r/p = 2/7$ is a proper fraction, and $p = 7$ is a prime number.

Arbitrarily, we let $H_1^*(1) = 0$ and employ the relation $H_i^*(n) = H_i^*(n - r) + F_i(n)$ to determine $H_1^*(3)$,

$$H_1^*(3) = H_1^*(1) + F_1(3)$$
$$H_1^*(3) = 0 + 2 = 2$$

Likewise, for $H_1^*(5)$

$$H_1^*(5) = H_1^*(3) + F_1(5)$$
$$= 2 + 2 = 4$$

Since $r = 2$, the next computation is for $H_1^*(7)$,

$$H_1^*(7) = H_1^*(5) + F_1(7)$$
$$= 4 - 4 = 0$$

Next, $H_1^*(9)$ is obtained; however, since $p = 7$, $H_1^*(9) = H_1^*(2)$. Hence,

$$H_1^*(2) = H_1^*(7) + F_1(2)$$
$$= 0 + 1 = 1$$

Likewise, for $H_1^*(4)$,

$$H_1^*(4) = H_1^*(2) + F_1(4)$$
$$= 1 + 2 = 3$$

Similarly,

$$H_1^*(6) = H_1^*(4) + F_1(6)$$
$$= 3 - 4 = -1$$

Hence, $\{H_1^*(n)\} = \{0, 1, 2, 3, 4, -1, 0\}$.

The values of $H_2^*(n)$ are obtained using the relation $H_2^*(n) = H_1^*(n) - f_1(n)$. Therefore,

$$\{H_2^*(n)\} = \{0, 1, 2, 3, 4, -1, 0\} - \{1, 1, 2, 2, 2, 1, 1\}$$
$$= \{-1, 0, 0, 1, 2, -2, -1\}$$

The value of c is found to be

$$c = \min_{i,n} H_i^*(n) = -2$$

for $i = 2$ and $n = 6$. Consequently, the sequences $\{H_i(n)\}$ are given by

$$\{H_1(n)\} = \{2, 3, 4, 5, 6, 1, 2\}$$
$$\{H_2(n)\} = \{1, 2, 2, 3, 4, 0, 1\}$$

and $B = \max_{i,n} H_i(n) = 6$, for $i = 1$ and $n = 5$. Therefore, each carrier must have sufficient capacity to accommodate 6 units of product.

As shown in Table 10.11 the amount of materials on the carriers will change with changing values of k. Additionally, the required carrier capacity B is affected by the value of k.

Table 10.11 Values of {H₁(n)} for Various Values of k in Example 10.30

n	$k=8$		$k=9$		$k=10$		$k=11$		$k=12$		$k=13$	
	$H_1(n)$	$H_2(n)$	$H_1(n)$	$H_2(n)$	$H_1(n)$	$H_2(n)$	$H_1(n)$	$H_2(n)$	$H_1(n)$	$H_2(n)$	$H_1(n)$	$H_2(n)$
1	2	1	2	1	5	4	4	3	6	5	9	8
2	3	2	3	2	2	1	7	6	5	4	8	7
3	5	3	4	2	5	3	5	3	5	3	7	5
4	7	5	5	3	7	5	3	1	4	2	5	3
5	9	7	6	4	4	2	6	4	3	1	3	1
6	5	4	1	0	1	0	3	2	2	1	1	0
7	1	0	2	1	3	2	1	0	1	0	5	4
	$B=9$		$B=6$		$B=7$		$B=7$		$B=6$		$B=9$	

The sequences $\{f_1(n)\}$ and $\{f_2(n)\}$ indicate, for example, that 5 units are removed from the sixth carrier at station 2 and, when it arrives at station 1, 1 unit is placed on the sixth carrier. When the sixth carrier passes station 1, its steady-state load $H_1(6)$ equals 1 unit of product; additionally, it becomes labeled the fifteenth carrier if $k = 9$.

When the fifteenth carrier reaches station 2 the action taken is given by $f_2(15) = f_2(8) = f_2(1) = 0$, since $p = 7$. Thus, nothing happens at station 2; however, at station 1, $f_1(1) = 1$ and 1 unit of product is added to the carrier, raising its total to $H_1(1) = 2$.

The carrier is now labeled the twenty-fourth carrier; when it reaches station 2, since $f_2(24) = f_2(3) = 0$, nothing happens at station 2. However, at station 1, $f_1(3) = 2$ and the content of the carrier is increased to 4 units. The carrier with 4 units arrives at station 2 numbered the thirty-third carrier. Since $f_2(33) = f_2(5) = 0$ nothing happens; however, at station 1, $f_1(5) = 2$, and the carrier's load is increased to 6 units.

When the carrier arrives at station 2 it is numbered the forty-second carrier. Since $f_2(42) = f_2(7) = -5$, the content of the carrier is reduced to 1 unit. When it arrives at station 1, $f_1(7) = 1$ and 1 unit is added to bring the total amount to 2 units.

The carrier is numbered the fifty-first carrier when it arrives at station 2. Since $f_2(51) = f_2(2) = 0$, nothing happens. At station 1 the content of the carrier is increased by 1 unit to total 3 units and it becomes the sixtieth carrier.

The sixtieth carrier arrives at station 2 with 3 units; since $f_2(60) = f_2(4) = 0$, nothing happens. At station 1, 2 units are added to the carrier and the number of the carrier is increased to 69. When the carrier arrives at station 2, $f_2(69) = f_2(6) = -5$, and 5 units are removed. Hence, the carrier arrives at station 1 empty, just as it did when it was labeled carrier six.

The above description for carrier 6 can be duplicated for each carrier on the conveyor by considering a period of time equal to $kp = 63$ time periods. The content of the carrier as it left station 1 followed the sequence $\{1, 2, 4, 6, 2, 3, 5\}$, which is a permutation of $\{H_1(n)\}$; likewise, the content of the carrier as it left station 2 followed the sequence $\{0, 1, 2, 4, 1, 2, 3\}$, which is the same permutation of $\{H_2(n)\}$ that generated the sequence for station 1.

If one tracks the content of any given carrier for kp time periods, it will be found that each carrier as it leaves station i will carry an amount of material given by a permutation of the sequence $\{H_i(n)\}$. Hence, each carrier will, at some time during the repeating cycle kp carry an amount of material equal to each element in the sequence $\{H_i(n)\}$ as it leaves station i. Consequently, a conveyor designed using Muth's results will satisfy the uniformity principle.

Example 10.31

Consider a conveyor system with one loading station and one unloading station. The material flow sequences are

$$f_1(n) = (0, 0, 2, 3, 1)$$
$$f_2(n) = (0, 0, 0, -2, -4)$$

If there are seven carriers on the conveyor, then $r = 2$. Solving for the sequences $\{H_i(n)\}$ for $i = 1, 2$ gives

$$\{H_1(n)\} = \{3, 2, 5, 3, 2\}$$
$$\{H_2(n)\} = \{3, 2, 3, 0, 1\}$$

Thus, $B = 5$.

Suppose the distance from station 2 to station 1 on the conveyor is changed, such that the loading and unloading sequences are given by

$$f_1(n) = (2, 3, 1, 0, 0)$$
$$f_2(n) = (0, 0, 0, -2, -4)$$

Solving for the sequences $\{H_i(n)\}$ gives

$$\{H_1(n)\} = \{3, 3, 4, 1, 0\}$$
$$\{H_2(n)\} = \{1, 0, 3, 1, 0\}$$

and $B = 4$.

Example 10.32

Suppose a closed-loop, irreversible conveyor with discretely spaced carriers is used to serve three workstations located as shown in Figure 10.21. The material flow sequences are assumed to be

$$f_1(n) = (4, 4, 0)$$
$$f_2(n) = (-3, -1, 0)$$
$$f_3(n) = (0, -2, -2)$$

and $k = 16$, such that $r = 1$. Solving for $\{H_i(n)\}$ gives

$$\{H_1(n)\} = \{4, 5, 3\}$$
$$\{H_2(n)\} = \{0, 1, 3\}$$
$$\{H_3(n)\} = \{3, 2, 3\}$$

with $B = 5$.

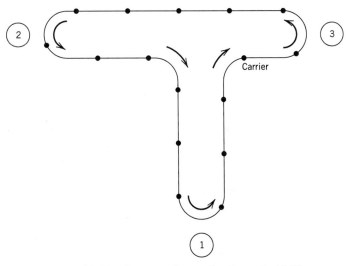

Figure 10.21   Conveyor layout for Example 10.32.

If the workstations are rearranged around the conveyor such that

$$f_1(n) = (0, -2, -2)$$
$$f_2(n) = (4, 4, 0)$$
$$f_3(n) = (-3, -1, 0)$$

then the appropriate computations establish that

$$\{H_1(n)\} = \{4, 5, 3\}$$
$$\{H_2(n)\} = \{4, 7, 5\}$$
$$\{H_3(n)\} = \{0, 5, 5\}$$

with $B = 7$.

The last two examples illustrate the effect on $\{H_i(n)\}$ of the locations of the workstations served by the closed-loop, recirculating conveyor. Not only can the capacity requirement for a carrier be affected by changing the number of carriers on the conveyor, but it also can be affected by the locations of the workstations around the conveyor.

Muth's modeling approach is descriptive, rather than prescriptive. Consequently, optimum conveyor designs are not obtained directly from the analysis. However, for a particular situation one can evaluate a number of alternative designs and compare the economic performance of each. As has been shown, the carrier capacity can be affected in a number of ways. Consequently, one would want to consider not only alternative conveyor lengths (measured in the number of carriers), but also the locations of stations around the conveyor.

Horsepower Calculations

In the previous section we considered a model of a recirculating trolley conveyor. The model allowed us to consider the effect on the carrying capacity of an individual carrier (trolley) of changes in loading and unloading sequences. Further, the model provided us with an indication of the amount of inventory on the conveyor during steady-state conditions.

We noted that the cost of the trolley conveyor is affected by the length of the conveyor and the number and size of carriers installed on the conveyor. In addition to the cost of the hardware represented by the trolleys and connecting chain, the installation cost is also a significant factor. Not to be overlooked, the cost of powering the conveyor is a function of its design. In particular, the horsepower requirement depends on the conveyor speed and the weight of the conveyor and the material it transports. The greater the horsepower requirement, typically, the greater the cost of the motor(s) used.

In this section, we consider the horsepower requirements for belt and roller conveyors used to transport unit loads, as well as the horsepower requirements for belt conveyors used to transport bulk materials. Drawing upon models provided in design manuals of various conveyor manufacturers, we present in this section approximate or quick-and-dirty models, rather than extremely precise models, of horsepower requirements.

Our purpose in considering the horsepower requirements that result from design decisions for a particular conveyor is to demonstrate that what might seem to be inconsequential design decisions can significantly impact horsepower requirements. We are not trying to make you an expert in determining the horsepower requirements for a particular conveyor; we realize that the accuracy of the assumptions made regarding material to be conveyed, as well as decisions regarding the placement of a motor on the conveyor, can affect final decisions regarding the size (and cost) of the motor installed.

Powered Unit and Package Conveyors

To determine the horsepower requirements for a belt conveyor or a roller conveyor to be used to convey packages, toteboxes, pallet loads, and so on, the following notation will be used:

HP = horsepower requirement, hp

S = speed of the conveyor, ft per minute (fpm)

L = load to be carried by the conveyor, in pounds

TL = total length of the conveyor, in feet

RC = spacing between centerlines of rollers (i.e., roller spacing, in inches)

WBR = width between rails, in inches (approximates belt width and/or roller length; rollers are generally at least 2.5″ longer than the width of the load being transported)

α = angle of incline, in degrees

LL_I = live load on incline, in pounds (weight of material on the conveyor section that is inclined)

BV = base value

FF = friction factor

LF = length factor

Belt Conveyors.   To estimate the horsepower requirement for a belt conveyor, BV can be approximated by multiplying WBR by two-thirds. To determine the value of FF, it is important to know how the belt will be supported: if it is roller supported, then $FF = 0.05$; if it is slider bed-supported, then $FF = 0.30$. The value of LF can be obtained from Table 10.12.

Table 10.12   *LF Values for Belt Conveyors, Given Specific RC and WBR Values*

RC Value	WBR =15″	WBR = 19″	WBR = 23″	WBR = 27″	WBR = 33″	WBR =39″
4.5″	0.41	0.50	0.61	0.72	0.85	1.00
6.0″	0.34	0.42	0.51	0.61	0.72	0.85
9.0″	0.29	0.35	0.43	0.52	0.62	0.73
12.0″	0.25	0.31	0.38	0.46	0.54	0.65
S.B.	0.50	0.70	0.80	1.00	1.20	1.50

S.B. denotes a slider bed-supported belt conveyor.

The horsepower requirement can be approximated as follows:

$$HP = [BV + LF(TL) + FF(L) + LL_1(\sin \alpha)](S)/14{,}000 \qquad (10.92)$$

Example 10.33

Consider a 100-ft-long roller-supported belt conveyor inclined at an angle of 10°, with a 6″ spacing between rollers and a WBR value of 27″. (Use roller beds, not slider beds, for inclines; 35° is the maximum angle of incline for most situations.) Suppose the conveyor is used to transport toteboxes of material, with each totebox being 18″ long and weighing 35 lb, loaded. (For stability of the load, it is recommended that a minimum of 2 rollers be under the load at all times; on an incline, 3 or more rollers should be under the load.) A spacing of 12″ is to be provided between toteboxes. The conveyor speed is planned to be 90 ft/min. What is the approximate horsepower requirement for the conveyor?

What is the maximum loading condition for the conveyor? To facilitate the determination, let a load-segment be defined as a totebox plus the clearance between toteboxes. In this case the length of the load-segment is equal to 30″. Hence, exactly 40 load-segments will be on the 100-ft-long conveyor at all times. Thus, the weight of the load will be 40(35) or 1400 lb.

In this case, $S = 90$ fpm, RC = 6″, WBR = 27″, BV = ⅔(27) = 18, LF = 0.61, TL = 100′, FF = 0.05, L = LL_1 = 1400 lb, and $\alpha = 10°$. Hence,

$$HP = [18 + 0.61(100) + 0.05(1400) + 1400(0.1737)](90)/14{,}000 = 2.52 \text{ hp}$$

Thus, a 2.75 hp or 3.0 hp motor should be adequate to meet the demands placed on the belt conveyor.

What if the spacing between consecutive toteboxes had been 8″, rather than 12″? Dividing the conveyor length by the length of the load-segment yields a value of 46, plus a fraction. Hence, there would be 46 complete load-segments, plus a partial load-segment. The 46 load-segments have a length of 46(18″ + 8″), or 1196″. Thus, the maximum weight condition would be 46 full toteboxes and 4″ of a partial totebox. If the weight of a totebox is uniformly distributed over its length, then the partial totebox contributes (4/18)(35), or 7.78 lb. Therefore, the maximum load would be 46(35) + 7.78, or 1617.78 lb, and the horsepower requirement would be 2.83 hp.

Example 10.34

Continuing the previous example, suppose it is desired to elevate the toteboxes 17.36 ft. Three alternatives are under consideration, using angles of incline of 10°, 20°, and 30°. Rounding off to the nearest foot, the resulting conveyor lengths would be 100 ft, 51 ft, and 35 ft in length. The horsepower requirement for the first alternative has already been determined, 2.52 hp.

For the case of $\alpha = 20°$ and TL = 51 ft, a calculation establishes that 20 complete load-segments and 12″ of a partial load-segment will be accommodated on the conveyor.

Thus, $L = LL_1 = 20(35) + (12/18)(35) = 723.33$ lb. The horsepower requirement will be approximately

$$HP = [18 + 0.61(51) + 0.05(723.33) + 723.33(0.3420)](90)/14,000 = 2.14 \text{ hp}$$

For the case of $\alpha = 30°$ and TL = 35 ft, a calculation establishes that 14 complete load-segments will be accommodated on the conveyor. Hence, $L = LL_1 = 14(35) = 490$ lb; hence, the horsepower requirement will be approximately

$$HP = [18 + 0.61(35) + 0.05(490) + 490(0.50)](90)/14,000 = 1.99 \text{ hp}$$

Depending on circumstances of the application and the overall layout of the facility, it appears that the steepest feasible angle of incline would be preferred since it requires a smaller horsepower motor and a shorter conveyor.

Example 10.35

In the initial example, suppose the toteboxes are to be transported horizontally over a distance of 100 ft, rather than elevated. Further, suppose a 2.5 hp motor is available and must be used. What is the maximum weight per totebox that should be transported by the conveyor?

From the initial example, S = 90 fpm, RC = 6″, WBR = 27″, BV = $\frac{2}{3}(27) = 18$, LF = 0.61, TL = 100 ft, FF = 0.05, and $\alpha = 0°$. Hence,

$$HP = [18 + 0.61(100) + 0.05L](90)/14,000 = 2.50 \text{ hp}$$

Solving for L yields a value of 6197.78 lb. Since 40 toteboxes will be on the conveyor at the same time (based on a 30″ load-segment), the maximum weight per totebox would be 154.94 lb.

Roller Conveyors. Using different values of FF, BV, and LF, the same formula can be used to determine the horsepower requirement for roller conveyors. The following values are used for the friction factor: FF = 0.10 when a flat belt is used to drive the rollers; FF = 0.085 when a zero-pressure accumulating belt is used to power the rollers; FF = 0.075 when a v-belt is used to drive the rollers; and FF = 0.05 when the rollers are chain driven. The base value for roller conveyors is given by the following relation:

$$BV = 4.60 + 0.445(WBR) \tag{10.93}$$

LF values are provided in Table 10.13 for various RC and WBR values.

Table 10.13 *LF Values for Roller Conveyors, Given RC and WBR Values*

RC Value	WBR = 15″	WBR = 19″	WBR = 23″	WBR = 27″	WBR = 33″	WBR = 39″
3.0″	1.3	1.5	1.7	2.0	2.4	2.6
4.5″	0.9	1.0	1.2	1.4	1.6	1.8
6.0″	0.7	0.8	0.9	1.0	1.2	1.5
9.0″	0.5	0.6	0.7	0.8	0.9	1.0

Example 10.36

Consider the conveying requirement given in Example 10.33, albeit without the requirement to incline the load. (Although it is possible to use a roller conveyor to elevate unit loads, belt conveyors are preferred due to the slippage between the load and the rollers on an incline. The friction of the belt prevents slippage of the load on belt conveyors.) Consider a flat belt-driven roller conveyor, with the rollers still on 6″ centers. (For stability of transport, at least three rollers should be under the load at all times; for soft loads, closer spacing is required.) Hence, for an 18″-long load, the largest roller spacing would be 6″.

For the example, $S = 90$ fpm, $RC = 6″$, $WBR = 27″$, $BV = 4.60 + 0.445(27) = 16.62$, $LF = 1.0$, $TL = 100$ ft, $FF = 0.10$, $L = 1400$ lb, and $LL_I = 0$ lb, since $\alpha = 0°$. Hence,

$$HP = [16.62 + 1.0(100) + 0.10(1400)](90)/14{,}000 = 1.65 \text{ hp}$$

Thus, a 1.75 hp or 2.0 hp motor should be adequate to meet the requirements to be placed on the roller conveyor.

What is the upper bound on the weight of a totebox if no more than 2.5 hp can be required? A calculation establishes that each totebox must not weigh more than 68 lb.

Example 10.37

In the previous example, suppose the toteboxes go directly from the roller conveyor operating at a speed of 90 fpm to a 100-ft-long, flat belt driven roller conveyor operating at 150 fpm. What would be the horsepower requirement for the second conveyor?

If 18″ toteboxes are moving at a speed of 90 fpm, with a clearance of 12″, what will be the spacing of toteboxes on the second conveyor? Be careful! Note that the rate at which toteboxes exit the first conveyor equals (90 ft/min)(12 in./ft)/(30 in./totebox), or 36 toteboxes/min. Since the rate at which they exit the first conveyor must equal the rate at which they enter the second conveyor, then the spacing of toteboxes on the second conveyor will be (150 ft/min)(12 in./ft)/(36 toteboxes/min), or 50 in./totebox. Since the toteboxes are 18″ long, then the clearance on the second conveyor will be 32″.

(Some might have thought that increasing the speed by two-thirds would cause the clearance to increase by two-thirds, to attain a value of 20″. Not so! If you have ever walked along an airport concourse and stepped onto a moving sidewalk, you have experienced the spacing differential that occurs with the toteboxes. Since the totebox is neither stretched nor shrunk as it moves from one conveyor to another operating at a different speed, the clearance must accommodate the expansion/contraction for the load-segment.)

On the second conveyor there will be exactly 24 complete load-segments. Hence, the following parameter values apply to the second conveyor: $S = 150$ fpm, $RC = 6″$, $WBR = 27″$, $BV = 16.62$, $LF = 1.0$, $TL = 100$ ft, $FF = 0.10$, and $L = 840$ lb. Therefore,

$$HP = [16.62 + 1.0(100) + 0.10(840)](150)/14{,}000 = 2.15 \text{ hp}$$

Example 10.38

From the two previous examples, we found that a 1.65 hp motor and a 2.15 hp motor would be required to transport the toteboxes over a distance of 200 ft by using two 100-ft-long conveyors operating at 90 fpm and 150 fpm. Suppose a single 200-ft-long conveyor is used. What would be the horsepower requirement if it is operated at a speed of 120 fpm and conveys the toteboxes at a rate of 36 toteboxes per minute?

A calculation establishes that the length of a load-segment is 40″ and 60 complete load-segments are accommodated on the conveyor. Thus, for this example, S = 120 fpm, RC = 6″, WBR = 27″, BV = 16.62, LF = 1.0, TL = 200 ft, FF = 0.10, and L = 2100 lb. Hence,

$$HP = [16.62 + 1.0(200) + 0.10(2100)](120)/14,000 = 3.66 \text{ hp}$$

Depending on the costs of the motors, the requirements of the particular application, and the costs of 100-ft and 200-ft conveyor sections, one might prefer having a single 200-ft conveyor to having two 100-ft conveyors.

Bulk Belt Conveyors

In designing a bulk belt conveyor it is important to understand the physical characteristics of the material being conveyed. Since bulk materials such as grain, coal, crushed stone, beans, steel trimmings, sand, earth, ashes, and salt have different densities and different flowability, different bulk materials will take on different shapes when conveyed at the same speed and the same angle of incline.

Based on physical experiments with a wide range of materials, maximum recommended belt speeds have been established for given belt widths (BW) for different categories of material. The following are examples of maximum belt speeds, in feet per minute:

$$S_{max} = \begin{cases} 16.667\text{BW} & \text{for fine abrasive material} \\ 100 + 8.333\text{BW} & \text{for fragile material and lumpy abrasive material} \\ 200 + 8.333\text{BW} & \text{for unsized material (e.g., coal, stone)} \\ 250 + 8.333\text{BW} & \text{for crushed stone and ore} \\ 200 + 16.667\text{BW} & \text{for grain and sand (wet or dry)} \end{cases}$$

where BW is expressed in inches.

Further, as shown in Table 10.14, different maximum angles of incline have been established for different bulk materials. Depending on the belt width used and the density of the material being conveyed, Table 10.15 provides values for the delivered capacity of the material in tons per hour per 100 ft/min of belt speed used.

To determine the horsepower requirements for conveying bulk material using a bulk belt conveyor, use Table 10.16 to determine the horsepower required to

Table 10.14 *Maximum Angle of Incline for Various Bulk Materials (in degrees)*

Bulk Material	Maximum Angle of Incline	Bulk Material	Maximum Angle of Incline
Ashes, dry	20°	Gravel, bank	18°
Ashes, wet	23°	Gravel, washed	12°
Briquettes	10°	Grain	20°
Coffee, green bean	10°	Sand, damp	20°
Cornmeal	22°	Sand, dry	15°
Coal, bituminous	20°	Salt, dry/coarse	20°
Coal, mine run	18°	Salt, dry/fine	11°
Earth, loose	20°	Steel trimmings	18°

Table 10.15   *Delivered Capacity in Tons Per Hour(tph) per 100 fpm of Belt Speed for Various Belt Widths (in inches) and Material Densities (in pounds per cubic feet)*

Belt Width (in.)	30 lb/ft^3 Density	50 lb/ft^3 Density	100 lb/ft^3 Density
18"	16.0	26.5	53.0
24"	31.0	52.0	104.0
30"	48.5	81.0	162.0
36"	68.0	113.5	227.0

Table 10.16   *Horsepower Required to Drive Empty Conveyor per 100 fpm of Belt Speed for Various Belt Widths (in inches) and Conveyor Lengths (in feet)*

Belt Width	50'	100'	150'	200'	250'	300'	400'	500'
18"	0.3	0.4	0.5	0.6	0.7	0.8	0.9	1.1
24"	0.5	0.6	0.7	0.8	0.9	1.0	1.3	1.5
30"	0.6	0.7	0.9	1.0	1.2	1.3	1.6	1.9
36"	0.7	0.9	1.1	1.3	1.5	1.6	2.0	2.4

Table 10.17   *Additional Horsepower Needed to Convey Material Horizontally at Any Speed Based on Various Conveyor Lengths (in feet) and Delivered Capacities (in tons per hour)*

Conveyor Length, ft	50 tph	100 tph	150 tph	200 tph	250 tph	300 tph	350 tph	400 tph	450 tph	500 tph
50'	0.3	0.6	0.9	1.2	1.5	1.8	2.1	2.4	2.7	3.0
100'	0.4	0.8	1.1	1.5	1.9	2.3	2.7	3.0	3.4	3.8
150'	0.5	0.9	1.4	1.8	2.3	2.7	3.2	3.6	4.1	4.5
200'	0.5	1.1	1.6	2.1	2.7	3.2	3.7	4.2	4.8	5.3
250'	0.6	1.2	1.8	2.4	3.0	3.6	4.2	4.8	5.5	6.1
300'	0.7	1.4	2.0	2.7	3.4	4.1	4.8	5.5	6.1	6.8
400'	0.8	1.7	2.5	3.3	4.2	5.0	5.8	6.7	7.5	8.3
500'	1.0	2.0	3.0	3.9	4.9	5.9	6.9	7.9	8.9	9.8

Table 10.18   *Additional Horsepower Required to Elevate Any Material at Any Speed for Various Handling Rates (in tons per hour) and Various Lifting Heights (in feet)*

Handling Rate (tph)	5'	10'	15'	20'	25'	30'	40'	50'
25	0.2	0.3	0.4	0.5	0.6	0.8	1.0	1.3
50	0.3	0.5	0.8	1.0	1.3	1.5	2.0	2.5
75	0.4	0.8	1.1	1.5	1.9	2.3	3.0	3.8
100	0.5	1.0	1.5	2.0	2.5	3.0	4.0	5.1
125	0.6	1.3	1.9	2.5	3.2	3.8	5.1	6.3
150	0.8	1.5	2.3	3.0	3.8	4.5	6.1	7.6
175	0.9	1.8	2.7	3.5	4.4	5.3	7.1	8.8
200	1.0	2.0	3.0	4.0	5.1	6.1	8.1	10.1
225	1.1	2.3	3.4	4.5	5.7	6.8	9.1	11.4
250	1.3	2.5	3.8	5.1	6.3	7.6	10.1	12.6
300	1.5	3.0	4.5	6.1	7.6	9.1	12.1	15.2
350	1.8	3.5	5.3	7.1	8.8	10.6	14.1	17.7
400	2.0	4.0	6.1	8.1	10.1	12.1	16.2	20.0
450	2.3	4.5	6.8	9.1	11.4	13.6	18.2	23.0
500	2.5	5.1	7.6	10.1	12.6	15.2	20.0	25.0

drive the empty conveyor; next, use Table 10.17 to determine the horsepower required to convey the material horizontally; third, use Table 10.18 to determine the horsepower required to elevate the material. Finally, the overall horsepower required is equal to the sum of the three components multiplied by 1.20, where the multiplier includes a 20% safety factor.

Example 10.39

Suppose we want to convey wheat having a density of 30 lb per cubic foot from a delivery truck to the top of a grain silo. From Table 10.14, we see that the maximum angle of incline for grain is 20°. If the vertical distance from the entry point to the exit point of the conveyor is 50 ft, then a trigonometric calculation establishes that the conveyor will need to be 146 ft long. (Assuming conveyor sections are available in 50-ft increments, a 150-ft conveyor would be used.)

If we use a belt that is 18 in. wide, the maximum belt speed is 500 ft/min. Further, at that speed, from Table 10.15, the wheat will be conveyed at a rate of (16.0)(500/100), or 80 tons per hour. On the other hand, if we used a 30″ belt, the speed could be 700 fpm and the delivered capacity would be (48.5)(700/100), or 339.5 tons per hour. (Assume a 30″ belt and the maximum belt speed are to be used to obtain a delivered capacity of 340 tph.)

Based on a conveyor length of 150 ft, a 30″ belt width, and a belt speed of 700 fpm, from Table 10.16 the horsepower required to drive the empty conveyor is (0.9)(700/100), or 6.3 hp. For a delivered capacity of 340 tph, the additional horsepower required to convey the material horizontally over the 150-ft length of the conveyor is found in Table 10.17 by interpolating between 300 tph and 350 tph; the result is 3.1 hp. To lift the wheat 50 ft at a rate of 340 tph, from Table 10.18, will require 17.2 hp. Hence, the total horsepower requirement is

$$HP = (6.3 + 3.1 + 17.2)(1.2) = 31.92 \text{ hp}$$

For this example, at least a 30 hp motor will be required if a speed of 700 fpm and a rate of 340 tph are to be realized. By slowing down the conveyor (and lengthening the time required to unload the delivery truck), the horsepower requirement can be reduced.

Suppose a rate of 300 tph is deemed feasible for transferring the wheat to the silo. With a 30″ belt, a belt speed of (100)(300/48.5), or 618.56 fpm, is required. At that speed, the horsepower requirement is reduced to

$$HP = (5.567 + 2.7 + 15.2)(1.2) = 28.16 \text{ hp}$$

Flow Path Design Models

In this section we focus on flow path design models since there are numerous types of material handling systems where system performance can be improved with an optimized flow path network design. They include in-floor towline conveyors, overhead trolley conveyors, automated guided vehicle systems, and others.

As discussed in Chapter 5, there are multiple alternative flow path configurations that can satisfy a given set of material handling requirements. In this section, we cover two configurations, namely, the conventional flow and tandem flow systems. The determination of which is the best flow path structure for specific applications requires more detailed study, and is beyond the scope of the discussion below.

1.   Conventional Flow Systems

The objective of the conventional flow path design problem is to find the direction of flow for each link in the flow path network. The direction of flow can be unidirectional or bidirectional. The flow path network represents the pickup/delivery stations, aisle intersections, and all usable aisles within a manufacturing or warehousing facility.

The conventional flow path design problem was first formulated by Gaskins and Tanchoco [25] as a zero-one integer linear programming problem. The model is formulated as a node-are network where the nodes represent pickup/delivery stations and aisle intersections and the arcs are the flow paths that connect the nodes. Kaspi and Tanchoco [35] presented an improved model. The formulation was developed in the context of developing guide paths for automated guided vehicle systems. The model assumes that the flow along each segment of the flow path network is unidirectional. The reader is referred to Kim and Tanchoco [37] for discussions on mixed models that include unidirectional as well as bidirectional flows within the same flow path network.

a. Analytical Model.   Let the nodes represent pickup points, delivery points, and aisle intersection points, and let the arcs represent possible direction of flow between two adjacent nodes. Each are is assigned a length equal to the distance (or travel time) between the nodes it connects. The flow rate between pickup and delivery stations is described by a from-to chart. Once the node-arc network and the from-to chart have been obtained, the following variables and parameters can be defined:

n = number of entries in the from-to chart

$f_{\ell m}$ = flow intensity from pickup node l to delivery node m

d_{ij} = length of arc $i\text{--}j$ (the distance from node i to an adjacent node j)

$Y_{\ell m}$ = path length from pickup node l to delivery node m

$$X_{ij\ell m} = \begin{cases} 1 & \text{if arc } i\text{--}j \text{ is included in the path from pickup node } l \text{ to delivery node } m \\ 0 & \text{otherwise} \end{cases}$$

$$Z_{ij} = \begin{cases} 1 & \text{if arc } i\text{--}j \text{ is directed from node } i \text{ to node } j \\ 0 & \text{otherwise} \end{cases}$$

The objective function for the unidirectional flow path design (FPD) problem is to minimize the total distance that vehicles have to traverse to satisfy the total transportation requirements, that is,

Minimize $\sum_{\ell,m} f_{\ell m} Y_{\ell m}$

The constraint set includes the following:

C1. Path length from pickup node l to delivery node m

$$\sum_{i,j} X_{ij\ell m} d_{ij} = Y_{\ell m} \qquad \forall l, m$$

C2. Ensure the path from pickup node l to delivery node m is feasible

$$X_{ijlm} \leq Z_{ij} \qquad \forall l, m \ \forall i, j$$

C3. Unidirectionality constraints

$$Z_{ij} + Z_{ji} \le 1 \qquad \forall i, j$$

C4. At least one input arc

$$\sum_i Z_{ij} \ge 1 \qquad \forall j$$

C5. At least one output arc

$$\sum_k Z_{jk} \ge 1 \qquad \forall j$$

C6. One output arc from pickup node l using the path from node l to delivery node m

$$\sum_k X_{jklm} = 1 \qquad \forall l, m$$

C7. One input arc to delivery station m using the path from pickup node l to node m

$$\sum_k X_{kmlm} = 1 \qquad \forall l, m$$

C8. Number of input arcs equal to the number of output arcs

$$\sum_i X_{ijlm} = \sum_k X_{jklm} \qquad \forall l, m \ \ \forall j$$

The main difficulty in finding the solution to the above problem is the large number of variables required for realistic size problems. For example, more than 10,000 variables are required for a 10 pickup/delivery station problem with a total of 30 nodes representing the pickup/delivery stations and aisle intersections. A natural approach to use is the branch-and-bound procedure.

b. Branch-and-Bound Approach.   The specific technique used is the branch and bound with depth-search first and backtracking rather than the jump-tracking type of approach. Using the backtracking method, a feasible complete solution (not necessarily optimal) is obtained very quickly and the required memory is much less than for the jump-tracking method. The proposed approach involves eight steps. Each of these steps is described below. But first, some additional definitions are needed.

UB = upper bound, that is, the current (known) best value of the objective function. The initial value of UB is set at infinity. Any time a feasible complete solution is obtained with a value less than UB, the value of the upper bound UB is updated.

LB_k = lower bound of branch k is the best value of the objective function with constraint C2 ignored for all arcs in $\{U\}$. The lower bound LB_k is used to label the branches in the search process. Any time a lower bound of a certain branch is greater than (or equal to) the upper bound UB, then this branch is bounded.

$\{D\}$ = the set of directed arcs.

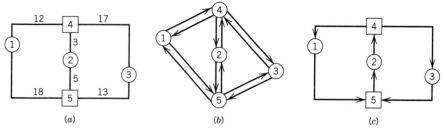

Figure 10.22   (*a*) The departmental layout, (*b*) the node-are network, and (*c*) the optimal flow path.

$\{U\}$ = the set of undirected arcs.
$\{A\}$ = the set of all the arcs, i.e., $\{U\} \cup \{D\} = \{A\}$, $\{U\} \cap \{D\} = \phi$.
$\{A'\}$ = the set of all the arcs connected to a pickup/delivery node.

c. Search Procedure.   For clarity's sake, the proposed branch-and-bound method is explained through a simplified numerical example. The departmental layout considered is shown in Figure 10.22*a* and the material flow from-to chart is given in Table 10.19. Note that an ε value equal to a very small number is entered as the flow value from node 2 to node 3 to ensure that each pickup/delivery station is reachable from any other pickup/delivery station.

Step 1. Initialization
The corresponding node-arc network is shown in Figure 10.22*b*. The procedure is initiated by determining set $\{A\}$. Initially, $\{A\} = \{1-4, 4-1, 1-5, 5-1, 2-4, 4-2, 2-5, 5-2, 3-4, 4-3, 3-5, 5-3\}$. Since all the arcs are currently undirected $\{U\} = \{A\}$ and $\{D\} = \phi$. The upper bound $UB = \infty$.

Step 2. Branching
The branching process is initialized using the arcs that are directly connected to pickup/delivery nodes (i.e., arc $i-j$, arc $j - i \varepsilon \{A'\}$). It is only after all arcs in $\{A'\}$ are exhausted that the other arcs in $\{A\}$ are considered.

Additionally, for the two branches connecting nodes i and j:
(i) $Z_{ij} = 1$, $Z_i = 0$ or $Z_{ij} = 0$, $Z_{ji} = 1$ where $Z_{ij}, Z_{ji} \varepsilon \{D\}$
(ii) $\{U\} = \{U\} - \{i - j, j - i\}$.

For the example, the branching is initialized with arcs $1-4$ and $4-1$ as shown in Figure 10.23. This initial branching corresponds to branch $k = 1$ ($Z_{14} = 1$, $Z_{41} = 0$) and branch $k = 2$ ($Z_{14} = 0$, $Z_{41} = 1$).

Table 10.19   *From-to Flow Chart*

		To		
	From	1	2	3
	1	—	10	15
	2	20	—	ε
	3	5	10	—

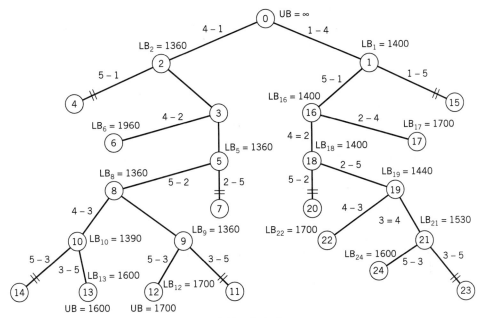

Figure 10.23   Search tree.

Step 3. Labeling

For each of the two new branches, the FPD problem is solved with constraint set C2 ignored for all arcs in $\{U\}$. Actually, the procedure looks for the shortest path Y_{lm} from each pickup node l to the delivery node m considering both the arcs in $\{D\}$ and $\{U\}$. Then the lower bound of each branch, $LB_k = \sum f_{lm} Y_{lm}$, is calculated.

For branch $k = 1$ ($Z_{14} = 1$, $Z_{41} = 0$), the shortest paths are: $Y_{12} = 17$, $Y_{13} = 29$, $Y_{21} = 23$, $Y_{31} = 31$, $Y_{32} = 18$, and the lower bound $LB_1 = 1400$. For branch $k = 2$ ($Z_{14} = 0$, $Z_{41} = 1$), the lower bound is $LB_2 = 1360$.

Step 4. Setting bounds

Branch k is bounded if:

(i) Lower bound $LB_k >$ upper bound UB;

(ii) No feasible path from a pickup node to a delivery node is obtained; or

(iii) For any node, all the arcs emanating from the node and all the arcs entering the node are in $\{D\}$, and constraint sets C4 and C5 are violated.

Step 5. Branch selection

Considering the two new branches (see the branching procedure), the one with the lowest LB is selected. The information on the other branch (LB, $\{U\}$, $\{D\}$) is recorded. If both branches are bounded, the backtracking procedure (step 7) is evoked.

The lower bound of branch $k = 1$ is $LB_1 = 1400$ and for branch $k = 2$, $LB_2 = 1360$. Thus, branch $k = 2$ is selected (see Figure 10.23). The branching process is continued with $LB_2 = 1360$, while branch $k = 4$ is bounded since $\{D\} = \{Z_{41} = 1, Z_{14} = 0, Z_{51} = 1, Z_{15} = 0\}$ violates constraint set C5.

Step 6. *Updating of upper bound*

When constraint set C2 is not violated, then the solution obtained in step 5 (branch selection) is a feasible complete solution to the FPL problem. If the value of the objective function for this solution, LB_k is less than UB (the current best value of the objective function), then UB is updated (i.e., $UB = LB$).

For this example, $UB_{12} = 1700$ and none of the arcs' direction violates any constraint, so it is a feasible complete solution. Until this point, UB has been equal to ∞, so UB is updated to be equal to 1700.

Step 7. *Backtracking*

The backtracking procedure is invoked any time a feasible complete solution is obtained. The backtracking procedure returns to the source branch. If a previously unselected branch of the source (i.e., a sibling branch) is available (i.e., it is not bounded and it has not been selected before), then the procedure continues through this branch. If the sibling branch is not available, then backtracking is performed again.

Referring to Figure 10.23, branch $k = 12$ represents a feasible complete solution, so the procedure returns to its source, branch $k = 9$. The sibling branch, branch $k = 11$, is bounded so the backtracking procedure continues to branch $k = 8$. Since branch $k = 10$ is available, it is selected; and the branching selection procedure (step 5) is executed.

Step 8. *Termination*

When backtracking reaches the root source ($k = 0$), and both branches $k = 1$ and $k = 2$ are no longer available, then the search is terminated. If UB is still infinite, then there is no feasible solution to FLP problem; otherwise, UB is the optimal solution.

The optimal flow path layout for the example problem is shown in Figure 10.22c.

2. *Tandem Flow Systems*

An alternative approach to flow path design is tandem flow systems, which was developed by Bozer and Srinivasan [10, 11] in the context of AGV systems. In tandem flow systems, the workstations to be served by the vehicles are divided into nonoverlapping, single-vehicle zones. Transfer stations are provided, as necessary, between adjacent zones in order to transfer a load from one zone to the next. Consider, for example, the eight workstations shown in Figure 10.24a. (The heavy black lines represent the aisles.) A possible two-zone tandem flow system is shown in Figure 10.24b, where the heavy black lines represent the guide paths for the two AGVs. The first zone consists of workstations {3, 4, 5, 6}, while the second zone consists of workstations {1, 2, 7, 8}. Workstations 2, 3, 6, and 7 are also utilized as transfer stations. The shaded lines in Figure 10.24b represent the (unidirectional) transfer conveyors that connect the transfer stations. Whether two pairs of conveyors and transfer stations are required depends on the direction of flow between the two zones. An alternative tandem flow system, again with two zones, is shown in Figure 10.24c.

In those cases where it is not possible to use the workstations as transfer stations, separate and additional transfer stations are set up between adjacent zones. For example, given the workstations shown in Figure 10.24a, two example tandem flow systems with two zones are shown in Figure 10.25a and 10.25b, where the

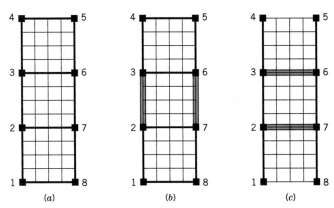

Figure 10.24 (*a*) Workstation locations; (*b*) and (*c*) alternative tandem flow systems with workstations used as transfer stations.

solid black line represents the guide path for the two AGVs, and the solid black diamond represents the transfer station. The number of loads that can be held at the transfer station is a design variable.

The examples we showed above have only two vehicles (and two zones). More vehicles (and more zones through further partitioning) will be required as the flow intensity in the system increases. As one might expect, the overall performance of the tandem flow system depends to a large extent on the partition used (i.e., the number of zones, the work/transfer stations in each zone, and the resulting flow patterns). Generally speaking, two factors seem to be important in partitioning. First, the workload among the vehicles (i.e., "how hard" each vehicle must work to meet the throughput requirement in its own zone) needs to be reasonably well-balanced. While a perfect workload balance among the vehicles is generally not going to be attainable, reducing the workload deviation among the vehicles is desirable so that a partition where a subset of the vehicles have to "work hard" while other vehicles are underutilized is avoided.

Second, the higher the flow intensity between stations i and j, the more incentive there is to place stations i and j in the same zone. For example, if the flow

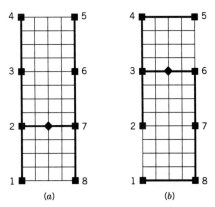

Figure 10.25 Alternative tandem flow systems with separate transfer stations.

intensity, say between stations 4 and 5 and between stations 1 and 8, is considerably high compared to the other flow values, the partition shown in Figure 10.25*b*
is going to be more desirable than the partition shown in Figure 10.25*a* from an
overall system performance point of view. Of course, if two stations have a high
flow intensity between them but they are located far from each other (such as stations 4 and 8 in Figure 10.25*a*), then placing them in the same zone may not be
possible. Such cases, that is, high-intensity flows over long distances, generally tax
the material handling system regardless of the type of flow system used.

Given a set of workstations (specified by their (x, y) coordinates in the plane),
and the flow intensity between the workstations, in [11] a heuristic partitioning procedure is presented for the case where there is no aisle structure imposed *a priori*.
This approach is more suitable if the aisle structure is relatively easy to change, or a
new facility is being planned and the "best" aisle structure for a tandem flow system
is sought. For those cases where the workstation locations are specified within a
fixed, user-defined aisle structure (such as a fixed departmental layout), a simulated-
annealing-based partitioning heuristic is presented in [49] for tandem flow systems.
(See Chapter 6 for simulated annealing as it applies to facility layout.)

A full description of either heuristic is beyond the scope of this section.
However, the partitioning model shown in [11] is reasonably straightforward to
solve with a basic optimization package that can handle 0/1 (binary) variables provided that a set of prospective zones with feasible vehicle workloads are identified
by the user. For example, given the workstation locations shown in Figure 10.24*a*,
the following set of prospective zones may be used as input for the partitioning
model: {1, 8}, {1, 2, 8}, {1 , 7, 8}, {1 , 2, 7, 8}, {2, 7}, {2, 3, 7}, {2, 6, 7}, {2, 3, 6, 7}, {3, 6},
{3, 4, 6}, and {3, 4, 5, 6}. Each prospective zone must be checked to ensure that the
vehicle in that zone meets the throughput requirement. Some allowance can be
made for load transfers in and out of each zone. Larger prospective zones such as
{1, 2, 3, 4}, {5, 6, 7, 8}, {1, 6, 7, 8}, and {2, 3, 4, 5} can be considered as long the vehicle in the zone is not overloaded. (Heuristic procedures to identify prospective
zones with feasible vehicle workloads are described in [11]; we refer the reader to
[11] for details.)

Given a set of prospective zones, let w_p denote the vehicle workload in zone
p. Let $a_{ip} = 1$ if workstation i appears in zone p, and 0 otherwise. Note that both
w_p and a_{ip} are parameters specified by the user. The w_p value for each zone must
theoretically be less than 1.0 but for practical purposes, a w_p value of 0.85 or less
is more desirable. The (binary) decision variable, x_p, is equal to 1 if prospective
zone p is used in the partition (i.e., it is part of the partition), and 0 otherwise.

To balance the workload among the vehicles, the objective in [11] is to minimize the maximum vehicle workload (denoted by z) in the partition:

Minimize z

The constraint set includes the following:

C1. Ensures that z is the maximum vehicle workload in the partition

$z \geq w_p x_p$ for all p

C2. Ensures that each workstation i is assigned to exactly one zone

$\sum_p a_{ip} x_p = 1$ for all i

C3. Allows the user to specify the number of zones (L) in the partition

$$\sum_p x_p = L$$

Small-to-medium instances of the above problem can be solved to optimality. (We will not define what "small" and "medium" are since they depend on the software and hardware used by the analyst.) For "large" instances, the partitioning model may be solved heuristically [11]. Also, the analyst should experiment with alternative values for L since L is the number of vehicles in the partition, by definition.

As the above partitioning model suggests, being able to check the vehicle workload in a given zone is an important part of developing and evaluating alternative partitions. We will next show a model one can use to compute the utilization (or the w_p value) of each vehicle in a given (user-specified) partition. Alternative or prospective partitions can then be manually generated and evaluated by the user by computing the vehicle utilizations for all the zones. The model we show here is based on the one developed in [17]. For simplicity, we treat each workstation as a combined pickup and deposit node. The model also applies in situations where the pickup and deposit nodes of a workstation are separate.

Consider an AGV serving a given set of stations in one zone. Each station is represented by a pickup/deposit (P/D) point; each P/D point corresponds to either an input/output (I/O) point (where loads enter and exit the zone) or a machine (where processing takes place). An empty trip occurs when the AGV has just delivered a load at station k and the next load it needs to move is at station i. A loaded trip occurs when the AGV picks up a load at station i and delivers it to station j. (A load is defined as a unit that the AGV moves in one trip.) When a load is ready to be moved by the AGV, we refer to it as a "move request."

Suppose the flow in the system (specified by the user) is expressed as a from-to chart, where f_{ij} denotes the flow intensity from station i to station j. (We assume that $f_{ii} = 0$.) Let Λ_i denote the rate at which loads must be delivered to station i (i.e., $\Lambda_i = \sum_j f_{ji}$); likewise, let λ_i be the rate at which loads must be picked up at station i (i.e., $\lambda_i = \sum_j f_{ij}$). We assume that flow is conserved at each machine (i.e., $\Lambda_i = \lambda_i$). The same is true for an I/O point if there is only one such point. Otherwise, flow is not necessarily conserved at each I/O point since the loads may enter the zone from one I/O point and exit from another. Even with multiple I/O points, however, we assume flow is conserved globally, that is, $\Lambda_T = \sum_i \Lambda_i = \sum_i \lambda_i = \lambda_T = \sum_i \sum_j f_{ij}$.

Let τ_{ij} and σ_{ij} denote the loaded and empty AGV travel time from station i to j, respectively. The τ_{ij} and σ_{ij} values are supplied by the user; they are computed based on the AGV travel distance and speed (including acceleration/deceleration), the load weight, and the guide path (such as number of turns the vehicle has to make in traveling from station i to j). Since the AGV does not share the guide path with other vehicles, we assume that the travel distance values are obtained by taking the shortest bidirectional route from station i to j. Note that τ_{ij} includes the load pickup time at station i, and the load deposit time at station j, by definition.

As long as it is up and running, the AGV is in one of three states at any given instant: (1) traveling loaded, (2) traveling empty, or (3) sitting idle at the last delivery point. (We will remark on station buffers at the end of this section.) Let α_f and α_e denote the fraction of time the AGV is traveling loaded and traveling empty,

respectively. Let $\rho = \alpha_f + \alpha_e$ denote the utilization of the AGV. Our goal is to estimate ρ for user-supplied values of f_{ij}, τ_{ij}, and σ_{ij}.

Computing α_f from the problem data is straightforward. That is,

$$\alpha_f = \sum_i \sum_j f_{ij}\tau_{ij} \tag{10.94}$$

In contrast, computing α_e is more difficult in general since its value depends on the
operating discipline, which is also known as "empty vehicle dispatching." A dispatching rule often used in industry is the *first-come, first-served* (FCFS) rule, where the
AGV, upon completing a loaded trip and becoming empty, is assigned to the "oldest"
move request in the system. If there are no move requests, the AGV becomes idle at
its last delivery point.

If we model the AGV system as an *M/G/1/FCFS* queue, and use Chow's model
[17], we first derive the "service time" (s_{kij}) for the AGV to serve one move request.
To serve a move request, the AGV travels empty from its last deposit point (say, station k) to the station where the move request is waiting (say, station i), picks up the
load, and then travels with the load to say, station j ($j \neq i$), where the load is deposit-
ed. (If $k = i$, empty travel time is zero since we assumed a combined deposit and
pickup node for each station; otherwise, the appropriate travel time from the deposit
to the pickup node of station i can be used for empty travel time.) The probability
that the AGV delivers a load to station k, say r_k, is equal to Λ_k/Λ_T. The probability
that a given move request must be picked up at station i and delivered to j, say r_{ij},
is equal to f_{ij}/λ_T. Hence, the probability that the AGV starts empty at station k, and
performs a loaded trip from station i to j, is given by $r_k r_{ij}$ (since the last delivery point
of the AGV and the origin of the next move request to be served are independent
under the FCFS rule).

The expected *service time*, say $\bar{s}$, is obtained as:

$$\bar{s} = \sum_k \sum_i \sum_{j \neq 1} r_k r_{ij} s_{kij} = \sum_k \sum_i \sum_{j \neq i} \left(\frac{\Lambda_k}{\Lambda_T}\right)\left(\frac{f_{ij}}{\lambda_T}\right)\sigma_{ki}\tau_{ij} \tag{10.95}$$

time units/move request. By definition, the device utilization, ρ, is equal to $\lambda_T\bar{s}$.
Note that, with the above approach, we can obtain α_e through a slight modification
of Equation 10.95. However, since we already obtained ρ, we can simply compute
α_e as $\alpha_e = \rho - \alpha_f$. The AGV will satisfy the workload requirement if $\rho < 1$. Of
course, one can also derive the second moment of the service time and obtain additional results such as the expected waiting time for a move request (averaged across
all the stations). We will now demonstrate the model through a simple example.

Example 10.40

Assuming the transfer station in Figure 10.25b is labeled station 9, consider the zone
defined by stations {1, 6, 7, 8, 9}. Suppose two job types (labeled A and B) are
handled through this zone. Assuming that stations 1 and 9 serve as I/O stations for
the zone, suppose the average hourly production volume and the production route
for the two job types are given as follows: Job A: 6 jobs/hour (1-7-9); Job B: 9 jobs/
hour (9-6-8-1). The resulting from-to chart (including the Λ and λ values) is shown
in Figure 10.26a.

	1	6	7	8	9	λ
1	0.0	0.0	6.0	0.0	0.0	6.0
6	0.0	0.0	0.0	9.0	0.0	9.0
7	0.0	0.0	0.0	0.0	6.0	6.0
8	9.0	0.0	0.0	0.0	0.0	9.0
9	0.0	9.0	0.0	0.0	0.0	9.0
Λ	9.0	9.0	6.0	9.0	6.0	39.0

	1	6	7	8	9
1	0.00	1.00	0.67	0.33	1.17
6	1.00	0.00	0.33	0.67	0.17
7	0.67	0.33	0.00	0.33	0.50
8	0.33	0.67	0.33	0.00	0.83
9	1.17	0.17	0.50	0.83	0.00

(*a*) From-To Matrix (Loads/hr)   (*b*) Empty Travel Time Matrix (mins/trip)

Figure 10.26   Flow data and travel time data for Example 10.40.

Suppose the empty AGV travel speed is fixed at 5 sec/grid. (A more elaborate travel time scheme, including AGV acceleration/deceleration and straight-travel speed versus turning speed, can be used instead of our fixed speed of 5 sec/grid.) The resulting empty travel time matrix (i.e., the σ_{ij} values expressed in min/trip) is shown in Figure 10.26*b*. Assuming the load pickup or deposit time is equal to 9 sec (0.15 min), the loaded AGV travel times (i.e., the τ_{ij} values) are obtained simply by adding the load pickup plus deposit time (0.30 min) to the corresponding σ_{ij} values. (Separate τ_{ij} values can be used if the loaded AGV travel speed differs significantly from the empty AGV travel speed.)

Using the above flow and travel time values in Equations 10.94 and 10.95, we obtain $\alpha_f = 0.4875$ and $\bar{s} = 0.02057$ hr (1.2342 min) per move request, respectively. Hence, $\rho = 0.8023$ and $\alpha_e = 0.3148$. Note that the busy vehicle would travel empty about 40% (0.3148/0.8023) of the time, which is a considerable portion of time, especially when empty travel is viewed as wasted travel or unproductive travel. Unfortunately, this is often the case with the FCFS dispatching rule since the location of the (empty) vehicle is not considered relative to all the move requests in the system when the vehicle is assigned to serve the next move request. Another dispatching rule used in industry, *shortest-travel-time-first* (STTF), assigns the empty vehicle to the closest move request as opposed to the oldest move request. STTF generally reduces empty vehicle travel but analytically it is more difficult to model and, therefore, system performance under STTF is often analyzed using simulation. (For a dynamic dispatching rule that further reduces empty vehicle travel in multivehicle AGV systems, the interested reader may refer to Bozer and Yen [14].) Both the FCFS and the STTF rules are "centralized" dispatching rules in that a central controller must keep track of all the move requests in a zone and dispatch the AGV to the appropriate station.

A simple dispatching rule that does not require a central controller, *first-encountered, first-served* (FEFS), was proposed and evaluated analytically by Bartholdi and Platzman [4] for closed-loop-based AGV systems. Under FEFS, when an AGV becomes empty, it simply follows a predetermined "polling sequence" to inspect each station one-at-a-time to see if there is a move request waiting to be served. Note that, under FEFS, the AGV never "sits idle." Even when it's empty, the AGV will continue to travel and poll the stations, looking for a move request. The reader may refer to Bozer and Srinivasan [10] for an analytical model to determine the workload of an AGV serving a set of stations under the FEFS dispatching rule.

The model we presented in this section is based on the assumption that there is sufficient buffer space at each station so the AGV does not get blocked when delivering a load. Also, a move request is placed only when a load is ready to be moved to the next machine (i.e., the idle AGV does not wait for a possible move request at any particular station or a "parking spot"). If these assumptions are relaxed, the possible states we defined earlier for the vehicle would have to be expanded.

10.9 WAITING LINE MODELS[5]

Among the many analytical approaches that can be used to aid in the design of facilities is that of **waiting line analysis** or **queueing theory.** As the name implies, waiting line analysis involves the study of waiting lines. Examples of waiting lines include the accumulation of parts on a conveyor at a workstation, pallet loads of material at receiving or shipping, in-process inventory accumulation, customers at a tool crib, customers at a checkout station in a grocery store, and lift trucks waiting for maintenance, to mention but a few.

Waiting line problems can be analyzed either mathematically or by using simulation. In this section, we present some quick-and-dirty methods for analyzing queueing problems mathematically. These mathematical methods are useful in preliminary planning and in many cases will yield results not significantly different from those obtained using simulation.

Because waiting lines occur in a variety of forms and contexts, it is important for the facilities planner to define clearly and accurately the waiting line situation under study. The following four elements are useful in defining a waiting line system: customers, servers, queue discipline, and service discipline.

The term **customer** can be interpreted very broadly. *Trucks* arriving at a receiving dock, *pallet loads* of materials arriving at a stretchwrap machine, *cases* arriving at a case sealing station, *employees* arriving at a tool crib, *patients* arriving at the hospital admissions window, and *airplanes* arriving at an airport are examples of customers. In brief, customers are the entities that arrive and require some form of service.

The term **server** can also be interpreted very broadly. For example, S/R machines, shrinkwrap tunnels, order pickers, industrial trucks, cranes, elevators, machine tools, computer terminals, inspector/packers, postal clerks, and nurses can be considered to be servers in certain contexts. Servers are the entities or combination of entities that provide the service required by customers.

The term **queue discipline** refers to the behavior of customers in the waiting line, as well as the design of the waiting line. In some cases, customers may refuse to wait; some customers may wait for a period of time, become discouraged, and depart before being served. Some waiting line systems have a limited amount of waiting space; some systems employ a single waiting line, while others have separate waiting lines for each server.

The term **service discipline** refers to the manner in which customers are served. For example, some systems have servers that serve customers singly, while others have servers that serve more than one customer at a time. Additionally, some systems serve customers on a first-come, first-serve basis, while others use a priority or random basis.

There exist a large number of different varieties of queueing problems. To facilitate the discussion of queueing problems, we will adopt the following classification scheme.

$$(x/y/z):(u/v/w)$$

---

[5]The material presented in this section draws heavily from the text by White, Schmidt, and Bennett [62].

where

x = the arrival (or interarrival) distribution

y = the departure (or service time) distribution

z = the number of parallel service channels in the system

u = the service discipline

v = the maximum number allowed in the system (in service, plus waiting)

w = the size of the population

The following codes are commonly used to replace the symbols x and y:

M = Poisson arrival or departure distributions (or equivalently exponential inter-arrival or service time distributions—M refers to the Markov property of the exponential distribution)

GI = General independent distribution of arrivals (or interarrival times)

G = General distribution of departures (or service times)

D = Deterministic interarrival or service times

The symbols z, v, and w are replaced by the appropriate numerical designations. The symbol u is replaced by a code similar to the following:

FCFS = first come, first served

LCFS = last come, first served

SIRO = service in random order

SPT = shortest processing (service) time

GD = general service discipline

Additionally, a superscript is attached to the first symbol if bulk arrivals exist and to the second symbol if bulk service is used. To illustrate the use of the notation, consider $(M^b/G/c):(\text{FCFS}/N/\infty)$. This denotes exponential interarrival times with b customers per arrival, general service times, c parallel servers, "first come, first serve" service discipline, a maximum allowable number of N in the system, and an infinite population.

Poisson Queues

The simplest waiting lines to study mathematically are those involving Poisson-distributed arrivals and services. In the case of an infinite customer population, if λ is the average number of arrivals per unit time, then the probability distribution for the number of arrivals during the time period T is given by

$$p(x) = \frac{e^{-\lambda T}(\lambda T)^x}{x!} \qquad x = 0, 1, 2, \ldots \qquad (10.96)$$

There exists an interesting, and important, relationship between the Poisson distribution and the exponential distribution. Namely, if arrivals are Poisson distributed, then the time between consecutive arrivals (interarrival time) will be exponentially distributed.

If services are Poisson distributed, then service times are exponentially distributed and can be represented by

$$f(t) = \mu e^{-\mu t} \qquad t > 0 \tag{10.97}$$

where μ is the service rate or expected number of services per unit time by a busy server, and $(1/\mu)$ is the expected service time.

In analyzing waiting line problems, we will let P_n denote the probability of n customers in the system (in service or waiting). Letting λ_n denote the arrival rate and μ_n denote the service rate when there are n customers in the system, it can be shown that

$$P_n = \frac{\lambda_0 \lambda_1 \ldots \lambda_{n-1}}{\mu_1 \mu_2 \ldots \mu_n} P_0 \tag{10.98}$$

where P_0 is the probability of 0 customers in the system.

If the system can hold not more than N customers in total, then $\lambda_N = 0$. Additionally, because the sum of the probabilities must be unity, then

$$P_0 + P_1 + \cdots P_N = 1 \tag{10.99}$$

From Equation 10.98 it is possible to express P_n in terms of P_0. From Equation 10.99 we then solve for P_0 and substitute the value of P_0 in Equation 10.98 to obtain the value of P_n for each value of n.

Example 10.41

In order to illustrate the approach suggested for analyzing a waiting line problem, suppose a workstation receives parts automatically from a conveyor. An accumulation line has been provided at the workstation and has a storage capacity for five parts ($N = 6$). Parts arrive randomly at the switching junction for the workstation; if the accumulation line is full, parts are diverted to another workstation. Parts arrive at a Poisson rate of 1/min; service time at the workstation is exponentially distributed with a mean of 45 sec. Hence,

$$\lambda_n = \begin{cases} 1 & n = 0, 1, \ldots, 5 \\ 0 & n = 6 \end{cases}$$

(*Note:* the system has a capacity of 6, the waiting line has a capacity of 5.)

$$\mu_n = \begin{cases} 0 & n = 0 \\ \frac{4}{3} & n = 1, \ldots, 6 \end{cases}$$

From Equation 10.98

$$P_1 = \frac{\lambda_0}{\mu_1} P_0 = \frac{1}{\frac{4}{3}} P_0 = \frac{3}{4} P_0$$

Likewise,

$$P_2 = \frac{\lambda_0 \lambda_1}{\mu_1 \mu_2} P_0 = \frac{(1)(1)}{\left(\frac{4}{3}\right)\left(\frac{4}{3}\right)} P_0 = \left(\frac{3}{4}\right)^2 P_0$$

Table 10.20 *Computation of P_n for Example 10.41*

n	λ_n	μ_n	P_n	P_n
0	1	0	$1P_0$	$\dfrac{4,096}{14,197} = 0.2885$
1	1	$\frac{4}{3}$	$\dfrac{\lambda_0}{\mu_1}P_0 = \dfrac{3}{4}P_0$	$\dfrac{3,072}{14,197} = 0.2164$
2	1	$\frac{4}{3}$	$\dfrac{\lambda_1}{\mu_2}P_1 = \left(\dfrac{3}{4}\right)^2 P_0$	$\dfrac{2,304}{14,197} = 0.1623$
3	1	$\frac{4}{3}$	$\dfrac{\lambda_2}{\mu_3}P_2 = \left(\dfrac{3}{4}\right)^3 P_0$	$\dfrac{1,728}{14,197} = 0.1217$
4	1	$\frac{4}{3}$	$\dfrac{\lambda_3}{\mu_4}P_3 = \left(\dfrac{3}{4}\right)^4 P_0$	$\dfrac{1,296}{14,197} = 0.0913$
5	1	$\frac{4}{3}$	$\dfrac{\lambda_4}{\mu_5}P_4 = \left(\dfrac{3}{4}\right)^5 P_0$	$\dfrac{972}{14,197} = 0.0685$
6	0	$\frac{4}{3}$	$\dfrac{\lambda_5}{\mu_6}P_5 = \left(\dfrac{3}{4}\right)^6 P_0$	$\dfrac{729}{14,197} = \dfrac{0.0513}{1.0000}$
			$\dfrac{14,197}{4,096}P_0 = 1$	

Table 10.20 summarizes the calculations involved in solving for P_n. Notice, in terms of P_0, the sum of the probabilities equals $(14,197/4,096)P_0$, thus, P_0 equals $4,096/14,197$. Given the value of P_0, the values of P_n are computed for $n = 1, \ldots, 6$.

From Table 10.20 it is seen that $P_0 = 0.2885$ and $P_6 = 0.0513$; hence, the workstation will be idle 28.85% of the time and the accumulation line will be full 5.13% of the time. The average number of parts in the accumulation (waiting) line L_q is given by

$$L_q = 0(P_0 + P_1) + 1P_2 + 2P_3 + 3P_4 + 4P_5 + 5P_6$$

or

$$L_q = 1(0.1623) + 2(0.1217) + 3(0.0913) + 4(0.0685) + 5(0.0513)$$
$$= 1.2101 \text{ parts in the accumulation line}$$

The average number of parts in the system, L, is given by

$$L = \sum_{n=0}^{N} nP_n$$

$$= 0\,(0.2885) + 1(0.2164) + 2(0.1623) + 3(0.1217) + 4(0.0913) + 5(0.0685) + 6(0.0513)$$

$$= 1.9216 \text{ parts in the system}$$

The rate at which parts actually enter the accumulation line is not 1/min because the accumulation line is sometimes full and parts are diverted elsewhere. The effective arrival rate $\bar{\lambda}$ is defined as the rate at which customers *enter* the system and is given by

$$\bar{\lambda} = (L - L_q)\mu \qquad (10.100)$$

or

$$\bar{\lambda} = (1.9216 - 1.2101)\left(\tfrac{4}{3}\right)$$
$$= 0.9487 \text{ parts per minute}$$

Alternately, in this case

$$\bar{\lambda} = (1 - P_N)\lambda$$
$$= 0.9487 \text{ parts per minute}$$

The percentage of arriving parts that do not enter the accumulation line is $P_N = 0.0513$. Suppose the production manager desires that no more than 2% of the arriving parts be diverted, how long should the accumulation line be? We will provide an answer to this question subsequently.

Example 10.42

The previous example involved a single server ($c = 1$). We consider now an example in which two servers are present ($c = 2$). Suppose in the previous situation the accumulation line supplies parts for two workstations, as depicted in Figure 10.27. Parts are removed from the accumulation line, worked on, and placed on the conveyor which delivers the parts to the packaging department. In this case, we assume $\lambda = 2$ parts per minute and let each server operate at a rate of $\frac{4}{3}$ parts per minute. Hence, $N = 7$ (waiting space for 5 parts, plus 2 being serviced) and

$$\lambda_n \begin{cases} 2 & n = 0, 1, \ldots, 6 \\ 0 & n = 7 \end{cases} \qquad \mu_n = \begin{cases} 0 & n = 0 \\ \frac{4}{3} & n = 1 \\ \frac{8}{3} & n = 2, 3, \ldots, 7 \end{cases}$$

Notice that if both servers are busy ($n \geq 2$), then the rate at which departures (services) occur is twice the service rate for a single server.

From Table 10.21 it is seen that $P_0 = 0.1613$, $P_1 = 0.2420$, $P_7 = 0.0430$. Hence, both workstations are idle 16.13% of the time, one workstation is busy 24.20% of the time, and the accumulation line is full 4.30% of the time. The average number of parts in the accumulation line, L_q and the average number of parts in the system, L, are found to be

$$L_q = 1P_3 + 2P_4 + 3P_5 + 4P_6 + 5P_7$$
$$= 1.0146 \text{ parts in the accumulation line}$$

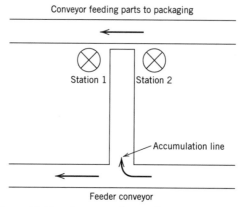

Conveyor feeding parts to packaging

Station 1   Station 2

Accumulation line

Feeder conveyor

Figure 10.27   Conveyor layout for Example 10.42.

Table 10.21   *Computation of P_n for Example 10.42*

n	λ_n	μ_n	P_n	P_n
0	2	0	$1P_0$	0.1613
1	2	$\dfrac{4}{3}$	$\dfrac{\lambda_0}{\mu_1}P_0 = \dfrac{6}{4}P_0$	0.2420
2	2	$\dfrac{8}{3}$	$\dfrac{\lambda_1}{\mu_2}P_1 = \dfrac{6}{4}\left(\dfrac{6}{8}\right)P_0$	0.1815
3	2	$\dfrac{8}{3}$	$\dfrac{\lambda_2}{\mu_3}P_2 = \dfrac{6}{4}\left(\dfrac{6}{8}\right)^2 P_0$	0.1361
4	2	$\dfrac{8}{3}$	$\dfrac{\lambda_3}{\mu_4}P_3 = \dfrac{6}{4}\left(\dfrac{6}{8}\right)^3 P_0$	0.1021
5	2	$\dfrac{8}{3}$	$\dfrac{\lambda_4}{\mu_5}P_4 = \dfrac{6}{4}\left(\dfrac{6}{8}\right)^4 P_0$	0.0766
6	2	$\dfrac{8}{3}$	$\dfrac{\lambda_5}{\mu_6}P_5 = \dfrac{6}{4}\left(\dfrac{6}{8}\right)^5 P_0$	0.0574
7	0	$\dfrac{8}{3}$	$\dfrac{\lambda_6}{\mu_7}P_6 = \dfrac{6}{4}\left(\dfrac{6}{8}\right)^6 P_0$ $(3{,}250{,}112/524{,}288)\,P_0 = 1$	0.0430 1.0000

and

$$L = \sum_{n=0}^{7} nP_n$$
$$= 2.4501 \text{ parts in the system}$$

In analyzing waiting lines a number of operating characteristics are typically used. Among them are the following:

1. L, expected number of customers in the system
2. L_q, expected number of customers in the waiting line
3. W, expected time spent in the system per customer
4. W_q, expected time spent in the queue per customer
5. U, server utilization
6. $\bar{\lambda}$, effective arrival rate

The values of L, L_q, W, and W_q are related as follows:

$$L = \sum_{n=0}^{N} nP_n \tag{10.101}$$

$$L_q = \sum_{n=c}^{N} (n - c)P_n \tag{10.102}$$

$$U = (L - L_q)/c \tag{10.103}$$

$$\hat{\lambda} = (L - L_q)\mu \tag{10.104}$$

$$W = L/\hat{\lambda} \tag{10.105}$$

$$W_q = L_q/\hat{\lambda} \tag{10.106}$$

$$L = L_q + \frac{\hat{\lambda}}{\mu} \qquad (10.107)$$

$$W = W_q + \frac{1}{\mu} \qquad (10.108)$$

Equations 10.104, 10.107, and 10.108 are based on the assumption that an individual server maintains the same service rate (μ) regardless of the number of customers in the system ($n > 0$). If a server has a changeable service rate then a more complicated situation exists and the values of W and W_q are not easily obtained.

Example 10.43

Consider a tool crib with two attendants ($c = 2$). Time studies indicate customers arrive in a Poisson fashion at a rate of 12 per hour so long as no more than one customer is waiting to be served. If two customers are waiting, the rate at which people enter the tool crib is reduced to 8 per hour. If three customers are waiting, customers enter at a rate of 4 per hour. No additional customers enter if four customers are waiting. Hence, the maximum number of customers in the system (N) equals 6. The time required to fill a customer's order is exponentially distributed with a mean of 10 min.

From Table 10.22 it is seen that $L = 2.7211$ and $L_q = 1.0923$. Because the service rate for an individual server, μ, is constant for $0 < n < 6$, then

$$U = \frac{(2.7211 - 1.0923)}{2}$$
$$= 0.8144 \text{ or } 81.44\%$$

$$\hat{\lambda} = (L - L_q)\mu$$
$$= (2.7211 - 1.0923)(6)$$
$$= 9.7728 \text{ customers per hour}$$

and

$$W = L/\hat{\lambda}$$
$$= \frac{2.7211}{9.7728}$$
$$= 0.2784 \text{ hr per customer}$$
$$= 16.706 \text{ min per customer}$$

Table 10.22 *Computation of P_n, L, and L_q for Example 10.43*

n	λ_n	μ_n	P_n	P_n	nP_n	$(n - c)P_n$
0	12	0	$1P_0$	0.0928	—	—
1	12	6	$2P_0$	0.1856	0.1856	—
2	12	12	$2P_0$	0.1856	0.3712	—
3	12	12	$2P_0$	0.1856	0.5568	0.1856
4	8	12	$2P_0$	0.1856	0.7424	0.3712
5	4	12	$\frac{4}{3}P_0$	0.1237	0.6185	0.3711
6	0	12	$\frac{4}{9}P_0$	0.0411	0.2466	0.1644
			$\frac{27}{9}P_0 = 1$	1.0000	$L = 2.7211$	$L_q = 1.0923$

$$W_q = \frac{L_q}{\lambda}$$

$$= \frac{1.0923}{9.7728}$$

$$= 0.1118 \text{ hr per customer}$$

$$= 6.706 \text{ min per customer}$$

(M/M/c):(G/N/∞) Results

We consider next the special case of an infinite population with Poisson arrivals and services having arrival rates and service rates defined by

$$\lambda_n = \begin{cases} \lambda & n = 0, \ldots, N-1 \\ 0 & n = N \end{cases} \qquad \mu_n = \begin{cases} n\mu & n = 0, \ldots, c-1 \\ c\mu & n = c, \ldots, N \end{cases}$$

In words, the arrival rate is constant and equal to λ until the system is full, at which time $\lambda_N = 0$. The service rate per busy server is μ for each server; hence, if n customers are present, then the overall service rate for the system is $n\mu$, up to a maximum value of $c\mu$ when all servers are busy. It can be shown that Equation 10.98 reduces to

$$P_n = \begin{cases} \dfrac{(c\rho)^n}{n!} P_0 & n = 0, 1, \ldots, c-1 \\[2mm] \dfrac{c^c \rho^n}{c!} P_0 & n = c, \ldots, N \\[2mm] 0 & n > N \end{cases} \qquad (10.109)$$

Table 10.23 *Summary of Operating Characteristics for the (M/M/c) Queue*

	$(M\|M\|c){:}(GD\|N\|\infty)^a$	$(M\|M\|c){:}(GD\|\infty\|\infty)^b$
L_q	$\dfrac{\rho(c\rho)^c P_0}{c!(1-\rho)^2}[1 - \rho^{N-c+1} - N - c + 1)(1-\rho)\rho^N]$	$\dfrac{\rho(c\rho)^c P_0}{c!(1-\rho)^2}$
L	$L_q + \dfrac{(c\rho)^c(1-\rho^{N-c+1})P_0}{(c-1)!(1-\rho)} + \displaystyle\sum_{n=0}^{c-1} nP_n$	$L_q + \dfrac{\lambda}{\mu}$
W_q	$\dfrac{(c\rho)^c[1 - \rho^{N-c+1} - (N-c+1)(1-\rho)\rho^N]}{c!c\mu(1-\rho)^2(1-P_N)} P_0$	$\dfrac{(c\rho)P_0}{c!c\mu(1-\rho)^2}$
W	$W_q + \dfrac{1}{\mu}$	$W_q + \dfrac{1}{\mu}$
U	$\rho(1-P_N)$	ρ
λ	$\lambda(1-P_N)$	λ
P_0	$\left[\dfrac{(c\rho)^c(1-\rho^{N-c+1})}{c!(1-\rho)} + \displaystyle\sum_{n=0}^{c-1}\dfrac{(c\rho)^n}{n!}\right]^{-1}$	$\left[\dfrac{(c\rho)^c}{c!(1-\rho)} + \displaystyle\sum_{n=0}^{c-1}\dfrac{(c\rho)^n}{n!}\right]^{-1}$

$^a\rho \neq 1.$
$^b\rho < 1.$

Table 10.24 *Summary of Operating Characteristics for the (M/M/l) Queue*

	$(M\|M\|1){:}(GD\|N\|\infty)^a$	$(M\|M\|1){:}(GD\|\infty\|\infty)^b$
L_q	$\dfrac{\rho^2[1 - \rho^N - N\rho^{N-1}(1 - \rho)]}{(1 - \rho)(1 - \rho^{N+1})}$	$\dfrac{\rho^2}{1 - \rho}$
L	$\dfrac{\rho[1 - \rho^N - N\rho^N(1 - \rho)]}{(1 - \rho)(1 - \rho^{N+1})}$	$\dfrac{\rho}{1 - \rho}$
W_q	$\dfrac{\rho[1 - \rho^N - N\rho^{N-1}(1 - \rho)]}{\mu(1 - \rho)(1 - \rho^N)}$	$\dfrac{\rho}{\mu(1 - \rho)}$
W	$\dfrac{[1 - \rho^N - N\rho^N(1 - \rho)]}{\mu(1 - \rho)(1 - \rho^N)}$	$\dfrac{\rho}{\mu(1 - \rho)}$
U	$\dfrac{\rho(1 - \rho^N)}{1 - \rho^{N+1}}$	ρ
$\bar{\lambda}$	$\dfrac{\lambda(1 - \rho^N)}{1 - \rho^{N+1}}$	λ
P_0	$\dfrac{1 - \rho}{1 - \rho^{N+1}}$	$1 - \rho$

$^a\rho \neq 1.$
$^b\rho < 1.$

where $\rho = \dfrac{\lambda}{c\mu}$. Solving for P_0 gives

$$P_0 = \begin{cases} \left[\dfrac{(c\rho)^c(1 - \rho^{N-c+1})}{c!(1 - \rho)} + \displaystyle\sum_{n=0}^{c-1}\dfrac{(c\rho)^n}{n!}\right]^{-1} & \rho \neq 1 \\ \left[\dfrac{c^c}{c!}(N - c + 1) + \displaystyle\sum_{n=0}^{c-1}\dfrac{c^n}{n!}\right]^{-1} & \rho = 1 \end{cases}$$

The quantity ρ is referred to as the traffic intensity for the system.

Formulas for the operating characteristics are given in Table 10.23. Also, given in Table 10.23 are the values of the operating characteristics when N is infinitely large, that is, there is no limitation on the number the system can accommodate. (Notice that it must be true that $\rho < 1$ for the case of N infinitely large.) Values of the operating characteristics are given in Table 10.24 for the special case of a single server ($c = 1$).

Example 10.44

Recall, in Example 10.41 it was given that $c = 1$, $\lambda = 1$ and $\mu = \frac{4}{3}$. It was desired to determine the length of an accumulation line (N) such that $P_N \leq 0.02$. From Table 10.24 it is seen that

$$P_0 = \frac{1 - \rho}{1 - \rho^{N+1}}$$

where $\rho = \lambda/\mu$ or 0.75. From Equation 10.109 it is seen that the value of N is to be determined such that

$$P_N = \frac{\rho^N(1 - \rho)}{1 - \rho^{N+1}} \leq 0.02$$

or

$$\rho^N - \rho^{N+1} \le 0.02 - 0.02\rho^{N+1}$$

$$\rho^N(1 - 0.98\rho) \le 0.02$$

$$\rho^N \le \frac{0.02}{1 - 0.98\rho}$$

$$(0.75)^N \le \frac{0.02}{1 - 0.98(0.75)}$$

Taking the logarithm of both sides gives

$$N \log(0.75) \le \log(0.07547)$$

$$N \ge 8.982$$

Thus, $N = 9$, and the accumulation line will have space to accommodate 8 parts. The operating characteristics for the system can be obtained using the results given in Table 10.24. Namely, for $N = 9$, $\lambda = 1$, $\mu = \frac{4}{3}$, $c = 1$, and $\rho = \lambda/c\mu = 0.75$,

$$L = \frac{\rho[1 - \rho^N - N\rho^N(1 - \rho)]}{(1 - \rho)(1 - \rho^{N+1})} = 2.403$$

$$L_q = \frac{\rho^2[1 - \rho^N - N\rho^{N-1}(1 - \rho)]}{(1 - \rho)(1 - \rho^{N+1})} = 1.668$$

$$W = \frac{1 - \rho^N - N\rho^N(1 - \rho)}{\mu(1 - \rho)(1 - \rho^N)} = 2.452$$

$$W_q = \frac{[1 - \rho^N - N\rho^{N-1}(1 - \rho)]}{\mu(1 - \rho)(1 - \rho^N)} = 1.702$$

$$U = L - L_q = 2.403 - 1.668 = 0.735$$

Values of L and L_q can be obtained using Figures 10.28 and 10.29 for the $(M/M/c):(GD/\infty/\infty)$ case. Notice, when $N = \infty$, then $U = \rho$. Figures 10.28 and 10.29 indicate an important concept when designing material handling systems. Namely, as equipment utilization (ρ) increases, the waiting lines for material to be moved increases exponentially. It is a common tendency for managers to evaluate material handling effectiveness on the basis of equipment utilization; in many cases, that is counterproductive for the overall system.

Example 10.45

A fleet of lift trucks is used to place pallet loads of materials in storage and to retrieve pallet loads from storage. Job requisitions are received in a Poisson fashion at a rate of 10 per hour. The time required to perform the required material handling for a requisition is exponentially distributed with a mean of 15 min. Management wishes for the lift trucks to be utilized at least 80% of the time; hence, $\rho \ge 0.80$ or $\lambda/c\mu \ge 0.80$, which means $c \le 10/4(0.80)$ or $c \le 3.125$. Because an integer number of lift trucks must be assigned, $c \le 3$.

The operating characteristics for the system having $\lambda = 10$, $\mu = 4$, $c = 3$, and $N = \infty$ are obtained from Table 10.23 as follows:

$$P_0 = \left[\frac{(c\rho)^c}{c!(1 - \rho)} + \sum_{n=0}^{c-1} \frac{(c\rho)^n}{n!} \right]^{-1} = \left[\frac{(2.5)^3}{6(0.1667)} + 1 + 2.5 + \frac{(2.5)^2}{2} \right]^{-1}$$

$$= 0.0449$$

$$L_q = \frac{\rho(c\rho)^c}{c!(1 - \rho)^2} P_0 = 3.5078 \text{ requisitions}$$

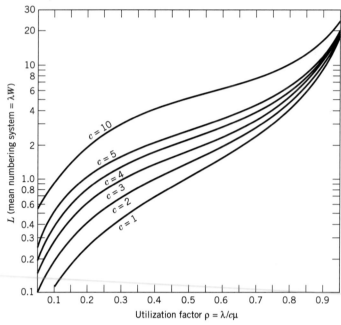

Figure 10.28 L (mean number in system $= \lambda W$) for different values of c versus the utilization factor ρ. (*From White et al. [62] with permission*)

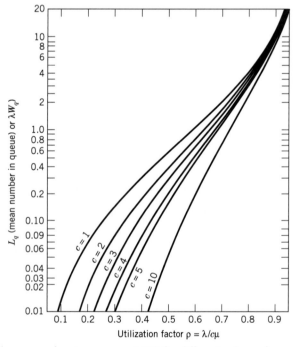

Figure 10.29 L_q (mean number in queue, $= \lambda W_q$) for different values of c versus the utilization factor ρ. (*From White et al. [62] with permission*)

$$L = L_q + \frac{\lambda}{\mu} = 6.0078 \text{ requisitions}$$

$$W_q = \frac{(c\rho)^c P_0}{c!c\mu(1 - \rho)^2} = 0.35078 \text{ hr} = 21.0468 \text{ min per requisition}$$

$$W = W_q + \frac{1}{\mu} = 21.0468 + 15.0 = 36.0468 \text{ minutes per requisition}$$

When the manager realizes that the average time a requisition is delayed before processing begins equals approximately 21 min, it is decided that the objective should be to have orders wait, on the average, less than 10 min. Hence, $W_q \le 10$ minutes per requisition

or

$$L_q = W_q \lambda \le \frac{10 \text{ min}}{\text{req.}} \left(\frac{10 \text{ req.}}{\text{hour}} \right) \left(\frac{\text{hour}}{60 \text{ min}} \right) = 1.667$$

From Figure 10.29 it is seen that for $c = 4$, $\rho = 0.626$, and $L_q \approx 0.60$. Hence, $c = 4$ should provide the level of service desired, yet the utilization of the lift trucks will be only 62.5% rather than the 80% figure desired initially.

$(M^b/M/1):(GD/\infty/\infty)$ Results

In a number of situations customers do not arrive singly, but in groups. Such a situation is termed a bulk arrival queueing system. Likewise, customers can often be serviced in groups rather than individually. In this section, we consider the bulk arrival situation under the usual Poisson assumptions; we also assume unlimited waiting space and a single server.

The arrival of b customers per arrival instant results in the following operating characteristics:

$$L = \frac{\rho(1 + b)}{2(1 - \rho)} \tag{10.111}$$

$$L_q = L - \frac{b\lambda}{\mu} \tag{10.112}$$

$$W = \frac{1 + b}{2\mu(1 - \rho)} \tag{10.113}$$

$$W_q = W - \frac{1}{\mu} \tag{10.114}$$

where $\rho = b\lambda/\mu$ is also the utilization of the single server. The probability the server will already be busy when the arrival occurs is ρ.

When the number of customers that arrive at a given arrival instant is a random variable, then the operating characteristics become

$$L = \frac{\lambda[V(b) + E^2(b) + E(b)]}{2[\mu - \lambda E(b)]} \tag{10.115}$$

$$L_q = L - \frac{\lambda E(b)}{\mu} \tag{10.116}$$

$$W = \frac{V(b) + E^2(b) + E(b)}{2E(b)[\mu - \lambda E(b)]}$$

(10.117)

$$W_q = W - \frac{1}{\mu}$$

(10.118)

where $E(b)$ and $V(b)$ are the expected value of b and the variance of b, respectively, and $\rho = \lambda E(b)/\mu$.

Example 10.46

Truckloads of material arrive at a receiving dock according to a Poisson process at a rate of two per hour. The number of unit loads per truck equals 10. The time required to remove an individual unit load from a truck is exponentially distributed with a mean of 2.4 min. A single lift truck is used in the receiving area for transporting unit loads from the truck to a conveyor that delivers the unit load to storage.

In this situation $b = 10$, $\lambda = 2$, $\mu = 25$, and $c = 1$. Hence, $\rho = 0.80$ and

$$L = \frac{\rho(1 + b)}{2(1 - \rho)} = \frac{0.8(11)}{2(0.2)} = 22 \text{ unit loads}$$

$$L_q = L - \frac{b\lambda}{\mu} = 22 - \frac{10(2)}{25} = 21.2 \text{ unit loads}$$

$$W = \frac{1 + b}{2\mu(1 - \rho)} = \frac{11}{2(25)(0.2)} = 1.1 \text{ hr per unit load}$$

$$W_q = W - \frac{1}{\mu} = 1.1 - 0.04 = 1.06 \text{ hr per unit load}$$

Example 10.47

Suppose the number of unit loads per truck is a random variable. Given the data from the previous example, consider the situation where b is binomially distributed with a mean of 10 and a variance of 5. Since $E(b) = 10$, $V(b) = 5$, $\lambda = 2$, $\mu = 25$, and $c = 1$, then $\rho = 0.80$ and

$$L = \frac{2[5 + 100 + 10]}{2[25 - 2(10)]} = 23 \text{ unit loads}$$

$$L_q = 23 - 0.80 = 22.20 \text{ unit loads}$$

$$W = \frac{5 + 100 + 10}{2(10)[25 - 2(10)]} = 1.15 \text{ hr per unit load}$$

$$W_q = W - \frac{1}{\mu} = 1.15 - 0.04 = 1.11 \text{ hr per unit load}$$

(M/M/c): (GD/K/K) Results

The previous analysis of waiting lines was based on an assumption that the population of customers is infinitely large. When the customer population is sufficiently large that the probability of an arrival occurring is not dependent on the number of customers in the waiting line, then the assumption of an infinite population is appropriate.

In many cases the number of customers is sufficiently small that the probability of an arrival is dependent on the number of customers in the queueing system. In such a situation, it is recommended that the arrival rate for an individual customer be determined. In particular, λ is defined as the reciprocal of the time between a customer completing service and requiring service the next time. If the number of customers is K and each customer has an identical arrival rate, then the arrival rate at the queueing system is given by

$$\lambda_n = (K - n)\lambda \qquad n = 0, 1, \ldots, K \tag{10.119}$$

Furthermore, since we are assuming service times are exponentially distributed, then

$$\mu_n = \begin{cases} n\mu & n = 0, 1, \ldots, c - 1 \\ c\mu & n = c, c + 1, \ldots, K \end{cases} \tag{10.120}$$

Substituting Equations 12.113 and 12.114 into Equation 12.92 gives

$$Pn = \begin{cases} \binom{K}{n}(c\rho)^n P_0 & n = 0, 1, \ldots, c - 1 \\ \binom{K}{n}\dfrac{n!\,c\rho^n}{c!}P_0 & n = c, c + 1, \ldots, K \\ 0 & n > K \end{cases} \tag{10.121}$$

where, as before, $\rho = \lambda/c\mu$ and

$$P_0 = \left[\sum_{n=0}^{c-1} \binom{K}{n}(c\rho)^n + \sum_{n=c}^{K} \binom{K}{n}\dfrac{n!\,c^c\rho^n}{c!} \right]^{-1} \tag{10.122}$$

Peck and Hazelwood [50] have developed extensive tables of computational results for the $(M/M/c) : (GD/K/K)$ queue which can be used to determine the operating characteristics for a given situation. A sample of the results obtained by Peck and Hazelwood is given in Table 10.25. To interpret these data, let the service factor X be defined as

$$X = \frac{\lambda}{\lambda + \mu} \tag{10.123}$$

and let

$$F = \frac{W - W_q + \lambda^{-1}}{W + \lambda^{-1}} \tag{10.124}$$

Peck and Hazelwood provide values of F for various combinations of X, K, and c. Based on the value of F, the following values are obtained.

$$L = K[1 - F(1 - X)] \tag{10.125}$$

$$L_q = K(1 - F) \tag{10.126}$$

$$W = \frac{1 - F(1 - X)}{\mu FX} \tag{10.127}$$

$$W_q = \frac{1 - F}{\mu FX} \tag{10.128}$$

Table 10.25 Results Obtained by Peck and Hazelwood for (M/M/c):(GD/K/K)

K	c	D	F	K	c	D	F	K	c	D	F
		X=0.05		12	1	0.879	0.764	10	1	0.987	0.497
4	1	0.149	0.992		2	0.361	0.970		2	0.692	0.854
5	1	0.198	0.989		3	0.098	0.996		3	0.300	0.968
6	1	0.247	0.985		4	0.019	0.999		4	0.092	0.994
	2	0.023	0.999	14	1	0.946	0.690		5	0.020	0.999
7	1	0.296	0.981		2	0.469	0.954	12	1	0.998	0.416
	2	0.034	0.999		3	0.151	0.992		2	0.841	0.778
8	1	0.343	0.977		4	0.036	0.999		3	0.459	0.940
	2	0.046	0.999	16	1	0.980	0.618		4	0.180	0.986
9	1	0.391	0.972		2	0.576	0.935		5	0.054	0.997
	2	0.061	0.998		3	0.214	0.988	14	2	0.934	0.697
10	1	0.437	0.967		4	0.060	0.998		3	0.619	0.902
	2	0.076	0.998	18	1	0.994	0.554		4	0.295	0.973
12	1	0.528	0.954		2	0.680	0.909		5	0.109	0.993
	2	0.111	0.996		3	0.285	0.983		6	0.032	0.999
14	1	0.615	0.939		4	0.092	0.997	16	2	0.978	0.621
	2	0.151	0.995		5	0.024	0.999		3	0.760	0.854
	3	0.026	0.999	20	1	0.999	0.500		4	0.426	0.954
16	1	0.697	0.919		2	0.773	0.878		5	0.187	0.987
	2	0.195	0.993		3	0.363	0.975		6	0.066	0.997
	3	0.039	0.999		4	0.131	0.995		7	0.019	0.999
18	1	0.772	0.895		5	0.038	0.999	18	2	0.994	0.555
	2	0.243	0.991	25	2	0.934	0.776		3	0.868	0.797
	3	0.054	0.999		3	0.572	0.947		4	0.563	0.928
20	1	0.837	0.866		4	0.258	0.987		5	0.284	0.977
	2	0.293	0.988		5	0.096	0.997		6	0.118	0.993
	3	0.073	0.988		6	0.030	0.999		7	0.040	0.998
25	1	0.950	0.771	30	2	0.991	0.664	20	2	0.999	0.500
	2	0.429	0.978		3	0.771	0.899		3	0.938	0.736
	3	0.132	0.997		4	0.421	0.973		4	0.693	0.895
	4	0.032	0.999		5	0.187	0.993		5	0.397	0.963
30	1	0.992	0.663		6	0.071	0.998		6	0.187	0.988
	2	0.571	0.963			X=0.20			7	0.074	0.997
	3	0.208	0.994	4	1	0.549	0.862		8	0.025	0.999
	4	0.060	0.999		2	0.108	0.988	25	3	0.996	0.599
		X=0.10			3	0.008	0.999		4	0.920	0.783
4	1	0.294	0.965	5	1	0.689	0.801		5	0.693	0.905
	2	0.028	0.999		2	0.194	0.976		6	0.424	0.963
5	1	0.386	0.950		3	0.028	0.998		7	0.221	0.987
	2	0.054	0.997	6	1	0.801	0.736		8	0.100	0.995
6	1	0.475	0.932		2	0.291	0.961		9	0.039	0.999
	2	0.086	0.995		3	0.060	0.995	30	4	0.991	0.665
7	1	0.559	0.912	7	1	0.883	0.669		5	0.905	0.814
	2	0.123	0.992		2	0.395	0.941		6	0.693	0.913
	3	0.016	0.999		3	0.105	0.991		7	0.446	0.963
8	1	0.638	0.889		4	0.017	0.999		8	0.249	0.985
	2	0.165	0.989	8	1	0.937	0.606		9	0.123	0.995
	3	0.027	0.999		2	0.499	0.916		10	0.054	0.998
9	1	0.711	0.862		3	0.162	0.985		11	0.021	0.999
	2	0.210	0.985		4	0.035	0.998				
	3	0.040	0.998	9	1	0.970	0.548				
10	1	0.776	0.832		2	0.599	0.887				
	2	0.258	0.981		3	0.227	0.978				
	3	0.056	0.998		4	0.060	0.996				

Source: From Peck and Hazelwood [50].

Example 10.48

A distribution center delivers material to the retail stores in its service region. Twelve trucks are used for delivery. The time required to load a truck is exponentially distributed with a mean of 40 min; the time required to deliver a load and return is exponentially distributed with a mean of 6 hr. There are two crews available for loading trucks. The distribution center operates continuously, that is, 24 hr per day and 7 days per week.

Based on the data for the distribution center $K = 12$, $\lambda = \frac{1}{6}$, $\mu = \frac{3}{2}$, and $c = 2$; hence, $X = 0.1$. From Table 10.25, it is found that $F = 0.970$. Therefore,

$$L = K[1 - F(1 - X)] = 12[1 - 0.970(1 - 0.1)]$$
$$= 1.524 \text{ trucks}$$

$$L_q = K(1 - F) = 12(1 - 0.970)$$
$$= 0.360 \text{ trucks}$$

$$W = \frac{1 - F(1 - X)}{\mu F X} = \frac{1 - 0.970(1 - 0.1)}{1.5(0.970)(0.1)}$$
$$= 0.873 \text{ hr per truck}$$

$$W_q = \frac{1 - F}{\mu F X} = \frac{1 - 0.970}{1.5(0.970)(0.1)}$$
$$= 0.206 \text{ hr per truck}$$

The results given above are based on the assumption that either of the two crews can load any truck. It has been suggested that trucks 1, . . . , 6 be serviced by crew 1 and trucks 7, . . . , 12 be serviced by crew 2. In such a case $K = 6$, $\lambda = \frac{1}{6}$, $\mu = \frac{3}{2}$, $c = 1$, and $X = 0.1$ for the first group of trucks. From Table 10.25, $F = 0.932$. Hence,

$$L = 6[1 - 0.932(1 - 0.1)] = 0.9672 \text{ trucks}$$
$$L_q = 6(1 - 0.932) = 0.408 \text{ trucks}$$
$$W = \frac{1 - 0.932(1 - 0.1)}{1.5(0.932)(0.1)} = 1.153 \text{ hr per truck}$$
$$W_q = \frac{1 - 0.932}{1.5(0.932)(0.1)} = 0.486 \text{ hr per truck}$$

Since there are two crews, then the average number of trucks in the total system is $L = 2(0.9672) = 1.9344$ trucks; likewise, the average number of trucks waiting to be loaded is $L_q = 2(0.408) = 0.816$ trucks. Under the proposed plan a complete delivery cycle by a truck will require $6 + 1.153 = 7.153$ hr; therefore, a truck can be expected to make $7(24)/7.153 = 23.49$ deliveries per week. Under the present plan a truck can be expected to make $7(24)/6.873 = 24.44$ deliveries per week.

Non-Poisson Queues

In general, when arrivals and/or services are not Poisson distributed, mathematical results are difficult to obtain and simulation is often used. However, there are some simple non-Poisson queues for which the operational characteristics are known. We will consider three non-Poisson queueing systems: $(M/G/1){:}(GD/\infty/\infty)$, $(D/M/1){:}(GD/\infty/\infty)$, and $(M/G/c) : (GD/c/\infty)$. The first case allows any general service time distribution; the second case is appropriate when the time between consecutive arrivals is deterministic (constant); and the third case allows any general service time distribution when no waiting is provided.

$(M/G/1) : (GD/\infty/\infty)$ Results

Assume arrivals are Poisson distributed and a single server is present. If the average service time is μ^{-1} and the variance of service time is σ^2, then the following expressions are available for determining the operating characteristics of the system:

$$L = \rho + \frac{\lambda^2\sigma^2 + \rho^2}{2(1 - \rho)} \qquad (10.129)$$

$$L_q = \frac{\lambda^2\sigma^2 + \rho^2}{2(1 - \rho)} \qquad (10.130)$$

$$W = \frac{1}{\mu} + \frac{\lambda\left(\sigma^2 + \dfrac{1}{\mu^2}\right)}{2(1 - \rho)} \qquad (10.131)$$

$$W_q = \frac{\lambda\left(\sigma^2 + \dfrac{1}{\mu^2}\right)}{2(1 - \rho)} \qquad (10.132)$$

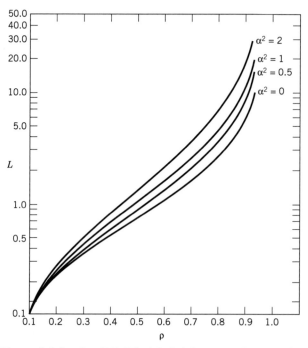

Figure 10.30   Values of L for the $(M/G/1):(GD/\infty/\infty)$ queue. (*From White et al. [62] with permission*)

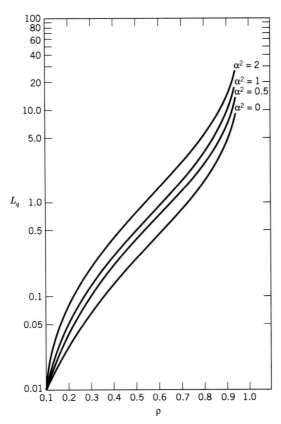

Figure 10.31   Values of L_q for the $(M/G/1):(GD/\infty/\infty])$ queue. (*From White et al. [62] with permission*)

where $\rho = \lambda/\mu$. Letting $\alpha^2 = \mu^2\sigma^2$, values of L and L_q are depicted in Figures 10.30 and 10.31 for various values of α^2. Notice that when service times are constant then $\alpha^2 = 0$ and when service times are exponentially distributed then $\alpha^2 = 1$.

Example 10.49

Unit loads arrive randomly at a shrinkwrap machine. An average of 15 unit loads arrive per hour at a Poisson rate. The time required for a unit load to be processed through the shrink wrapping operation is a constant of 2.5 min. The operating characteristics are obtained using Equations 10.129 through 10.132 by letting $\lambda = 15$, $\mu = 24$, $\sigma^2 = 0$, and $\rho = 0.625$.

$$L = 0.625 + \frac{(15)^2(0) + (0.625)^2}{2(1 - 0.625)} = 1.1458 \text{ unit loads}$$

$$L_q = \frac{(15)^2(0) + (0.625)^2}{2(1 - 0.625)} = 0.5208 \text{ unit loads}$$

$$W = \frac{L}{\lambda} = 0.07639 \text{ hr per unit load}$$

$$= 4.583 \text{ min per unit load}$$

$$W_q = W = \frac{1}{\mu} = 2.083 \text{ min per unit load}$$

$(D/M/1):(GD/\infty/\infty)$ *Results*

When the time between consecutive arrivals is deterministic (constant) and services are Poisson, then for the single server queue the following results are available:

$$L = \frac{\theta}{1-\theta} \qquad (10.133)$$

$$L_q = \frac{\theta^2}{1-\theta} \qquad (10.134)$$

$$W = \frac{1}{\mu(1-\theta)} \qquad (10.135)$$

$$W_q = \frac{\theta}{\mu(1-\theta)} \qquad (10.136)$$

where θ is a fractional-valued parameter $(0 < \theta < 1)$ satisfying the relation

$$\theta = e^{-(1-\theta)/\rho} \qquad (10.137)$$

with $\rho = \lambda/\mu$ and the time between consecutive arrivals is λ^{-1}. Values of θ are given in Figure 10.32 for various values of ρ.

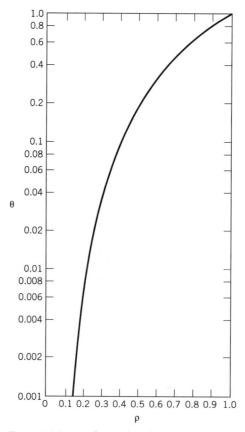

Figure 10.32 Relationship between θ and ρ.

Since a more accurate estimate of θ might be needed, the following numerical procedure can be used to refine the value obtained from Figure 10.32. Let θ_0 be the value obtained from Figure 10.32. An updated estimate, θ_1, can be obtained from the relation,

$$\theta_1 = e^{-(1-\theta_0)/\rho}$$

More generally, the kth estimate of θ can be obtained from

$$\theta_k = e^{-(1-\theta_{k-1})/\rho} \tag{10.138}$$

The iterative process can be continued until successive estimates of θ are less than, say, 0.001.

Example 10.50

Parts are supplied by a conveyor to an inspection station from a numerically controlled milling machine. Because of the insignificant variation in machining time and conveyor speed, parts arrive at the inspection station in a deterministic fashion at a rate of 6 per minute. Inspection is performed manually and the inspection times are exponentially distributed with a mean of 7.5 sec. In this situation, $\lambda = 6$, $\mu = 8$, $c = 1$, and $\rho = 0.75$. From Figure 10.32, $\theta = 0.55$. Therefore, if an estimate for θ is desired such that $|\theta_k - \theta_{k-1}| < 0.001$ then, for the example with $\rho = 0.75$, the following iterative process can be used:

$$\theta_k = e^{-(1-\theta_{k-1})/0.75}$$

Calculations show that if $\theta_0 = 0.55$, then $\theta_1 = 0.5488116$, $\theta_2 = 0.5479427$, and $\theta_3 = 0.5473083$. Therefore, substituting 0.5473083 for θ in Equations 10.133 through 10.136 yields

$$L = \frac{0.5473083}{0.4526917} = 1.209 \text{ parts}$$

$$L_q = \frac{0.5473083}{0.4526917} = 0.662 \text{ parts}$$

$$W = \frac{1}{8(0.4526917)} = 0.276 \text{ min per part}$$

$$W_q = \frac{0.5473083}{8(0.4526917)} = 0.151 \text{ min per part}$$

$(M/G/c):(GD/c/\infty)$ Results

When no waiting space is provided and arrivals are Poisson distributed, the situation gives rise to a formula called *Erlang's loss formula*,

$$P_c = \frac{(\lambda/\mu)^c/c!}{\sum_{k=0}^{c} (\lambda/\mu)^k/k!} \tag{10.139}$$

which gives the probability all servers are busy in a $(M/G/c):(GD/c/\infty)$ queue. Values of P_c are provided in Figure 10.33 for selected values of c and λ/μ. (Multiplying both the numerator and denominator of Equation 10.139 by exp-(λ/μ) gives an interesting result. Namely, the numerator is the probability of a Poisson

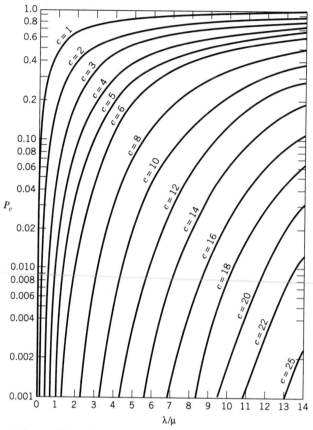

Figure 10.33   Values of Erlang's loss formula. (*From White et al. [62] with permission*)

distributed random variable with an expected value of λ/μ equaling c; the denominator is the probability of c or less occurrences for the Poisson distributed random variable. Hence, by consulting tables of the Poisson distribution, one can easily compute the value of P_c.)

The operating characteristics of the $(M/G/c):(GD/c/\infty)$ queue are easily obtained since $L_q = 0$, $W_q = 0$, and $W = 1/\mu$. The expected number in the system is given by

$$L = \lambda(1 - P_c)/\mu \qquad\qquad (10.140)$$

Example 10.51

A conveyor belt is used to feed parts to 10 workstations located along the conveyor belt. For convenience it is assumed that travel along the belt between workstations is instantaneous. Parts arrive at the first workstation at a Poisson rate of 12 per minute. If the first station is busy, then the part moves to the second station, and so forth, until, finally, if all 10 workstations are busy, then the part passes all workstations and is accumulated until the second-shift when they are processed by other workstations.

If the service times at the individual workstations are normally distributed with a mean of 45 sec and a standard deviation of 5 sec, what percentage of the parts will overflow the system? In this case $\lambda = 12$, $\mu = \frac{4}{3}$, $c = 10$, and from Figure 10.33, $P_c = 0.17$ or approximately 17% of the parts will pass all 10 workstations. The average number of busy workstations is equal to L, or $12(.83)(0.75) = 7.47$ busy workstations.

Having presented a number of waiting line models for use in analyzing facilities planning requirements for buffer spaces, waiting lines, and in-process storage, we would be negligent if we did not admit that many situations exist that do not fit the assumptions underlying the models we presented. In such a case, what should you do? Three alternatives come to mind: utilize a more advanced waiting line model, one that more closely fits your situation; utilize one of the simple models we presented, recognizing that the solutions obtained are, at best, approximations; and utilize a simulation model, rather than a waiting line model.

10.10 SIMULATION MODELS

Considerable improvements have occurred in the past 10 to 15 years in developing simulation packages for use in modeling material flow problems. Indeed, simulation technology has evolved to the point that serious facilities planners will use simulation one or more times during the facilities planning cycle. Relatively sophisticated simulations can be performed today using microcomputers; hand-held pocket computers can be used to perform rudimentary simulations; spreadsheet software typically includes random number generators for use in performing simulations. Advances in software graphics allow simulation results to be displayed using animated representations of systems.

Due to the explosion of options available for performing simulations of flow processes, we cannot provide more than a brief introduction to simulation. Because we can only scratch the surface, it was tempting to omit coverage of simulation. However, its power and increasingly important role in facilities planning mandated that we provide at least an introduction for those unfamiliar with simulation. So, for those who are knowledgeable about simulation, our apologies for taking up space in your book!

Simulation involves building a model of a system and experimenting with the model to determine how the system reacts to various conditions. Simulation does not provide an optimum solution, since it is a descriptive model. It simply provides us with a mechanism to use in understanding and, perhaps, predicting the behavior of a system. By asking enough "what if" questions, the configuration of the system that best satisfies some criteria can be chosen.

Some of the major reasons for using simulation are:

1. When a mathematical solution cannot be obtained easily or at all.
2. Selling the facilities plan to management.
3. Explaining to operating personnel how a proposed system will function.
4. Testing the feasibility of a proposed system.
5. Developing throughput and storage requirements.

6.   Validating mathematical models.

7.   Predicting the impact of a change in the physical system, the environment, or operating procedures.

Simulation can result in improved understanding of the facilities plan. Some ways in which this can occur are as follows:

1.   The process of creating a simulation model requires a detailed understanding and documentation of the activity being simulated. For example, the flow of information through a warehouse typically seems straightforward at first glance. Once a simulation of this information flow is developed, however, many exceptions and alternative information flows are identified.

2.   The teaching of some concepts is quite difficult because of the complex inter-relationship among variables. Simulation can be used in a "gaming" sense to relate these complex interrelationships. For example, determining the number and location of distribution centers to serve a market is based on the customer service that can be provided, the transportation cost, and the cost of carrying inventories.

3.   The orientation of employees on a system can often create significant problems. Simulation can be used to orientate existing employees to new systems or new employees to existing systems. For example, the pattern of loading parts onto a power-and-free conveyor system has a significant impact on the balance of effort in different paint booths. If parts are loaded incorrectly, excessive queues in front of some paint stations stop the conveyor while other paint stations are idle. This imbalance impacts not only the paint booths but also drying and assembly sta-tions. In an effort to gain experience in loading the conveyor prior to actually loading parts, an operator may be trained on a simulator where the only impact of loading errors will be a greater understanding of the system.

It is virtually impossible to plan a facility of any magnitude without conduct-ing some type of simulation. The question is not whether a simulation is to be done, but whether it is to be done formally or informally, and what will be the scope of the simulation. If the simulation is to be done informally, such as a mental simula-tion of a lift truck picking up a load and traveling down an aisle, a detailed design obviously will not be required. Formal simulation experiments requiring detailed designs are often an important portion of the facilities planning process.

The scope of facilities planning simulation experiments may include an indi-vidual workstation, a piece of equipment, an entire facility, or a whole series of facilities. A very brief example of the use of simulation is given in the example below. For one desiring to understand the process of simulation, this example illus-trates the process. For one interested in learning to develop simulation models, enough examples cannot be cited to properly present the "art" of simulation. The more serious reader is referred to [3], [41], [52], and [55], among others.

Example 10.52

An existing warehouse is 50 years old and is to be abandoned as it requires considerable renovation to promote efficient warehousing practice. A warehouse is to be built adjacent to the existing site to handle all existing business. The question that has been asked is,

Table 10.26   *Truck Arrivals Between 8:00 A.M. and 10:00 A.M., 11:00 A.M. and 2:00 P.M., and 3:00 P.M. and 5:00 P.M.*

Time Between Arrivals (Hours)	Relative Frequency	Cumulative Frequency
0− .25	0.02	2
0.251− .50	0.07	9
0.501− .75	0.19	28
0.751−1.00	0.34	62
1.001−1.25	0.26	88
1.251−1.50	0.08	96
1.501−1.75	0.03	99
1.751−2.00	0.01	100

"How many dock bays should be constructed?" The first thought is that waiting line analysis could be used to answer this question. Unfortunately, a surge of vehicles arrives in the middle of the morning and the middle of the afternoon. This prevents defining arrival rates in a manner acceptable for mathematical analysis. The best method of determining the number of dock bays appears to be simulation.

A time study is made of the arrival of trucks to obtain the data given in Tables 10.26 and 10.27. The time required to load and unload vehicles does not vary with the time of day. A time study of these activities results in the data given in Table 10.28.

The truck spotting time is observed to be a constant of 0.1 hr. The existing warehouse has three dock bays. Considerable knowledge is available with respect to "typical" operations for three docks. Hence, the simulation model can be validated with three dock bays. The logic underlying the simulation model is as follows:

1. Initialize the model. Begin with three dock bays and add one bay each time it is reinitialized.
2. Generate a series of random numbers from 0 to 99.

Table 10.27   *Truck Arrivals Between 10:00 A.M. and 11:00 A.M., and 2:00 P.M. and 3:00 P.M.*

Time Between Arrivals (Hours)	Relative Frequency	Cumulative Frequency
0−0.25	0.36	36
0.251−0.50	0.41	77
0.501−0.75	0.23	100

Table 10.28   *Truck Loading and Unloading Times*

Unloading Time (Hours)	Relative Frequency	Cumulative Frequency
0− .5	0.01	1
0.51−1.0	0.10	11
1.01−1.5	0.22	33
1.51−2.0	0.20	53
2.01−2.5	0.20	73
2.51−3.0	0.18	91
3.01−3.5	0.06	97
3.51−4.0	0.02	99
4.01−4.5	0.01	100

3. Transform the random numbers to a series of truck interarrival times by relating random numbers to the cumulative frequencies in Tables 10.26 and 10.27. For example, in Table 10.27, random numbers 0 to 35 would result in an interarrival time of 0.125 hr, random numbers 36 to 76 would result in an interarrival time of 0.375 hr and random numbers 77 to 99 would result in an interarrival time of 0.625 hr.

4. Generate a series of random numbers from 0 to 99.

5. Transform the random numbers to a series of truck loading and unloading times by relating the random number to the cumulative frequencies in Table 10.28.

6. Assign trucks to dock bays, or if unavailable, to a queue. Unload the trucks, dispatch waiting trucks, and assign spotting time. Perform the truck loadings and unloadings for the entire day and maintain statistics.

7. Determine whether steady state is reached. If so, and six dock-bays have been considered, print all statistics and terminate the model.

8. If steady state is reached and less than six dock bays have been considered, return to step 1.

9. If steady state is not reached, return to step 6 and simulate another day's operation.

By running this model, the types of data that are available are

Factor	Number of Bays			
	3	4	5	6
Average truck waiting time (minutes)	46.3	19.6	3.2	1.6
Longest truck waiting time (minutes)	60.4	26.1	4.3	2.0
Truck waiting time variance (minutes^2)	4.1	1.9	.2	.07
Average time truck spent at warehouse (minutes)	167.4	139.7	124.2	116.3
Average dock bay utilization (percentage)	82%	61%	49%	41%

The data the model generated for three dock bays is seen to be consistent with what is experienced with the present operation. Therefore, the model is considered to be a valid representation of the truck being loaded and unloaded for various numbers of dock bays. The data presented for four, five, and six dock bays can be evaluated by management and a decision reached with respect to the number of dock bays to be built for the new facility.

Simulation Software

Simulation models are often implemented using a specialized simulation language, rather than more general programming languages such as BASIC, C, PASCAL, or FORTRAN. Indeed, there are several simulation languages that have been designed with material flow and facilities planning targeted as the principal application focus.

Among the simulation languages that are either designed for or well-suited for facilities planning and simulation of material handling/manufacturing systems are ARENA, AutoMod, eM-Plant, Factory Explorer, GPSS/H (and SLX), GPSS World for Windows, MAST Simulation Environment, ProModel 2001, Quest, Simscript II.5, Simul8, Taylor ED, and Witness. For information on some of the above packages, the reader may refer to [41] or annual listings published by *IIE Solutions* and *OR/MS Today*.

In recent years, considerable improvements have occurred in the development of animation and 3-D modeling software to accompany the simulation packages cited above. Indeed, most of the software cited above includes animation and/or 3-D modeling capabilities.

As noted in Chapter 6, several software packages have been developed to facilitate facilities planning. Likewise, a number of simulation models have been developed for material handling applications [5], [15], [19], [21], [33], [36], [38], [58]. Among the material handling technologies commonly targeted for simulation studies are automated guided vehicles (AGV) and automated storage and retrieval systems (AR/RS); however, it is safe to assert that simulation applications have included all material handling technologies in use [7], [26], [28], [34], [51], [53], [57], [59], [60], [63].

10.11 SUMMARY

In this, the longest chapter, we attempted to provide you with a wide range of quantitative/analytical models that can be used in facilities planning. Some of the models have broad application, whereas others are constrained to very specific applications. Likewise, some are relatively simple representations of complex applications, while others more accurately represent the situation being studied.

Although the range of problems addressed in the chapter is sizeable, we make no claim to having covered all the pertinent models. Indeed, we are unable to assert that we included the most important and the most relevant. Why? Because new models are being developed even as we write the chapter.

Our objective for this chapter was threefold: (1) within one chapter, expose you to the richness of the world of quantitative/analytical modeling in facilities planning; (2) provide you with a large buffet of relatively simple models to launch you on your facilities planning journey and to whet your appetite for more; and (3) present models that can provide you with additional insights regarding trade-offs, sensitivities, and data requirements in facilities planning. We hope we accomplished our objective *as it relates to you!*

BIBLIOGRAPHY

1. Anstreicher, K. M., and Brixius, N. W., "A New Bound for the Quadratic Assignment Problem Based on Convex Quadratic Programming," *Mathematical Programming, Series A*, vol. 89, 2001, pp. 341–357.

2. Askin, R. G., and Standridge, C. R., *Modeling and Analysis of Manufacturing Systems*, Wiley, New York, 1993.

3. Banks, J., Carson, J. S. II, and Nelson, B. L., *Discrete-Event System Simulation*, 2nd ed., Prentice-Hall, Englewood Cliffs, NJ, 1996.

4. Bartholdi, J. J., and Platzman, L. K., "Decentralized Control of Automated Guided Vehicles on a Simple Loop," *IIE Transactions*, vol. 21, no. 1, 1989, pp. 76–81.

5. Basnet, C. B., Karacal, S. C., and Beaumariage, T. G., "Experiences in Developing Object-Oriented Modeling Environment for Manufacturing Systems," *1990 Winter*

Simulation Conference Proceedings, Association for Computing Machinery, New York, 1990, pp. 477–481.

6. Bazaraa, M. S., and Elshafei, A. N., "Exact Branch-and-Bound Procedure for the Quadratic-Assignment Problem," *Naval Research Logistics,* vol. 26, no. 1, 1979, pp. 109–121.

7. Benjaafar, S., "Intelligent Simulation for Flexible Manufacturing Systems: An Integrated Approach," *Computers and Industrial Engineering,* vol. 22, no. 3, 1992, pp. 297–311.

8. Bozer, Y. A., "Optimizing Throughput Performance in Designing Order Picking Systems," Ph.D. dissertation, Georgia Institute of Technology, Atlanta, 1985.

9. Bozer, Y. A., Schorn, E. C., and Sharp, G. P., "Geometric Approaches to Solve the Chebyshev Traveling Salesman Problem," *IIE Transactions,* vol. 22, no. 3, 1990, pp. 238–254.

10. Bozer, Y. A., and Srinivasan, M. M., "Tandem Configuration for AGVS and the Analysis of Single Vehicle Loops," *IIE Transactions,* vol. 23, no. 1, 1991, pp. 72–82.

11. Bozer, Y. A., and Srinivasan, M. M., "Tandem AGV Systems: A Partitioning Algorithm and Performance Comparison with Conventional AGV Systems," *European Journal of Operational Research,* vol. 63, 1992, pp. 173–191.

12. Bozer, Y. A., and White, J. A., "Travel-Time Models for Automated Storage/Retrieval Systems," *IIE Transactions,* vol. 16, no. 4, 1984, pp. 329–338.

13. Bozer, Y. A., and White, J. A., "Design and Performance Models for End-of-Aisle Order Picking Systems," *Management Science,* vol. 36, no. 7, 1990, pp. 852–866.

14. Bozer, Y. A., and Yen, C., "Intelligent Dispatching Rules for Trip-based Material Handling Systems," *Journal of Manufacturing Systems,* vol. 15, no. 4, 1996, pp. 226–239.

15. Carrie, A. S., Moore, J. M., Roczniak, R., and Seppanen, J. J., "Graph Theory and Computer Aided Facilities Design," *OMEGA, International Journal of Management Science,* vol. 6 no. 4, 1978, pp. 353–361.

16. Çela, E., *The Quadratic Assignment Problem: Theory and Algorithms,* Kluwer, 1998.

17. Chow, W. M., "Design for Line Flexibility," *IIE Transactions,* vol. 18, no. 1, 1986, pp. 95–103.

18. DeMars, N. A., Matson, J. O., and White, J. A., "Optimizing Storage System Selection," in *Proceedings of the 4th International Conference on Automation in Warehousing,* Tokyo, Japan, 1982.

19. Drolet, J. R., Moodie, C. L., and Montreuil, B., "Object Oriented Simulation with Smalltalk-80: A Case Study," *1991 Winter Simulation Conference Proceedings,* eds. B. L. Nelson W. D. Kelton, and G. M. Clark, Association for Computing Machinery, New York, 1991, pp. 312–321.

20. Eilon, S., Watson-Gandy, C. D. T., and Christofides, N., *Distribution Management: Mathematical Modelling and Practical Analysis,* Hafner Publishing, New York, 1971.

21. Eom, J. K., *Selection, Design, Control of Material Handling Storage Systems: An Object-Oriented and Knowledge-based Approach,* Ph.D. dissertation, Department of Industrial Engineering, North Carolina State University, Raleigh, NC, 1992.

22. Foley, R. D., and Frazelle, E. H., "Analytical Results for Miniload Throughput and the Distribution of Dual Command Travel Time," *IIE Transactions,* vol. 23, no. 3, 1991, pp. 273–281.

23. Francis, R. L., McGinnis, L. F., and White, J. A., *Facility Layout and Location: An Analytical Approach,* 2nd ed., Prentice-Hall, Englewood Cliffs, NJ, 1992.

24. Francis, R. L., and White, J. A., *Facility Layout and Location: An Analytical Approach,* 1st, ed., Prentice-Hall, Englewood Cliffs, NJ, 1974.

25. Gaskins, R. L., and Tanchoco, J. M. A., "Flow Path Design for Automated Guided Vehicle Systems," *International Journal of Production Research,* vol. 25, no. 5, 1987, pp. 667–676.

26. Gobal, S. L., and Kasilingam, R. G., "A Simulation Model for Estimating Vehicle Requirements in Automated Guided Vehicle Systems," *Computers and Industrial Engineering,* vol. 21, nos. 1–4, 1991, pp. 623–627.

27.   Goetschalckx, M. P., and Ratliff, H. D., "Sequencing Picking Operations in a ManAboard Order Picking System," *Material Flow*, vol. 4, no. 4, 1988, pp. 255–264.

28.   Gonzalez, C. J., *A Design Procedure for Microload Automated Storage/Retrieval Systems*, Project Report, Department of Industrial Engineering, North Carolina State University, Raleigh, NC, 1990.

29.   Hahn, P. M., and Grant, T., "Lower Bounds for the Quadratic Assignment Problem Based upon a Dual Formulation," *Operations Research*, vol. 46, 1998, pp. 912–922.

30.   Hall, R. W., "Distance Approximations for Routing Manual Pickers in a Warehouse," *IIE Transactions*, vol. 25, no. 4, 1993, pp. 76–87.

31.   Heragu, S. S., "Recent Models and Techniques for Solving the Layout Problem," *European Journal of Operations Research*, vol. 57, 1992, pp. 136–144.

32.   Heragu, S. S., and Kusiak, A., "Machine Layout in Flexible Manufacturing Systems," *Operations Research*, vol. 36, no. 2, 1988, pp. 258–268.

33.   Hollinger, D., and Bell, G., "An Object-Oriented Approach for CIM Systems Specification and Simulation," *Computer Applications in Production and Engineering*, Elsevier Science Publishers B. V., New York, 1987.

34.   Jeyabalan, V., and Otto, N. C., "Simulation Models of Material Delivery Systems," *1991 Winter Simulation Conference*, eds. B. L. Nelson, W. D. Kelton, and G. M. Clark, Association for Computing Machinery, New York, 1991, pp. 356–364.

35.   Kaspi, M., and Tanchoco, J. M. A., "Optimal Flow Path Design of Unidirectional AGV Systems," *International Journal of Production Research*, vol. 28, no. 6, 1990, pp. 1023–1030.

36.   Kim, K. S., *Object-Oriented Design and Simulation for AGV Systems*, Ph.D. dissertation, Department of Industrial Engineering, North Carolina State University, Raleigh, NC, 1993.

37.   Kim, K. H., and Tanchoco, J. M. A., "Economical Design of Material Flow Paths," *International Journal of Production Research*, vol. 31, no. 6, 1993, pp. 1387–1407.

38.   King, C. U., and Fisher, E. L., "Object-Oriented Shop-Floor Design, Simulation, and Evaluation," *Proceedings of the Fall IIE Conference*, Atlanta, 1986.

39.   Kwo, T. T., "A Theory of Conveyors," *Management Science*, vol. 5, no. 1, 1958, pp. 51–71.

40.   Kwo, T. T., "A Method for Designing Irreversible Overhead Loop Conveyors," *Journal of Industrial Engineering*, vol 11, no. 6, 1960, pp. 459–466.

41.   Law, A. M., and Kelton, W. D., *Simulation Modeling and Analysis*, 2nd ed., McGraw-Hill, New York, 1991.

42.   Lawler, E. L., Lenstra, J. K., Rinnooy K., A. H., Shmoys, D. B., (eds.), *The Traveling Salesman Problem*, Wiley, 1985.

43.   *Matlab User's Guide*, The MathWorks, South Natick, MA, 1989.

44.   Mayer, H., "Introduction to Conveyor Theory," *Western Electric Engineer*, vol. 4, no. 1, 1960, pp. 43–47.

45.   Murty, K. G., *Operations Research: Deterministic Optimization Models*, Englewood Cliffs, NJ, Prentice Hall., 1995.

46.   Muth, E. J., "Analysis of Closed-Loop Conveyor Systems," *AIIE Transaction*, vol. 4, no. 2, 1972, pp. 134–143.

47.   Muth, E. J., "Analysis of Closed-Loop Conveyor Systems: The Discrete Flow Case," *AIIE Transactions*, vol. 6, no. 1, 1974, pp. 73–83.

48.   Muth, E. J., "Modelling and Systems Analysis of Multistation Closed-Loop Conveyors," *International Journal of Production Research*, vol. 13, no. 6, 1975, pp. 559–566.

49.   Park, J., "A Cost-Driven Partitioning Algorithm for Tandem Trip-based Material Handling Systems," Ph.D. dissertation, University of Michigan, Ann Arbor, 1995.

50.   Peck, L. G., and Hazelwood, R. N., *Finite Queueing Tables*, Wiley, New York, 1958.
      Pegden, C. D., Shannon, R. E., and Sadowski, R. P., *Introduction to Simulation Using SIMAN*, McGraw-Hill, New York, 1990.

51. Prasad, K., and Rangaswami, M., "Analysis of Different AGV Control Systems in an Integrated IC Manufacturing Facility Using Computer Simulation," *1998 Winter Simulation Conference Proceedings,* Association for Computing Machinery, New York, 1988, pp. 568–574.

52. Pritsker, A. A. B., *Introduction to Simulation and SLAM II,* 3rd ed., Wiley, New York, 1986.

53. Quiroz, M. A., *Simulation of a Microload Automated Storage/Retrieval System,* M.S. thesis, School of Industrial and Systems Engineering, Georgia Institute of Technology, Atlanta, 1986.

54. Ratliff, H. D., and Rosenthal, A. S., "Order-Picking in a Rectangular Warehouse: A Solvable Case of the Traveling Salesman Problem," *Operations Research,* vol. 31, no. 3, 1983, pp. 507–521.

55. Shannon, R. E., *Systems Simulation: The Art and Science,* Prentice-Hall, Englewood Cliffs, NJ, 1975.

56. Taha, H. A., *Operations Research: An Introduction,* 4th ed., New York, Macmillan, 1987.

57. Thomasma, T., and Ulgen, O. M., "Modeling of a Manufacturing Cell Using a Graphical Simulation System Based on Smalltalk-80," *1987 Winter Simulation Conference Proceedings,* Association for Computing Machinery, New York, 1987, pp. 683–691.

58. Ulgen, O. M., "Simulation Modeling in an Object-Oriented Environment using Smalltalk-80," *1986 Winter Simulation Conference Proceedings,* Association for Computing Machinery, New York, 1986, pp. 474–484.

59. Ulgen, O. M., and Kedia, P., "Using Simulation in Design of a Cellular Assembly Plant with Automatic Guided Vehicles," *1990 Winter Simulation Conference Proceedings,* Association for Computing Machinery, New York, 1990, pp. 683–691.

60. Webster, R. L., and Foster, D. F., "Building Flexible AGV and AS/RS System Models for Facility Design Phase Applications," *1990 Winter Simulation Conference Proceedings,* Association for Computing Machinery, New York, 1990, pp. 692–698.

61. White, J. A., and Muth, E. J., "Conveyor Theory: A Survey," *AIIE Transaction,* vol. 11, no. 41, 1979, pp. 270–277.

62. White, J. A., Schmidt, J. W., and Bennett, G. K., *Analysis of Queueing Systems,* Academic Press, New York, 1975.

63. Xu, Z., and Pirasteh, R. M., "Airline-Catering Plant Material Handling System Analysis with Simulation and Scaled Automation," *1991 Winter Simulation Conference Proceedings,* eds. B. L. Nelson and G. M. Clark, Association for Computing Machinery, New York, 1991, pp. 402–410.

64. Zollinger, H. A., "Planning, Evaluating, and Estimating Storage Systems," presented at the *Institute of Material Management Education First Annual Winter Seminar Series,* Orlando, February 1982.

65. *Considerations for Planning an Automated Storage/Retrieval System,* Material Handling Institute, Charlotte, North Carolina, 1982.

66. *Warehouse Modernization and Layout Planning Guide,* Department of the Navy, Naval Supply Systems Command, NAVSUP Publication 529, Washington, D. C., 1978.

PROBLEMS

10.1 Let four existing facilities by located at $P_1 = (0,10)$, $P_2 = (5,10)$, $P_3 = (5,15)$, and $P_4 = (10,5)$ with $w_1 = 15$, $w_2 = 20$, $w_3 = 5$, and $w_4 = 30$. Determine the optimum location for a single new facility when cost is proportional to rectilinear distance. Construct a contour line passing through the point having coordinates (10, 10).

10.2 The XYZ Company has six retail sales stores in the city of Raleigh. The company needs a new warehouse facility to service its retail stores. The location of the stores and the expected delivery per week from the warehouse to each store are:

Store	Location (miles)	Expected Deliveries
1	(1,0)	4
2	(2,5)	7
3	(3,8)	5
4	(1,6)	3
5	(−5,−1)	8
6	(−3,−3)	3

Assume that travel distance within the city of Lafayette is rectilinear and that after each delivery the delivery truck must return to the warehouse. If there are no restrictions on the warehouse location, where should it be located?

10.3 A new back-up power generator is to be located to serve a total of six precision machines in a manufacturing facility. Separate electrical cables are to be run from the generator to each machine. The locations of the six machines are $P_1 = (0,0)$, $P_2 = (30,90)$, $P_3 = (60,20)$, $P_4 = (20,80)$, $P_5 = (70,70)$ and $P_6 = (90,40)$. Determine the location for the generator that will minimize the total required length of the electrical cable. Assume rectilinear distance.

10.4 Six housing subdivisions within a city area are targeted for emergency service by a centralized fire station. Where should the new fire station be located such that the maximum rectilinear travel distance is minimized? The centroid locations (in miles) and total value of the houses in the subdivisions are as follows:

Subdivision	X-Coordinate	Y-Coordinate	Total Value
A	20	15	50 mil.
B	25	25	120 mil.
C	13	32	100 mil.
D	25	14	250 mil.
E	4	21	300 mil.
F	18	8	75 mil.

How does the answer change if all houses are valued equally?

10.5 The city council of Fayetteville has decided to locate an emergency response unit within the city. This unit is responsible for four housing sectors (A) and three major street intersections (P) as shown in the figure below. Assume that the weights are uniformly distributed over the housing sectors.

a. Determine the minimum location based on the weights given in the table below.
b. Determine the minimax location based on the weights given in the table below.

Housing Sector	Weight	Intersection	Weight
A1	10	P1	30
A2	15	P2	15
A3	20	P3	5
A4	30		

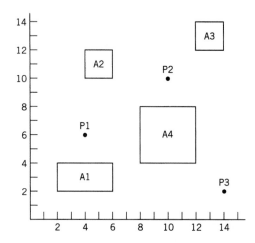

10.6   A new elementary school is needed in a suburban area of Detroit, Michigan. After extensive research, the school board has narrowed down its choice to three possible sites. The
locations for the current residential areas, expected students from each residential area,
and possible locations for the school are shown in the tables below. Determine the
optimal location for the new elementary school such that the total distance the students have to travel is minimized. It is fair to assume that the construction cost for all
three sites are similar and distance is measured rectilinearly.

Residences	X-Coordinate	Y-Coordinate	Weight
A	20	25	600
B	36	18	400
C	62	37	500
D	50	56	300
E	25	0	200

Possible sites for the new elementary school:

Possible Sites	X-Coordinate	Y-Coordinate
1	50	50
2	30	45
3	65	28

10.7   The Lafayette Exponent plans to rent building space for a new print shop within the
city limits. The locations for current distribution centers, expected deliveries, and possible locations for the facility are shown in the tables and figures below.
a. Determine the optimal location for the new print shop.
b. Rank the alternative locations in order of preference using contour lines.
Current Distribution Centers:

Center	X-Coordinate	Y-Coordinate	Weight
A	5	10	200
B	50	15	400
C	25	25	500
D	35	5	300
E	15	20	400
F	30	30	600

Possible locations for the new print shop:

Building	X-Coordinate	Y-Coordinate
1	20	20
2	40	25
3	25	35

10.8 Plot the contour line passing through the point $(1,5)$ for the following formula:

$$f(X, Y) = 6|X - 6| + 3|X - 4| + 9|Y - 1|$$

10.9 A small machine shop has five existing machines (M1 through M5) located at coordinate locations $P_1 = (10, 25)$, $P_2 = (10, 15)$, $P_3 = (15, 30)$, $P_4 = (20, 10)$, and $P_5 = (25, 25)$. Two new machines (N1 and N2) are to be located in the shop. It is anticipated that there will be four trips per day between the new machines. The number of trips per day between each machine and each existing machine is:

M/C	M1	M2	M3	M4	M5
N1	8	6	5	4	3
N2	2	3	4	6	6

a. *Formulate* the objective function assuming that rectilinear distance is used.
b. *Formulate* the objective function assuming that Euclidean distance is used.

10.10 Three new facilities are to be located relative to six existing facilities. The weighting factors are [v_{jk} = flow intensity between new machines j and k, and w_{ji} = flow intensity between new machine j and existing machine i];

$v_{12} = 5,$ $v_{13} = 1,$ $v_{23} = 4,$ $w_{11} = 4,$ $w_{12} = 0,$
$w_{13} = 2,$ $w_{14} = 0,$ $w_{15} = 5,$ $w_{16} = 5,$ $w_{21} = 0,$
$w_{22} = 0,$ $w_{23} = 0,$ $w_{24} = 0,$ $w_{25} = 10,$ $w_{26} = 1,$
$w_{31} = 0,$ $w_{32} = 0,$ $w_{33} = 10,$ $w_{34} = 10,$ $w_{35} = 2$
$w_{36} = 5$

The coordinate locations of the existing facilities are $P_1 = (0,10)$, $P_2 = (5,0)$, $P_3 = (10,0)$, $P_4 = (5,10)$, $P_5 = (20,10)$, and $P_6 = (5,20)$.
a. *Formulate* the location problem assuming rectilinear distance.
b. *Formulate* the location problem assuming Euclidean distance.

10.11 An enterpreneur plans to locate coffee shops in a newly constructed office building. The current office tenants are located at $P_1 = (20,70)$, $P_2 = (30,40)$, $P_3 = (90,30)$, and $P_4 = (50,100)$. 50 persons per day are expected to visit the first office, 30 in the second office, 70 in the third office, and 60 in the last office. Seventy percent of the visitors are expected to drop by the coffee shop. For each additional unit distance a customer has to travel, the coffee shop is expected to lose $0.25 in revenue. The daily operating cost for n coffee shops is $C(n) = 5000n$. Determine the optimal number of coffee shops to construct and specify their locations.

10.12 Five special-purpose machines are located in a plant at the point $(0,0)$, $(0,10)$, $(30,25)$, $(15,10)$, and $(20,20)$. The machines require maintenance at expected frequencies of 10, 16, 8, 5, and 12 times per month, respectively. Due to the nature of the maintenance, all machine maintenance must be performed at the maintenance center. The annual cost of owning and operating a maintenance center is $5,000. The cost of moving machine to the maintenance center is estimated to be $10 per unit distance.

a. What are the total cost and location of the maintenance center if *one* maintenance center is the optimum number?

b. What are the total cost and locations of the maintenance centers if *two* maintenance centers is the optimum number?

10.13 Amco Frozen Corp. is planning on locating its food distribution centers in Dallas, Texas, to serve the southwestern part of the country. After extensive research, the company has narrowed down its choice to five existing facilities. The monthly cost of meeting the customer's demands and rental cost for each of these five facilities are summarized in the figure below. Determine which facilities are the optimal site for the distribution center such that the overall cost is minimized.

Existing Facility Sites

Customer	A	B	C	D	E
1	2,500	500	3,600	10,000	8,000
2	1,800	6,000	5,400	1,200	7,200
3	5,000	5,000	4,700	1,500	6,500
4	2,800	2,600	4,800	6,000	7,000
5	6,000	1,500	9,000	8,000	6,300
Rental Cost	5,000	7,000	6,000	2,000	8,000

10.14 A warehouse company plans to locate its regional warehouse in Miami, Florida, to serve the southeastern part of the country. The company has four sites to choose from. The *monthly* costs of meeting the customers' demands and *annual* rental cost for each of these four facilities are summarized below. Determine the optimal size for the distribution center such that the overall cost is minimized.

Existing Facility Sites

Customer	A	B	C	D
1	12,000	20,000	15,000	24,000
2	18,000	20,000	10,000	15,000
3	25,000	20,000	12,000	32,000
4	16,000	18,000	10,000	24,000
5	52,000	48,000	25,000	58,000
6	32,000	30,000	10,000	55,000
Rental Cost	50,000	75,000	72,000	45,000

10.15 A manufacturing plant is to be constructed in the city of Cleveland, Ohio, in order to provide a faster service to its major customers in the surrounding area. There are three possible locations the company is considering. The distances (in miles) from these three locations to the four major customers are listed in the table below. The transportation cost is approximately $1.00/mile. The annualized fixed cost of constructing the plant and the customers' demand (number of trips required per year) are also given in the table. Find the best location for this plant.

Possible Locations

Customer	No. of Trips	A	B	C
1	1,000	25	15	20
2	2,400	30	45	40
3	4,200	10	10	50
4	3,200	25	25	35
Construction Cost		120,000	75,000	70,000

10.16 Five departments are involved in the processing required for the products in a manufacturing company. A summary of the material flow matrix (between departments) and distance matrix (between sites in feet) can be found in the matrices below. The material handling cost is directly proportional to the distance traveled.

a. Determine a lower bound for the material handling cost.

b. Determine the material handling cost for the current layout design.

$$V = \begin{bmatrix} 0 & 4 & 8 & 6 & 5 \\ 4 & 0 & 3 & 7 & 9 \\ 8 & 3 & 0 & 2 & 1 \\ 6 & 7 & 2 & 0 & 6 \\ 5 & 9 & 1 & 6 & 0 \end{bmatrix} \quad D = \begin{bmatrix} 0 & 70 & 45 & 62 & 35 \\ 70 & 0 & 57 & 47 & 91 \\ 45 & 57 & 0 & 28 & 71 \\ 62 & 47 & 28 & 0 & 62 \\ 35 & 91 & 71 & 62 & 0 \end{bmatrix}$$

10.17 Six cells are to be located in a two by three facility as shown in the figure below. The material flow matrix is given in a separate table. The material handling cost is directly proportional to the distance traveled (from centroid to centroid).

a. Determine a lower bound for the material handling cost.

b. Determine an initial cell layout and the total material handling cost.

c. Use the improvement procedure to determine a better layout and calculate its total material handling cost.

A	C	E
B	D	F

Flow	1	2	3	4	5	6
1	—	100	200	150	0	0
2		—	20	10	150	0
3			—	30	250	150
4				—	0	50
5					—	10
6						—

10.18 Four manufacturing cells are served by an automatic guided vehicle (AGV) on a linear bidirectional track as shown below. The product routing information and required production rates are given in the table below. Determine a cell arrangement based on an improvement procedure. (Assuming that the pickup/delivery stations are located along the AGV track at the midpont of the cell edge.)

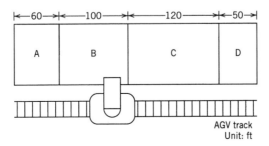

AGV track
Unit: ft

Product	Processing Sequence	Weekly Production
1	A B D C D	1,000
2	B A D C A	700
3	B D A	300
4	A B C A	400

10.19 Six machines are located on either side of a bidirectional conveyor system as shown in the figure below. For the flow information given in the table, how can the machines be rearranged to minimize the time products spend on the conveyor? Assume the load transfer station for each machine is located in the midpoint of the machine edge facing the conveyor. Distances are given in feet.

Bidirectional conveyor system

	A	B	C	D	E	F
A	—	100	200	20	100	100
B		—	150	250	250	75
C			—	0	50	0
D				—	125	225
E					—	300
F						—

10.20 Consider an array of storage locations as illustrated below. The array represents storage bays. Rectilinear travel is used and is assumed to originate and/or terminate at the centroid of the storage bay. Products are received through one of the I/O points (docks). The storage activity is divided equally between the docks.

Define the following parameters and variable:

m = number of items

A_i = number of grid squares required by item $i = 1, \ldots, m$.

Assume $n = \sum A_i$.

d_{kj} = distance between dock k (= 1, ... , p) and the centroid of grid j
  (j = 1, ... , n).

w_{ik} = cost/unit distance incurred in transporting item i between dock k and its
  storage region

X_{ij} = if item i is assign to grid square j, 0 otherwise.

a. Calculate the average distance item i travels between dock k and its storage region.

b. Calculate the average cost of transporting item i between dock k and its storage region.

c. Formulate an integer programming model which minimizes the total transportation cost.

10.21 Consider a warehouse illustrated by the figure below. Thirty-two bays (25 ft. × 25 ft.) are available for storage. Four different types of products (A, B, C, and D) are to be stored. Each type of product cannot share a bay with any other type of product. Products are received from Dock 1 and are shipped out in equal proportion from docks 2 and 3. The area requirement and weekly load rate for each of these four products are shown in the table below. Determine the layout that will minimize the average distance traveled per week.

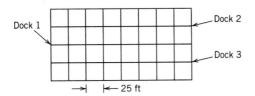

Product	Area (ft²)	Weekly Load Rate
A	4,375	500
B	7,500	600
C	1,500	700
D	6,250	400

10.22 An existing warehouse will be used for the storage of six product families. The warehouse consists of storage bays of size 20 ft × 20 ft. Dock 1 has been designated as the receiving dock, while dock 2 is used as the shipping dock. The area requirement and monthly load rate for each product family are shown in the following table. Determine the layout that will minimize the total expected travel distance.

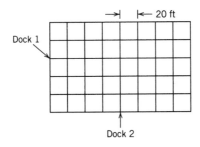

Product Family	Area (ft^2)	Load Rate
1	2,400	600
2	3,200	400
3	2,000	800
4	2,800	400
5	4,000	400
6	1,600	800

10.23 Three classes of products (A, B, and C) are to be stored in the warehouse depicted in the figure below. Product storage requirements are 15 bays for A, 5 bays for B, and 16 bays for C. Fifty percent of the shipment has to go through dock 1, and the other shipment are evenly distributed between docks 2 and 3. All products require the same number of trips from/to storage (dock) per day. Recommend a layout which maximizes throughput.

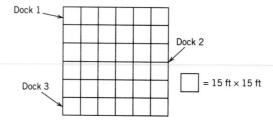

10.24 Shown in the figure below is a layout of an existing warehouse. One-way aisles are 8 ft wide, while two-way aisles are 16 ft wide. Each storage bay is approximately 20 ft × 20 ft. Three products (X, Y, and Z) are to be stored. Door A and B serve as the receiving and shipping docks, respectively. The area requirements and load rates are shown in the table below. Determine the layout that will minimize the expected travel distance.

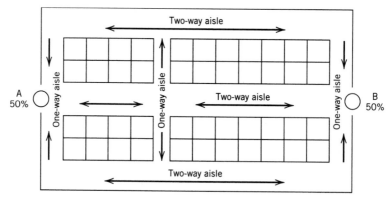

Product	Area	Load Rate
X	6,400	400
Y	8,800	400
Z	2,400	600

10.25 Four equal-sized (20 ft × 20 ft) machines are grouped into a manufacturing cell in a linear layout. The material handling device is a two-way shuttle cart system. The pick-up and delivery station(s) for each type of machine is shown in the figure below. The material flow information is given in the table.

a. Determine the machine location that will minimize the loaded travel distance.

b. Calculate the total loaded travel distance (assuming the clearance between each pair of machines if 8 ft).

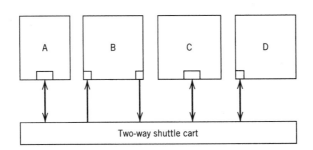

M/C	A	B	C	D
A	—	100	150	100
B	100	—	300	50
C	150	300	—	0
D	100	50	0	—

10.26 Block stacking is used for the storage of unit loads of paper products. The dimensions of the unit load are 48 in. × 52 in. A clearance of 8 in. is provided between storage rows. The storage aisle is 12 ft wide. The product is stacked four levels high in the warehouse. Sixty loads of a particular product have been received.

   a. Determine the depth of the storage rows that will minimize the average amount of floor space required. Assume uniform depletion of product over the life of the forty loads. Use enumeration to obtain the solution.

   b. By using a continuous approximation, what row depth would minimize the average amount of floor space?

10.27 Deep lane storage is used for storing a soft drink product. The dimensions of the unit load are 42 in. × 48 in. The flue clearance is 12 in., the rack is 6 in. wide, the clearance between a load and the rack is 4 in., and the aisle width is 8 ft. Seven tiers of storage are used. Three hundred pallet loads of the product are to be stored. Determine the following values:

   a. x, using continuous approximation.

   b. x, when the quantity of pallet loads to be stored is reduced to 35 and using enumeration.

10.28 Block stacking is used for the storage of 40 in. × 48 in. unit loads. A counterbalanced lift truck is used to store and retrieve the unit loads, which are stacked five-high. The storage aisle is 13 ft. wide. A clearance of 8 in. is used between storage rows.

   a. If 300 loads of a particular product are received and are to be stored, what storage depth will minimize the average floor space required during the length of time the product is stored? Assume the product is withdrawn from storage at a uniform rate. Use continuous approximation and round off the answer to the nearest integer value.

   b. Suppose the lot size in a) had been 50 unit loads. Using enumeration, determine if the optimum row depth is 3, 4, or 5.

   c. Suppose the lot size in a) had been 50 unit loads and the inventory levels for the product were distributed as follows during its life cycle.

Inventory	Percent	Inventory	Percent
50	20	25	5
45	20	20	5
40	15	15	5
35	10	10	5
30	10	5	5

Determine the expected number of storage rows required using a storage row depth of 3.

    d. In (c), determine the expected amount of floor space required using a storage row depth of four.

10.29 Thirty 48 in. × 40 in. unit loads of a particular product are to be stored in storage racks. The warehouse manager cannot decide whether to store the product in single deep or double deep storage racks. Seven tiers of storage are used; the storage aisle is 8 ft wide. A clearance of 4 in. is used between a load and the 4 in. upright truss and a clearance of 4 in. side-to-side between loads. Two loads are stored side-by-side on a load beam. The flue space has a width of 12 in. between loads. Withdrawals occurs at a uniform rate over the life cycle of the 30 loads.

    a. Determine the expected amount of floor space required to store the product using single deep storage rack.

    b. Determine the expected amount of floor space required to store the product using double deep storage rack.

    c. Suppose a scissors type reach attachment will allow 3-deep storage in the storage rack. As with double deep storage, 2 loads are stored side-by-side on a load beam. The same clearance apply. Based on assumptions similar to those used for double deep lane storage, determine the expected amount of floor space required to store the product.

10.30 Block stacking is used to store unit loads, 40″ × 48″; each unit load is 54″ high, including the pallet. A clearance of 10″ is provided between storage rows. The storage aisle is 13 ft. wide; storage is assumed to occur on both sides of the aisle. Unit loads are stacked 3-high.

    a. Given $Q = 30$, using enumeration determine x to minimize S.

    b. Suppose $Q = 300$. Using continuous approximation, how deep should the storage rows be?

    c. Suppose $Q = 300$ and $s = 30$ loads. Use continuous approximation to determine x.

    d. Suppose $Q = 30$ and $s = 5$ loads. Use enumeration to determine x.

10.31 An appliance manufacturer stores its finished goods using block stacking and clamp attachments on counterbalanced lift trucks. Leased warehouse space is currently being used to store freezers. The outside dimensions of the freezers are 36″ × 60″ × 48″, with the latter being the height of the freezer. The unit can be stored in either the 36″ × 60″ configuration or the 60″ × 36″ configuration. Units can be stacked 3-high in storage. If the 36″ × 60″ configuration is used, then a clearance of 12″ between storage lanes is required; if the 60″ × 36″ configuration is used, then a clearance of 24″ is required. A 13 ft. storage aisle is used.

    a. If the 36 in. × 60 in. storage configuration is used, what row depth will minimize the average amount of floor space required over the life of a storage lot of size 30, assuming uniform withdrawals from storage? (Use enumeration to obtain you answer).

    b. Given the conditions of part (a), what row depth is indicated when using continuous approximation with a lot of size 300?

    c. Given the conditions of part (a), what row depth is indicated when using continuous approximation with a lot of size 300 and a safety stock of 30?

   d. For a lot of size 30 and uniform withdrawal, what is the resulting minimum square footage for the best configuration, that is, the minimum value of S for either 36 in. $\times$ 60 in. $\times$ 48 in. or 60 in. $\times$ 36 in. $\times$ 48 in.?

10.32 The Most Delicious Ice Cream Company used block stacking to store pallet loads of its finished goods in frozen storage. The dimensions of the pallet load are 36 in. $\times$ 48 in. Each load consists of 40 half-gallon cartons. Eight cartons to a layer, and five layers to a pallet. Using walkie stacker lift trucks, pallet loads of ice cream are stacked 3-high. The storage aisle is 8 ft wide, and a clearance of 12 in. is to be used between storage rows. A production run of 30-pallet loads of Hot Fudge Delight occurs. It is expected that the product will be withdrawn from storage at a uniform rate of 5/day; however, the first withdrawal will not occur immediately. Specifically, the inventory level for day k is given to be

$I_k = 30$ for $k = 1, \ldots, 4$

$I_k = 50\text{-}5k$ for $k = 5, \ldots, 9.$

   a. If loads are stacked 2-deep in a row, what will be the average amount of floor space required in the freezer over the life of the production lot?
   b. If loads are stacked 3-deep in a row, what will be the average amount of floor space required in the freezer over the life of the production lot?

10.33 A storage system is to be designed for storing 60 in. $\times$ 38 in. pallet loads of a wide variety of products. The storage alternatives to be considered are block stacking, deep lane storage, single-deep pallet rack, double-deep pallet rack, and triple-deep pallet rack.
   In the case of block stacking, a counterbalanced lift truck will be used. The resulting design parameters are 13-ft storage aisles, 10 in. clearance between storage rows, and 4-high storage of products.
   With deep lane storage, an S/R machine, will be used. Also, the following design parameters are to be used: a 6-ft aisle, 12-in. flue space, 3-in. clearance between the load and the rack, 4-in. rack member, and 8-high storage.
   With single-deep, double-deep and triple-deep pallet rack, a narrow-aisle lift truck will be used. Two loads are stored side-by-side on a common load beam. The following design parameters are to be used: an 8-ft aisle, 12-in. flue space, 5-in. clearance between loads side-by-side on the load beam, 5-in. clearance between the load and the rack, 5-in. rack member, and 6-high storage.
   a. Based on a lot size of 30, determine the average amount of floor space required for block stacking with the row depth that minimizes the average amount of floor space, measured in square feet. Solve for the row depth using enumeration.
   b. Solve part (a) for the case of deep-lane storage, but determine the lane depth using a continuous approximation; round-off the lane depth to the nearest integer value.
   c. Based on a lot size of 30, determine the average amount of floor space required to store the product using single-deep pallet rack. Express answer in square feet.
   d. Solve part (c) using double-deep pallet rack.
   e. Solve part (c) using triple-deep pallet rack.

10.34 Deep-lane storage is used to store 36 in. $\times$ 48 in. unit loads of Captain Crunchy Goober Butter. The clearance between the load and the upright rack member is 4 in.; the upright rack member is 4 in. wide; a 12-in. flue space is provided at the end of the deep lane, the storage aisle is 5 ft. wide; there are 15 storage levels in the deep lane system. The system consists of 5-deep lanes, 10-deep lanes, and 15-deep lanes.
   a. Suppose 250 unit loads are received. Which lane depth is the worst choice for storing the Captain Crunchy Goober Butter in terms of minimizing the average square foot-age of floor space required? What is the resulting square footage? Assume the 250 units loads are withdrawn uniformly.

b. In part (a), suppose a safety stock of 50 units exists. Which of the three lane depths is the best in terms of minimizing the average square footage of storage space? What will be the resulting square footage?

c. In part (b), suppose double-deep storage rack is used for the product. With two loads stored side-by-side on a load beam, 4 in. clearances between the loads and the 4-in. upright trusses and between loads side-by-side, a 12 in. flue space, and an 8-ft. aisle, what would be the average square footage required if the rack contained 6 storage levels?

10.35 60 in. × 48-in. loads are block stacked 5-high using 10-ft. wide aisle. Two hundred unit loads of a product are received; a safety stock level of 20 unit loads is desired. In this case, suppose 20 units loads of the the previous lot of the product are in the warehouse at the time of receipt of the current lot. Due to the random nature of demand for the product, the probability distribution for the amount of the particular lot of product in inventory is given as follows:

Inventory Level	Probability	Inventory Level	Probability
200	.10	100	.04
190	.02	90	.04
180	.02	80	.06
170	.02	70	.06
160	.02	60	.06
150	.02	50	.08
140	.02	40	.08
130	.04	30	.08
120	.04	20	.08
110	.04	10	.08

Determine the lane depth for block stacking that minimizes the average amount of floor space over the lifetime of a lot of the particular product in question.

a. Use continuous approximation based on uniform demand.

b. Using enumeration, determine a local minimum near the solution obtained in part (a).

10.36 An AS/R system is being designed for 48 × 42 in. unit loads that are 45 in. high. One of the design alternatives involves 12 aisles, no transfer cars, 1440 openings per aisle, 12 levels and 60 columns of storage on each side of the aisle, and sprinklers. The S/R will travel horizontally at an average speed of 320 fpm; simultaneously, it travels vertically at an average speed of 75 fpm. The time required to pick up or deposit a load equals 0.25 min. Each S/R will have to store and retrieve at a rate of 15 storages per hour and 15 retrievals per hour. It is felt that 70% of the storages and retrievals will be performed on a dual-command cycle. What will be the utilization of each S/R? What will be the cost of the building if a 25-ft-high building costs $20/ft²? What will be the cost of the AS/RS if a central computer console is used and if each unit load weighs 2500 lb?

10.37 Five aisles of storage are provided with an S/R machine dedicated to each aisle. Randomized storage is used. The S/R machine travels simultaneously horizontally and vertically at speeds of 400 fpm and 80 fpm, respectively. Each aisle is 300 ft long and 50 ft high. The time required to pick up or put down a load is 0.30 min. Over an 8-hr period a total of 400 single-command cycles and 400 dual-command cycles are performed. Determine the system utilization.

10.38 An AS/R system is being designed for 42 × 48 in. unit loads that are 42 in. high. One of the design alternatives involves eight aisles, no transfer cars, 1100 openings per aisle, 11 levels and 50 columns of storage on each side of the aisle, and sprinklers.

The S/R will travel horizontally at an average speed of 300 fpm; simultaneously, it travels vertically at an average speed of 70 fpm. The time required for the shuttle to perform either a pickup (P) or a deposit (D) operation is 0.35 min. Each S/R will have to store and retrieve at a rate of 18 storages per hour and 18 retrievels per hour. Due to the uncertainty concerning the proportion of the operations that will be single command versus dual command, you are to determine the maximum percentage of single-command operations so that utilization of the S/R does not exceed 85%. Also, the cost of the building, rack, and S/R machines are to be determined. Assume 2000-lb loads, off-board computer controls (but not a central console), and a $20/ft^2 building cost for a 25-ft-high building.

10.39 An AS/RS is being designed; randomized storage is to be used. One alternative being considered is to have each aisle be 70 ft tall and 310 ft long. The S/R has a horizontal speed of 425 fpm and a vertical speed of 80 fpm. It requires 0.40 min to pick up or deposit a load. The activity level for the system varies during each shift. The heaviest operating period is a time during which a large number of retrievals must be performed. During peak activity each S/R must be capable of performing 30 retrievals/per hour and 10 storages per hour. It is expected that 100% of the storages will be performed with dual-command cycles.

　　a. Can the system handle the required workload? Why or why not?

　　b. If during the peak period the number of retrievals is 28 per hour and the number of storages is six per hour, what will be the utilization for each S/R?

10.40 An AS/RS is to be designed for a 40 × 48-in. unit load that is 48 in. high. There are eight aisles; each aisle is 12 loads high and 50 loads long. There are to be sprinklers and no transfer cars. It takes 0.30 min to perform a P/D operation. The S/R travels horizontally at a speed of 300 fpm and travels vertically at a speed of 60 fpm. Thirty percent of the retrievals and 30% of the storages are single-command operations.

　　a. How many storages per hour and retrievals per hour can each S/R handle without being utilized more than 90%?

　　b. Estimate the cost of racks, S/R machines, and building. Assume it costs $25/ft^2 to build a 25-ft-high building, loads weigh 3000 lb, central console, and conventional structure (not rack supported).

10.41 An AS/R system is to be designed to store approximately 12,000 loads. Due to space limitations in the existing facility, it is known that the system will have eight tiers of storage. The choice has been reduced to having either (a) 10 aisles, 75 openings long, (b) 9 aisles, 84 openings long, or (c) 8 aisles, 94 openings long. The S/R will travel horizontally at an average speed of 450 fpm; simultaneously, it travels vertically at an average speed of 80 fpm. 48 × 40-in. unit loads, 44 in. high and weighing 2500 lb are to be stored. The P/D station is located at the end of the aisle, at floor level. The time required to pick up (P) or to deposit (D) a load is 0.30 min. The thoughout requirement on the system is to perform a total of 180 storages per hour and 180 retrievals per hour. Randomized storage is used. Forty percent single-command operations are anticipated. A maximum utilization of 95% can be planned for an individual S/R. A central computer console is to be used to control the S/Rs. Determine the least-cost alternative that satisfies the throughout constraint. Consider only rack cost and S/R machine cost; do not consider building cost. (Assume sprinklers and no transfer cars are to be used.)

10.42 An AS/RS is to be designed for a 48 × 54 in. unit load that is 50 in. high. There are 10 aisles; each aisle is 14 loads high and 60 loads long. There are to be sprinklers and no transfer cars. It takes 0.40 min to perform a P/D operation. The S/R travels horizontally at a speed of 425 fpm and travels vertically at a speed of 80 fpm. During peak activity each S/R in the system must be capable of performing 24 retrievals/hr. and 6 storages/hr. During peak activity it is expected that 100% of the storages will be performed with dual-command cycles. The maximum utilization of an S/R during peak activity must be less than or equal to 98%. Does the system satisfy the utilization constraint?

10.43 If a 42 × 46-in. pallet load (62 in. high including the pallet) is to be stored in an AS/RS, what building dimensions are required to accommodate six aisles? Assume each aisle has 40 columns and 10 levels of storage along each side of the aisle. Do not allow for transfer cars; do consider sprinklers.

10.44 Based on a peak rate of 100 storages per hr. and 100 retrievals per hr, determine the minimum number of S/R machines required without having greater than 75% utilization of the equipment. Assume 60% of the operations are performed on a dual-command basis. The shuttle on the S/R requires 15 sec to perform a P/D operation. The average horizontal speed of the S/R is 400 fpm; the average vertical speed of the S/R is 80 fpm; horizontal and vertical movement is simultaneous. Each aisle is 300 ft long and 60 ft high.

10.45 For an AS/RS, suppose it has been determined that A sq.ft. of rack space is required on either side of each aisle. Given that $A = L H$ and assuming that b_v and v_v are fixed, determine the rack shape (i.e., the Q value) for which the expected S/R machine travel time for a single command cycle is minimized.

10.46 In deriving the cycle time equations, the P/D station was assumed to be located at the lower-left-hand corner of the rack. Based on the particular type of hardware involved, however, the exact location of the P/D station may not coincide with the lower-left-hand corner of the rack. In fact, suppose the P/D station is located 12 ft from the lower-left-hand corner of the rack as shown below:

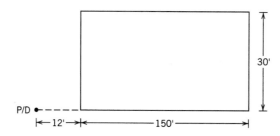

Further assume that the storage rack is small; say, 150 ft long and 30 ft high. The S/R machine travels at 400 fpm and 80 fpm in the horizontal and vertical directions, respectively. Calculate the expected S/R machine travel time for single- and dual-command cycles for the above case. Compare your results with the expected travel times obtained from Equations 10.70 and 10.72 (show the percent difference). Note: In traveling from the P/D station to a location in the rack (and vice versa), the S/R machine is **not** required to pass through the lower-left-hand corner of the rack.

10.47 Consider a normalized storage rack Q units long in the vertical direction. The P/D station is located at the lower-left-hand corner of the rack, and each trip originates and terminates at the P/D station. Given that the rack will be serviced by a straddle truck, it is assumed that the truck will travel according to the *rectilinear* metric.
a. Derive (in closed-form) the normalized expected travel time for a single-command cycle.
b. Derive (in closed-form) the normalized expected travel time for a dual-command cycle.

10.48 An AS/RS is to be designed for the storage of 60 in. × 48 in. unit loads. The S/R machine will travel horizontally at a speed of 500 fpm; simultaneously, it will travel vertically at a speed of 80 fpm. The storage aisle will be 400 ft long and 60 ft tall. The pickup and deposit station will be located at the end of the aisle and will be elevated

20 ft above floor level. Assume randomized storage will be used and the storage region served by the S/R is a continuous, rectangular region 400 ft × 60 ft. The P/D time will be 0.30 min.

a. Determine the single-command cycle time, including P/D times.

b. Determine the dual-command cycle time, including P/D times.

c. Suppose three classes of products are to be stored in the storage aisle. Fast movers will be stored in a rectangular region defined by the following coordinates: (0,0), (0,40), (125,40), and (125,0). Medium movers will be stored in an L-shaped region between the fast and slow movers. Slow movers will be stored in a rectangular region defined by: (250,0), (250,60), (400,60), and (400,0). Fast movers represent 60% of the storage and retrieval operations performed; medium movers account for 30% of the activity; and slow movers represent the balance. Determine the average single command cycle time, including P/D times.

10.49 Consider a unit load AS/RS rack that is 400 ft long and 80 ft high. The S/R machine travels at 450 fpm and 90 fpm in the horizontal and vertical directions, respectively. The P/D station is located at the lower-left-hand corner of the rack. Each trip starts and ends at the P/D station. Randomized storage is used. Suppose two types of products, labeled A and B, are stored in the rack. As shown below, product type A is stored in a 40 ft by 200 ft rectangular region located at the center of the rack. The remainder of the rack is used for storing product type B.

a. Determine the expected *single command* travel time (in minutes) for handling product type A.

b. Determine the expected *single command* travel time (in minutes) for handling product type B.

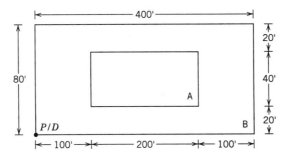

10.50 Consider a unit load AS/RS rack that is 500 ft long and 45 ft high. The S/R machine travels at 400 fpm and 120 fpm in the horizontal and vertical directions, respectively. The P/D station is located at the lower-left-hand corner of the rack. As shown by the shaded area below, the S/R machine will **not** be allowed to use one-third of the rack during the first shift. Compute the expected dual command travel time (in minutes) to perform a dual command trip assuming that the first operation is in region A and the second operation is in region B. (Each trip starts and ends at the P/D station.)

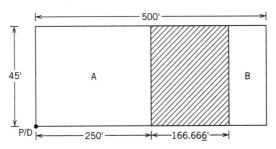

10.51 Consider a single-aisle AS/R system where the S/R machine travels at 450 fpm and 70 fpm in the horizontal and vertical directions, respectively. The pickup station is located at the lower-left-hand corner of the rack while the deposit station is located at the lower-right-hand corner. Loads to be stored must be picked up at the pickup station. Loads that are retrieved must be delivered to the deposit station. The rack is 400 ft long and 70 ft high. All the operations are performed on a dual command basis where the storage operation is assumed to be always performed before the retrieval.

Currently the system operates only for 8 hours through the first shift. An IE recently hired by the firm has proposed to use the S/R machine to move loads **within** the rack during the second shift (8 hr). Since no loads enter or leave the system during the second shift, the S/R machine can move loads with virtually no human intervention. Suppose the rack is divided into two equal pieces with an "imaginary" line as shown below:

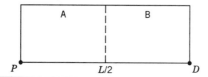

According to the proposal, during the second shift the S/R machine will move each load one-at-a-time from region A to region B. That is, the S/R machine will pick up a load from region A and deposit the load in an empty opening in region B. Subsequently, it will return to region A and pick up the next load to be moved to region B and so on. The above activity will continue for 8 hours as long as region A is not empty and region B is not full. The purpose is to move each load closer to the deposit station while providing more empty openings close to the pickup station for storages made during the *first* shift.

a. For the second shift, suppose the point where a load is picked up is randomly located within region A. Likewise, the point of deposit is assumed to be randomly located within region B. Determine the one-way expected S/R machine travel time (in minutes) to move a load from region A to region B. Compare your answer with the expected travel between time if the above two points were randomly located within the entire rack.

b. If the above proposal is implemented, the *first* shift operations are estimated to be affected as follows:

Pr(storage in region A) = 0.75 Pr(retrieval from region A) = 0.20

Pr(storage in region B) = 0.25 Pr(retrieval from region B) = 0.80

where $Pr(k)$ denotes the probability of event k. On the other hand, if the proposal is not implemented, then the storage and retrieval locations for the first shift will remain randomly distributed over the entire rack. If implemented, would the proposal made by the new IE reduce the expected S/R machine travel time during the first shift? Why or why not?

10.52 Consider an in-the-aisle order picking system based on the person-on-board AS/RS. Suppose the storage rack is 400 ft long and 80 ft tall. The S/R machine travels at a constant speed of 450 fpm and 90 fpm in the horizontal and vertical direction, respectively. On each trip, the picker performs eight picks on the average.

a. Assuming that the pick points are sequenced in an optimum fashion, compute the expected travel time required to complete a trip with eight picks.

b. Solve part (a) assuming that the two-band heuristic is used.

c. Solve part (a) assuming that the four-band heuristic is used.

10.53 Suppose in Problem 10.52 it takes the picker, on the average, 20 sec to complete a pick. Further suppose the picker spends 0.80 min at the P/D station between successive trips.

a. Assuming that the pick points are sequenced in an optimum fashion, compute the throughput capacity of the system in picks/hr.

b. Solve part (a) assuming that the two-band heuristic is used.

c. Solve part (a) assuming that the four-band heuristic is used.

10.54 Suppose the system in Problem 10.53 does not meet the required throughput capacity. Two alternative solutions have been proposed: (a) increase the S/R machine travel speed by 10% both in the horizontal and vertical direction, or (2) reduce the time per pick from 20 sec to 17 sec by improving the packaging of the items. Assuming that the pick points are sequenced in an optimum fashion, which one of the above two alternatives would increase the thoughput capacity of the system more than the other?

10.55 Consider an in-the-aisle order picking system based on the person-on-board AS/RS. The user would like to store 7000 loads in the system, where each load requires 18 ft^2 of rack surface (including necessary clearances). The system must be capable of performing 215 picks/hr. On each trip, the picker performs eight picks; each pick requires 1.5 min. The S/R machine travels at a speed of 400 fpm and 100 fpm in the horizontal and vertical direction, respectively. Last, the picker spends 1.20 min at the P/D station between successive trips.

a. Assuming that the pick points are sequenced in an optimum fashion, determine the "minimum" number of pickers and storage aisles required.

b. Determine the rack dimensions for the "minimum" number of storage aisles.

c. What fraction of the system's throughput capacity will be utilized?

10.56 Consider an in-the-aisle order picking system based on the person-on-board AS/RS. Total storage space required (expressed in square footage of storage rack) is equal to 300,000 ft^2. On each trip, the picker performs 10 picks; each pick requires 0.90 min. The S/R machine travels at a speed of 400 fpm and 100 fpm in the horizontal and vertical direction, respectively. The picker spends 1.40 min at the P/D station between successive trips. Assuming that the pick points are sequenced in an optimum fashion:

a. Determine the throughput capacity of a three-aisle system.

b. Determine the throughput capacity of a four-aisle system.

c. Determine the throughput capacity of a five-aisle system.

d. What can you say about the relationship between the throughout capacity of the system and the number of storage aisles?

10.57 Consider a walk-and-pick system where the total aisle length to be provided is fixed at 1440 ft. That is, if there are 10 aisles ($M = 10$), then each aisle must be 144 ft long; if there are 20 aisles, each aisle must be 72 ft long, and so on. Including shelving on either side, and the aisle clearance itself, suppose the aisle width is equal to 10 ft/aisle. For example, for $M = 10$, the warehouse width will be 100 ft. Using the results given for walk-and-pick systems, determine the optimum number of aisles (M) that would minimize the total expected walking distance per order.

10.58 The results we derived for walk-and-pick systems are based on a P/D point located at the midpoint of the front of the warehouse. Assuming the traversal policy is used, and that each pick is equally likely to be located anywhere in the warehouse, derive the expected walking distance per trip for the case where the P/D point is located at the left-hand side of the front of the warehouse (i.e., the P/D point is located at the lower left-hand corner of the warehouse). The number of aisles (M), the length (y) and width (x) of the warehouse, and the number of picks per trip (n) are given and fixed.

10.59 Consider an end-of-aisle order picking system based on the miniload AS/RS. The P/D station is located at the lower-left-hand corner of the rack, and each trip originates and terminates at the P/D station. One aisle is assigned to each picker. Furthermore, the S/R machine travels at 400 fpm and 100 fpm in the horizontal and vertical directions, respectively. It takes 0.10 min to pick up or deposit a container. The user would like to store approximately 7000 loads in the system where each load requires 18 ft^2 of rack surface (including the necessary clearances). Last, the system must be capable of performing at least 215 picks/hr. The pick time is constant at 1.5 min/pick.

   a. For $b = 0.90$, determine the near minimum number of pickers required to meet the throughput and storage space requirements.
   b. Assuming that the vertical dimension of the rack yields the value of Q, determine the resulting rack length and height in feet.
   c. Determine the *effective* values of the expected picker and S/R machine utilizations for the resulting system.

10.60 Consider one aisle of a miniload AS/R system used by a book distributor for order picking. The P/D station is located at the lower-left-hand corner of a rack which is 360 ft long and 63 ft tall. The S/R machine travels at 450 fpm and 90 fpm in the horizontal and vertical directions, respectively. It takes 0.15 minutes to pick up or deposit a tray. On the average, the picker picks 8 books from each tray. Time and motion studies indicate that, including the necessary paperwork, on the average it takes the picker 15 sec to pick one book. It has also been determined that the total time required to make the necessary picks from a tray is exponentially distributed.

   a. Compute the expected dual command cycle time for the above system.
   b. Suppose the expected dual command cycle time is equal to 2.30 min/cycle. Compute the *number of books picked per hour* by the above picker.

10.61 Consider a miniload AS/RS with two pick positions per aisle. It is assumed that the P/D station is located at the lower left-hand corner of the rack. Randomized storage is used. The S/R machine travels at 400 fpm and 80 fpm in the horizontal and vertical direction, respectively. The *total* load handling time on a dual command trip is equal to 0.42 min. The total storage area required is equal to 80,000 ft^2. Last, the rack is required to have a shape factor of 0.90. Assuming that the pick time is *constant* (i.e., deterministic) at 1.50 min/pick, determine the minimum number of aisles required to obtain a picker utilization of 100%.

10.62 Consider a miniload AS/RS used for end-of-aisle order picking. It is assumed that the P/D station is located at the lower-left-hand corner of the rack and that there are two pick positions. Consider a single-aisle system with one picker. Under the current mode of operation, the containers *within a given order* are retrieved on a closest-container-first basis. That is, the container closest to the P/D station (in travel time) is retrieved first and so on. Ties are broken randomly. (Recall that no other container can be retrieved before all the containers needed for filling the current order have been retrieved.) Furthermore, the parts handled by the system are small enough so that multiple stock keeping units (SKUs or part numbers) can be stored in each container (or tray).

   Currently, the system is unable to meet the required throughput level. Moreover, it has been established that:

   (i) It is impossible to further reduce the mean and/or variance of the time required to pick an SKU.
   (ii) It is impossible to further increase the travel velocity (or acceleration and deceleration) of the S/R machine (in either direction, horizontal or vertical). Likewise, the container handling time (or pickup/deposit time) is fixed.
   (iii) The space requirement is expected to remain at its present level. Therefore, the rack size may not be reduced.
   (iv) Management is unwilling to purchase another miniload aisle. However, they are willing to support reasonable modification(s) to the current system.

Given the above situation, how would you attempt to improve the throughput performance of the system? Explain and discuss at least three conceptually different alternatives. For each alternative, you must carefully argue why, how, and under what conditions you would expect the throughput to show a reasonable improvement. The economic comparison of the alternatives and the relative amount of improvement offered by each alternative is **not** within the scope of the question. However, your suggestions must be realistic and reasonable from an implementation standpoint.

10.63 A trolley conveyor serves four workstations. There are 543 carriers equally spaced around the conveyor. The material flow patterns for the work stations are as follows:

$\{f_1(n)\} = (0,0,3,2,0,0,0,3,2,0,0,0,3,2, \ldots)$

$\{f_2(n)\} = (-3,0,-3,0,-3,0-3,0-3,0-3,0, \ldots)$

$\{f_3(n)\} = (3,0,4,0,2,0,4,0,2,0,4,0, \ldots)$

$\{f_4(n)\} = (0,-3,0,0,-3,0,0,-3,0,0,-3,0,0, \ldots)$

a. Given the patterns, what is the smallest period?
b. Can general sequences be accommodated? If so, why? If not, why not?
c. Can steady-state operations be achieved? If so, why? If not, why not?
d. Is conveyor compatibility guaranteed for all values of k not an integer multiple of the period of the material flow patterns? If so, why? If not, why not?

10.64 A trolley conveyor is used to transport parts among four production stations. There are 32 carriers equally spaced around the conveyor, with 10-ft separation between carriers. The material flow patterns for the work stations are

$\{f_1(n)\} = (-5,0,0); \{f_2(n)\} = (0,-3,0); \{f_3(n)\} = (0,0,4); \{f_4(n)\} = (4,0,0)$

a. Determine the value of B.
b. Suppose the number of carriers on the conveyor can be decreased by one or two without affecting the cost of the conveyor. Consider each of the three options and specify the number of carriers that will minimize the amount of inventory on the conveyor. Use as an estimate of inventory on the conveyor the sum of the $H_i(n)$ values for all i and n.
c. Suppose the unloading patterns at stations 1 and 2 can be changed. Consider each of the options shown below and specify the patterns (including the original patterns) that will minimize the amount of inventory on the conveyor. Use as an estimate of inventory on the conveyor the sum of the $H_i(n)$ values for all i and n.
    (1) $\{f_1(n)\} = (0,0,-4); \{f_2(n)\} = (-4,0,0)$
    (2) $\{f_1(n)\} = (-4,0,0); \{f_2(n)\} = (0,-4,0)$

10.65 A trolley conveyor is used to transport parts among four production stations. There are 33 carriers equally spaced around the conveyor, with 10-ft separation between carriers. The material flow patterns for the work stations follows:

$\{f_1(n)\} = (-2,0,-2,0);$ $\{f_2(n)\} = (0,-2,0,-2);$

$\{f_3(n)\} = (0,0,4,0);$ $\{f_4(n)\} = (4,0,0,0);$

a. Can steady-state operations be achieved? If so, why? If not, why not?
b. Can general sequences for $\{F_i(n)\}$ be accommodated? If so, why? If not, why not?
c. Determine the value of $H_3(2)$.
d. Determine the value of B.
e. Suppose the number of carriers on the conveyor can be increased by one or two without affecting the cost of the conveyor. Consider each of the three options and specify the number of carriers that will minimize the amount of inventory on the conveyor. Use as an estimate of inventory on the conveyor the sum of

the $H_i(n)$ values for all i and n. Specify the value obtained for inventory on the conveyor.

f. Suppose the loading patterns at stations 3 and 4 can be changed. Consider each of the options shown below and specify the patterns (including the original patterns) that will minimize the amount of inventory on the conveyor. Use as an estimate of inventory on the conveyor the sum of the $H_i(n)$ values for all i and n.

(1) $\{f_3(n)\} = (0,0,2,2)$ $\{f_4(n)\} = (2,2,0,0)$

(2) $\{f_3(n)\} = (2,0,2,0)$ $\{f_4(n)\} = (0,2,0,2)$

10.66 A trolley conveyor is used to transport parts between three production stations. There are 62 carriers equally spaced around the conveyor, with 10-ft separation between carriers. The material flow patterns for the work stations follow:

$\{f_1(n)\} = (0,-4,0)$

$\{f_2(n)\} = (3,3,3)$

$\{f_3(n)\} = (0,-5,0)$

a. Determine the value of B.

b. Suppose the unload sequence for station 3 can be modified by delaying its start time or location. Which of the following permutations of the unloading sequence for station 3 will minimize the value of B:

$\{f_3(n)\} = (0,-5,0);$

$\{f_3(n)\} = (-5,0,0);$ or

$\{f_3(n)\} = (0,0,-5)?$

c. In (b), what permutation of the unloading sequence for station 3 will minimize the cumulative amount of inventory on the conveyor?

10.67 An overhead trolley conveyor is used to transport assemblies within an assembly department. There are three loading stations in the assembly department and one unloading station. The trolley conveyor has 83 carriers equally spaced around the conveyor, with 20-ft separation between carriers. The material flow patterns for the work stations follow:

$\{f_1(n)\} = (1,0,0);$ $\{f_2(n)\} = (0,2,0);$ $\{f_3(n)\} = (0,0,1);$ and

$\{f_4(n)\} = (0,-4,0)$

a. Determine the value of $H_3(2)$.

b. Determine the value of B and the amount of inventory on the conveyor. Estimate the inventory on the conveyor by summing the $H_i(n)$ values for all i and n.

c. Suppose the unload sequence can be changed to $\{f_4(n)\} = (0,-a,-b)$, where $a + b = 4$. What integer values of a and b will minimize the value of B?

d. Suppose the work performed by the first and third loading stations is combined to yield two loading stations and one unloading station with the following sequences:

$\{f_1(n)\} = (1,0,1);$ $\{f_2(n)\} = (0,2,0);$ and $\{f_3(n)\} = (0,-4,0)$

What, if anything, will be the impact on the value of B?

10.68 Cartons are to be conveyed over a distance of 300 ft using a combination of roller and belt conveyors. Movement over the first 200 ft will be via a flat belt driven roller conveyor with WBR dimensions of 33 in. rollers on 9 in. centers, and a speed of 200 fpm. Each tote box is 24 in. long and weighs 25 lb loaded. The spacing between tote boxes on the roller conveyor is 4 in. The tote boxes are transferred automatically from the roller conveyor onto a 100 ft. long roller supported belt conveyor operating at a speed of 300 fpm. The rollers are also on 9-in. centers and the WBR dimension is also 33 in. For the roller conveyor and the belt conveyor, determine the values of LF, BV, and L.

10.69 A chain driven roller conveyor is used to move pallet loads 150 ft. from storage to shipping. The pallet is 48 in. × 36 in. For smooth movement over the rollers, the pallet is oriented so that the stringer is aligned with the rollers (i.e., the stringer is perpendicular to the conveyor frames and parallel to the roller length.) The WBR dimension is 39 in. Each pallet load of product weighs 1200 lb. The spacing between pallets is 24 in. Rollers have a 9 in. spacing. The conveyor speed is 90 fpm. As the pallets near shipping, they are transferred directly from the roller conveyor to a roller supported belt conveyor.
   a. What are the values of BV, LF, FF, and L for the roller conveyor?
   b. If a 36-in. spacing is desired between pallets on the belt conveyor, what must be its speed?

10.70 A V-belt driven roller conveyor is used to convey cartons from the order picking area in a warehouse to the order accumulation and packing area in the warehouse. The conveyor is 250 ft long. Rollers are 2.5 in. in diameter and are spaced on 7.5 in. centers. A WBR dimension of 30 in. is used. A variety of cartons are placed on the conveyor. The cartons are equally likely to be 12 in., 18 in., 24 in., 30 in., or 36 in. long; they are equally likely to weight 5 lb, 10 lb, 15 lb, 20 lb, or 25 lb. The length and the weight are statistically independent. The minimum design clearance between cartons is 6 in.
   a. Determine the BV, LF, and FF values.
   b. Determine the average load on the conveyor.
   c. Determine the "worst case" load on the conveyor.

10.71 A flat belt driven roller conveyor is used to convey tote boxes of material from the receiving area in a warehouse to the storage area. The conveyor consists of two sections, 200 and 100 ft long. Rollers are on 6-in. centers. A WBR dimension of 27 in. is used. The tote boxes are 24 in. long and are equally likely to weight 10 lb, 20 lb, 30 lb, 40 lb, or 50 lb. The design clearance between tote boxes on the first conveyor (i.e., the 200-ft section is 6 in.
   a. The first conveyor operates at a speed of 200 fpm. What should be the speed of the second conveyor if the spacing between tote boxes is to be 8 in.
   b. What would be the horsepower requirement for the longest conveyor based on an average load per tote box? Would you recommend this horsepower be used? If not, what loading condition would you use to specify the horsepower requirement?

10.72 Two conveyors having 6-in. RC values are used to transport cartons horizontally 150 ft in a distribution center from receiving to storage. The first conveyor is a 100-ft-long roller supported belt conveyor with a speed of 300 fpm. Cartons are fed directly from the belt conveyor to a belt driven roller conveyor. Cartons weight 35 lb each, they are 30 in. long. The clearance between cartons on the first conveyor is 24 in.
   a. It is desired to establish the speed of the second conveyor that will result in a spacing of 12 in. between cartons on the second conveyor. What must be the speed of the second conveyor to achieve the desired clearance?
   b. Suppose the WBR dimensions for the conveyors are 33 in. What is the horsepower requirement for the first conveyor?
   c. Given the information in part (b), what is the horsepower requirement for the second conveyor? Assume the second conveyor is a 50-ft-long belt driven roller conveyor with the same WBR and RC values as the first conveyor.

10.73 Two roller supported belt conveyors are used to convey tote boxes of material 250 ft from the assembly department to the packing department. The first conveyor is 150 ft long and has a speed of 300 fpm; the second is 100 ft long and has a speed to be determined. The tote boxes are 24 in. in length. Rollers are 2.5 in. in diameter and are spaced on 10-in. centers. The WBR dimension is 27 in. The design clearance between tote boxes on the first conveyor is 12 in. It is reasonable to assume that the weight of consecutive tote boxes on the conveyor is statistically independent. The weight of the loaded tote boxes varies according to the following distribution.

Weight (lb)	Percentage
10	10
20	20
30	40
40	20
50	10

a. Tote boxes are conveyed from the 150-ft-long conveyor directly to the 100-ft-long conveyor. What should be the speed of the second conveyor in order for the spacing between tote boxes to be 6 in.?
b. Determine the lightest, average, and heaviest load on the first conveyor.

10.74 Tote boxes are conveyed over a distance of 200 ft using a combination of two 100-ft-long belt conveyors. Both are roller supported using 2.5 in. rollers weighing 5 lbs. each and installed on 6-in. centers. The WBR dimension is 33 in. for both conveyers. A conveyor speed of 150 fpm is used on the first conveyor, which feeds the tote boxes onto the second conveyor operating at a speed of 250 fpm. Each tote box is 20 in. long and weighs 40 lbs loaded. The spacing between tote boxes on the first conveyor is 6 in.
a. Determine the load on the first conveyor.
b. Determine the load on the second conveyor.
c. What should be the maximum roller center for stability of the tote boxes?

10.75 Cartons are to be conveyed using a 200-ft-long roller-supported belt conveyor with WBR dimensions of 27 in. and rollers on 6 in. centers. A conveyor speed of 100 fpm is to be used. The conveyor is inclined at an angle of 15°. Five types of cartons are to be conveyed on the belt. Their weights and lengths are as follows:

Carton	Weight	Length
A	15 lb	12 in.
B	20 lb	18 in.
C	30 lb	18 in.
D	40 lb	24 in.
E	30 lb	30 in.

For purposes of computing horsepower, the load on the conveyor should be based on cartons being distributed in the following sequence along the conveyor (A,A,B,B,C,C,C,D,D,D,E) with a 6 in. space between cartons. Assume the sequence moves along the conveyor with A's followed by B's followed by C's, etc. Determine the following values:
a. *L*, based on the heaviest condition.
b. *L*, based on the lightest condition.

10.76 A bulk belt conveyor is to be used to convey grain horizontally over a distance of 150 ft. The grain weighs 50 lb/cu ft. It is required that the conveyor deliver grain at a rate of 250 tph. The available belt widths are 18 in., 24 in., 30 in., and 36 in.
a. Determine the belt width and speed that delivers exactly the required capacity at the lowest horsepower to drive the empty conveyor.

    b. Suppose the cost of the 150 ft. conveyor is a function of the belt width and horse-power; specifically, suppose it costs $300 per in. of belt width and $5,000 per increment of horsepower. Determine the belt width that yields the least cost combination of belt width and horsepower.

10.77 A bulk belt conveyor is to be used to convey dry sand from a storage hopper to a cement mixer. The conveyor is to be 75 ft long. Using a 30-in.-wide belt, the sand is to be elevated a distance of 24 ft at a speed of 600 fpm.

    a. Is the angle of incline feasible?

    b. Is the speed feasible?

    c. Assuming the angle of incline and speed are feasible, what will be the delivered capacity if the sand has a density of 50 lb/cu ft?

10.78 A bulk belt conveyor is to be used to convey lumpy, abrasive steel trimmings from a basement area beneath a machine shop to a dumpster at ground level. The conveyor is to be 50 ft long. Using a 30-in.-wide belt, the waste material is to be elevated a distance of 20 ft at a speed of 400 fpm.

    a. Is the angle of incline feasible?

    b. Is the speed feasible?

    c. Assuming the angle of incline and speed are feasible, what will be the delivered capacity if the steel trimmings have a density of 100 lb/cu ft?

10.79 A bulk belt conveyor is used to transport washed gravel having a density of 75 lb/cu-ft over a horizontal distance of 232.55 ft. Also, the unsized gravel is to be elevated a vertical distance of 59.33 ft. A conveyor speed of 400 fpm is to be used in combination with a 24-in.-wide belt.

    a. Is the angle of incline feasible?

    b. Is the speed feasible?

    c. Assuming the angle of incline and the speed are feasible, what will be the delivered capacity in tph?

    d. Assuming feasible angle of incline and speed, how much horsepower is required to drive the empty conveyor?

10.80 A bulk belt conveyor is to be used to convey material having a density of 60 lb/cu-ft. It will be conveyed over a horizontal distance of 200 ft and at an incline of $10°$. The material is to be conveyed at a rate of 100 tons/h.

    a. Using a 24-in.-wide belt, what conveying speed should be used?

    b. Using a 24-in.-wide belt, what is the horsepower requirement to drive the empty conveyor?

    c. Using a 24-in.-wide belt, what is the incremental horsepower requirement for conveying the material horizontally?

    d. Using a 24-in.-wide belt, what is the additional horsepower requirement for lifting the material?

    e. Suppose the 24-in.-wide belt is conveyed at its maximum recommended belt speed. What is the capacity, in tons/hr, for material of the same density?

10.81 A bulk belt conveyor is to be used to convey coffee beans to a storage silo. The conveyor will have an incline of $10°$; the material is to be elevated 30 ft. In determining maximum belt speed, coffee beans are classified as grain. Coffee beans have a density of 30 lb/cu-ft.

    a. Using a 24-in. belt width, what is the maximum delivered capacity for the coffee beans?

    b. Using a 24-in. belt width, what speed delivers the beans at a rate of 200 tph?

    c. Using a 24-in. belt width and a speed of 500 fpm, what horsepower is required to drive the empty conveyor, excluding the safety factor?

    d. Using a 24-in. belt width and a delivered capacity of 200 tph, what additional hp is required to convey the load horizontally, excluding safety factor?

e. Using a 24-in. belt width and a delivered capacity of 200 tph, what additional horsepower is required to elevate the coffee beans, excluding safety factor?

10.82 A bulk belt conveyor is used to convey oats from ground-level to the top of a grain silo for storage. The oats are conveyed an an angle of 15° to a height of 50 ft. The oats have a density of 30 lb/cu-ft.

a. Using a belt that is 30 in. wide and a belt speed of 400 fpm, what is the horsepower requirement to drive the empty conveyor, excluding the safety factor?

b. For the conditions in part (a), what additional horsepower is required to elevate the oats 50 ft. to the top of the silo, including the horsepower required to move the oats horizontally and vertically, but excluding the safety factor?

c. Suppose a belt speed of 400 fpm can be used for any feasible belt width and belts can be purchased in 1-in. increments in widths ranging from 18 in. to 36 in. What is the smallest belt width that will deliver the oats at a rate of 200 tph?

d. In part (c), suppose the maximum belt speed (in fpm) is given by $S_{max} = 200 + 10$ BW where BW denotes the belt width in inches. What is the smallest integer-valued belt width that will deliver the oats at a rate of 200 tph?

10.83 Consider the tandem flow system shown earlier in Figure 10.25b. In Example 10.40 we focused on zone I defined by stations {1,6,7,8,9}. Let's now focus on zone II defined by stations {2,3,4,5,9}, where stations 2 and 5 are I/O stations. (Recall that station 9 is a transfer station, which is also treated as an I/O station for the zone.) Suppose Job A, which is delivered by the AGV in zone I at a rate of 6 jobs/hr at station 9, must be handled in zone II as follows: 3 jobs/hr following the route 9–3–2 (say, Job A'), and 3 jobs/hr following the route 9–4–5 (say, Job A"). Job B, on the other hand, must be handled through the route 5–4–3–9 at a rate of 9 jobs/hr. (Once a load for Job B is delivered at station 9, it is picked up by the AGV in zone I.) Suppose the same data given for the example earlier applies to the AGV in zone II. That is, the empty AGV travels at a speed of 5 sec/grid, the load pickup or deposit time is equal to 9 seconds (0.15 min), and the loaded AGV travel times are obtained simply by adding the load pickup plus deposit time (0.30 min) to the corresponding empty travel times. Using the model shown for tandem flow systems, compute the values of α_f and α_e for the AGV in zone II. Is the AGV in zone II able to meet the throughput requirement? If so, what fraction of the time would the AGV be traveling empty in zone II while it is busy?

10.84 Cartons arrive at a workstation via a roller conveyor spur. The arrivals are Poisson distributed with a mean of 0.5/min. The time required to process the carton at the work station is exponentially distributed with a mean of 0.8 min. How long must the accumulation line be such that it will not be full more than 5% of the time? (*Note:* When the accumulation line is full, cartons are diverted to a separate accumulation area.)

10.85 A conveyor delivers parts to an inspection station, with a part arriving every 20 sec. The time required to inspect a part is exponentially distributed with a mean of 15 sec. On the average, how many parts will be waiting to be inspected?

10.86 Parts arrive for packaging at a Poisson rate of 30/hr. The time required for packaging is normally distributed with a mean of 3 min and a standard deviation of 1 min. If the system operates as an Erlang loss system, how many packaging stations should be provided in order to have no greater than 5% loss?

10.87 Unit loads arrive randomly at a quality control (QC) station. An average of 10 unit loads arrive per hour at a Poisson rate. The time required to perform the QC check is normally distributed with a mean of 4 min and a standard deviation of 1 min. On the average, how long will a unit load have to wait at the QC station before the inspection begins?

10.88 Cartons arrive at a workstation at a Poisson rate of 10/min. The time required to process a carton is exponentially distributed with a mean of 5 sec. On the average, how many cartons will be waiting for processing at the work station?

10.89 Unit loads arrive randomly at a quality control station. Two inspectors are located at the station; the inspection time is exponentially distributed with a mean of 6 min. Loads arrive at a Poisson rate of 18/hr.
a. What is the probability of both inspectors being idle?
b. What is the average number of units loads waiting to be inspected?

10.90 An accumulation conveyor is to be provided at a work station. When the conveyor is full, parts are diverted to another area for processing. Parts arrive at a Poisson rate of 1.0/min. The time required to process a part at the work station is exponentially distributed with a mean of 0.80 min per part. It is desired to provide an accumulation line sufficiently long such that less than 2% of the arriving parts will be diverted to another area of processing. What is the minimum number of *waiting* spaces that will satisfy the objective?

10.91 A tote box of parts arrives at an inspection station. The inspector selects a random number of parts for inspection; they are inspected individually. The tote boxes arrive at a Poisson rate of 5/min. The number of parts selected for inspection is equally likely to be 1, 2, 3, or 4. The inspection time per part is exponentially distributed with a mean of 2.5 sec. On the average, how many seconds will parts be waiting to be inspected? Recall, $\text{Var}(X) = E(X^2) - [E(X)]^2$.

10.92. A tote box of parts arrives at an inspection station. The inspector selects a random number of parts for inspection; they are inspected individually. Suppose tote boxes arrive at a Poisson rate of 6/min. Also, suppose the number of parts selected for inspection is distributed as follows:

Number of Parts	Probability
1	0.25
2	0.25
3	0.25
4	0.25

Furthermore, suppose the inspection time per part is exponentially distributed with a mean of 2.5 sec. On the average, how many parts will be waiting to be inspected?

10.93 Consider a queueing situation involving a single server. Suppose service time is exponentially distributed with an expected value of 5 min/customer.
a. Suppose arrivals occur in a Poisson fashion and there is unlimited waiting capacity for the system. Determine the maximum arrival rate that will be acceptable if L can be no greater than 4.
b. Suppose arrivals occur deterministically. If it is desired that L be no greater than 4.0, what is the maximum arrival rate for the customers?
c. Suppose arrivals occur in a Poisson fashion, but with the following pattern.

#Customers in System	Arrival Rate
0	15/hr
1	10/hr
2	5/hr
3	0

What will be the value of L?

10.94 Consider a queueing situation involving a single server. Service time is exponentially distributed with an expected value of: 6 min/customer when there is one customer in the system; 5 min when there are two in the system; 4 min when there are three in the system; 3 min when there are four in the system; and 2 min/customer when there are five in the system.

Suppose arrivals occur in a Poisson fashion at an average rate of 20/hr when no more than one customer is in the system. If one customer is waiting for service, then the arrival rate drops to a level of 15/hr. If two are waiting, then arrivals occur at a rate of 10/hr; if three are waiting, then the arrival rate is 5/hr; if four are waiting, then arrivals no longer occur.

a. Determine the probability of an idle server.

b. If the probability of an idle server equals 0.1, for the arrival and service rates given above, what is the probability of exactly one customer in the system?

10.95 Consider a distribution center that uses a fleet of eight battery powered automated guided vehicles for material delivery. Due to the wide variety of demands placed on an AGV, the life of a battery, before recharging is required, is a random variable. Suppose the time from recharging a battery for an AGV until it requires recharging again is exponentially distributed with a mean value of 10 hr. Also, suppose the AGV is idle during battery recharging. Finally, suppose the time required to recharge a battery is also exponentially distributed with a mean of 2 hr. There are two battery recharging machines, and the distribution center operates 24 hr/day and 7 days/week.

a. Determine the probability of at least one battery recharging machine being idle.

b. Determine the average number of AGVs *waiting* to be recharged. (Don't include those being recharged.)

10.96 Consider a manufacturing situation involving the use of six robots. There is one robot attendant who changes end-of-arm tooling for the robots, programs the robots, and performs setup changes for the robots. The time between requirements for an attendant for an individual robot is exponentially distributed with an average value of 2 hr. The time required to tend to the robot is also exponentially distributed and has an average value of 0.40 hr.

a. What is the probability of an idle attendant?

b. Determine the average number of robots waiting on the attendant to begin performing a service.

10.97 Draw a flow chart of a simulation experiment to determine how many receiving docks should be provided at a distribution center. Assume three kinds of receipts occur: small parcel delivery services (e.g., Federal Express, UPS, U.S. Postal Service); commercial carrier (J.B. Hunt, Schneider, Yellow Freight); and private carriers (your firm's and suppliers' trucks). Because of the differences in dock design needed to accommodate small parcel delivery (SPD) trucks, over-the-road (OTR) trailers, and your firm's (OWN) delivery trucks, three different types of customers exist. The simulation is to examine the impact of having general purpose docks versus having three different categories of docks.

10.98 Given the situation in Problem 10.97, suppose the time each category of customer occupies a dock space has the following distribution:

Time at the Dock (min)	Frequency (%)		
	SPD	OTR	OWN
0–4	60	0	10
5–9	30	10	20
10–14	10	20	35
15–19		40	25
20–24		20	10
25–29		10	

Can theoretical distributions be used in place of the data? If so, what distributions appear to be likely candidates? How should you decide which theoretical distributions to use?

10.99 Consider the situation described in Problem 10.97. Suppose the times (min) between consecutive arrivals of trucks of each category are given by the following distributions:

Time between Arrivals (min)	Frequency (%)		
	SPD	OTR	OWN
0–4	22	0	0
5–9	17	0	0
10–14	14	5	0
15–19	12	10	0
20–24	10	20	10
25–29	8	30	20
30–34	6	15	40
35–39	4	10	20
40–44	3	5	10
45–49	2	5	0
50–54	1	0	0
55–59	1	0	0
>59	0	0	0

Can theoretical distributions be used in place of the data? If so, what distributions appear to be likely candidates? How should you decide which theoretical distributions to use?

10.100 Given the information in Problems 10.97, 98, and 99, perform a simulation to determine the number of docks required to provide the following level of service: The probability an arriving truck has to wait for a dock is no greater than 0.15. Consider the following design alternatives: designated docks for each category of truck and undesignated docks. In the latter case, any truck can use any dock. What is the difference in the total number of docks required to provide the service level specified? Are there other design alternatives that should be considered?

10.101 Suppose the data provided in Problems 10.98 and 10.99 were obtained by studying for 15 working days the arrival patterns during the 8-hr work shift of arriving trucks and the "dwell times" at the docks for each category of truck. What concerns might you have about the data that were collected? Suppose the distribution center is used to store products that are highly seasonal or have a strong cyclic behavior during the year, during a month, during a week, and during a day.

10.102 A facility has two bridge cranes operating on a single bridge crane runway. To increase the amount of crane capacity, it has been suggested that a third bridge crane be placed on the runway. Given it is structurally feasible, describe the steps to use in developing and using a simulation model to determine the desirability of installing a third bridge crane.

10.103 Consider the facility location models and layout models in Sections 10.2, 10.3, and 10.4. When might simulation be used to aid in facility location and layout decisions?

10.104 Consider the storage models in Section 10.5. How might simulation be used in actually designing a warehouse based on the results from the models?

10.105 Consider the AS/RS and order picking models in Section 10.6 and 10.7. How might simulation be used in designing an AS/RS or order picking system? What role, if any, would the models presented in Sections 10.6 and 10.7 play if simulation is used?

10.106 Consider the conveyor models presented in Section 10.8. What role might simulation play in such analyses? When should simulation be used, rather than the deterministic models?

10.107 Perform a literature search (paying particular attention to the Winter Simulation Conference Proceedings); identify and describe simulation software not cited in the chapter that can be used in facilities planning.

10.108 Perform a literature search and describe five facilities planning simulation applications for a distribution center or warehouse not referenced in the text.

10.109 Perform a literature search and describe five facilities planning simulation applications for a manufacturing facility not referenced in the text.

Part Five

EVALUATING,
SELECTING,
PREPARING,
PRESENTING,
IMPLEMENTING,
AND MAINTAINING

11

EVALUATING AND SELECTING THE FACILITIES PLAN

11.1 INTRODUCTION

The preceding discussion focused on defining facilities requirements (Part One) and developing alternative facilities plans (Parts One, Three, and Four). Part Five addresses the evaluation, selection, preparation, presentation, implementation, and maintenance of the facilities plan. In this chapter, evaluation and selection of the facilities plan are considered.

As we noted in the early chapters of the book, the process of developing facilities plans contains elements of both art and science. The artist's dependence on *creativity, synthesis,* and *style* combined with the scientist's use of *analysis, reduction,* and *deduction* is the essence of facilities planning.

Analysis is a dissection process; synthesis is a combining or creating process. Through the application of quantitative models, including computer-aided layout models, the facilities planner explores many different solution spaces for the facilities plan; through the application of synthesis, the facilities planner combines the quantitative and qualitative aspects of the plan into a set of alternative facilities plans to be evaluated. Throughout, the facilities planner is engaged in a design process.

Recall that the design process was depicted in Chapter 1 as a six-step process:

1. Define the problem.
2. Analyze the problem.
3. Generate alternative solutions.
4. Evaluate the alternatives.
5. Select the preferred solution.
6. Implement the preferred solution.

677

It is recommended that the discussion on developing layout alternatives in Section 6.10 be read again before proceeding further. Special attention should be given to the treatise on style in design. In some sense the process of developing alternative facilities plans is best described as a "groping" process. At times it seems to be a search through a maze with few clear directions as to the way out and few, if any, indications whether progress is being made.

This statement is not intended to discourage, but rather as preparation for the realities of facilities planning. In particular, it should be noted that the six-step process is probably a simplistic representation of the facilities planning process for any but the simplest problems. However, we do believe it is very valuable to follow the six steps when possible to do so. At the same time, it is recognized that the "real" design process is often an iterative one, with many steps. Also, it is typically the case that the process is applied to subsets of the design problem.

Perhaps a more accurate representation of the design process would be similar to that given below.

1. Define the design problem and review with management to ensure the scope, objectives, schedule, and budget for the design effort are acceptable.

2. If management accepts the problem definition, then proceed to step 3; otherwise, redefine the problem and repeat the process until acceptance is obtained.

3. Analyze the problem; develop the material flow requirements and the database to be used as the foundation for the design.

4. Review the database with management and operating personnel to ensure acceptance of designs based on the database.

5. If the database review reveals changes are needed in the definition of the problem, adjust the definition; if the database review reveals gaps exist in the database, gather additional data.

6. Break up the problem into subproblems; often, departmental boundaries will serve as a logical basis for subdivision. As an example, if the facility being planned is a warehouse, the problem might be divided into receiving, moving to storage, small parts storage, unit load storage, order picking/retrieval, order accumulation, packing, moving to shipping, and shipping.

7. Generate alternative solutions to each subproblem; macrodesigns or concepts are developed focusing on generic categories of alternatives. Included in the set of alternatives is an improved present method, when an existing system exists.

8. Where appropriate, develop mathematical and/or simulation models of the alternatives to test their feasibility and perform sensitivity tests; if the data collected previously are not sufficient to support the modeling effort, collect additional data.

9. Review the feasible alternatives with operating personnel to verify the reasonability of the concepts; modify the alternatives as appropriate. Often, additional alternatives will be developed at this point; if so, return to step 8.

10. Evaluate the alternatives in terms of the criteria identified in step 1, focusing on tangible and intangible areas of concern. Rank the alternatives.

11. Aggregate the solutions to the subproblems. Pay particular attention to the interfaces to ensure the integration of individual solutions results in a reasonable overall solution. Where it can be justified, develop a simulation model of the overall system to verify it will meet the throughput requirements for the system and that bottlenecks do not develop at the interfaces.

12. In many cases, alternative solutions will exist for subproblems that are close contenders for selection. If so, incorporate them into the aggregated design to obtain alternative facility plans. (It might occur that the total system will perform best by using one or more solutions to subproblems that are not best when considered alone.)

13. Evaluate alternative aggregate facility plans. Determine the economic impact of each alternative, focusing on capital investment requirements and operating expenses for budget purposes. Perform sensitivity analyses and breakeven analyses, focusing on changes in production volumes, product mix, and technological changes.

14. Rank the alternatives and present the ranking to management for selection of the preferred alternative; obtain management approval to proceed with the detailed design for the preferred alternative.

15. Develop a detailed design for the preferred facility plan. Determine equipment locations; locate columns, doors, aisles, walls; determine floor loadings, facility service requirements, and so on. At this point the architect/engineering/construction plans are initiated. Utilize project management techniques to control the project through implementation.

16. Develop detailed bid specifications for equipment and facilities construction; develop hardware and software specifications for the computer control systems.

17. Send bid specifications to qualified suppliers.

18. Receive and review bid packages; obtain management participation in the selection of the suppliers/contractors; perform any required modifications to the facility plan.

19. Perform the required construction, install the equipment, train personnel, debug the system, and turn over the system to the operating personnel.

20. Perform audits of the facility plan. When changes in condition and/or requirements justify an additional facilities planning effort, return to step 1.

This chapter is concerned with the fourth and fifth steps of the six-step design process: the evaluation of alternatives and selection of a facilities plan. It is assumed that steps 1 through 3 have been completed and that step 6 will follow the completion of the fourth and fifth steps. The input into the fourth step should be a collection of feasible facilities plans. Unfortunately, there exists a tendency to pre-empt the evaluation of alternatives by prejudging feasible plans. Such an approach often eliminates the plan that may have been most promising. To guard against the possibility, a broad cross-section of feasible plans should be included in the initial evaluation and selection steps. For example, the initial evaluation and selection of solutions to a problem involving the transport of parts to an assembly line may include a lift truck alternative, a trolley conveyor alternative, a belt conveyor alternative, a roller conveyor alternative, and an AGVS alternative. To eliminate any of the generic alternatives prior to the initial evaluation and selection might bias the solution. Conversely, to include in the initial evaluation and selection several specific models and brands of lift trucks, trolley conveyors, belt conveyors, roller conveyors, and AGVS would probably be a poor use of time. Detailed data and competitive bidding are not required to perform an initial screening. Once the initial screening is made, specific brands and vendors may be evaluated to determine the most suitable solution.

11.2   EVALUATING FACILITIES PLANS

The evaluation process includes the assessment of each alternative in terms of the criteria identified previously. *Evaluating* is not the same as *selecting!* For example, if the least-cost alternative is to be selected, then the costs for each alternative must be evaluated.

Hopefully, by separating the evaluation of alternatives from selection of the preferred alternative, the selection will be made more objectively. If one is forced to formally evaluate the performance of each alternative, it is less likely that someone's "favorite" (based on subjective factors) will be selected without understanding the impact of the selection.

If the criteria are easily quantified, the process of evaluation is easier to perform. However, it is typically the case in facilities planning that both quantitative and qualitative considerations are employed in evaluating alternatives. Few would disagree with the claim that the evaluation and selection steps are the most difficult steps of the six-step design process.

Among the techniques used in evaluating alternative facilities plan are the following:

1.   List the positive and negative aspects of each alternative.
2.   Rank the performance of each alternative against each of several enumerated criteria.
3.   Perform weighted factor comparison of the alternatives by assigning a numerical weight to each criterion (factor), ranking numerically each alternative against each criterion, and summing the weighted rankings over all criteria to obtain a total weighted factor for each alternative.
4.   Determine the economic performance of each alternative over a specified planning horizon.

Listing Advantages and Disadvantages

The listing of advantages and disadvantages of each alternative provides a simple way to evaluate facilities plans. Of the evaluation techniques listed, it is probably the easiest to perform. However, it is also difficult to obtain an accurate, balanced, and objective evaluation of alternative facilities plans by simply listing the positive and negative aspects of each. Biases are easily introduced when using this approach; to guard against the possibility that someone might be favoring one alternative over the others, it is wise to engage a team of individuals in identifying the advantages and disadvantages of each alternative.

Ranking

Ranking the alternatives has the benefit of requiring that all alternatives be compared against a common set of factors. Furthermore, it forces an explicit consideration of the several factors that will influence the selection decision. However, there is no guarantee that all important factors will be considered. Additionally, the ranking process may yield too much information; it may overwhelm the decision makers and confuse the selection process. After using the ranking procedure, the results must be integrated to allow a selection to be made. The integration can be either explicitly or implicitly performed; either way, it occurs! Again, to guard against possible bias, engage a team in performing the ranking.

Weighted Factor Comparison

Similar to the prioritization matrix described in Chapter 3, the weighted factor comparison method provides an explicit method for integrating the rankings. Numerical values or weights are assigned proportionally to each factor based on their degrees of importance. A numerical score is then assigned to each alternative based on its performance against a particular factor. The scores are multiplied by the weights and the products are summed over all factors to obtain a total weighted score. Although there are some very obvious scaling problems associated with the technique, it is quite popular.

Both the ranking approach and the weighted factor comparison approach involve a comparison of facilities planning alternatives for each factor. In an attempt to make the comparison as objective as possible, a paired comparison is recommended for each factor.

To illustrate the approach, suppose there are four facilities planning alternatives (W, X, Y, and Z). Further, suppose the following preferences are obtained by comparing the alternatives two at a time for a particular factor.

$$W < Y \qquad X > Y$$
$$W < X \qquad X < Z$$
$$W < Z \qquad Y < Z$$

where $W < Y$ means W ranks lower than Y and $X > Y$ means X ranks higher than Y. Combining the paired comparisons yields the following ranking.

$$Z > X > Y > W$$

for the factor considered. For some other factor an entirely different ranking would probably occur.

By performing the paired comparisons, inconsistencies in the rankings will be revealed. For example, suppose in the previous example a ranking of $Y > Z$ had occurred. Obviously, it is inconsistent to state that

$$X > Y \qquad Y > Z \qquad Z > X$$

Even though the ranking procedure indicates preferences, it does not indicate the strength of the preference. For example, there is no distinction between

$$Z >>> X >> Y > W$$

and

$$Z > X >> Y >>> W$$

Suppose values could be accurately assigned to a factor in direct proportion to its preference. Then $W = 1$, $Y = 3$, $X = 6$, and $Z = 10$ would receive the same ranking ($Z > X > Y > W$) as would $W = 1$, $Y = 2$, $X = 9$, and $Z = 10$.

In using ranking and weighted-factor comparison approaches, care is required to avoid the halo effect. The halo effect is a phenomenon that occurs when a high ranking on one factor carries over and influences the ranking on other factors. If one alternative clearly dominates all other alternatives on all or almost all factors, then it is likely that the halo effect is present.

Among the factors that are considered in using the ranking procedure and weighted factor comparison are the following:

1. Initial investment required
2. Annual operating costs
3. Return on investment (ROI)
4. Payback period
5. Flexibility or ease of changing or rearranging the installed system
6. Integration with and ability to serve the process operations
7. Versatility and adaptability of the system to accommodate fluctuations in products, quantities, and delivery times (ease of changing or rearranging the installed system)
8. Ease of future expansion
9. Limitations imposed by handling methods on the flexibility and ease of expansion of the layout and/or buildings
10. Space utilization
11. Safety and housekeeping
12. Working conditions and employee satisfaction
13. Ease of supervision and control
14. Availability of trained personnel
15. Frequency and seriousness of potential breakdowns
16. Ease of maintenance and rapidity of repair
17. Volume of spare parts required to stock
18. Interruption or disruption of production and related confusion during the installation period
19. Quality of product and risk of damage to materials
20. Ability to pace, or keep pace with, production requirements
21. Effect on in-process time
22. Personnel problems—available workers with proper skills, training capability, disposition of redundant workers, job description changes, union contracts, or work practices
23. Availability of equipment needed
24. Availability of repair parts
25. Tie-in with scheduling, inventory control, paper work
26. Effect of natural conditions—land, weather, sun, temperature
27. Compatibility with the operating organization
28. Potential delays from required synchronization and peak loads
29. Supporting services required

30. Integration with other facilities
31. Tie-in with external transportation
32. Time required to get into operation—installation, training, and debugging
33. Degree of automation
34. Software requirements
35. Promotional or public relations value

A form that is useful in performing a weighted factor comparison is given in Figure 11.1. The factors that are relevant for a particular evaluation and the weights

WEIGHTED FACTOR COMPARISON FORM

Company _____ Prepared by _____ Date _____

Facility Plan for _____

Sheet _____ Of _____

Alternatives

Factor	Weight	A		B		C		D		E	
		Rt.	Sc.	Rt.	Sc.	Rt.	Sc.	Rt.	Sc.	Rt.	Sc.
1.											
2.											
3.											
4.											
5.											
6.											
7.											
8.											
9.											
10.											
11.											
12.											
13.											
14.											
15.											
16.											
17.											
18.											
19.											
20.											
Totals											

Figure 11.1 Weighted factor comparison form.

of these factors should be listed in the first two columns. Once the factors and weights have been determined, each of the alternatives should be rated (Rt) and the score (Sc) determined for each factor by multiplying the rating times the weight. The overall evaluation of each alternative is obtained by adding the scores for all the factors. It is convenient to assign factor weights such that their sum equals 100 and to select rating values between zero and ten.

When using weighted factor comparison, it is important to remember that quantifying subjective factors can be tricky business. The underlying process is one of ranking; hence, the data are ordinal data, not ratio data. As a result, arithmetic results from rank-ordered data do not have the same degree of accuracy as arithmetic operations with ratio data such as weight, distance, and cost.

For further discussion of scaling difficulties when using ordinal data, see [7]. Likewise, for further treatment of multiattribute decision making, see [1], [3], [5], [6], [9], and [10]. Software has been developed to facilitate the assignment of numerical values to each alternative for each factor when using weighted factor comparison [2].

Example 11.1

Four layout alternatives (A, B, C, and D) are to be evaluated for BWT², Inc. Five factors are to be used in evaluating the layout alternatives: annual material handling cost, construction cost, ease of expansion, employee's preference, and access to railroad siding. The following weights are assigned to the evaluation factors:

1.	Annual material handling cost	25
2.	Construction cost	15
3.	Ease of expansion	20
4.	Employee's preference	30
5.	Rail siding access	10
	Total	100

Each layout alternative was ranked on the basis of each factor, with the following results:

	Layout Alternatives			
Factor	**A**	**B**	**C**	**D**
1	10	7	5	2
2	8	4	10	9
3	7	8	10	9
4	6	8	7	10
5	10	6	5	7

The weighted factor comparison for the four alternative layouts is summarized in Figure 11.2. From the results of the analysis, alternative A has the highest score. If the weights and ratings truly reflect management's feelings, then alternative A would be recommended. From the example, it is easy to see how sensitive the selection decision is to the assignment of ratings. For example, if the factor 4 ratings for alternatives C and D had been swapped, the totals would have been 790, 695, 825, and 645, resulting in alternative C being recommended.

WEIGHTED FACTOR COMPARISON FORM

Company: BWT², Inc.   Prepared by: J. Bo Whitankins   Date April 1, 2003
Description of investment: Assembly plant

Factor	Wt.	A		B		C		D		E	
		Rt	Sc	Rt	Sc	Rt	Sc	Rt	Sc	Rt	Sc
1. Annual material handling cost	25	10	250	7	175	5	125	2	50		
2. Construction cost	15	8	120	4	60	10	150	9	135		
3. Ease of expansion	20	7	140	8	160	10	200	9	180		
4. Employee's preference	30	6	180	8	240	7	210	10	300		
5. Rail siding access	10	10	100	6	60	5	50	7	70		
6.											
7.											
8.											
9.											
10.											
11.											
12.											
13.											
14.											
15.											
16.											
17.											
18.											
19.											
20.											
Totals	100		790		695		735		735		

Figure 11.2   Weighted factor comparison for Example 11.1.

Economic Comparison

Justifying facilities planning investments is of intense interest. Because some have had difficulty justifying their favorite technologies, discounted cash flow (DCF) methods have been criticized and the adoption of new justification approaches have been advocated. Such reactions are akin to shooting the messenger because you don't like the message. Generally, it isn't the justification methods that are inadequate, it is the way they are applied!

In the early 1900s, Henry Ford noted, "If you need a new machine and don't buy it, you pay for it without ever getting it." Cost savings opportunities are often missed due to poorly executed economic justifications.

In performing financial justifications of facilities plans, the following Systematic Economic Analysis Technique (SEAT) can be used to justify investments *that deserve to be justified:*

1.   Specify the feasible alternatives to be compared.
2.   Define the planning horizon to be used.
3.   Estimate the cash flows for each alternative.
4.   Specify the discount rate to be used.
5.   Compare the alternatives using a discounted cash flow (DCF) method.
6.   Perform supplementary analyses.
7.   Select the preferred alternative. [30]

Specifying the Feasible Alternatives

Specifying the set of feasible alternatives to be compared is the most important step in justifying a facilities plan. For, if an alternative is not included in the comparison, it will not be selected! Likewise, it is fruitless to include in the comparison those alternatives that will not meet the facilities planning requirements.

A frequently used alternative is the "do-nothing" alternative, since it often serves as the baseline against which the "do-something" alternatives are compared. However, no business situations stand still. Hence, doing nothing is seldom feasible. For while your firm is doing nothing, your competition will be doing something. "Business as usual," if perpetuated, is a failing business strategy. Yet, situations will arise in which "standing pat" is preferred to making a capital investment in a new or improved facility.

An objective assessment of "doing nothing" must include both cost and revenue impacts. Too often, a myopic approach is taken and only those costs that arise *inside* a distribution facility or manufacturing plant are considered. Since investments in facilities often enhance revenues, due to cost reductions, increased quality, and more timely responses to customer demands, marketing representatives should be invited to assess the impact of doing nothing. Many have found facilities planning investments easier to justify with marketing on their side.

In addition to focusing on the do-nothing alternative, a range of alternatives should be considered. Unfortunately, attempts to justify the acquisition of the ultimate in technological sophistication frequently occur by comparing its economic performance to that of a very poor current method. No consideration is given to intermediate alternatives, since one might be the better economic choice.

Widespread biasing of justifications toward particular solutions led the treasurer of a Fortune 100 company to admit that he did not trust engineers' justifications. When "important" decisions had to be made, he depended on his intuition and judgment of what was best for the firm, rather than the numbers submitted by the engineers! Care must be taken to ensure that technological preferences and egos do not cloud one's judgment of what is best for the firm over the planning horizon.

A related phenomenon is the engineers' attempts to thwart the use of "size gates" in the approval process, where investments less than a particular size require plant manager approval only, those within another range of sizes require divisional approval, and those above a specified upper limit require corporate approval. When companies delegate approval authority on the basis of the size of the investment, engineers often subdivide the investment into smaller pieces in order to obtain approval at lower levels. Rather than propose an integrated system, several "islands of automation" are defined and one or more are proposed each year, depending on the investment requirements and the size gates used.

Making piecemeal investments can postpone the benefits of an integrated system for years. Also, in trying to "eat the elephant a bite at a time," one or more smaller "bites" might not be sold. Instead of proposing small increments, many firms believe the best strategy is to "design the whole, sell or justify the whole, and then implement the pieces." In particular, the total system is justified and sold to management. When capital is limited or size gates are in use, a phased implementation plan is used and individual components of the total system are implemented over time [9].

Defining the Planning Horizon

The second step in the SEAT approach is the definition of the planning horizon to be used. (For those unfamiliar with the term, the planning horizon is the period of time over which the economic performance of an investment will be measured and evaluated.)

In general, U.S. firms use too small a planning horizon, often less than five years. Japanese firms, on the other hand, typically think in terms of 10 to 20 years. One manifestation of short-sighted management is the use of short planning horizons.

The planning horizon defines the width of a window through which we will look in evaluating each investment alternative. An important economic justification principle is that the same window be used in evaluating all alternatives. Hence, a 3-year window should not be used for one alternative and a 10-year window used for another alternative. If a 10-year window is allowed, then each alternative should be evaluated over the 10-year period; if alternatives exist that are not expected to last 10 years, then explicit consideration should be given concerning the intervening investment decisions over the planning horizon.

The planning horizon should be distinguished from the working life of equipment and the depreciable life of equipment, since it might have no relationship to the other two periods of time. The planning horizon is simply the time frame to be used in comparing the alternatives and should realistically represent the period of time over which reasonably accurate cash flow estimates can be provided.

Some firms use standard planning horizons, which, as noted above, are too short. This was the case for a major U.S. textile firm which was losing market share. A major modernization was proposed. However, to realize the full benefits from the investment, the system had to be debugged, all personnel had to be trained, and lost customers had to be regained. It was impossible to accomplish this within the 2-year period required. Faced with the economic consequences of maintaining the status quo, the firm elected to adopt a longer planning horizon.

Planning horizons also can be too long; a balance is needed. Based on our experience in justifying facilities planning investments, we prefer at least a 10-year planning horizon [18].

Estimating the Cash Flows

The third step in justifying a facilities plan is estimating the cash flow profiles for each alternative. What is a cash flow profile? Year-by-year estimates of money received (receipts, revenues, income, benefits or savings) and money spent (disbursements, expenses or costs) constitute a cash flow profile. Both the obvious and the not-so-obvious cash flows should be included. For example, investments in new or improved facilities often reduce inventory, increase quality, reduce space, increase flexibility, reduce lead time, increase throughput, reduce scrap, and increase morale. Though real, the economic benefits of each are difficult to quantify. Hence, many call them *intangible benefits* and do not include them in economic justifications.

Robert Kaplan, Harvard Business School accounting professor, noted, "Although intangible benefits are difficult to quantify, there is no reason to value them at zero in a capital expenditure analysis. Zero is, after all, no less arbitrary than any other number. Conservative accountants who assign zero values to many intangible benefits prefer being precisely wrong to being vaguely right. Managers need not follow their example" [8]. In the end a rough estimate of intangible benefits is better than no estimate at all! Intangibles can be effective competitive weapons in increasing revenues and decreasing operating costs.

Since the investment is for the future not the past, estimates of the cash flows for each alternative should reflect *future* conditions not *past* conditions. Best estimates are needed of the annual costs and incomes anticipated over the planning horizon, including a liquidation or salvage value estimate at the end of the planning horizon. Although it is not necessary to include estimates of cash flows that will be the same for all alternatives, it is important to capture the differences in the alternatives. Figure 11.3 provides a cost determination form that may prove useful in accumulating the annual investment and operating costs for each investment alternative.

A number of different approaches can be used to develop cost estimates. Among the estimation techniques commonly used are the following: guesstimates, historical standards, converting methods, ratio methods, pilot plant approaches, and engineered standards.

A **guesstimate** is an educated guess or estimate based on experience. Commonly, when guesstimates are to be used, several individuals are consulted independently and requested to provide estimates; these are screened and pooled to develop a figure to be used.

Historical standards are developed through years of experience in designing facilities. The more refined and accurate historical standards are those which are expressed in small "chunks" rather than an aggregated standard. As an illustration, equipment installation costs can be expressed as a percentage of acquisition costs. However, if the standard percentage is an aggregate for a wide variety of equipment, then it will not be as accurate an estimate as one that is developed specifically, for, say, cantilever racks.

The **converting method** is closely related to the use of historical standards. Using the converting method to estimate production space requirements, the present space would be converted to square footage per unit produced. Consequently, if the production rate is to be doubled, it would be estimated that twice as much production space is required. Obviously, the method is subject to error and caution should be exercised when it is used.

COST DETERMINATION FORM

Company _____ Prepared by _____ Date _____
Handling System for _____
Sheet _____ of _____

	Alternative ___ Year ___	Alternative ___ Year ___	Alternative ___ Year ___	Alternative ___ Year ___	Alternative ___ Year ___
Equipment type					
Model					
Vendor					
Date of quote					
Investment					
Invoice-price					
Installation charge					
Maintenance facilities					
Power facilities					
Alterations of facilities					
Freight chrges					
Consulting charges					
Supplies					
Other					
Total Investment Cost					
Operating cost					
Interest on Investment					
Taxes					
Insurance					
Supervision					
Clerical					
Maintenance labor					
Power costs					
Other					
Total Operating Cost					

Figure 11.3 Cost determination form.

Ratio methods are related to the converting method and involve the development of ratios of, say, number of rack openings to some other factor that can be measured and predicted for the proposed layout. Examples include number of rack openings per pallet load received; square footage of floor space per rack opening; number of receiving employees per ton of materials received; square footage of aisle space per square footage of production space; annual utility cost per square foot of space; annual maintenance cost per dollar invested in equipment; annual operating cost per dollar invested in equipment; annual inventory carrying cost per dollar invested annually in inventory; annual pilferage cost per dollar invested annually in inventory; total investment cost per rack opening in an automated storage system; conveyor investment cost per foot of conveyor; AGV operating cost per foot of guide path; in-process handling cost per dollar spent in production; and material handling cost per dollar spent in production.

The **pilot plant** approach involves the operation of the proposed system on a significantly reduced scale. As a substitute for a pilot plant, scale models and simulations can be used to develop reasonable estimates for the "real thing."

Engineered standards can be developed by systematically analyzing the situation, identifying cause—effect relationships, and developing prediction equations. Using simulation, work measurement, waiting line, inventory control, production planning, and forecasting methods, as well as other analytical models, the engineer can normally develop accurate estimates of many of the important benefit and cost elements.

An example of the use of engineered standards in estimating AS/RS equipment costs was described in Chapter 10. Three components were considered: S/R machine, storage racks, and building. The S/R machine cost was represented as a function of the height of the AS/RS, the weight of the unit load, and the type and location of the control logic. The storage rack cost was given as a function of the dimensions of the unit load, the weight of the unit load, and the height of the rack. The building cost was expressed as a function of building height.

"Ballpark" estimates of equipment costs frequently can be obtained from equipment vendors; however, such estimates should not be interpreted as firm quotes or bids. As an example, the ratio method is often used to estimate the acquisition cost of storage rack. Based on estimates obtained from four different firms, the costs per 2500-lb load given in Table 11.1 were obtained for various storage methods. (The costs given include the cost of the pallet and storage rack.)

Acquisition cost alone does not provide an accurate estimate of the cost of storage rack. Freight cost ($3 to $4/100 lb of weight) and installation cost should be estimated as well. These, and other, "add-ons" can substantially increase the cost of installation.

Many who evaluate facilities planning investment alternatives fail to recognize the impact such an investment will have on revenues. Consideration of the competition and the market environment are seldom included in the evaluation, but they should be! Often, it is blindly assumed that the only cash flows associated with replacing existing equipment with state-of-the-art equipment are the costs associated with each alternative. The fact that installing the newest equipment will influence customers to

Table 11.1 *Storage Rack Cost Estimates*

Storage Method	Cost per Pallet Load (dollars)
Bulk storage	$10–15
Portable stacking rack	50–100
Drive-in rack	50–75
Pallet flow rack	125–175
Selective pallet rack	35–60
AS/RS rack	125–200

select your product rather than the competitor's product is often overlooked. Too frequently, the adverse economic consequences of the "do nothing" alternative are underestimated. Maintaining the status quo can have a substantial impact on both costs and revenues. If the competition is doing "something" while you are doing "nothing," your firm's revenue stream could be reduced significantly [19].

Specifying the Discount Rate

The next step in justifying facilities planning investments is specifying the discount rate to be used. Variously called a hurdle rate, interest rate, return on investment, and minimum attractive rate of return (MARR), the discount rate is essential to the discounted cash flow (DCF) methods of economic justification.

Underlying all DCF methods is the recognition that *money has time value!* Since money has time value, you would prefer receiving $100,000 today to receiving $100,000 a year from today. By owning the money one year sooner, you can rent (loan) the money to someone who needs it and charge interest for its use. By receiving the money sooner, you can increase its value over time.

If you choose to invest the money in, say, material handling equipment, then you should charge interest against the investment to reflect the loss of income available to you by "loaning the money" or investing it elsewhere and earning a particular return. Using such an approach, the discount rate reflects the foregone opportunity to invest elsewhere and is called an *opportunity cost* discount rate.

The discount rate is used to convert the mix of cash flows occurring throughout the planning horizon to a meaningful economic measure that can be used rationally to compare facilities planning investment alternatives. Due to its impact on the comparative advantages of the investment alternatives, the discount rate used should reflect the real opportunity cost of alternative uses of the investment capital.

An argument given frequently for higher hurdle rates in the United States is the cost of capital. However, as Kaplan [8] pointed out, over a 60-year period the real cost of capital in the United States was less than 8%. He believes that the use of an overly large discount rate is a primary reason U.S. industry has underinvested in capital improvements. If a firm is using an after-tax, inflation-free discount rate greater than 10%, a careful examination should be performed to ascertain its negative impact on the long-term viability of the firm.

In our treatment of the economic justification of facilities planning investments, the discount rate will be referred to as the MARR and its value will be based on the opportunity cost concept. Throughout, we will argue that an investment should not be made if it will not earn a return at least as great as the MARR, since the MARR should be based on the opportunities for alternative investment at comparable risk [20].

Comparing the Investment Alternatives

Comparing the economic performances of facilities planning investment alternatives is the fifth step in the justification process. Because money has time value the following rules apply when comparing investment alternatives that might involve cash flows of different magnitudes occurring at different points in time: monies can be added and/or subtracted only at the same point(s) in time; to move a cash flow forward in time by one time period, multiply the magnitude of the cash flow by the quantity $(1 + i)$, where i is the time value of money expressed in decimal form; to move a cash flow backward in time by one time period, divide the magnitude of the cash flow by the quantity $(1 + i)$. The process of multiplying and dividing cash flows by $(1 + i)$ is called compounding and discounting, respectively. The overall process of converting cash flows to single-sum equivalents (called present values and future values) or equivalent annual series (called annual value) is called a *discounted cash flow process* (DCF).

Although many discounted cash flow (DCF) methods exist, it is likely that your employer will have a perferred method of comparison. The most popular DCF methods are net present value (NPV) and internal rate of return (IRR); others include net annual value (NAV), net future value (NFV), and discounted payback period (DPP). The NPV method involves discounting all cash flows over the planning horizon to a common point in time called the present. The IRR, as noted, involves the determination of the discount rate that yields an NPV equal to zero; since trial and error can be time consuming, the use of financial calculators or spreadsheet software programs is highly recommended. (A note of caution: consult an authority on the subject if the net cash flows over the planning horizon change signs multiple times.) The NAV method involves converting the NPV into an equal annual series of cash flows over the planning horizon. The NFV method involves compounding all cash flows over the planning horizon to the end of the planning horizon. The DPP method involves the determination of the point during the planning horizon when the cumulative NPV equals zero or becomes positive. There are a host of very fine books devoted to the economic comparison of investment alternatives. For detailed treatments of alternative DCF methods, see any of the economic justification texts [4], [30].

Using Spreadsheets. Rather than repeat what is presented in engineering economics texts, we will focus on the use of readily accessible software packages to compare the investment alternatives. A variety of software packages are available to facilitate the performance of SEAT's fifth and sixth steps. Of particular significance is the development of spreadsheet software (e.g., Excel®, Lotus 1-2-3®, and Quattro® Pro). To show how a spreadsheet can be used to facilitate both financial calculations and the performance of sensitivity analyses, note that a spreadsheet can be developed by using a single formula: $V_0 = V_t(1 + i)^{-t}$, where V_t denotes the value of a cash flow occurring at time t, V_0 denotes the value of the cash flow at time zero (the present value), and i denotes the time value of money.

Consider the spreadsheet given in Table 11.2. The columns are identified alphabetically and the rows are identified numerically. The interest rate is entered, using decimal format, in cell B1. The magnitudes of cash flows are entered in column B, beginning with row 4.

As shown in Table 11.2, when using Excel[1], the net present value of a series of cash flows can be easily obtained. Enter **=npv(b1,b5:b9)+b4** in cell **B10**. To determine the net annual value, use Excel's PMT function, which converts a net present value to a uniform series of cash flows; as shown in Table 11.2, **=pmt(b1, a9,−b10)** is intered in cell **B11** to obtain the NAV; the parameters required for the PMT function are the discount rate, the number of years, and the negative of the value obtained for NPV. To obtain the net future value, use Excel's FV function, which converts a uniform series into its equivalent net future value; as shown in Table 11.2, **=fv(b1,a9,−b11)** is entered in cell **B12** to obtain the NFV; the parameters required

---

[1]Microsoft® Excel 2000, Microsoft Corporation, 2000.

Table 11.2   *Sample Spreadsheet*

R\C	A	B
1	Discount Rate =	(enter interest rate)
2		
3	Time	Cash Flow
4	0	A_1
5	1	A_2
6	2	A_3
7	3	A_4
8	4	A_5
9	5	A_6
10	NPV	=npv(bl,b5:b9)+b4
11	NAV	=pmt(bl,a9,−b10)
12	NFV	=fv(bl,a9,−b11)
13	IRR	=irr(b4:b9)

for the FV function are the discount rate, the number of years, and the negative of the value obtained for NAV. Excel also has a financial function for determining the internal rate of return for an investment. Entering **=irr(b4:b9)** in any cell in the spreadsheet will yield the value of the discount rate that will result in a net present value of zero for the cash flows B4 through B9.

Example 11.2

A conveyor is purchased for $10,000. Its application yields net annual cost savings of $3,000 per year for a 5-year period. What are the net present value, the net annual value, the net future value, and internal rate of return for the investment if a discount rate of 10% is used and a zero salvage value for the conveyor is assumed?

As shown in Table 11.3, the net present value for the investment is $1,372.36. *It is important to note that the NPV function computes the present value of a series of cash flows as of one period prior to the first value in the series.* Using Excel's NPV financial function,

Table 11.3   *Spreadsheet Solution for Example 11.2*

R\C	A	B
1	Discount Rate =	0.10
2		
3	Year	Cash Flow
4	0	($10,000.00)
5	1	$3,000.00
6	2	$3,000.00
7	3	$3,000.00
8	4	$3,000.00
9	5	$3,000.00
10	NPV	$1,372.36
11	NAV	$362.03
12	NFV	$2,210.20
13	IRR	15.24%

the net present value is determined for the cash flows entered in cells B5 through B9; hence, to obtain the present value for the entire cash flow series it is necessary to add the entry in cell B4 to the value obtained by using the NPV function over the range of cells from B5 through B9.

Also, as shown in Table 11.3, the NAV equals $362.03, the NFV equals $2,210.20, and the IRR equals 15.24% for an investment of $10,000 that returns $3,000 per year for 5 years.

When faced with choosing the most economic alternative from among several mutually exclusive investment alternatives, rank them on the basis of their NPVs and recommend the one having the largest value. However, when using the IRR method, it is not necessarily the case that the one with the greatest rate of return is preferred. Hence, it is recommended that NPV analysis be used to determine the preferred alternative and, then, its IRR be computed.

Example 11.3

To illustrate the DCF process for a more realistic material handling investment, suppose $1,000,000 is to be invested in a distribution center to improve the material handling system. A planning horizon of 10 years is used, along with an inflation-free, after-tax minimum attractive rate of return of 10%. Let the net annual after-tax cash flow resulting from the investment be a positive $200,000 over the 10-year period, plus a salvage value of $250,000. Hence, the cash flow profile for the investment consists of a negative single sum of $1,000,000 occurring at time zero, a uniform series of $200,000 occurring annually at the end of years 1 through 9, and a single sum of $450,000 occurring at the end of year 10.

As shown in Table 11.4, the net present value for the investment is $325,295, which can be interpreted as the excess above a 10% return on the invested capital. The internal rate of return is 17.98%.

Table 11.4   *Spreadsheet Solution for Example 11.3*

R\C	A	B
1	Discount Rate =	0.10
2		
3	Year	Cash Flow
4	0	($1,000,000.00)
5	1	$200,000.00
6	2	$200,000.00
7	3	$200,000.00
8	4	$200,000.00
9	5	$200,000.00
10	6	$200,000.00
11	7	$200,000.00
12	8	$200,000.00
13	9	$200,000.00
14	10	$450,000.00
15	NPV	$325,299.24
16	IRR	17.98%

When investments are being made in facilities and in manufacturing and warehousing equipment, it is important to consider the effects of income taxes on the economic viability of capital investments. To illustrate the effect of income taxes on a decision, suppose you are considering using a rack-supported warehouse versus a conventional column-supported warehouse. It might well be that a substantial portion of the capital investment in the rack-supported warehouse would qualify as equipment. The tax treatment, alone, could make a substantial difference in the after-tax NPV of the warehouse investment.

Another factor to consider is inflation. In the United States, depreciation charges do not change as a result of inflation. Consider the case in which you must choose between two investment alternatives. The first has high labor and annual operating costs, but low capital investment. The second alternative involves an investment in automation and has low labor and operating costs, but a high capital investment. In the absence of inflation, suppose the automation alternative has the largest NPV of the two. If modest-to-high inflation rates occur during the investment period, it could well reverse the economic preference between the two investment alternatives.

Other DCF methods yield the same recommendations as the net present value method. However, some do not guarantee to provide the same recommendation as the net present value method—chief among them is the payback period method, which is commonly applied using a 2-year payback requirement. It calls for a determination of the length of time required to fully recover the initial investment without regard to the time value of money.

As shown in Table 11.5, for Example 11.3, it would take 5 years to recover the initial investment. Even though it is a simple method to use, beware the payback period method! The alternative that is the first to recover the initial investment is

Table 11.5 *Payback Period and Discounted Payback Period Solutions for Example 11.3*

R\C	A	B	C	D	E
1	Discount Rate =	0.10			
2					
3	Year	Cash Flow	Net Cash Flow	Cumulative NPV	Cumulative NPV
4	0	($1,000,000.00)	($1,000,000.00)	=b4	($1,000,000.00)
5	1	$200,000.00	($800,000.00)	=npv(b1,b5:b5)+b4	($818,181.82)
6	2	$200,000.00	($600,000.00)	=npv(b1,b5:b6)+b4	($652,892.56)
7	3	$200,000.00	($400,000.00)	=npv(b1,b5:b7)+b4	($502,629.60)
8	4	$200,000.00	($200,000.00)	=npv(b1,b5:b8)+b4	($366,026.91)
9	5	$200,000.00	$0.00	=npv(b1,b5:b9)+b4	($241,842.65)
10	6	$200,000.00	$200,000.00	=npv(b1,b5:b10)+b4	($128,947.86)
11	7	$200,000.00	$400,000.00	=npv(b1,b5:b11)+b4	($26,316.24)
12	8	$200,000.00	$600,000.00	=npv(b1,b5:b12)+b4	$66,985.24
13	9	$200,000.00	$800,000.00	=npv(b1,b5:b13)+b4	$151,804.76
14	10	$450,000.00	$1,250,000.00	=npv(b1,b5:b14)+b4	$325,299.24
15	NPV	$325,299.24			
16	IRR	17.98%			

the preferred choice; hence, a $1,000,000 investment promising payback of $500,000 per year for two years and $5,000 per year for the next eight years would be preferred to the alternative under consideration.

The payback period method is short-sighted and ignores the financial performance of an investment beyond its payback period; yet it is probably the most popular method in use among U.S. firms. Why? Because it is easy to compute, use, and explain; it provides a rough measure of the liquidity of a project; and it tends to reflect management's attitude under limited capital conditions. However, due its shortcomings, it is recommended that the payback period be used as a tie-breaker only when DCF methods are used [21].

The discounted payback period method can be used to determine how long it will take for the initial investment to be recovered without ignoring the time value of money. As shown in Table 11.5, the discounted payback period is the time required for the cumulative net present value to equal zero. In the case of the example, slightly more than 7 years is required for the initial investment to be fully recovered based on a time value of money equal to 10%.

Performing Supplementary Analyses

The sixth step, performing supplementary analyses, though surely the most neglected, often proves to be the most important step. Engineers often resist using formal justification processes due to the absence of accurate data. The notion of "garbage-in, garbage-out" comes to mind. While the quality of the recommendation can be no greater than the quality of the inputs to the process, it is also true that supplementary analysis can provide valuable insights concerning the economic viability of an investment alternative *in the absence of perfect information.*

Three kinds of supplementary analyses are available: *break-even analysis, sensitivity analysis,* and *risk analysis.* Break-even analysis is useful when you have no clue as to the true value of one or more parameters, but you can decide if the value is greater than or less than some *break-even value.* Break-even analysis, for example, might prove useful when you are unable to quantify exactly the value of reductions in space requirements, cycle times, and inventory levels, as well as increases in quality levels, market share, throughput levels, and flexibility. In each case, the answer to the following question might prove useful. How much must the reduction (or increase) in _____ be worth for the investment to be justified? If the answer is $5,000,000 for the value of, say, increased flexibility and you know that such a value is unreasonable, then the investment should not be undertaken. Next, suppose an investment in material handling will reduce from eight to two weeks the time required to fill a customer's order; if at least a 2% increase in customer orders have to occur annually to justify the material handling system, then the investment appears attractive.

Even if your firm does not require supplementary analysis in its economic justification package, give strong consideration to doing it—if for no one else's benefit than yours! By addressing many what-if questions, supplementary analysis serves as an "insurance policy" and increases your confidence in your recommendation to management. In summary, do more than *just what is required.* Supplementary analysis provides a means of *going the extra mile* in ensuring that a firm's scarce capital is invested wisely.

In justifying facilities planning investments, remember the axiom: *don't wade in rivers that on the average are two feet deep!* Economic justifications typically use best estimates or "average values" of the parameters embodied in the analysis. Yet, it is seldom the case that the "average condition" will occur. Managers are seldom as concerned with average conditions as they are with rare or unusual conditions! Many people have drowned in rivers that are on the average two feet deep; such rivers can have very deep holes and swift currents. The purpose of supplementary analysis is to anticipate the "deep holes and swift currents" and ascertain their impact on the viability of the facilities planning investments under consideration.

The three supplementary analysis techniques differ in the degree of knowledge you must have concerning the "true values" of the "uncertain parameters." Break-even analysis requires knowing if the "true value" of a parameter is above or below the break-even value and sensitivity analysis requires knowledge of the range or possible values of each parameter. Risk analysis, on the other hand, requires knowing not only the possible values of the parameters but also their probabilities of occurrence.

Typical of the parameters addressed through the use of sensitivity analysis are the discount rate used, the length of the planning horizon, the magnitude of the initial investment, the annual savings realized, and the operating costs required. Sensitivity analysis has as its objective the determination of the sensitivity of the decision to incorrect estimates of the values of one or more parameters. One approach often used is to specify minimum, maximum, and most likely estimates for each parameter in question. By determining the economic performance of each alternative for each combination of values for the "uncertain" parameters, it is anticipated that a better-informed decision will result.

One of the greatest powers of spreadsheets is their use in performing sensitivity analyses. As an illustration, recall Example 11.2. If the net annual savings had been only $2,500, then, as shown in Table 11.6, the resulting NPV value would have been −$523.03. Likewise, as shown in Table 11.7, if the interest rate in Table 11.3 is changed to 16% and the net annual savings remain $3,000, the net present value changes to −$177.12.

Risk analysis requires the greatest knowledge concerning the future values of the parameters. Simulation is generally used to convert the probability distributions

Table 11.6 *Using a Spreadsheet to Analyze Changes in Cash Flows*

R\C	A	B
1	Discount Rate =	0.10
2		
3	Time	Cash Flow
4	0	($10,000)
5	1	$2,500
6	2	$2,500
7	3	$2,500
8	4	$2,500
9	5	$2,500
10	NPV	−$523.03
11	IRR	7.93%

Table 11.7   *Using a Spreadsheet to Analyze Changes in Interest Rate*

R\C	A	B
1	Discount Rate =	0.16
2		
3	Time	Cash Flow
4	0	($10,000.00)
5	1	$3,000.00
6	2	$3,000.00
7	3	$3,000.00
8	4	$3,000.00
9	5	$3,000.00
10	NPV	($177.12)
11	IRR	15.24%

for the parameters to a simulated probability distribution for the measure of economic worth. Based on the results obtained, an estimate can be provided of the probability of the investment yielding a positive outcome. (A more sophisticated version utilizes probability distributions of probability distributions, but such an approach is not likely to be merited; those who undertake this approach are in danger of succumbing to a terminal case of *paralysis of analysis!*)

Although it is easy to become enamored of the supplementary analysis techniques, it is important to remember why they are used—to enhance the quality of the ultimate investment decision. One test of this is to ask if you would make the investment if your money were involved. That, indeed, is the bottom line! [22].

Selecting the Preferred Alternative

At this point, you have available a set of facilities planning investment alternatives, as well as extensive information concerning their financial and operational performance. Additionally, you should have available information concerning the reasons for undertaking the investment, analyses of the impacts of the investment for each alternative, and a recommended preferred alternative. In some instances, only the recommended investment will be submitted for approval; in other cases, you will have to submit the "top candidates."

In preparing economic justification proposals, it is important to *go the extra mile.* Examples include performing sensitivity analyses and performing a competitive benchmarking analysis. The latter should assess the impact of "doing nothing" while the competition is "doing something"; throughout, the impact of the facilities planning investment on the revenue stream should be considered explicitly.

Recognize that the final selection can be quite different from the recommendation you make. Management decisions are based on many considerations, not all of which were included in your analysis. To improve the chances of your recommended action being accepted, consider the following "lessons learned":

- Obtain the support of the users of the recommended system.
- Pre-sell your recommendation!
- Speak the language of the listener.

- Do not oversell the technical aspects of the facilities plan; technical aspects seldom convince management to make the required investment.

- The decision makers' perspectives are broad; know their priorities and tailor the economic justification package accordingly.

- Relate the proposed investment to the well-being of the firm; show how the investment relates to the firm's strategic plan and the stated corporate objectives.

- Your proposal will be only one of many submitted and many will not be funded.

- Failure to fund your proposal does not mean management is stupid; management's decision to fund your proposal does not necessarily mean they are brilliant.

- Do not confuse unfavorable results with destiny.

- Timing is everything!

- Profit maximization is not always the "name of the game," but "selling" is!

- A firm's ability to finance the proposal is as important as its economic merit.

- You get what you pay for; don't be penny-wise and dollar-foolish.

- Remember the Golden Rule: those with the gold make the rules!

- The less you bet the more you stand to lose in case you win! [23], [24]

Example 11.4

A furniture company is considering three alternative methods of meeting its order picking requirements. The first alternative method of order picking is very basic and is designed to maximize space utilization rather than operator efficiency. All products are picked from storage locations, regardless of location or level. The order picker travels on a standup reach truck to the pick location along a storage aisle. The pick-to-pallet is placed on the floor; the pick-from-pallet is removed from the rack and placed on the floor adjacent to the pick-to-pallet. The operator removes the necessary cartons from the pick-from-pallet and stacks them on the pick-to-pallet; the pick-from-pallet is returned to the storage location; and the operator travels to the next picking location. Product is stored in a mix of single-deep and double-deep pallet racks, as well as bulk storage.

An alternative being considered is the use of an order picker truck that will raise the operator with the pallet and allow cartons to be removed from the pick-from-pallet and placed directly on the pick-to-pallet. While this option is ideal for single-deep pallet rack, it eliminates the use of double-deep pallet racks, but requires additional handling with bulk storage.

A second alternative is to create an order picking area that is separate from the replenishment storage area. In this case, one or more pallet loads of product to be picked will be placed on the floor in a centralized area. As cartons are removed, a replenishment pallet will be delivered to the picking area. Replenishment product will be stored in single-deep and double-deep pallet racks, as well as bulk storage.

An analysis of the labor requirements, equipment investments, space requirements, and operating costs for each alternative yielded the after-tax, inflation-adjusted cash flow profiles shown in Table 11.8. Using a 10-year planning horizon, a 37.2% combined state and federal income tax rate, and a 20% after-tax minimum attractive rate of return, the net present value for the reach truck alternative (RT) was found to be −$16,550,219. For the alternative involving the use of order picker trucks (OPT), the net present value equalled −$18,525,927. The net present value for the separate order picking area alternative (SOP) was found to be −$16,230,944.

Table 11.8   *Cash Flow Profiles for Example 11.4*

R\C	A	B	C	D	E
1	Discount Rate =	0.2			
2					
3	Year	CF(RT)	CF(OPT)	CF(SOP)	CF(SOP-RT)
4	0	−$13,790,000	−$16,306,000	−$14,251,000	−$461,000
5	1	−$659,309	−$560,711	−$498,841	$160,468
6	2	−$542,018	−$411,455	−$368,499	$173,519
7	3	−$574,741	−$441,908	−$395,411	$179,330
8	4	−$608,610	−$473,427	−$423,265	$185,345
9	5	−$643,664	−$506,049	−$452,094	$191,570
10	6	−$679,944	−$539,813	−$481,932	$198,012
11	7	−$717,495	−$574,759	−$512,814	$204,681
12	8	−$851,606	−$712,347	−$632,627	$218,979
13	9	−$891,831	−$749,781	−$665,709	$226,122
14	10	−$933,464	−$788,526	−$699,948	$233,516
15	NPV	−$16,550,219	−$18,525,927	−$16,230,944	$319,275
16	IRR				37.35%

The difference between the cash flows for the reach truck alternative and the separate order picking area alternative are also shown in Table 11.8. As shown, the internal rate of return on the incremental investment of $461,000 is 37.35%.

While many factors will need to be considered in making the final selection of the order picking method, based on the economic analysis performed, the separate order picking area would be recommended.

Economic Value Added (EVA[2])

An example of "speaking the language of the listener" is provided by the emergence of *economic value added* (EVA).

Since the mid-1980s, a management tool called economic value added has been used by an impressive set of firms to make investment decisions. It focuses management's attention on an important objective: adding value for the shareholders. In fact, it has been a principal tool used by upper management within AT&T, Briggs & Stratton, Coca-Cola, CSX, Eastman Chemical, Quaker Oats, and Wal-Mart Stores, and a host of other firms use EVA to ensure that their investments are adding value for the shareholders.

EVA is used to facilitate decisions regarding major capital investment, as well as acquisitions and divestitures. It has also been used to analyze products and operating units to determine the economic dogs and cash cows within the firm's portfolio of products and businesses.

What is EVA? Basically, EVA is a management tool that examines the difference between the net operating profit after taxes and the cost of capital, which includes the cost of both debt and equity capital. Hence, the interest charges and bond rates

---

[2]Stern Stewart & Co., a corporate financial advisory services firm located at 450 Park Avenue, New York, 10022, is generally credited with developing EVA and claims ownership of the trademark on the use of EVA [31], [33].

that contribute to the cost of debt capital are combined with the cost to the share-holders of providing the firm with equity capital (by purchasing its stock) [29].

As noted in *Fortune,* the cost of equity capital needs to include the opportu-nity cost to the shareholders, who can choose where they are going to invest their money [12]. An analysis of the cost of capital typically reveals that debt capital is far less expensive than equity capital.

Many firms that adopted EVA found that few of their managers knew how much capital was tied up in their business units. Moreover, few managers had a firm grasp on the true cost of capital. EVA seeks to remedy this by focusing attention on adding value for the shareholder through more effective use of capital. The result of an increased emphasis on effective use of capital will result in lower inventories, fewer warehouses, and so on [29].

Stern Stewart argues that there are four ways to create value for the shareholder:

1. Increase profitability without using additional capital (e.g., increase profit mar-gins and increase sales without using additional capital).
2. Invest in projects that earn more than the cost of capital.
3. Free-up capital that earns less than the cost of capital.
4. Use debt to reduce the cost of capital [31].

Real estate, equipment, facilities, working capital, and inventories are exam-ples of capital being used within the firm. Other, not-so-obvious examples of cap-ital are investments in training and in research and development. The investments in training result in increased value of the firm's human capital. While it is not easy to quantify the value of capital in R&D and in training, they should not be com-pletely overlooked in the quest for value.

Calculating Economic Value Added

The following three examples illustrate the EVA calculation.

Example 11.5

Two firms (A and B) are being considered for acquisition. The assets of the firms are $100 million and $200 million, respectively. Both are debt free; hence, the equity equals the assets. The annual operating profits for the firms are $40 million and $70 million, respectively. Taxes equal 40% of operating profit. Consequently, the annual net incomes for the firms are $24 million and $42 million, respectively. Dividing the net income by equity yields return on equity of 24% and 21%, respectively. The cost of capital is 12%. Which firm is best from a shareholder value point of view? Some would choose A because it has the greatest ROE (return on equity) or ROA (return on assets). However, B maximizes value for shareholders.

To determine the EVA, subtract the cost of capital from net operating profit after taxes. For A, this yields: $24 million − 0.12($100 million) = $12 million. For B, the EVA equals $42 million − 0.12($200 million) = $18 million. (As was the case with the IRR method, one cannot choose the alternative that has the greatest return on equity; instead, incremental net operating profits should be compared with the cost of capital.)

In 1924, Donaldson Brown, chief financial officer of General Motors Corporation, recognized that maximizing return on investment was the wrong objective. He noted, "The object of management is not necessarily the highest rate of return on capital, but ... to assure profit with each increment of volume that will at least equal the economic cost of additional capital required." [31]

Example 11.6

There are several ways to compute EVA. The method used in the previous examples was:

	Firm C	Firm D
Capital Invested	$100	$200
Operating Profits	$25	$35
Taxes (40%)	$10	$14
Net Operating Profit After Taxes (NOPAT)	$15	$21
Cost of Capital (12%)	$12	$24
EVA	+$3	−$3

Alternatively, EVA can be computed as follows:

Return on Assets (ROA)	15%	10.5%
Cost of Capital	12%	12%
Difference	+3%	−1.5%
EVA = Difference × Assets	+$3	−$3

Example 11.7

The relationship between EVA and cash flow approaches can be illustrated by considering the data for Firm C in Example 11.6. To simplify the analysis, consider a $100 million investment that returns $15 million annually, forever. With a 12% MARR, the net present value (NPV) of the NOPAT can be obtained by dividing the annual NOPAT by the MARR to obtain $125 million. Subtracting the initial investment yields a NPV of $25 million. Alternately, the $100 million investment yields an annual EVA of $3 million. Dividing the EVA by the MARR yields a NPV of $25 million. The NPV for the projected EVA is the same as the NPV of the projected cash flow.

From Example 11.7, since using DCF approaches yields the same recommendation as using discounted EVA, why use discounted EVA? Stern Stewart note that when using DCF, after an investment is made there is no explicit "accounting" for it since it is a "sunk cost." As a result, "It apparently does not carry an ongoing performance obligation." They go on to say, "The incremental profits from the project, on the other hand, make an ongoing contribution to earnings growth, thereby stimulating investments that may not make good economic sense. Because cash flow cannot be used for any other purpose—not for expressing goals or appraising performance, for example—the analysis is apparently disconnected from the rest of the financial management system" [33, p. 5].

In support of their argument for using EVA, they note, "The result is the same as discounting cash flow, an important equivalence. The obligation to earn an attractive return on capital is explicit and enduring. The EVA performance targets required to create value fall right out of the analysis. Unlike cash flow, EVA can be used as a performance target and measure. If EVA is used for determining bonuses, managers will see clearly how prospective business decisions will affect their compensation, an important link" [33, p. 5].

11.3 SELECTING THE FACILITIES PLAN

As mentioned previously, the selection of the facilities plan tends to be a "satisficing" process, as opposed to an optimizing process. The selection decision typically requires a compromise over several important criteria. Furthermore, the decision maker's value system is not necessarily the same as the facilities planner's.

The facilities planner has a responsibility to do a thorough, accurate, and objective job of developing several good facilities plans. The facilities planner also has a responsibility to perform a thorough, accurate, and objective evaluation of the alternative plans. Furthermore, the facilities planner is responsible for presenting professionally the results of the facilities planning effort to management and making a recommendation concerning the preferred plan. However, it is management's responsibility either to approve or disapprove the recommendation or to make the selection decision.

The selection process is critically dependent on the facilities planner selling the facilities plan. Not only must management be sold, but also the end users of the system must be sold.

The sales job for management must address the long-term benefits of the facilities plan. The flexibility, reliability, and adaptability of the system must be addressed. Management may take a strategic, offensive (as opposed to defensive) position and may want to know what is going to be the return on investment, increased services, or other benefits of pursuing the recommended system.

To the contrary, the sales job for users must address the short-term implications of the new system. Noise, safety, and impacts on the user of the system are the key factors. The user often takes a contingency, defensive position, and wants to know how this system will affect the people involved.

Not understanding to whom the system must be justified is a major mistake. Once a survey indicated that 80% of the people who had installed a major material handling system were unhappy with the results. Of these unhappy people, 80% said the reason for their unhappiness was not the equipment aspects of the system, but instead, the people aspects. Systems must be explained and sold to both managers and users and the approach taken to these two groups should be different.

Because of the importance of preparing and selling facilities plans, the subject is treated separately. Chapter 12 is devoted to the preparation of the facilities plan and the development of a sales presentation.

11.4 SUMMARY

In summary, the evaluation of alternative facilities plans was considered; four evaluation techniques were described and a seven-step economic justification procedure was presented. The importance of presenting the justification package in a form that "communicates" to management was illustrated by describing the Economic Value Added approach. The use of spreadsheets to facilitate the computational aspects of the systematic approach was also illustrated. Further treatment of the selection of the facilities plan is reserved for Chapter 12, where the preparation and selling aspects of the facilities plan are covered.

BIBLIOGRAPHY

1. Belton, V., "A Comparison of the Analytic Hierarchy Process and a Simple Multi-Attribute Value Function," *European Journal of Operational Research,* vol. 26, 1986, p. 7.
2. Boucher, T. O., and MacStravic, E. L., "Multiattribute Evaluation Within a Present Worth Framework and Its Relation to the Analytic Hierarchy Process," *Engineering Economist,* vol. 37, no. 1, Fall 1991, pp. 1–32.
3. Canada, J. R., and Sullivan, W. G., *Economic and Multiattribute Evaluation of Advanced Manufacturing Systems,* Prentice-Hall, Englewood Cliffs, NJ, 1989.
4. Canada, J. R., Sullivan, W. G., and White, J. A., *Capital Investment Decision Analysis for Management and Engineering,* 2nd ed., Prentice-Hall, Englewood Cliffs, NJ, 1996.
5. Dyer, J. S., "Remarks on the Analytic Hierarchy Process," *Management Science,* vol. 36, no. 3, 1990, p. 249.
6. Falkner, C. H., and Benhajla, S., "Multiattribute Decision Models in the Justification of CIM Systems," *Engineering Economist,* vol. 35, no. 2, Winter 1990, p. 91.
7. Francis, R. L., and White, J. A., *Facility Layout and Location: An Analytical Approach,* Prentice-Hall, Englewood Cliffs, NJ, 1974.
8. Kaplan, R. S., "Must CIM be Justified by Faith Alone?," *Harvard Business Review,* vol. 64, no. 2, March–April 1986, pp. 87–95.
9. Lane, E. F., and Verdini, W. H., "A Consistency Test for AHP Decision Makers," *Decision Sciences,* vol. 20, 1989, pp. 575–582.
10. Saaty, T. L., *The Analytic Hierarchy Process,* McGraw-Hill, New York, 1980.
11. Sellers, P., "Corporate Reputations," *Fortune,* January 29, 1990, p. 50.
12. Stern, J. M., "The Mathematics of Corporate Finance—or EVA = $NA[RONA-C]," transcript of a speech given in Atlanta, 1993.
13. Stewart, G. B., *The Quest for Value,* HarperCollins, New York, 1991.
14. Tompkins, J. A., and White, J. A., *Facilities Planning,* Wiley, New York, 1984.
15. Tully, S., "The Real Key to Creating Wealth," *Fortune,* September 20, 1993, pp. 38–50.
16. White, J. A., "How to Justify Equipment Investments," *Modern Materials Handling,* vol. 44, no. 1, January 1989, p. 39.
17. White, J. A., "Specifying the Investment Alternatives," *Modern Materials Handling,* vol. 44, no. 2, February 1989, p. 27.
18. White, J. A., "Defining Your Planning Horizon," *Modern Materials Handling,* vol. 44, no. 4, March 1989, p. 29.
19. White, J. A., "Developing a Cash Flow Profile," *Modern Materials Handling,* vol. 44, no. 5, April 1989, p. 27.
20. White, J. A., "The Cost of Opportunity," *Modern Materials Handling,* vol. 44, no. 6, May 1989, p. 29.
21. White, J. A., "Comparing Investment Options," *Modern Materials Handling,* vol. 44, no. 7, June 1989, p. 27.
22. White, J. A., "Justifying with Break-Even Analysis," *Modern Materials Handling,* vol. 44, no. 8, July 1989, p. 29.

23. White, J. A., "Making Informed Investment Decisions," *Modern Materials Handling,* vol. 44, no. 9, August 1989, p. 27.
24. White, J. A., "Winning Investment Approval," *Modern Materials Handling,* vol. 44, no. 10, September 1989, p. 27.
25. White, J. A., "Leveling the Playing Field," *Modern Materials Handling,* vol. 44, no. 12, October 1989, p. 25.
26. White, J. A., "The Money Value of Time," *Modern Materials Handling,* vol. 44, no. 13, November 1989, p. 27.
27. White, J. A., "Investment, Taxes, and Inflation," *Modern Materials Handling,* vol. 44, no. 14, December 1989, p. 27.
28. White, J. A., "Manage to Better Bottom Line," *Modern Materials Handling,* vol. 45, no. 13, November 1990, p. 29.
29. White, J. A., "Three Ways to Build Economic Value," *Modern Materials Handling,* vol. 49, no. 7, June 1994, p. 31.
30. White, J. A., Case, K. E., Pratt, D. B., and Agee, M. H., *Principles of Engineering Economic Analysis,* 4th ed., Wiley, New York, 1998.
31. *EVA*^{TM}, presentation by Stern Stewart & Co., 450 Park Avenue, New York.
32. "Managing," *Fortune,* April 22, 1991, p. 86.
33. *The Quest for Value: How Value Is Created and How to Create Value,* presentation by Stern Stewart & Co., 450 Park Avenue, New York.

PROBLEMS

11.1 List the advantages and disadvantages of leasing production equipment versus purchasing the equipment.

11.2 List the advantages and disadvantages of providing on-site storage of materials and supplies versus using off-site leased storage.

11.3 List the advantages and disadvantages in building a conventional (25- to 30-ft-tall) warehouse versus building a high-rise (55- to 75-ft-tall) warehouse.

11.4 List 10 factors to be used in evaluating the following horizontal handling alternatives for moving pallet loads: lift truck, tractor-trailer train, automated guided vehicle, in-floor towline conveyor, roller conveyor for pallets.

11.5 List 10 factors to be used in evaluating site location alternatives for a corporate headquarters building.

11.6 List 10 factors to be used in evaluating site location alternatives for a distribution center.

11.7 Two material handling alternatives are being considered for a new distribution center. The first involves a smaller capital investment, but larger operating costs. Specifically, alternative A involves an initial investment of $200,000 and has annual operating expenses of $500,000. Alternative B involves an initial investment of $1,000,000 and has annual operating expenses of $300,000. Both alternatives are assumed to have negligible salvage values at the end of the 10-year planning horizon. A minimum attractive rate of return of 10% is used for such analyses.
   a. Using NPV, NAV, and NFV analyses, which is preferred?
   b. What is the IRR on the $800,000 incremental investment required for alternative B?
   c. Determine the payback period and the discounted payback period for the $800,000 incremental investment required for alternative B.

11.8 In Problem 11.7, suppose there is considerable uncertainty regarding the annual operating expenses for alternative B. What is the break-even value?

11.9 A warehouse modernization plan requires an investment of $8 million in equipment; at the end of the 10-year planning horizon, it is anticipated the equipment will have a salvage value of $750,000. Savings in operating and maintenance costs due to the modernization are anticipated to total $1.75 million per year. A MARR of 10% is used by the firm. Perform a sensitivity analysis to determine the effects on the economic viability of the plan due to errors in estimating the initial investment required, as well as the magnitude of the annual savings.

11.10 True or False: The NPV and the IRR methods, when applied correctly, always yield the same recommendation.

11.11 True or False: The payback period method will always yield the same recommendation as the NPV method.

11.12 True or False: The use of "size gates" in the authorization of capital expenditures can result in major projects being broken into several seemingly independent investments, some of which might not be able to stand alone in the economic justification process.

11.13 True or False: The "do-nothing" alternative is always the lowest cost alternative and serves as the baseline against which other alternatives are compared.

11.14 Given the following financial data for a firm, determine its EVA.

Earnings Before Income Taxes	$680	
Taxes	$316	
Net Operating Profit After Taxes		$364
Investment in:		
Total Assets (Avg)	$6,276	
Payables (Avg)	$2,434	
Net Investment		$3,842
Cost of Capital	10%	

11.15 Consider two firms, C and D, each with a 10% cost of capital Based on the financial data for the two firms (in $M), which has the greatest EVA?

	Firm C	Firm D
Equity	$100	$200
Annual Operating Profit	$25	$40
Taxes (40%)	$10	$16

11.16 Three investment alternatives (A, B, and C) are under consideration. The discounted present worths (PW) are $30,000, $24,000, and $18,000, respectively. The three alternatives have very different rankings on time required to fill a customer's order (T) and flexibility (F). The following weights have been assigned to the three factors (PW, T, and F): 40, 35, and 25. On a scale from 1 to 10, the following ratings have been assigned to the three alternatives for the three factors:

	A	B	C
Present worth (PW)	10	8	6
Fill time (T)	7	10	8
Flexibility (F)	5	7	10

Using the weighted factor comparison method, which investment alternative would be recommended?

11.17 Which investment alternative would be recommended in Problem 11.16, if
a. the weightings for the three factors are 50, 10, and 40?
b. the rating values for fill time for A, B, and C had been 5, 7, and 10, respectively?

c. another factor is introduced, safety (**S**), the weights assigned to the factors (**PW, T,F,** and **S**) are 30, 20, 10, and 40 and the rating values for **S** are 10, 8, and 9 for **A, B,** and **C,** respectively?

11.18 Recall Example 11.1. Determine the preferred alternative with the following weights assigned to factors 1 through 5, respectively: 20, 25, 15, 20, and 20.

11.19 Solve Problem 11.18 using the following weights: 15, 25, 15, 30, and 15.

11.20 Solve Problems 11.18 and 11.19 using the following ratings for each alternative.

Layout Alternatives

Factor	A	B	C	D
1	8	10	2	5
2	8	4	9	10
3	7	8	10	9
4	10	8	7	6
5	10	6	5	8

11.21 Solve Problems 11.18 and 11.19 using the following ratings for each alternative.

Layout Alternatives

Factor	A	B	C	D
1	7	10	6	5
2	8	7	9	10
3	7	8	10	9
4	10	8	7	9
5	10	6	7	8

11.22 From solving Problems 11.16 through 11.21, what conclusions do you draw regarding the use of the weighted factor comparison method?

12

PREPARING, PRESENTING, IMPLEMENTING, AND MAINTAINING THE FACILITIES PLAN

12.1 INTRODUCTION

In the previous chapter, we considered the evaluation and selection of the facilities plan; here, we complete our consideration of facilities planning by addressing the physical preparation of the facilities plan, its presentation to management, the implementation of the selected facilities plan, and its maintenance over time. Obviously, the previous 11 chapters have little practical value unless the resultant facilities plan is properly prepared, presented, and sold. However, neither will its value be fully realized if it is not properly implemented and maintained.

12.2 PREPARING THE FACILITIES PLAN

If the facilities plan is not sold properly, the probability of acceptance by management is very low no matter now high its quality. As such, it is critical that careful attention be given to the "lessons learned" in justifying capital investments provided in the previous chapter under the seventh step in the SEAT process.

The facilities plan records the results of the entire facilities planning process. It is during the preparation of the facilities plan that decisions must be made concerning several extremely important details. The facilities plans resulting from the analysis of department arrangements yield an overall facility configuration and a department layout. To convert the overall facility configuration into a functioning facility requires the specification of details with respect to the total plot of land. To convert the department layout into a functioning facility, the details of each department must be specified. These plot and layout details may be specified on a plot plan and layout plan, respectively. This section describes plot plans and layout plans.

There are three components required to successfully sell a facilities plan. The first is a high-quality facilities plan that has been prepared in a clear and accurate manner. The second component is a written report describing the benefits of a facilities plan and documenting why the particular facilities plan selected is best. (A description of how to prepare a facilities planning written report is contained in Section 12.4.) The third component, an oral presentation, is the focal point of an entire facilities planning project. It is after the oral presentation that a decision will be reached to either implement or not implement a facilities plan. This critical component is described in Section 12.5.

In preparing the facilities plan, it is important for both a plot plan and a layout plan to be prepared. Each is described in the following sections.

Plot Plans

A plot plan (Figure 12.1) is a drawing of the facility, the total site, and the features on the property that supports the facility. Features that may be included on a plot plan include:

* Access roads, driveways, and truck aprons
* Railways, waterways, aircraft runways, and heliports
* Yards and storage tanks and areas
* Parking lots and sidewalks
* Landscaping and recreation areas
* Utilities including water, sewage, fire hydrants, gas, oil, electric, and telephone

A plot plan should be drawn to a scale that is consistent with the size of the plot to be drawn. Common plot scales include 1/16 in. or 1/32 in. equal to 1 ft, or 1 in. equal to 50 ft (or 100 ft for larger areas).

Factors that must be considered when developing a plot plan include:

1. *Expansion.* A plot plan should be constructed while projecting space requirements 10 years into the future. Facility expansion should be planned in at least two directions and particular attention should be given to the expansion of functions that will be difficult to relocate. Sites should be planned so that plant services, utilities, docks, and railways need not be altered when expansion occurs.

2. *Flow patterns.* Separate traffic patterns for materials, employees, and visitors. Plan all flow patterns while considering security, ease of access, and integration of internal flow patterns with external flow patterns.

3. *Energy.* Whenever possible, have the building face south and have the shorter dimension running north to south and the longer east to west. Consider the use of underground structures or at least consider the use of large amounts of

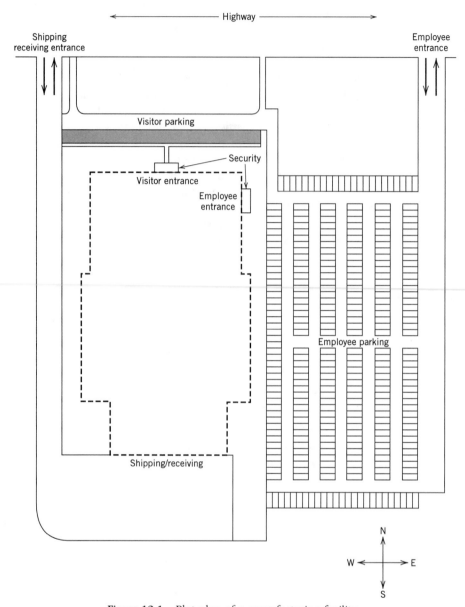

Figure 12.1   Plot plan of a manufacturing facility.

back-fill on northern and western walls. Avoid asphalt or concrete areas around the building as much as possible, and use deciduous trees for shade during the summer while still allowing sunlight to penetrate in the winter.

4.   *Aesthetics.* Locate the facility on the site to obscure unsightly activities. Plan landscaping and land use to blend into the environment and to promote an attractive setting.

Layout Plans

Layout plans are detailed representations of facilities. The representations may be either two-dimensional and drawn by hand, constructed with templates, or drawn by computer, or be three-dimensional and consist of modular building blocks or detailed models. No matter what type of representation is used, the same procedure should be followed to create the layout plan.

Constructing a Layout Plan

Prior to beginning the construction of the layout plan, the following data must be gathered.

1.   A department layout that has been developed using appropriate layout techniques
2.   The department and area requirement sheets for each department
3.   The material handling planning charts for each product to be manufactured

Once the above data are collected, the layout plan should be constructed using a systematic procedure. The systematic procedure for a manufacturing facility is:

1.   *Select the scale.* A scale should be selected so that the overall facility configuration, the department layout, and the available layout planning equipment fit appropriately. Whenever feasible, a scale of ¼ in. equals 1 ft should be used. If possible, use the same scale used by the architect, construction engineer, or other professionals working on the facility.

2.   *Decide on the method of representation.* In general, the selection of the method of representation for layout plans should be based on a combination of clarity and economics. Three-dimensional representations should be used when the third dimension plays an important role in design decisions. Each situation should be treated individually and a decision made as to which approach will result in the most effective layout plan while minimizing the life cycle cost for the layout plan.

3.   *Obtain layout plan supplies and/or hardware and software.* Depending on the method selected in step 2, possibly only paper and pencil are needed. Alternatively, many widely available computer-aided drafting software packages may be used with a PC and plotter or laser printer. Whatever supplies and/or equipment are needed, they should be obtained before proceeding with the construction of the layout plan.

4.   *(For an existing facility) Locate all permanent facilities on the layout plans.* All columns, windows, doors, walls, ramps, stairs, elevators, sewers, cranes, and other permanent fittings should be the first elements placed on the layout plan for an existing facility. Also, when appropriate, floor loading and ceiling heights should be recorded on layout plans for existing facilities.

5.   *Locate the exterior wall that includes the receiving function.* It is necessary to begin making decisions concerning the locations of functions in the facility.

One approach that works well is to first locate the receiving function and then locate functions along the primary material flow paths until all manufacturing departments are included within the layout plan. Clearly, other starting points may also be selected but as long as the overall department layout, the department and area requirement sheets, and the material handling planning charts are followed, no significant difference should exist in the resulting layout plan.

6. *(For nonexistent facilities) Locate all columns.* The size, span, and location of columns must be among the initial planning decisions when a facility is to be constructed. Architectural or construction engineers should be consulted for the column requirements. This is done to avoid the interference of columns with material flow or manufacturing equipment. The column spacing decisions must often be made after analyzing the economic tradeoff between facility construction costs and material flow costs.

7. *Locate all manufacturing departments and equipment.* Beginning with receiving, each department should be tentatively located in the layout plan based on the department layout. Aisles between departments should be included as appropriate. The details of each department should be developed based on the data recorded on the department area and requirements sheet. When necessary, department boundaries and aisles should be altered to allow department details to be shaped into the required department areas.

8. *Locate all personnel and plant services.* Alterations should be made to the manufacturing layout to include all personnel and plant services. Efforts should be made to ensure uniform aisle spacing and to integrate the service functions into the layout to maximize the use of production space.

9. *Audit the layout plan.* Prior to finalizing the layout plan, it should be audited from both the material and personnel perspective. The material audit should involve tracing all material flows through the facility. Material handling planning charts should be used as guides and each material move, storage, operation, and inspection should be traced through the facility. The personnel audit should involve tracing the tasks performed by every person to be employed within the facility. For example, a production worker's efforts should be traced through at least the following activities:

    (1) Park automobile.

    (2) Walk into facility.

    (3) Hang up coat and store lunch.

    (4) Sign in.

    (5) Talk with supervisor.

    (6) Report to work station.

    (7) Begin production.

    (8) Handle machine breakdown.

    (9) Handle machine setup.

  (10) Go to restroom.

  (11) Get a drink of water.

  (12) Take a break.

  (13) Go to lunch.

(14) Receive first aid.

(15) Attend production meeting.

(16) Meet with human resources representative.

(17) Perform housekeeping activity at workstation.

(18) Interact with quality control.

(19) Interact with production control.

(20) Interact with material handling personnel.

(21) Report production difficulties.

(22) Report production milestones.

(23) Wash hands.

(24) Retrieve coat and lunch box.

(25) Sign out.

(20) Return to automobile.

These types of audits will often result in a change in workstation orientation and/or spacing.

10. *Finalize the layout plan.* Once the layout plan is fully audited, it should be finalized by permanently locating everything on the layout plan and recording appropriate headings and clarifying notations. A detailed legend should be included, and all areas should be clearly labeled and, when appropriate, color-coded.

Although a detailed procedure has been given for developing a layout plan, it should not be inferred that development is an easy or straightforward task. To the contrary, while developing the layout plan the majority of data used and assumptions made will be questioned repeatedly. Many iterations will be necessary and many trial-and-error attempts will be made before obtaining an acceptable layout plan.

Alternative Methods of Representing Layout Plans

The alternative methods of representing two-dimensional layout plans are:

1. *Drawings.* Figure 12.2 illustrates a handmade drawing, which serves as a rough sketch of an area. Drawings are quickly and conveniently made to illustrate an alternative layout plan. In addition, drawings may be the best approach for layout plans of small areas. However, handmade drawings are much too expensive to produce and to alter for use as final layout plans for large areas. Therefore, handmade drawings should be used either as rough sketches or as final layout plans for small areas.

2. *Templates and tapes.* This was the most often used method, prior to the 1990s, of creating layout plans. Templates may be homemade or purchased; typically, they are made from cardboard, mylar, adhesive-backed mylar, or magnetized plastic. Templates may be either block templates or contour templates. A block template is simply a labeled rectangle representing the maximum length and width of equipment. A contour template illustrates the contour and the clearances for the movable portions of the machine. A broad range of tapes are available for representing walls, partitions, belt conveyors, roller conveyors,

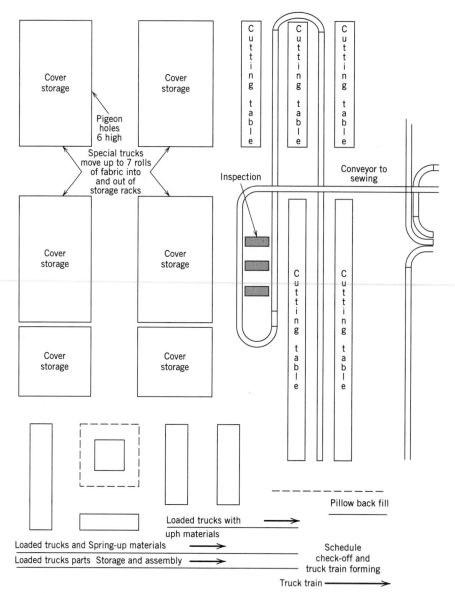

Figure 12.2   Handmade layout drawing. (*Courtesy of Furniture Design and Manufacturing.*)

windows, overhead monorails, and aisles. In addition, different-colored tapes and arrows are available for a wide range of uses. The flexibility and low cost of producing template and tape layouts makes it a popular approach. Figure 12.3 illustrates a contour adhesive-backed Mylar template and tape layout. Figure 12.4 illustrates a magnetic template layout.

3.   *Computer-aided drafting.* A convenient approach for combining the ease of hand-drawn layouts with the quality of a template and tape layout is computer-aided drafting (CAD). A CAD system consists of an input terminal, a central

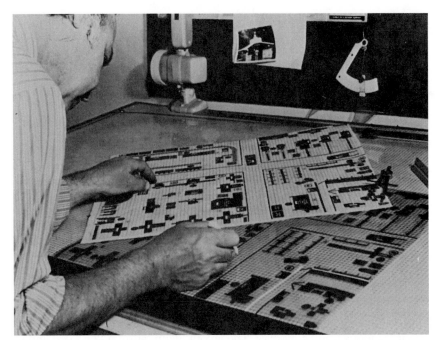

Figure 12.3 A contour adhesive-backed Mylar template and tape layout. (*Courtesy of A.D.S. Company*)

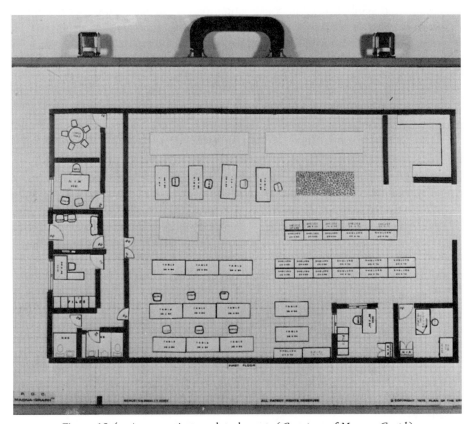

Figure 12.4 A magnetic template layout. (*Courtesy of Magna-Graph*)

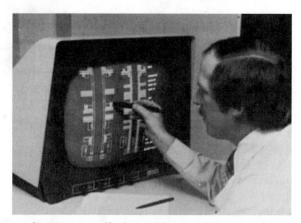

Figure 12.5   The use of a CAD system for layout planning. (*Courtesy of Computer Graphics, Inc.*)

processor, and output peripherals. The input terminal typically consists of a keyboard, a mouse, and a monitor. The layout planner uses a keyboard and mouse to provide instructions to the CAD system. The monitor is used to graphically display the layout being developed. The central processor (PC) translates the instructions from the layout planner into graphics layouts and stores the templates and symbols used in constructing the layout. The output peripherals can include line printers, impact and laser printers, and/or plotters. Figure 12.5 illustrates the use of a CAD system for layout planning. A layout may be developed by calling for specific templates and symbols from the central processor. Once a layout is complete, the layout planner may have the layout reproduced using any one of many output peripherals. In addition to the ease and quality of the resulting layouts, CAD has the following useful features:

a. The capability to view a layout from various perspectives. An entire facility, a department, a workstation, or a portion of a workstation may be viewed simply by requesting a different perspective.

b. The scale of layouts can be quickly and easily changed. This allows greater consistency between different users of layout plans.

c. Animated simulations can be performed to test the layout and demonstrate the operation of the system. Congestion points can be tested and evaluated by subjecting the design to peak and changing requirements. Animated 2-D and 3-D representations of the flow of people, material, equipment, information, and paper through the facility can be viewed, providing a degree of realism not possible otherwise. Transforming a static representation of the facilities plan into a dynamic representation, simulation is useful in testing and evaluating a design, as well as selling the design to management and other individuals who will be affected.

The use of overlays often adds considerable clarity to two-dimensional layout plans. Overlays are rarely used with drawings but are very common with either template and tape layout plans or CAD layout plans. Overlays for template and tape layouts are typically made of transparent sheets of mylar or acetate and are placed over the layout plan to illustrate planned features. CAD overlays are easily added and deleted. Plumbing, electrical, heating, and lighting overlays can provide considerable insight into the complexities involved with integrating plant services in a layout. An additional overlay that is extremely useful in a manufacturing facility is a material flow overlay. Material flow overlays facilitate evaluating, improving, and presenting the layout plan. All material flows within the facility should be recorded on the overlay. The color of the flow lines may indicate the product or the type of material being moved, including raw materials, in-process materials, or finished products. In addition, the volume and method of movement may be specified by changing the width of the flow lines or by using various symbols on the flow lines.

Three-dimensional models are the clearest and most easily understood method of developing layout plans. However, because of the cost of three-dimensional models they often cannot be justified. Nevertheless, whenever the third dimension plays a critical role in the layout plan, or when a detailed presentation is necessary for a non-technical group, three-dimensional models should be considered. There are three methods of establishing three-dimensional models.

1. *Modular block models.* The use of modular block models to develop a three-dimensional model is illustrated in Figure 12.6. The advantage of modular block models is that special models of each machine need not be constructed. The block models may be made of modular building blocks. This significantly reduces the cost and difficulty of making three-dimensional models without significantly reducing the quality of the layout plan.

2. *Scale models.* Like templates, scale models may be either block models or contour models. Contour models show the actual machine contour rather than the machine travel. Figure 12.7 illustrates a three-dimensional, contour model

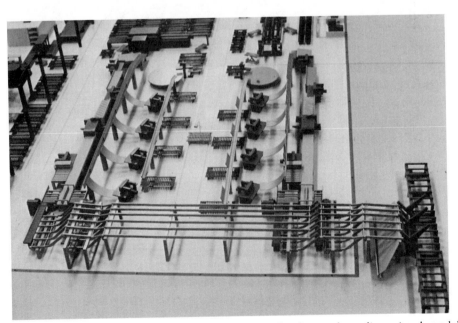

Figure 12.6 The use of modular block models to develop a three-dimensional model. (*Courtesy of Model Display Studios*)

Figure 12.7   Contour model three-dimensional layout plans. (*a*) An office layout. (*b*) An automated storage/retrieval system (AS/RS) layout. (*Courtesy of Visual Industrial Products, Inc.*) (*c*) A layout of the waste treatment system for a wool-combing plant. (*d*) A two-level air-conditioning system layout. (*Courtesy of Burlington Industries, Inc.*)

Figure 12.7   (continued)

layout plan. Although these models are the most easily understood, their cost
and difficulty to duplicate and transport limit their use.

3. *Computer models.* Computerized three-dimensional (3-D) representations of
    facilities can be produced, allowing the designer to rotate the facility, "enter
    the door," "walk through the facility," and make changes as needed.

In addition to using modular block models, scale models, and 3-D computer-
generated representations of a facility, photographs may be used to provide a 3-D

representation of facilities in 2-D space. The quality of the 3-D representation has improved in recent years. Photography is often used to portray 3-D models when it is not feasible to take the physical model to the client.

Given the level of research activity underway in holography and "virtual reality," the computer will play a greater role in facilities planning in the future in design, evaluation, and sales. Regardless of the method used to represent the facilities plan, it must have a professional appearance. Color, overlays, tapes, templates, photographs, simulations, multimedia presentations, 2-D and 3-D models—all possibilities should be considered. The final choice will depend on the audience, as well as the costs (and benefits) associated with the alternative methods of presenting the facilities plan.

12.3   PRESENTING THE FACILITIES PLAN

The 2-D or 3-D depiction of the recommended facilities plan is generally accompanied by a written report. Additionally, an oral presentation is typically used to communicate the plan to management in order to provide an opportunity for dialogue regarding the plan and any alternatives or modifications that might be of interest to management. Both the written report and the oral presentation are considered in the following sections.

The Written Report

The objective of most facilities planning written reports is to describe the benefits of a facilities plan and document why the particular facilities plan selected is the best. It is not the objective of most written reports to describe the detailed procedures required to establish a facilities plan. Hence, it is surprising and disappointing to read facilities planning reports that dwell on the process of establishing the facilities plan and do little to sell it.

The process of developing the written report should include the following steps:

1. *Define the objective of the report.* The objective should state what conclusions should be reached by anyone reading the report and should be used as a guide while writing the report.
2. *Establish a table of contents.* The table of contents serves as an outline for the written report. It forces one to organize the material in the report and ensures that a logical train of thought exists from the introduction to the conclusion.
3. *Determine the audience.* The audience will dictate the style, level of detail, and length of the report. Clearly, a report written to the chief facilities planner would differ from a report written for a chief executive officer.
4. *Write the report.* The report should be brief but should accurately communicate the reasons for implementing the facilities plan. It should be assumed that the reader will believe the results presented. Therefore, the reader should not be forced to read through detailed justifications or explanations. Justifications

and explanations should be included in tables or figures. Full documentation of tables and figures should be placed in appendixes.

5. *Document the report.* Prepare appendixes that define the sources of data and assumptions included in the report. Include within them data and procedural details describing exactly how the conclusions reached in the report were determined. Provide sufficient information in the appendixes so that the reader can recreate the conclusions reached in the body of the report.

6. *Edit the report.* Before finalizing the report, allow a few days to pass and then critically review it. Focus your attention on the objective of the report, and the need for brevity and clarity.

It is difficult to specifically describe what should be contained in a facilities planning written report. If possible, review previously successful reports that have been prepared for other organizations. A successful format could include:

- Letter of transmittal
- Cover page
- Executive summary
- Table of contents
- Introduction
- Body
- Implementation plan
- Conclusions
- Appendixes

The **letter of transmittal** is the document that introduces the reader to the report. It sets the stage for the reader by stating what is in the report and why it is being presented. In addition, the letter of transmittal may be used to recognize persons who contributed to the report and to describe future reports or activities that pertain to the report being transmitted.

The **cover page** includes the title of the report, the audience, the author, and the report date.

The **executive summary** lists the major findings of the report. It should not exceed a page and should describe all conclusions reached in the report.

The **table of contents** for a report is like the table of contents for a textbook.

The **introduction** describes the background of the report. The objectives, scope, and boundaries of the report are given in the introduction.

The **body** of the report is broken down into major sections. It leads the reader through a brief justification and the benefits of the facilities plan.

Calculations and listings of data appear in the **appendixes,** not in the body. The appendixes should be referenced in the body.

The **implementation plan** is the most important section in many reports. It describes the specific activities and the dates for completing the activities required to fully implement the facilities plan described in the report.

The **conclusion** reiterates the information present in the introduction, body, and implementation plan. The conclusion differs from the executive summary in that it presents a narrative summary of the entire report, whereas the executive summary is simply a listing of the major findings. As previously stated, the appendixes contain all supporting data and all backup information. If the appendixes are very long, it may be appropriate to bind them under separate cover.

At least two days before the oral presentation, the written report should be given to all persons who will be attending the oral presentation. After distributing the report, it may be wise to check with a few key persons who will attend the oral presentation and learn their reactions. Their reactions may be very helpful in preparing the oral presentation.

The Oral Presentation

The first step in preparing an oral presentation is to determine the audience. Analyze the perspective of each person attending the presentation to determine their expectations. Fully understanding the audience and viewing the presentation from its perspective allows the presenter to predict what questions will be asked and what benefits to emphasize. Next, decide what information should be presented. Two critical factors are brevity and a logical progression from *what* to *why* to *how*.

The oral presentation should be brief. For most projects, a one-hour oral presentation is all that is necessary. If an oral presentation can be done effectively in 15 minutes, then plan for a 15-minute presentation.

Oral presentations should begin by stating *what* should be done, describing *why* it should be done, and *how* to do what is recommended. One should not attempt to train the attendee in *how* to establish a facilities plan. To the contrary, an oral presentation should dwell on *what* facilities plans should be implemented, *who* will be impacted, *why* the facilities plan should be implemented, and *how* to implement it. Following are some helpful guidelines for preparing and presenting an oral presentation.

1. Do not oversell. Be realistic and make certain that all statements can be proven.
2. Refer to systems that are similar to the one recommended. Be certain the audience knows that what is recommended was successful elsewhere.
3. Before concluding the presentation, describe the implementation plan. Explain how to accomplish what is recommended.
4. Conclude the presentation with a tabulation of results. Show cost savings and improvements in space utilization.
5. Rehearse the oral presentation. This boosts self-confidence, which makes the presentation easier for the audience to accept.
6. Prepare quality visuals. Do not place too much information on any one slide. Do not use reproductions from the report.
7. Dress appropriately. Look successful and your presentation will be believable.
8. Begin on time. Have visual-aid equipment ready and focused.
9. Address all objections and questions. Be aware and communicate to the audience that objections and questions are good to get out in the open. Listen carefully and allow objections and questions to be fully voiced. Stay open-minded.
10. If you do not know the answer to a question, respond: "I do not know, but I will find out," and do find out!

12.4 IMPLEMENTING THE FACILITIES PLAN

Because of a failure to implement the system effectively, many excellent systems designs fail to deliver their promised benefits. Most implementation failures are technical problems or people-related issues. Experience suggests that the more frequent and formidable problems are often the people problems.

Invariably a new, improved facilities plan will require that changes be made, and people tend to resist change. The process of managing change and dealing with resistance to change has been well documented. It is recommended that works [1], [3], [5] and [6] in this chapter's bibliography be consulted.

In the final analysis, the ability to sell the facilities plan is dependent upon one's ability to deal effectively with resistance to change. Krick [10] lists a number of causes of resistance to change including:

1. Inertia
2. Uncertainty
3. Failure to see the need for the proposed change
4. Fear of obsolescence
5. Loss of job content
6. Personality conflict
7. Resentment of implied criticism
8. Poor approach by the facilities planner
9. Inopportune timing
10. Resentment of outside help

Some methods of minimizing resistance to change are recommended by Krick [10] including:

*   Explaining the need for the change thoroughly and convincingly
*   Involving many people in the development of the facilities plan
*   Tactfully introducing the proposal, carefully planning the timing
*   Introducing major changes in stages
*   Emphasizing those aspects of the change that are most beneficial to the individuals involved
*   Letting others "think of the answer" and receive the credit
*   Showing genuine interest in the feelings and reactions of others
*   Having changes announced and introduced by the immediate supervisors of those affected

When implementing the facilities plan, it is important for the facilities planner to be cognizant of the sensitivities of all parties involved. Selling the facilities plan does not end with top management; each employee must believe the new system will work as well. Managers may be the decision makers, but operations personnel are generally the decision wreckers. It is critical to prepare employees for the new plan.

Though the steps of defining, analyzing, generating, evaluating, and selecting a facilities plan are very important, the real payoff comes from implementation. Improvement occurs *if and only if* the plan is implemented properly [2].

Implementation is a tedious, often frustrating "fire-fighting" activity that must be pursued with vigor. Someone who has been intimately involved with the selection of the solution and who has the capability and time to coordinate and manage diverse activities must carry it out. The implementation process includes the following phases [8]:

1. *Document the solution.* Develop a series of manuals describing the systems operation, standard operating procedures, systems staffing, maintenance policies, and managerial responsibilities.

2. *Plan for solution implementation.* Develop a Gantt chart or a critical-path network of the activities required for the system to become operational.

3. *Select vendors.* Prepare functional specifications, mail specifications to qualified vendors, and select the vendors to participate in the systems implementation.

4. *Purchase equipment.* Place an order for all components, supports, controls, attachments, and containers required for the system to become operational.

5. *Supervise installation.* Maintain control of the implementation plan and exercise judgment when modifications become necessary.

6. *Make ready.* Establish and install a training program and oversee employee acceptance.

7. *Start up.* Begin operation.

8. *Debug.* Work with employees to ensure the system meets *their* expectations. Check with all shifts; identify and rectify difficulties.

9. *Follow up.* Perform a post-installation audit and verify that the system is operating as *you* expected. If not functioning as expected, determine the causes and rectify.

10. *Maintenance.* Routinely oversee system operation and maintenance to verify the system's continual success.

In discussing the importance of implementation in the context of automated material handling and storage systems, Searls [23] claims it is the key to a successful system. In defending his claim, he notes that such systems typically involve many different interacting elements. Specifically, fire protection, mechanical and electrical components associated with guided vehicles, conveyors, and storage/retrieval machines, storage racks, and computers must be integrated. If any one component is installed incorrectly, it will not perform correctly. Though "forgiveness" may have been designed into the system, if the components do not function correctly, neither will the system. Furthermore, if the system does not function properly, then the planned (and promised) benefits will not be realized.

Searls [23] addresses the implementation problem from the perspective of a storage rack supplier; however, his remarks are pertinent to the overall issue of implementing facilities plans. For this reason, the following is excerpted from his presentation:

> Today, most automated storage and retrieval systems utilize a control scheme based on XYZ coordinates, and require relative precision in the location of rack elements. The rack must be plumbed to a tolerance that the S/R machines manufacturer's control method and his equipment can tolerate. A prerequisite then to a successful rack installation is a complete and accurate understanding of the overall equipment operating requirements.

It is the responsibility of the system supplier to interpret these overall requirements, and if there is a separate contract for rack, transmit them to his rack supplier who, in turn, must convey them to the installer.

Before the installation contractor can offer a meaningful quotation for installation work, it will be necessary for him to visit the site to determine what constraints affect his ability to perform efficiently. He must determine if there are access roads to the site, if there is sufficient area to store his material near the work area, and he needs to determine the availability of water, power, and other required utilities. In the case of freestanding installations, he needs to check for overhead obstructions and interferences. Most important of all, he must have a clear understanding of the conditions of the floor on which he is to set his rack.

Assuming the scope of work is clear, site conditions understood, and an acceptable proposal made, a signed contract serves to initiate the work.

The construction contract should establish responsibility for insurance to be carried by the several subcontractors on site, and identify those areas where the general contractor has responsibility. Some states allow contractors to self-insure; others require third party insurers. Some construction sites require that all contractors use the same insurance carrier, so that disputes as to responsibility are avoided. It is important to establish when title passes to the owner, as this will establish responsibility for insurance, security, and liability. The construction contract will recognize permits that are required and establish who has the responsibility to obtain them—it will also provide for access to the work by the owner or his representative—for inspecting progress and the quality of work.

Contractors should recognize the importance of maintaining an agreeable working relationship with one another on the construction site. The construction site has confusions that arise with the intermingling of different contractors and trades, and a certain amount of give and take is required to get the job done. Those areas where there is a direct interface require particular attention—these would include the fire protection contractor, who will be hanging piping from the rack, and the electrical contractors, who may be attaching conduits and conductor bars to the different parts of the working system.

A good contract contains a clear statement of the work to be done, along with responsibilities of each contractor and of the owner. It will include detailed schedules with easily identifiable milestones so that the interested parties can monitor progress. It is essential that there is a clear understanding of responsibilities and reporting lines. Generally, the systems supplier bears responsibility for conveying all the requirements of the owner, as well as his own, to the subcontractors. On occasion, although it is not recommended for the inexperienced, the owner will contract for rack, crane, etc. separately, and act as his own systems coordinator. He may also elect to retain a consultant to monitor progress or, perhaps, even to assume overall responsibility.

Such arrangements generally increase the risk of the individual sub-system suppliers. If the owner, or his representative, is not totally competent and experienced in design and implementation, system integration errors are likely to result, and attempts will be made to pass responsibility through to the individual suppliers. In such instances, the equipment suppliers' management has to be strong enough and knowledgeable enough to withstand efforts to expand his scope of work to cover the integration errors.

Even with great care in overall planning, changes to the scope of work are likely. It is important that the mechanism for handling changes to the scope of work, and the conditions under which added or revised work will be done, be clearly spelled out in the contract. Quite often, there can be tremendous pressure on the installation supervisor to make adjustments, just to keep the work moving. He may have a site crew of 50 or 100 people, with a payroll as high as $50,000 a day, and he is pressed to keep the work moving. Work that is essential, but is not anticipated by contractual arrangements can force expensive delays. Therefore, it is essential that the installation contractor is cognizant not only of the work to be accomplished today, but tomorrow and next week's activity, and identifies the entire work scope prior to site activities.

The contract should provide for the possibilities of the parties reaching an impasse on a disagreement. Generally, this is provided for by the contract calling for the establishment of an arbitration procedure.

Once a proposal has been made and accepted, and a contract executed, "Project Management"—the business of getting on with the work in a manner that will accomplish time and cost objectives—requires a very careful and well-executed plan. Receiving and storing material, the flow of that material through any required on-site fabrication, and through the erection process itself is an industrial engineering feat every bit as complex as that of a manufacturing plant. The magnitude of the feat can be appreciated when it is recognized that the site installation management must organize and execute his work plan with a force that is temporary and of varying levels of skill, and which is seldom motivated by loyalty to the employer. The conditions at the construction site, the independent nature of the job force, and the pressures to complete schedules on time generate problems for installation site management, that are similar but often more intense than those that are faced by their counterparts in manufacturing.

Control of the job site is essential to productivity, and once lost is almost impossible to regain. Therefore, foremen and site superintendents must be quick to discipline, and quick to remove nonproductive or disruptive personnel.

The decision to use union or non-union labor depends on many factors. In some areas, this option is not open. In other areas, particularly right to work states, non-union construction is possible and feasible. If you are a union-installer, it may be possible to work in a non-union area. If you are a non-union installer, you cannot work on job sites where other contractors are union, and where union trades are on site. In areas where the unions are particularly powerful, it may be impossible to even ship material to the site if the material was not fabricated with union labor. These conditions vary from location to location and from time to time, and help to contribute to the uncertainties that are part of the installation business.

There are advantages to the installer in using union construction. A well-run, well-disciplined union can provide a well-trained and disciplined work force. Unions will also enforce work rules with consistency, which allows the knowledgeable installation contractor to estimate his costs accurately. A pre-job meeting with the appropriate business agents will help prepare the way for a trouble-free operation. Problems with local business agents can be referred to the national level, where national agreements are in operation, which is a further assurance of a predictable working situation.

Most rack installations are handled with a composite crew of millwrights and ironworkers, with representation worked out by the two trades. Perhaps the worst experience that the installer can have is to be involved in a jurisdictional dispute between two trades.

The effectiveness of labor on the job site can be enhanced by the careful selection and application of erection equipment. Job trailers should be in good condition and kept neat and organized to set the tone for the performance expected on the job site. Cranes, fork trucks, welders, and handling equipment should be provided with the objective of utilizing manpower as effectively as possible. Lock boxes and cribs must be provided along with procedures for checking out tooling so pilferage is discouraged.

The site superintendent on an installation project must work very closely with the project engineer to obtain interpretation of drawings and resolutions of questions so as to keep the project moving. Even a well-designed system requires input from the project engineer. The worst possible situation is design or redesign at the job site, with engineering details being resolved while project crews wait.

Quality control and inspection are essential to avoiding costly rework and repair situations. Mistakes found in the field are rectified with cutting torches and sledgehammers. They are costly.

Administration at the site must provide for timely and accurate records to meet the legal requirements of the construction business. A laborer cannot be terminated without being paid—so there must be a high degree of flexibility in administrative systems being utilized on the construction site. Any subcontracts, based on time and material, need to be tracked very carefully to avoid excessive costs. Local, state and national safety requirements must be observed, and records kept of all injuries, so as to avoid fines and unplanned shutdowns.

Relations between the system supplier's project team and the user need to be close and cooperative. A good understanding of the overall system requirements and a close knowledge of daily operations can enhance relationships. Specifications can sometimes be relaxed in certain areas to accommodate construction problems if working customer/contractor relationships are fully developed.

It is essential that installation site management track their progress against a planned schedule, and report that progress to their own management, as well as to the customers. Problems that are recognized early are most easily dealt with. Re-planning is a continuing activity on-site. On large projects, critical path scheduling may be employed, and is essential when the project is too complex to yield to intuition.

Some projects, which are well designed and carefully executed, still face problems toward the end of the installation contract. If workers fear no new work is available, productivity can fall off markedly as the job nears completion.

Rack installation projects (large ones) obtain a high degree of efficiency of certain specified movements, such as the standing of frames. As this work is completed, crews are cut back and the less glamorous functions, such as plumbing and leveling, can become disorganized. This last 5 percent of the job can eat up a disproportionate share of the resources unless the site superintendent plans the completion of the project just as carefully as he planned the rest of the work. He must be very familiar with the contractual requirements, so there are no unpleasant surprises at the end of a project. Correcting a 100,000 opening rack for any sort of systematic error can use a lot of manhours.

Even assuming that all rack specifications have been met, the installation is not complete until after the system startup. There will be instances of interference between the stacker cranes, shuttle forks, and sprinkler heads, or support arms.

Finally, the installation contractor should be very careful about leaving the site in good condition. Long months of quality workmanship can be blemished by the impressions left by an inadequate cleanup job. People tend to think in exceptions, and the customer's last impression should be a good one.

Installation is a risky business. Problems on the job site are expensive and costs escalate quickly. In automatic storage retrieval systems, the most common problem is dimensional out-of-tolerance. Recognizing that tolerance is a frequent problem, we should anticipate it and approach the design and installation so as to minimize it.

The advent of the rack-supported building presents new requirements on design and installation. An exposed rack structure is more susceptible to wind loading than the completed building by a factor of 2-1/2. This means that rack-supported building structures must be well guyed during construction to prevent damage that would not be a problem once the building is enclosed. The same is true of paint finishes. The exposure of the rack to sun, wind, and rain for extended periods of time, places a more severe requirement on the finish than it will ever see in use. The system supplier and ultimate customer must be made aware of the results of this exposure so that controversy over oxidized paint or rust spots can be avoided. Once the rack structure is enclosed, it is doubtful that any paint is needed if warehouse temperatures are maintained above the dew point.

Strikes invariably increase administrative expenses, and attention should be given to scheduling rack installation projects to avoid periods when contracts will be under negotiation.

Picket lines thrown up by unions other than the construction trades may nonetheless shut down a project. Generally, these problems can be averted by opening up a different "construction gate," and obtaining an injunction against that gate being picketed. Expect at least two or three days of production to be lost if you are faced with this problem. Jurisdictional problems cannot be avoided completely, but it pays to carefully monitor previous practices in the construction area. When problems do arise, it is essential to get in contact with the appropriate business agents, and force the problem to the highest level in the union organization as quickly as possible.

Theft can be a serious problem on a construction site. The high turnover of personnel with a great freedom of access suggests that small tools and equipment be locked up in secure cribs and major equipment bunched together in a protected and lighted area. Common sense precautions are probably the best protection against loss in this area.

Fire can be a problem if trash is allowed to accumulate at the site, or if construction is going on in or adjacent to an existing building. On-site motor fuels, of course, demand special care.

Government regulations can be a problem if not adhered to, and fines and shutdowns can result. The best solution to this problem is communication and education.

12.5 MAINTAINING THE FACILITIES PLAN

Once the facilities plan has been implemented, it would be nice to conclude that the planning effort is finished. However, as we noted in Chapter 1, the facilities planning cycle is continuous. Changing conditions and requirements necessitate that the facilities plan be audited periodically to ascertain the need for modifications and revisions.

The facilities planning audit involves a comprehensive, systematic examination of the location of the facility, the layout and material handling system, and the building design. The utilization of space, equipment, personnel, energy, and financial resources; the performance of the facility under current conditions; and the capacity of the facility to meet future requirements are to be assessed.

In performing the facilities planning audit, the following areas should be examined [10]:

1. *Layouts,* to verify the locations of equipment, storage areas, offices, doors, aisles, and service areas are documented accurately

2. *Installation,* to verify the hardware and software systems installed were, in fact, what was designed and/or purchased

3. *System performance,* to verify satisfactory performance is being provided, focusing on throughput, personnel, software, uptime, return on investment, costs, and productivity

4. *Requirements,* to verify sufficient capacity exist to meet the requirements for space, equipment, and personnel

5. *Management of the system,* to verify the policies, procedures, controls, and reporting channels support effective management of the system

The audit should be performed in order to learn from the past, to assess the need for change, to develop a database, and to establish or improve credibility within the organization. Despite the fact many benefits can be obtained from formal system audits, very few firms appear to perform them.

Some of the more common reasons for firms failing to conduct audits are [10]:

1. *Fear.* Being afraid promises weren't kept, benefits weren't realized, and resources were wasted.

2. *Mobility.* The lead times for planning, designing, and installing the system are such that the persons responsible have either been promoted or transferred, or have left the company.

3. *Lack of accountability.* Typically the persons responsible for planning, designing, and selling the system aren't responsible for installing, operating, and maintaining the system.

4. *Time pressures.* Other needs exist and a "There's never time to do it right" attitude exists.

5. *Changing requirements.* Due to internal and external factors, by the time the system is installed the need has changed.

6. *Lack of baseline.* The current status of the system cannot be compared with anything; neither relative nor absolute comparisons can be made.

7. *Hindsight attitude.* With the belief that "Anyone can look back, but the real pros in the business are those who only look ahead," such an attitude discourages audits.

Unfortunately, audits are associated with the dichotomy of failure and success. Yet, the terms *success* and *failure* are generally related to outcomes, not decisions. It is important to recognize the differences in decisions and outcomes. Good decisions can have good or bad outcomes; likewise bad decisions can have good or bad outcomes. The merit of the decision is often related to the extent to which it was objective, analytical, comprehensive, repeatable, and explainable.

An audit should always be performed onsite. Too often, quick judgments (based on little data) are made concerning systems; conclusions are drawn from comments made by persons lacking objectivity. Disgruntled suppliers, jealous users, and politically motivated employees can cast doubts on a system that are difficult to overcome.

The audit should be performed at a level within the organization that will allow objectivity. The report should be submitted at least at the funding approval level.

The audit should be designed when the system is designed. Too often, audits are designed *after* the system is installed. Such an approach is similar to playing a game and *then* deciding how the scoring will be performed, or submitting bid packages, receiving vendor responses, and *then* deciding what factors will be used to select the supplier. One should know what things are important from an audit point of view when the system is being designed.

The initial audit is best performed approximately six months after the user has accepted it. To do it sooner is to invite "buyer's remorse." Immediately following startup and during the debugging, the user typically undergoes a period of discouragement that can bias attempts to assess the system.

Not only can the audit be performed too soon, it can also be performed too late. If the initial audit is delayed, then bad habits will have been developed, firm judgments will have been made, and opportunities for improvements will be forgone.

After the initial audit, annual audits are appropriate for some installations. For others, periodic audits with indefinite timings are preferred.

To facilitate the audit, a checklist should be prepared during the design phase. Performance measures and/or productivity ratios can provide useful information concerning the performance of the system on a continuing basis. A number of such ratios is given in [9]. As examples, consider the following:

$$\text{Aisle space percentage} = \frac{\text{Cube space occupied by aisles}}{\text{Total cube space}}$$

$$\text{Damaged loads ratio} = \frac{\text{Number of damaged loads}}{\text{Number of loads}}$$

$$\text{Energy utilization index} = \frac{\text{BTUs consumed per day}}{\text{Cubic space}}$$

$$\text{Manufacturing cycle efficiency} = \frac{\text{Total time spent on machining}}{\text{Total time spent in productive system}}$$

$$\text{Order picking productivity ratio} = \frac{\text{Equivalent lines or orders picked per day}}{\text{Labor hours required per day}}$$

$$\text{Slot occupancy ratio} = \frac{\text{Number of occupied storage slots}}{\text{Total number of storage slots}}$$

$$\text{Storage space utilization} = \frac{\text{Cube space of material in storage}}{\text{Total cube space required}}$$

Care must be taken to ensure that the performance measures used are reasonable and do reflect positive performance. Additionally, the data required for calculating the performance measure must be readily available.

To be effective, the audit should be comprehensive. Hence, managers, operators, maintenance engineers, designers, system suppliers, and functions being served by the system should be included in the audit. System suppliers can frequently provide recommendations for change or modification.

It is also important that the audit be performed formally and that comparisons be made of:

1. Promises versus deliveries
2. Expectations versus realizations

3. Forecasts versus actual occurrences

4. Assumptions versus outcomes

5. Concepts versus installations

Where differences exist, they must be explained.

The results of the audit should be reported formally. They should be reviewed with those affected before being submitted. It is important that feedback occur to ensure that opinions or perceptions are minimized and facts are maximized. Actions for change or continuation should be recommended, and the impact of the recommendations must be assessed.

Management should provide the right climate in order for audits to be productive. An adversarial relationship must not exist; otherwise the audit process will be doomed to failure. Differences in decisions and outcomes must be recognized. Management must be patient. Complex systems must be debugged. The period of time following startup will tend to be frustrating; management support during this time is crucial.

Engineers should be accountable for their designs, avoid "pride of authorship," and recognize that changes in a design might be needed. Assumptions, objectives, alternatives considered, and rationale used should be documented throughout the design process for the audit process to be effective. Operating personnel must be involved in the design process. To the extent it is possible to do so, the design should be based on facts rather than guesses; it must be rooted in analysis, not hypotheses.

Operations personnel also need to be accountable. It is important that they follow through on commitments. If reductions in personnel, equipment, and/or space were promised, then they should be made. Operations personnel need to be a part of the solution, not a part of the problem. Resistance to change, lack of confidence in the new system, and a lack of patience with the system do not create the proper climate for an effective audit.

12.6 SUMMARY

The preparation of a facilities plan is an important step in developing a workable facility. Several important details must be resolved and considerable fine-tuning should take place during plan development. A facilities plan consists of a plot plan and a layout plan. The critical factors to be considered in developing a plot plan are expansion, flow patterns, energy, and aesthetics. Establishing a layout plan should follow a systematic course of action. Different approaches are available for presenting the plan, with the computer playing an increasingly important role in both the plan development and presentation.

Written and oral reports are important parts of a successful facilities planning project. As much care should be spent on selling a facilities plan using written and oral reports as is spent on developing it. The planner's mission is not to make plans for facilities, but rather to implement facilities plans. A well-written report and a good oral presentation are the initial steps in bringing about successful implementation.

It is during the implementation phase that the results of a design project become evident. By maintaining the plan, the benefits gained from it will be lasting, rather than transitory.

BIBLIOGRAPHY

1. Apple, J. M., *Plant Layout and Material Handling,* 3rd ed., John Wiley & Sons, New York, 1977.
2. Ballinger, R. A., *Layout and Graphic Design,* Van Nostrand Reinhold, New York, 1970.
3. Blair, E. L., and Miller, S., "Interactive Approach to Facilities Design Using Microcomputers," *Computers and Industrial Engineering,* vol. 9, no. 1, January 1985, pp. 91–102.
4. Brickner, W. H., *Managing Change,* Lansford Publishing, San Jose, CA, 1974.
5. Francis, R. L., McGinnis, L. F., and White, J. A., *Facility Layout and Location: An Analytical Approach,* Second Edition, Prentice-Hall, Englewood Cliffs, NJ, 1992.
6. Gupta, R. M., "Flexibility in Layouts: A Simulation Approach," *Material flow,* vol. 3, no. 4, June 1986, pp. 243–250.
7. Hales, H. L., *Computer-Aided Facilities Planning,* Marcel Dekker, New York, 1984.
8. Irwin, P. H., and Langham, F. W. Jr., "The Change Seekers," *Harvard Business Review,* vol. 44, no. 1, pp. 81–92, January-February 1966.
9. Konz, S., *Facility Design,* John Wiley & Sons, New York, 1985.
10. Krick, E. V., *Methods Engineering,* Wiley, New York, 1962.
11. Kusiak, A., and Heragu, S. S., "The Facility Layout Problem," *European Journal of Operational Research,* vol. 29, 1987, pp. 229–251.
12. Lawrence, P. R., "How to Deal with Resistance to Change," *Harvard Business Review,* vol. 47, no. 1, pp. 4–12, January–February 1969.
13. Lefferts, R., *Elements of Graphics: How to prepare Charts and Graphs for Effective Reports,* Harper & Row, New York, 1981.
14. Malouf, L. G., "Initiating Change Within a System Framework—How Do You Obtain Employer and Employee Commitment and Improve Results?," Presentation to the National Material Handling Forum, Material Handling Institute, Pittsburgh, PA, 1975.
15. Mullins, C. J., *The Complete Writing Guide to Preparing Reports, Proposals, etc.,* Prentice-Hall, Englewood Cliffs, NJ, 1980.
16. Muther, R., *Practical Plant Layout,* McGraw-Hill, New York, 1955.
17. Muther, R., *Systematic Layout Planning,* Industrial Education Institute, Boston, 1961.
18. Muther, R., and Hales, H. L., *Systematic Planning of Industrial Facilities,* vol. I, Management & Industrial Research Publications, Kansas City, MO, 1979.
19. Muther, R., and Hales, H. L., *Systematic Planning of Industrial Facilities,* vol. II, Management & Industrial Research Publications, Kansas City, MO, 1980.
20. O'Brien, C., and Abdel Barr, S. E. Z., "An Interactive Approach to Computer Aided Facility Layout," *International Journal of Production Research,* vol. 18, no. 2, 1980, pp. 201–211.
21. Reed, R., *Plant Layout: Factors, Principles, and Techniques,* Richard D. Irwin, Homewood, IL, 1961.
22. Reed, R., *Plant Location, Layout, and Maintenance,* Richard D. Irwin, Homewood, IL, 1967.
23. Searls, D. B., "Installation: The Key to a Successful Operation," Presentation to the 1980 Automated Material Handling and Storage Systems Conference, April 1980.
24. Sims, R. E., Jr., "*Selling Your Solution to Management,*" Presentation to the 28th Annual Material Handling Management Course, American Institute of Industrial Engineers; Airlie, VA, June 1981.
25. Souther, J. W., *Technical Report Writing,* 2nd ed., Wiley, New York, 1977.
26. Sule, D. R., *Manufacturing Facilities: Location, Planning, and Design,* 2nd ed., PWSKent Publishing Company, Boston, 1994.
27. Tompkins, J. A., "Justification and Implementation: The Critical Link," Presentation to the 1981 Automated Material Handling and Storage Systems Conference, Philadelphia, PA, September 1981.

28.   White, J. A., *Yale Management Guide to Productivity,* Eaton Corp., Philadelphia, PA, 1979.

29.   White, J. A., "Auditing the System After Installation," Presentation to the 1980 Automated Material Handling and Storage Systems Conference, April 1980. Subsequently Published in *Industrial Engineering,* vol. 12, no. 6, pp. 20–21, June 1980.

PROBLEMS

12.1   Draw a plot plan of the building housing the classroom for this course. How should the building be expanded if the demand for classrooms doubles? Could the facility have been designed to accommodate expansion better? Please explain.

12.2   Why should a facilities planner be concerned with aesthetics?

12.3   How could computer-aided layout techniques be used to study flow patterns outside a facility?

12.4   Describe specific situations where two-dimensional layouts would be superior to three-dimensional layouts and vice versa.

12.5   Describe specific situations where different two-dimensional approaches would be used.

12.6   Describe specific situations where different three-dimensional approaches would be used.

12.7   How could computer-aided layout techniques be integrated with computer graphics? Would this be useful?

12.8   Describe specific situations where block templates or models should be used instead of contour templates and models and vice versa.

12.9   List five reasons for people resisting change, other than those cited in the chapter.

12.10 List five additional ways resistance to change can be overcome.

12.11 Visit a manufacturing facility and prepare a written audit report documenting your findings.

12.12 Visit a nonmanufacturing facility and prepare a written audit report documenting your findings.

12.13 Perform an audit of the university, focusing on the location of facilities and personnel flow.

12.14 How might multimedia presentations be used to sell a facilities plan?

12.15 Consider the year 2020. What changes in technology do you believe will significantly impact the preparation and presentation of facilities plans?

INDEX